D1218274

Nonlinear Optics in Telecommunications

Springer
Berlin
Heidelberg
New York
Hong Kong
London
Milan
Paris
Tokyo

Advanced Texts in Physics

This program of advanced texts covers a broad spectrum of topics which are of current and emerging interest in physics. Each book provides a comprehensive and yet accessible introduction to a field at the forefront of modern research. As such, these texts are intended for senior undergraduate and graduate students at the MS and PhD level; however, research scientists seeking an introduction to particular areas of physics will also benefit from the titles in this collection.

Thomas Schneider

Nonlinear Optics in Telecommunications

With 199 Figures, 14 Tables, and 70 Problems

Solutions for Instructors available online
at springeronline.com/3-540-20195-5

Springer

Professor Dr. Thomas Schneider
Deutsche Telekom AG
Fachhochschule Leipzig
Gustav-Freytag-Str. 43-45
04277 Leipzig, Germany
E-mail: schneider@fh-telekom-leipzig.de

ISSN 1439-2674
ISBN 3-540-20195-5 Springer-Verlag Berlin Heidelberg New York

Cataloging-in-Publication Data

Schneider, Thomas, 1965-
Nonlinear optics in telecommunications / Thomas Schneider.
p.cm. – (Advanced texts in physics, ISSN 1439-2674)
Includes bibliographical references and index.
ISBN 3-540-20195-5 (alk. paper)
1. Optical communications. 2. Nonlinear optics. I. Title. II. Series.
TK5103.59.S36 2004
621.382'7–dc22

Springer-Verlag is a part of Springer Science+Business Media

springeronline.com

Typesetting: Camera-ready copy from the author using a Springer LaTeX macro package
Cover design: *design & production* GmbH, Heidelberg

Printed on acid-free paper SPIN 10934142 56/3141/jl 5 4 3 2 1 0

Preface

In today's backbone networks, data signals are almost exclusively transmitted by optical carriers in fibers. The continents and the whole world come closer together by a large number of optical fibers lying on the bottom of the oceans. Applications like the internet demand an increase of the available bandwidth in the network and reduction of costs for the transmission of data. Linear optics is, in principle, a special case of the much wider field of nonlinear optics. Nonlinearity can only be neglected if the optical intensities are small. With the increasing data rates, the optical amplification of the signals, and the incorporation of techniques like dense wavelength division multiplexing (DWDM) nonlinear optical effects become very important for system design since these effects can cause severe problems. On the other hand, the same effects offer the possibility for very fast all-optical signal processing.

This book tries to give an introduction to the fascinating field of optical telecommunications. In order to write a useful text for students, technicians and researchers working in the field, the technical aspect as well as the physical background of the different effects is described in detail. In most cases the easiest way to explain the physics of nonlinear optics was chosen. Nevertheless, an attempt was always made to integrate the latest research results.

Although the whole text was carefully written and corrected several times it could be possible that some mistakes were overlooked. I apologize for this and I am very thankful for everybody who informs me about mistakes.

First of all I want to thank my family, Kerstin and Clara for their support and understanding. Special thanks go to R. Fischer for the relief he gave me on important showplaces beside. I am deeply grateful to Prof. W. Glaser from the Humboldt University in Berlin and to Prof. J. Reif from the Brandenburgische Technische Universität in Cottbus who gave me the opportunity to engage in the research of nonlinear optics.

Special thanks to all the colleagues who discussed special subjects with me and who always had an open ear for any kind of problems: Dr. R. Schmid (BTU Cottbus), W. Ruhland and M. Messollen (FH Telekom Leipzig).

I wish to acknowledge the colleagues who have read parts of the book manuscript and given their comments and suggestions to the English translation: M. Sams (FH Telekom Leipzig), Dr. M. Ammann and Dr. A. Schwarzbacher (Dublin Institute of Technology).

Last but not least I want to thank G. Biermann, from the library of the FH Telekom Leipzig, who provided me with the required literature, as well as the editors and production team from the Springer Verlag for their kind support.

Leipzig, October 2003 *Thomas Schneider*

Contents

Part III. Applications of Nonlinear Effects in Telecommunications

1. Introduction

In the early days of optical telecommunications the attenuation and dispersion of the fiber were the most important aspects that determined the bit rate and the maximum distance of the links in communication systems. New fiber designs and a more or less intelligent dispersion management could rectify these linear effects. The nonlinearity of an optical fiber is very weak; it depends on the intensity of the transmitted signals and the interaction length. Due to the fact that only rather small distances between a transmitter and receiver were bridged optically, nonlinear effects played only a marginal role in special fields such as long haul undersea cable systems.

In the last few years the picture has completely changed. Responsible for this fact are especially two reasons: the ever increasing demand of bandwidth due to "packet oriented" applications like the internet, and the invention of optical amplifiers. This prediction will be substantiated in the following in more detail.

The bit rate in the communication systems in the 1980s was 155 Megabit per second (Mbit/s), ten years later it increased to 2.5 Gigabit per second (Gbit/s). Although 10 Gbit/s is being used today, the change for 40 Gbit/s is already planned and laboratories have show data rates up to a few Terabit per second (Tbit/s). Since the middle of the 1990s, optical telecommunication devices have been transmitted more than one wavelength over a fiber span to better exploit the usable bandwidth of optical fibers. The technique, dense wavelength division multiplexing (DWDM), transmits different channels at the same time in one fiber with different carrier wavelength.

Historically speech signals were transmitted via a virtual link between the transmitter and receiver through the network. The link lasted as long as the transmission was continued and a very high percentage of telephone conversations took place over a rather small distance. Contrary to this, the packets of the internet are transmitted over the communication networks world wide. Therefore not only the data rate of the systems increased, the distance over which the data signals had to be transmitted increased as well.

In order to reduce the costs for the transmission, it was necessary to minimize expensive elements like transmitters, receivers and regenerators in the optical network. At the same time the maximum data rate and transmission distance had to be increased drastically. This goal could be reached with the invention of the erbium-doped fiber amplifier (EDFA). Since this device allowed the amplification of the signal in the optical domain, the transformation into the electrical domain was no longer necessary. Hence a receiver, transmitter, and regenerator were spared with every optical amplifier.

On the other hand, an amplifier adds noise to the signal. With the increasing transmission distance, bit rate and number of amplifiers, the noise in the system will therefore increase. Since the bit error rate (BER) of the system depends on the signal to noise ratio, the signal power had to be increased. With use of the optical amplifiers, the drastically increased intensity and the interaction length together with the higher bit rates of current systems lead to a much higher impact of nonlinear effects onto the system design.

Another reason for the rise of nonlinear effects in optical communication systems is the very small attenuation the signal experiences in modern fibers together with the rather small core area in which the signal is guided. Hence, the high powers and very high intensities required for a particular BER will decrease very slowly and are amplified again with each EDFA. Assume, for instance, a DWDM system with 100 WDM channels and an input power of 3 mW each, propagating in a single mode fiber with an effective core area of 80 μm^2 and an attenuation constant of $\alpha = 0.2$ dB/km. The intensity in the core of the fiber is then 3.75×10^9 W/m^2, and will still be $\approx 3 \times 10^9$ W/m^2 after a distance of 5 km.

The nonlinearity is a fundamental physical property of the fiber. The highest bit rate that can theoretically be transmitted without an error over a classical (linear) link – like a copper wire or a wireless channel – depends only on the usable bandwidth and the signal to noise ratio. In optical fibers this bit rate is limited instead by the nonlinearity of the fiber.

A variety of different nonlinear effects can occur in optical fibers, all of them with different influences on the communications link. The self-phase modulation (SPM), for instance, leads to a change in the dispersion behavior in high-bit-rate transmission systems; the cross-phase modulation (CPM or XPM) and the Raman and Brillouin scattering implicate a decreasing of the signal to noise ratio; and four-wave-mixing (FWM) in dispersion-shifted fibers will increase the cross talk between adjacent channels.

On the other hand, the same nonlinear effects offer a variety of possibilities for ultrafast all-optical switching, amplification and regeneration. Currently, the whole spectrum of signal processing in the electrical domain and much more can, in principle, be carried out in the optical domain with the help of nonlinear effects. With solitons, for instance, it is possible to transmit optical pulses over extremely large distances without distortion. The FWM offers the possibility for a pure optical wavelength conversion. With an ultrafast optical grating that can switch optical signals in a few femtoseconds (10^{-15} s), the so called mid span spectral inversion can compensate the distortions of the optical pulses completely. Nonlinear effects like the FWM and the Raman and Brillouin-scattering are able to amplify optical signals in areas

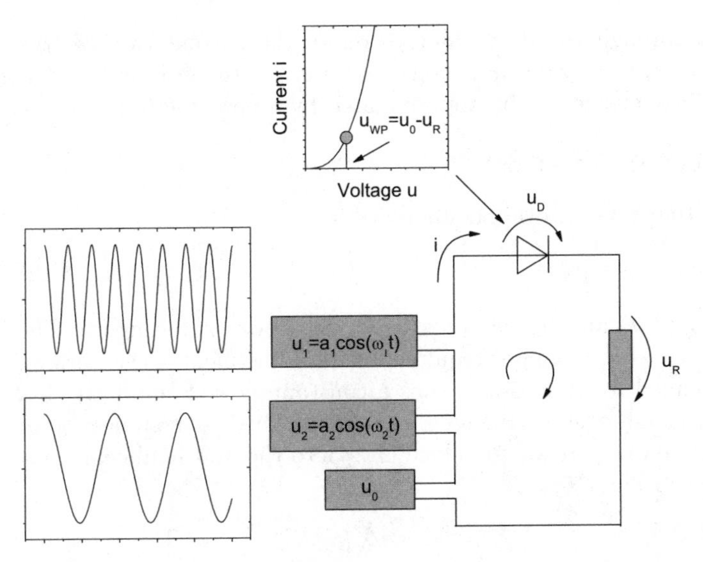

Fig. 1.1. Current diagram with a diode as nonlinear element

that can never be reached by EDFA's. Recent world records where data rates of a few Terabit per second (10^{12} bit/s) were transmitted over distances of thousands of kilometers would have been unthinkable without distributed Raman amplifiers.

With compact nonlinearities offered by highly nonlinear fibers, semiconductor optical amplifiers, and nonlinear photonic crystal microdevices [1], it may be possible to integrate some of these functionalities into small compact devices.

Nonlinear optics is the area of optics in which the response of a material system on an applied optical field is not linear and, hence, the superposition rule no longer holds. In other words, this means that different optical waves propagating in a nonlinear material are not independent of each other. They can not be simply added, like in the linear case, but the generation of harmonics and the mixing between them has to be considered.

Nonlinearity is not only responsible for the behavior of electromagnetic waves in the visible part of the spectrum, nonlinear dependence exists in electrical devices installed in circuits as well. Let's assume two alternating currents with different frequencies are propagating in a wire. If the frequencies are not too close together – so that fading can be neglected – and the intensities are not too high, both currents are propagated independently of each other in the medium. The two currents will only experience an attenuation, and the frequencies are the same at both ends of the wire. If, instead, a device like a simple diode is installed into the wire then the behavior is much more complicated. Figure 1.1 shows the dependence between the voltage and the current of a theoretical diode and the circuit diagram.

The current through the diode and therefore the current through the resistor depends nonlinearly on the voltage, as shown in the inset of Fig. 1.1. This nonlinear dependence can be approximated by a power series:

$$i = c_0 + c_1 u_\mathrm{D} + c_2 u_\mathrm{D}^2 + c_3 u_\mathrm{D}^3 + \ldots \tag{1.1}$$

The voltage that falls off over the diode is:

$$u_\mathrm{D} = u_1 + u_2 + u_0 - u_\mathrm{R} \tag{1.2}$$

with $u_1 = a_1 \cos(\omega_1 t)$ and $u_2 = a_2 \cos(\omega_2 t)$, as shown in the figure. The voltage difference $u_0 - u_\mathrm{R} = u_\mathrm{WP}$ is required for the adjustment of the working point and is assumed to be constant, c_0 is a constant as well, and $c_1 u_\mathrm{D}$ is a linear term that is only responsible for attenuation. The first nonlinear term in (1.1) is therefore $c_2 u_\mathrm{D}^2$. If we introduce (1.2) into the first nonlinear term of (1.1), the result is:

$$c_2 u_\mathrm{D}^2 = c_2 \begin{bmatrix} \frac{1}{2}a_1^2 + \frac{1}{2}a_2^2 + u_\mathrm{WP}^2 \\ +2a_1 u_\mathrm{WP} \cos\omega_1 t + 2a_2 u_\mathrm{WP} \cos\omega_2 t \\ +a_1 a_2 \cos((\omega_1 + \omega_2)t) + a_1 a_2 \cos((\omega_1 - \omega_2)t) \\ +\frac{1}{2}a_1^2 \cos 2\omega_1 t + \frac{1}{2}a_2^2 \cos 2\omega_2 t \end{bmatrix} \tag{1.3}$$

In the first row of (1.3), there are constant terms with $\omega = 0$, whereas the second row contains the original frequencies. In the third row the sum and difference frequencies will be found and the last row contains frequencies twice the input frequencies representing an effect called the second harmonic. The second nonlinear term of (1.1) will deliver:

$$c_3 u_\mathrm{D}^3 = c_3 \begin{bmatrix} f(\omega = 0) \\ +f(\omega_1) + f(\omega_2) \\ +f(\omega_1 + \omega_2) + f(\omega_1 - \omega_2) \\ +f(2\omega_1 + \omega_2) + f(2\omega_1 - \omega_2) \\ +f(\omega_1 + 2\omega_2) + f(\omega_1 - 2\omega_2) \\ +f(2\omega_1) + f(2\omega_2) \\ +f(3\omega_1) + f(3\omega_2) \end{bmatrix} \tag{1.4}$$

Due to the nonlinear element in the electric circuit, the two frequencies are no longer independent of each other and the current will consist of sum, difference, and the harmonics of the original frequencies. This behavior is always the same whenever nonlinearities come into play, for instance due to the nonlinear behavior of gutters or aerial masts for the mobile communication signals, or as a result of the nonlinear function of an amplifier. For the radio frequency range these phenomena are called mixing or intermodulation, whereas for the visible range of the electromagnetic spectrum the term nonlinear optics will be used.

This book is divided into three parts, the first part is an introduction to linear and nonlinear optics as well as optical telecommunications. The second part deals with various optical effects and their impact on communication systems in detail, whereas in the third part an overview of different applications of nonlinear optical effects for signal processing, amplification and distortion compensation is given.

Part I

Fundamentals of Linear and Nonlinear Optics

2. Overview of Linear Optical Effects

If an electromagnetic wave, or a photon, hits atoms or molecules of a material, these particles can react in two distinct ways. If the photons were absorbed, an excitation of the particles onto a higher energy level is the result. In this case the band gap of the material is smaller or equal to the energy of the photons ($\hbar\omega$). After the absorption, the excitation of the atom can decay in many different ways. The most important, in the context of optical telecommunications, is the spontaneous emission; in this case the material emits on its part photons with a different energy. This spontaneous emission can in some materials be transmuted into a stimulated emission and is the basis of lasers and the erbium-doped fiber amplifier (EDFA).

If the energy of the photons is smaller than the band gap, no excitation into a higher energy level of the atom takes place.[1] Therefore, the photon leads only to a disturbance of the distribution or motion of the internal charges of the atom. The charges will be accelerated, depending on the electrical field of the wave. Accelerated charges are on their part sources of electromagnetic waves. These secondary waves – emitted by the diodes – superimpose with the original wave. Contrary to the first possibility of photon absorption, the waves or photons emitted by the molecules have the same frequency or energy as the original wave; only their phase or momentum will be different. This is the origin of every kind of linear optical effect, like reflection, diffraction and scattering.

Optical materials, such as silica glass used for optical waveguides, have a very high band gap in comparison to the photon energy used for optical communications. Therefore, only an excitation in the form of charge acceleration can take place. This nonresonant possibility will be the focus of the following chapters.

The chapter starts with the classical mathematical description of electrodynamics, the Maxwell equations. From these equations we will derive the linear wave equation. The model of the harmonic oscillator is introduced to describe important properties of the interaction between an optical wave

[1] If the intensity is very high, i.e., a very large number of photons hits the particle in a short time, then a multiphoton absorption will be possible. In this case the particle can absorb two or more photons simultaneously. Consequently, photons with a smaller energy than the bandgap can excite the molecules.

and matter. From the oscillator model, the refractive index, attenuation, and susceptibility of an optical material are then derived.

The wavelength dependence of the refractive index leads to different velocities of the components of an optical pulse consists of and, hence, the definition of different velocities. This chapter declares the important concepts of phase velocity, group velocity, and dispersion.

2.1 The Wave Equation for Linear Media

First we will start with a classical description of the interaction between an optical wave and a medium. The Maxwell equations form the mathematical basis of this description. These equations, published by James Clerk Maxwell (1831-1879) in his book "A Treatise on Electricity and Magnetism," are as follows:

$$\nabla \times \boldsymbol{E} = -\frac{\partial \boldsymbol{B}}{\partial t}, \tag{2.1}$$

$$\nabla \times \boldsymbol{H} = \boldsymbol{j} + \frac{\partial \boldsymbol{D}}{\partial t}, \tag{2.2}$$

$$\nabla \cdot \boldsymbol{D} = \rho, \tag{2.3}$$

and

$$\nabla \cdot \boldsymbol{B} = 0, \tag{2.4}$$

where $\boldsymbol{E}$ and $\boldsymbol{H}$ are the electric and magnetic field vectors, respectively, with $\boldsymbol{D}$ and $\boldsymbol{B}$ as the corresponding flux densities; $\boldsymbol{j}$ indicates the current density, a result of the motion of carriers, ρ is the carrier density, and ∇ is the Nabla operator (see Appendix).

The physical background of these equations is the following: according to (2.1) the origin of the electric field vortices ($\nabla \times \boldsymbol{E}$) is the time dependent change of the magnetic field ($-\partial \boldsymbol{B}/\partial t$). Whereas, due to (2.2), the magnetic field vortices ($\nabla \times \boldsymbol{H}$) could be a result of the current density in the material ($\boldsymbol{j}$) or of the time dependent change of the electric flux density ($\partial \boldsymbol{D}/\partial t$). The third and fourth equation (2.3) and (2.4) indicate that the origin of the electric field ($\nabla \cdot \boldsymbol{D}$) could be the existence of sources (ρ), while this would be impossible for the magnetic field ($\nabla \cdot \boldsymbol{B} = 0$). In other words, the magnetic field will always be free of sources.

These equations enabled the first mathematical description of the connection between the known electrical and magnetical phenomena. Maxwell predicted the existence of electromagnetic fields. An ether was not necessary for the propagation of these waves.[2] This prediction was verified some years later (1887–1889) in an experiment by Heinrich Hertz.

[2]Contrary to this behavior, the propagation of sound waves would be impossible in vacuum. Sound waves are dependent on a medium like the surrounding air.

First, we will consider the propagation of electromagnetic waves in vacuum. In this case the electric flux density ($\boldsymbol{D}$) is connected to the electric field vector ($\boldsymbol{E}$) by the vacuum permittivity (ε_0); $\boldsymbol{D} = \varepsilon_0 \boldsymbol{E}$. In analogy to this, the magnetic flux density ($\boldsymbol{B}$) is connected to the magnetic field vector ($\boldsymbol{H}$) via the permeability (μ_0); $\boldsymbol{B} = \mu_0 \boldsymbol{H}$.

In vacuum there is neither a carrier density (ρ) nor a current density ($\boldsymbol{j}$). Hence, (2.3) is zero and the first term on the right side of (2.2) will vanish. If the magnetic flux density ($\boldsymbol{B}$) and the electric field vector ($\boldsymbol{E}$) are used instead of $\boldsymbol{H}$ and $\boldsymbol{D}$ in (2.2), then the Maxwell equations in vacuum become

$$\nabla \times \boldsymbol{E} = -\frac{\partial \boldsymbol{B}}{\partial t} \tag{2.5}$$

and

$$\nabla \times \boldsymbol{B} = \mu_0 \varepsilon_0 \frac{\partial \boldsymbol{E}}{\partial t}. \tag{2.6}$$

If (2.6) is introduced into (2.5) assuming that no sources are present, the wave equation can be derived[3]:

$$\Delta \boldsymbol{E} = \frac{1}{c^2} \frac{\partial^2 \boldsymbol{E}}{\partial t^2}, \tag{2.7}$$

with Δ as the Laplace operator. This equation describes the propagation of electromagnetic waves in vacuum. Due to (2.7), all waves have the same velocity $c = 1/\sqrt{\varepsilon_0 \mu_0}$. This is the so called velocity of light in vacuum ($\approx 3 \times 10^8$ m/s).

On the other hand, if the waves propagate in a medium, the material obviously has an influence on the fields and the wave equation. If an insulator is placed into an electric field of a condenser, then the charges within the atoms of the insulator shift due to the force the field exercises on them. This phenomenon is called polarization. In consequence, the electrical flux density $\boldsymbol{D}$ will get an additional factor, the relative permittivity (ε_r):

$$\boldsymbol{D} = \varepsilon_\mathrm{r} \varepsilon_0 \boldsymbol{E}. \tag{2.8}$$

The relative permittivity depends on the medium, the frequency, and the polarization of the waves. In an insulator the external field will superimpose with the internal field which is a consequence of the internal charge distribution. As a result the field strength in the medium will decrease. The material dependent relative permittivity (ε_r) is dependent on the dielectric susceptibility (χ):

$$\varepsilon_\mathrm{r} = 1 + \chi. \tag{2.9}$$

---

[3]The Laplace operator is $\Delta \boldsymbol{E} = \nabla (\nabla \cdot E) - \nabla \times \nabla \times \boldsymbol{E}$. Since the electric field in vacuum is free of sources, $\nabla \cdot \boldsymbol{E} = 0$ and $\Delta \boldsymbol{E} = -\nabla \times \nabla \times \boldsymbol{E}$.

The susceptibility is a measure of the polarizability of the material. With the equations 2.8 and 2.9, the electric flux density can be written as

$$\boldsymbol{D} = (1 + \chi)\varepsilon_0 \boldsymbol{E} = \varepsilon_0 \boldsymbol{E} + \varepsilon_0 \chi \boldsymbol{E} = \varepsilon_0 \boldsymbol{E} + \boldsymbol{P}, \tag{2.10}$$

where $\boldsymbol{P} = \varepsilon_0 \chi \boldsymbol{E}$ is the polarization of the material. In linear optics the polarization is always linearly proportional to the electric field strength. If it is assumed that in an isolator no current density is present, introducing (2.10) into the Maxwell equation (2.2) yields

$$\nabla \times \boldsymbol{B} = \varepsilon_0 \mu_0 \frac{\partial \boldsymbol{E}}{\partial t} + \mu_0 \frac{\partial \boldsymbol{P}}{\partial t}. \tag{2.11}$$

Analogously to the derivation of the wave equation in vacuum, one yields from the equations (2.5) and (2.11) the wave equation in insulators:

$$\Delta \boldsymbol{E} = \frac{1}{c^2} \frac{\partial^2 \boldsymbol{E}}{\partial t^2} + \frac{1}{\varepsilon_0 c^2} \frac{\partial^2 \boldsymbol{P}}{\partial t^2}. \tag{2.12}$$

According to (2.12), the wave in the material consists of two parts: the primary wave (first term) and the secondary wave (second term). The origin of the secondary wave is the polarization of the material due to the external field. The secondary wave has the same frequency but a phase shift relative to the primary wave. Both parts will superimpose during propagation in the material.

If the polarization in (2.12) is replaced by $\boldsymbol{P} = \varepsilon_0 \chi \boldsymbol{E}$ and if (2.9) is considered, then the wave equation in insulators is as follows

$$\Delta \boldsymbol{E} = \frac{\varepsilon_{\mathrm{r}}}{c^2} \frac{\partial^2 \boldsymbol{E}}{\partial t^2}. \tag{2.13}$$

In general, the material-dependent relative permittivity is complex (see Sect. 2.3) and a tensor of second rank. It is a scalar [2] only for isotropic materials. The square root of the permittivity is the complex refractive index of the material ($\hat{n}$), which is considered in detail in Sect. 2.3:

$$\hat{n}^2 = (n + \mathrm{j}\kappa)^2 = \varepsilon_{\mathrm{r}}. \tag{2.14}$$

Equation (2.13) can be used to describe all linear optical phenomena like refraction, reflection, scattering, diffraction and absorption. A comparison between the wave in vacuum (2.7) and the wave in a material (2.13) shows that the wave will propagate slower in the material (c/n). This behavior follows from the phase shift between primary and secondary wave. While the velocity of the waves in a vacuum is independent of frequency and is equal to the velocity of light, the velocity of electromagnetic waves in a medium depends on the properties of the material and on the frequency of the waves.

The origin of this behavior can be found in the frequency dependence of the susceptibility and therefore the linear part of the refractive index,

$n = n(\omega)$, which is treated in Sect. 2.6.1. Out of the range of material resonances the linear part of the refractive index increases with the frequency, this behavior is called normal dispersion. If the frequency is near the resonances in the material, the dependence of the refractive index on the frequency can be either positive or negative.

2.2 Solution of the Linear Wave Equation

Equation (2.13) is a vector equation where the vector $\boldsymbol{E}$ can consist of three parts. Assuming, for simplicity, a Cartesian coordinate system with the axes x, y and z the component of the field vector $\boldsymbol{E} = (E_x, E_y, E_z)$ in x-direction is represented by

$$\frac{\partial^2 E_x}{\partial x^2} + \frac{\partial^2 E_x}{\partial y^2} + \frac{\partial^2 E_x}{\partial z^2} = \frac{\hat{n}^2}{c^2}\frac{\partial^2 E_x}{\partial t^2}. \tag{2.15}$$

For the other components similar equations can be derived. Plane monochromatic waves depend only on one coordinate. Assume for instance a plane transverse wave propagating in the z direction. The transverse notation follows from the fact that the electric field vector $\boldsymbol{E} = (E_x, E_y, 0)$ is perpendicular to the propagation direction[4] z. In every plane perpendicular to z ($z = const$), the wave has the same phase everywhere. Therefore, these waves are called plane transverse waves. Due to the constant phase, the components of the field perpendicular to z are constant and hence

$$\frac{\partial \boldsymbol{E}}{\partial x} = \frac{\partial \boldsymbol{E}}{\partial y} = 0. \tag{2.16}$$

Therefore, the wave equation becomes

$$\frac{\partial^2 \boldsymbol{E}}{\partial z^2} = \frac{\hat{n}^2}{c^2}\frac{\partial^2 \boldsymbol{E}}{\partial t^2}. \tag{2.17}$$

According to (2.17), the solution of the wave equation for a plane, transverse, monochromatic wave propagating in the z-direction is

$$\boldsymbol{E}(z,t) = \frac{1}{2}\left(\hat{E}\mathrm{e}^{\mathrm{j}(\hat{n}k_0 z - \omega t)} + \text{c.c.}\right)\boldsymbol{e}_i = \left|\hat{E}\right| \cos\left(\hat{n}k_0 z - \omega t + \varphi_0\right)\boldsymbol{e}_i, \tag{2.18}$$

with

$$\hat{E} = \mathrm{Re}\left(\hat{E}\right) + \mathrm{j}\,\mathrm{Im}\left(\hat{E}\right) = a + \mathrm{j}b,$$

---

[4]Contrary to the longitudinal sound waves, electromagnetic waves are always transverse, because the electric and magnetic field vectors are always perpendicular to the wave vector.

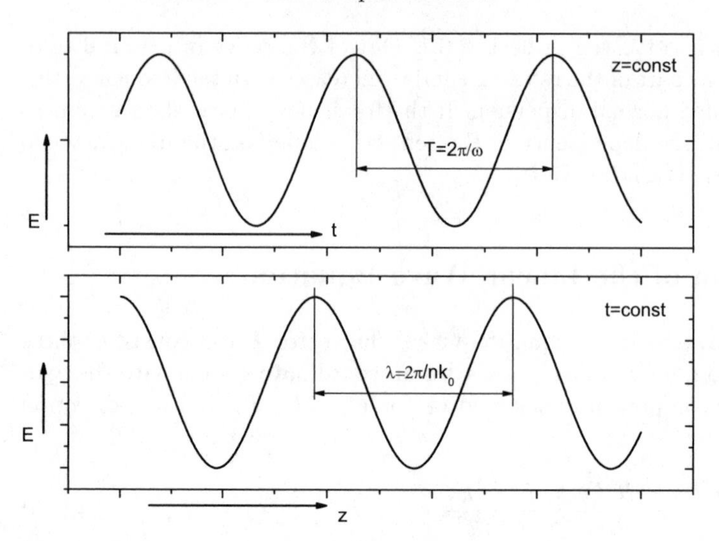

Fig. 2.1. Electric field of an electromagnetic wave for a constant time (*bottom*) and a constant position (*top*)

$$\left|\hat{E}\right| = \sqrt{a^2 + b^2},$$

and

$$\varphi_0 = \arctan \frac{b}{a},$$

where c.c. denotes the complex conjugate, Re and Im represent the real and imaginary part, respectively. The term $\hat{E}$ indicates the complex amplitude of the wave and contains the phase for $z = 0$, (φ_0):

$$\hat{E} = E_0 e^{j\varphi_0}, \tag{2.19}$$

and $\boldsymbol{e}_i$ denotes the unit vector for an arbitrary i-direction. For this particular case, it can show in a direction in the x-y plane. Equation (2.18) describes a monochromatic wave with the angular frequency $\omega = 2\pi f$ and the wavenumber $k_0 = 2\pi/\lambda$ propagating in the z-direction, as shown in Fig. 2.1. The wavenumber corresponds to the absolute value of the wave vector in vacuum $(\boldsymbol{k}_0)$. If the direction of the wave propagation corresponds to one axes of the coordinate system the vector equals its absolute value; $\boldsymbol{k}_0 = (0, 0, k_z = k_0)$.

As already mentioned, the superposition rule no longer holds in nonlinear optics, i.e., not only the sums and differences of the fields have to be considered, but products as well. Therefore, if the waves are expressed in a complex representation, their complex conjugate always has to be included in calculations. This follows from the fact that the real part of the product of two complex numbers z is not equal to the product of the real parts alone, i.e., $\mathrm{Re}(z_1 \times z_2) \neq \mathrm{Re}(z_1) \times \mathrm{Re}(z_2)$.

2.3 The Harmonic Oscillator Model

According to (2.12), the wave in the medium consists of two parts: (1) the primary wave and (2) the secondary wave that superimposes the primary one. The origin of the secondary wave is the polarization of the material. On the other hand, the polarization predicates something about the shift of charges inside an atom due to the external field. The external field changes with time and optical frequencies are rather high (a wavelength of 1.5 μm in free space corresponds to a frequency of 200 THz) so that the relatively heavy nucleus of the atom can not follow them. Hence, the primary wave will only excite the elastically bound valence electrons of the atom.

Due to the fact that the electrons are rather light, they are able to follow the fast fluctuating field. They will oscillate with the frequency of the external field around their equilibrium state. The electrons will be accelerated, but accelerated charges can emit electromagnetic waves by their own. Hence, the ensemble of an atom and oscillating electron can be seen as an irradiating dipole. The emission of the dipole has the same frequency as the external field and is the origin of the second term in (2.12).

Assuming the material consists of a huge number of identical dipoles or oscillators, ingrained in a host medium and oscillating with the frequency of the external field, the propagating wave in the material is then a consequence of the superposition of all emitted waves from the dipoles and the primary wave. The superposition can be described according to Huygens principle.

Figure 2.2 shows one of these oscillators; the light valence electron is represented by a mass m, the stationary atom is depicted by the fixed plate in Fig. 2.2, whereas the forces between atom and valence electron are represented by a spring.

The equation of motion of this oscillator is a consequence of the equilibrium between the inertial force, the frictional force the reset force on one side, and the force of the external electric field on the other. The inertial force

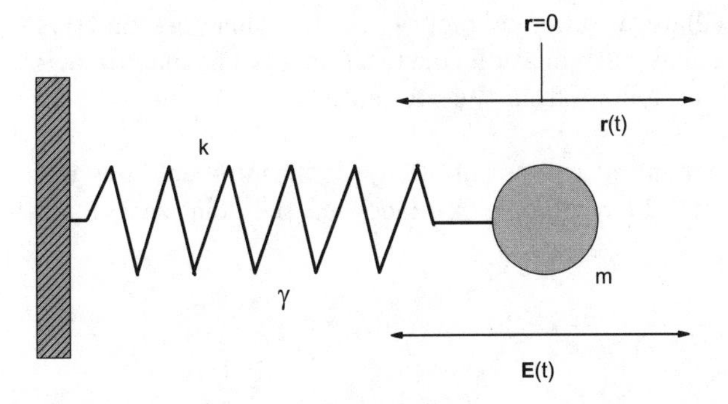

Fig. 2.2. Schematic illustration of a harmonic oscillator as a mechanical model (m=mass, k=spring constant, γ=attenuation constant, $\boldsymbol{r}$=space coordinate, $\boldsymbol{E}$=electrical field)

is proportional to the acceleration, where the constant of proportionality is the mass m. The frictional force is, for small excursions, proportional to the velocity with the constant of proportionality (γ), and the reset force is (for small excursions) – via the spring constant (k) – linearly proportional to the excursion of the oscillator.

The oscillator experiences an external electric field, $\boldsymbol{E}(t) = \boldsymbol{E}_0 \mathrm{e}^{\mathrm{j}\omega t}$, which sets it into oscillations around $\boldsymbol{r} = 0$. The proportionality constant for the electric force is the charge of the electron (q). Therefore, the equation of motion for the linear oscillator is

$$m \frac{\mathrm{d}^2 \boldsymbol{r}(t)}{\mathrm{d}t^2} + \gamma \frac{\mathrm{d}\boldsymbol{r}(t)}{\mathrm{d}t} + k\boldsymbol{r}(t) = q\boldsymbol{E}(t). \tag{2.20}$$

If (2.20) is divided on both sides by the mass and the resonance frequency of the oscillator is $\omega_0 = \sqrt{k/m}$, then

$$m\left(\ddot{\boldsymbol{r}}(t) + \gamma_{\mathrm{m}}\dot{\boldsymbol{r}}(t) + \omega_0^2 \boldsymbol{r}(t)\right) = q\boldsymbol{E}_0 \mathrm{e}^{\mathrm{j}\omega t}, \tag{2.21}$$

with $\gamma_m = \gamma/m$. The position coordinate $\boldsymbol{r}$ will periodically change with the frequency of the electrical field ω, this time dependence can be written as $\boldsymbol{r}(t) = \boldsymbol{r}_0 \mathrm{e}^{\mathrm{j}\omega t}$. Hence, (2.21) yields

$$m\left(\omega_0^2 - \omega^2 + \mathrm{j}\gamma_{\mathrm{m}}\omega\right)\boldsymbol{r}_0 = q\boldsymbol{E}_0. \tag{2.22}$$

The dipole moment of the oscillator will be changed periodically as well:

$$\boldsymbol{p}(t) = q\boldsymbol{r}(t) = \boldsymbol{p}_0 \mathrm{e}^{\mathrm{j}\omega t}, \tag{2.23}$$

with $\boldsymbol{p}_0 = q\boldsymbol{r}_0$, which according to (2.22) can be written as

$$\boldsymbol{p}_0 = \frac{q^2 \boldsymbol{E}_0}{m} \frac{1}{(\omega_0^2 - \omega^2) + \mathrm{j}\gamma_{\mathrm{m}}\omega}. \tag{2.24}$$

The whole medium consists of a large number of molecules or oscillators. Every molecule can oscillate in different modes and has therefore different resonance frequencies (ω_i) and attenuation constants ($\gamma_{\mathrm{m}i}$). The macroscopic polarization of the medium follows from the summation over all the different oscillators.

With l as the number of vibratory systems and f_i as the number of possible oscillating states each with the resonance frequency ω_i, the polarization ($\boldsymbol{P}$) of the medium is

$$\boldsymbol{P} = \frac{lq^2}{m} \sum_i \frac{f_i}{(\omega_i^2 - \omega^2) + \mathrm{j}\gamma_{\mathrm{m}i}\omega} \boldsymbol{E}(t) = \varepsilon_0 \chi(\omega) \boldsymbol{E}(t), \tag{2.25}$$

where f_i is also the so called oscillator strength that is proportional to the transition probability between different quantum states of an atom. The electrical susceptibility (χ) of the material in (2.25) is

$$\chi(\omega) = \frac{\boldsymbol{P}}{\varepsilon_0 \boldsymbol{E}(t)} = \frac{lq^2}{\varepsilon_0 m} \sum_i \frac{f_i}{(\omega_i^2 - \omega^2) + \mathrm{j}\gamma_{\mathrm{m}i}\omega} = \chi'(\omega) - \mathrm{j}\chi''(\omega). \quad (2.26)$$

$\sqrt{\frac{lq^2}{\varepsilon_0 m}}$ is defined as the plasma frequency ω_{p}. According to (2.26), the susceptibility is complex. The real and imaginary parts are then

$$\chi'(\omega) = \omega_{\mathrm{p}}^2 \sum_i \frac{f_i\left(\omega_i^2 - \omega^2\right)}{\left(\omega_i^2 - \omega^2\right)^2 + \left(\gamma_{\mathrm{m}i}\omega\right)^2} \quad (2.27)$$

and

$$\chi''(\omega) = \omega_{\mathrm{p}}^2 \sum_i \frac{\gamma_{\mathrm{m}i}\omega f_i}{\left(\omega_i^2 - \omega^2\right)^2 + \left(\gamma_{\mathrm{m}i}\omega\right)^2}. \quad (2.28)$$

According to (2.9) the material-dependent relative permittivity is

$$\varepsilon_{\mathrm{r}}(\omega) = 1 + \chi(\omega) = 1 + \omega_{\mathrm{p}}^2 \sum_i \frac{f_i}{(\omega_i^2 - \omega^2) + \mathrm{j}\gamma_{\mathrm{m}i}\omega} = \varepsilon_{\mathrm{r}}'(\omega) - \mathrm{j}\varepsilon_{\mathrm{r}}''(\omega). \quad (2.29)$$

The real and imaginary parts of the permittivity are then

$$\varepsilon_{\mathrm{r}}'(\omega) = 1 + \omega_{\mathrm{p}}^2 \sum_i \frac{f_i\left(\omega_i^2 - \omega^2\right)}{\left(\omega_i^2 - \omega^2\right)^2 + \left(\gamma_{\mathrm{m}i}\omega\right)^2} = 1 + \chi'(\omega) \quad (2.30)$$

and

$$\varepsilon_{\mathrm{r}}''(\omega) = \omega_{\mathrm{p}}^2 \sum_i \frac{\gamma_{\mathrm{m}i}\omega f_i}{\left(\omega_i^2 - \omega^2\right)^2 + \left(\gamma_{\mathrm{m}i}\omega\right)^2} = \chi''(\omega). \quad (2.31)$$

In Fig. 2.3, the real and imaginary parts of the material-dependent permittivity and susceptibility (+1) are shown for two resonance frequencies, the infrared and ultraviolet resonances of optical materials, for example.

The refractive index of the material is defined as the square root of the material dependent permittivity, hence

$$\varepsilon_{\mathrm{r}}(\omega) = \hat{n}^2(\omega). \quad (2.32)$$

The refractive index $\hat{n}^2(\omega)$ is complex with the real part n and the imaginary part κ:

$$\hat{n}(\omega) = n(\omega) + \mathrm{j}\kappa(\omega). \quad (2.33)$$

The imaginary part (κ) is called the extinction coefficient. According to the equations (2.33) and (2.32), one yields

$$\varepsilon_{\mathrm{r}}(\omega) = \hat{n}^2(\omega) = n(\omega)^2 - \kappa(\omega)^2 + 2\mathrm{j}n(\omega)\kappa(\omega). \quad (2.34)$$

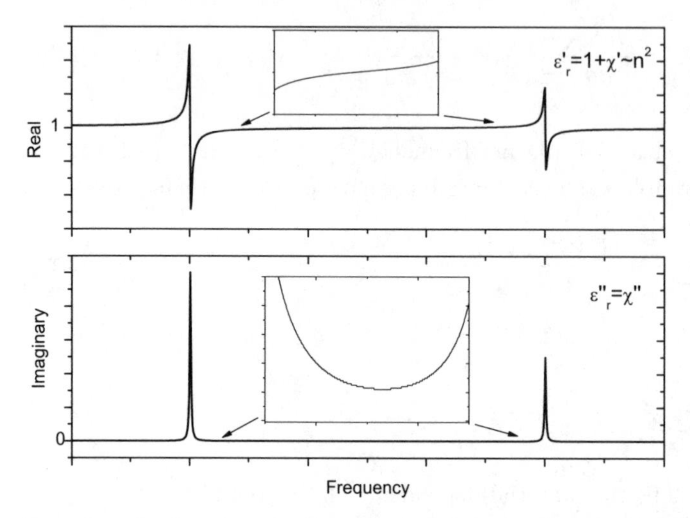

Fig. 2.3. Real and imaginary parts of the permittivity and the susceptibility for two resonance frequencies of the harmonic oscillators

Therefore, it follows with the equations (2.34), (2.32) and (2.29) that

$$n(\omega)^2 - \kappa(\omega)^2 = \varepsilon_{\mathrm{r}}'(\omega) \tag{2.35}$$

and

$$2n(\omega)\kappa(\omega) = \varepsilon_{\mathrm{r}}''(\omega) \tag{2.36}$$

If the extinction coefficient (κ) is much smaller than the real part of the refractive index one yields for the refractive index and the extinction coefficient:

$$n(\omega) \approx \sqrt{1 + \omega_{\mathrm{p}}^2 \sum_i \frac{f_i\left(\omega_i^2 - \omega^2\right)}{\left(\omega_i^2 - \omega^2\right)^2 + \left(\gamma_{\mathrm{m}i}\omega\right)^2}} = \sqrt{1 + \chi'(\omega)} \tag{2.37}$$

$$\kappa(\omega) = \frac{\varepsilon_{\mathrm{r}}''(\omega)}{2n(\omega)} = \frac{\omega_{\mathrm{p}}^2}{2n(\omega)} \sum_i \frac{\gamma_{\mathrm{m}i}\omega f_i}{\left(\omega_i^2 - \omega^2\right)^2 + \left(\gamma_{\mathrm{m}i}\omega\right)^2} = \frac{\chi''(\omega)}{2n(\omega)} \tag{2.38}$$

In this case, the real part of the refractive index is approximately equal to the square root of the real part of the material-dependent permittivity. The refractive index is responsible for a phase shift of the wave in the material, as will be shown in Sect. 2.6. According to Fig. 2.3 the refractive index, and therefore the phase shift of the secondary waves that were emitted by the oscillators, is a function of the frequency. Hence, this phase shift between primary and secondary wave depends on the frequency difference between the external field ω and the resonance frequency of the oscillator ω_i. Near a resonance the refractive index can increase or decrease, but between two resonances it will always increase with the frequency, or decrease with the wavelength. This wavelength dependence of the refractive index is the origin of the so called dispersion.

The extinction coefficient is, according to (2.38), coupled with the imaginary part of the material-dependent permittivity. According to Fig. 2.3, this imaginary part is especially large if the frequency of the external field is near a resonance of the medium, i.e., $\omega \approx \omega_i$.

The imaginary part leads to an absorption of the wave in the material. If the frequency of the electromagnetic wave that excites the oscillators is far enough from resonances in the material, then the real part will be much stronger than the imaginary.

2.4 Attenuation

If the complex refractive index according to (2.33) is introduced into the solution of the wave equation (2.18) the wave propagating in an insulator is then

$$\begin{aligned} \boldsymbol{E}(z,t) &= \frac{1}{2}\left(\hat{E}\mathrm{e}^{\mathrm{j}((n+\mathrm{j}\kappa)k_0 z-\omega t)} + c.c.\right)\boldsymbol{e}_i \\ &= \frac{1}{2}\left(\hat{E}\mathrm{e}^{-\kappa k_0 z}\mathrm{e}^{\mathrm{j}(nk_0 z-\omega t)} + c.c.\right)\boldsymbol{e}_i . \end{aligned} \tag{2.39}$$

The exponential function $\hat{E}\mathrm{e}^{-\kappa k_0 z}$ describes the decrease of the amplitude of the wave with distance z. The intensity of the wave is proportional to the square of its amplitude. Hence, the result for the intensity will be

$$I = \frac{1}{2}\varepsilon_0 c n \left|\hat{E}\right|^2 , \tag{2.40}$$

with

$$I(z) = I_0 \mathrm{e}^{-2\kappa k_0 z} = I_0 \mathrm{e}^{-\alpha z} , \tag{2.41}$$

where $\alpha = 2\kappa k_0 = \frac{4\pi}{\lambda_0}\kappa$ is the so called attenuation constant in km^{-1}. In the wavelength range of telecommunications, optical fibers will only absorb a very small part of the electromagnetic wave. Ordinary glass for windows has for instance an attenuation constant of $\alpha = 11500\ \mathrm{km}^{-1}$, whereas the attenuation constant of silica glass used for optical fibers is only $\alpha \approx 0.04 - 0.07\ \mathrm{km}^{-1}$ for a wavelength of 1.5 µm.

Assuming a lightwave with a power of 10 mW and using the attenuation constant in ordinary glass, the power of the wave after passing through a 6 cm thick window is 5 mW. On the other hand, if the wave propagates in an optical fiber, this power loss will not be reached before a distance of 15 km.

Nevertheless, the power losses in optical fibers are not negligible because optical fibers for transmission systems are very long. A transatlantic link between the west coast of Ireland and Boston has, for instance, a direct length of $\approx$ 4600 km. Without amplifiers for the compensation of the attenuation this distance can not be bridged even with the best optical fibers of today.

The attenuation in optical fibers has in principle two origins: first, the scattering of light into all directions at density fluctuations or due to other impurities in the material; and second, the absorption.

The absorption has its origin in the electronic resonances of the oscillators in the media. For optical fibers these oscillators are built by the molecules of the basic material (silica glass) and the impurities in the fiber. According to Fig. 2.3 the attenuation constant is especially large near the resonance frequencies of the oscillators. This happens because in the case of a resonance the strongly oscillating molecules detract optical power from the external wave for their motion. The power of the electromagnetic field will be converted into the kinetic energy of the oscillators and therefore the temperature of the material will increase. This will not only happen near the resonances, but also if the frequency of the external field is near a harmonic of the fundamental resonance.

The electronic resonances of the basic material of an optical fiber are in the UV-range for wavelengths smaller than 0.4 μm and in the infrared for $\lambda > 2\mu$m. In the frequency range that is used for telecommunications, harmonics from resonances of impurities, especially OH ions, are much more important. These ions are residuals of the water required for the production of the fiber. The basic stretching vibration of the O-H bond is centered at $f_{\mathrm{OH}} = 2.73$ μm, hence the first harmonic is near 1.365 μm ($2f_{\mathrm{OH}}$) and the third near 0.91 μm ($3f_{\mathrm{OH}}$). Due to a mixing with a resonance of the silica glass at $f_{\mathrm{SiO}} = 12.5$ μm, there are additional resonance near 1.24 μm ($2f_{\mathrm{OH}} + f_{\mathrm{SiO}}$) and a much smaller near 1.13 μm ($2f_{\mathrm{OH}} + 2f_{\mathrm{SiO}}$) (see Fig. 2.4). In new developments in the production process of optical fibers, no water is required, hence the above resonances are not present in these fibers.

The other source of attenuation in the fiber is the scattering of the optical wave due to density fluctuations or impurities. The density fluctuations have their origin in the thermal motions of the molecules in the molten glass used as a preform to draw the fiber. These density fluctuations are frozen during the production process of the fiber and can not be avoided. The impurities are atoms like germanium and phosphor, used to change the refractivity and therefore to adjust a particular refractive index regime in the fiber. This process is called doping.

The resulting scattering is called Rayleigh scattering (Sect. 10.1) and is proportional to the fourth power of the frequency of the lightwave. Hence, light with a short wavelength is scattered much stronger than light with a longer wavelength. The attenuation due to scattering in an optical fiber is shown in Fig. 2.4 (dashed line). The attenuation for a wavelength of 0.9 μm is more than 1 dB/km, whereas this attenuation is reduced to $\approx$ 0.15 dB/km for a wavelength of 1.5 μm.

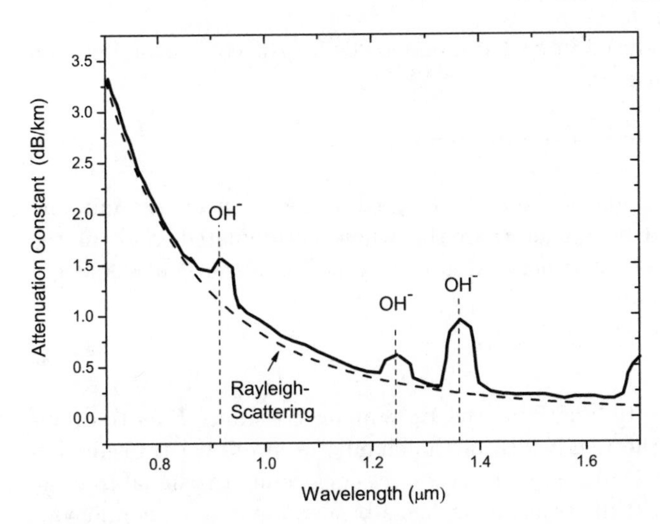

Fig. 2.4. Attenuation in a fiber made of silica glass, the dashed line depicts the Rayleigh scattering (OH^- hydroxyl ions)

The intensity of a wave is its power in relation to the area the wave will propagate through:

$$I = \frac{P}{A_{\text{eff}}}, \tag{2.42}$$

with A_{eff} as the effective area of the fiber core (see Sect. 4.8). If the effective area remains constant, not only the intensity, but the power of the wave is proportional to the amplitude squared as well. Hence, for the decrease of the power at a distance z, a similar rule as for the intensity (2.41) must hold. Therefore, the power of the wave in the material is

$$\frac{\mathrm{d}P(z)}{\mathrm{d}z} = -\alpha P(z) \tag{2.43}$$

For the length $z = L$, (2.43) yields

$$P(L) = P(0)\mathrm{e}^{-\alpha L}, \tag{2.44}$$

where $P(0)$ depicts the power at the fiber input. From (2.44), the attenuation constant is

$$\alpha_{\text{km}^{-1}} = -\frac{1}{L}\ln\frac{P(L)}{P(0)}. \tag{2.45}$$

In optical telecommunications, the attenuation constant related to the logarithm with base 10 ($\alpha_{\text{dB/km}}$) is more usable:

$$\alpha_{\text{dB/km}} = -\frac{10}{L}\log\frac{P(L)}{P(0)}. \tag{2.46}$$

With (2.46) introduced into (2.45), one yields for the conversion between both attenuation constants:

$$\alpha_{\mathrm{km}^{-1}} = \frac{\alpha_{\mathrm{dB/km}}}{10} \ln 10 \approx 0.23026\ \alpha_{\mathrm{dB/km}}. \tag{2.47}$$

For a secure detection of the optical signal at the receiver, the minimal optical power has to be higher than the whole accumulated noise in the system.[5] Beside the thermal noise, due to the motion of molecules and the electrons in the detector,

$$N = kTB, \tag{2.48}$$

with $k = 1.380658 \times 10^{-23}\mathrm{J/K}$ as the Boltzmann constant, T as the temperature, and B as the bandwidth of the channel – in optical transmission systems there are some additional noise sources degrading the signal to noise ratio. The laserdiode of the transmitter has a basic noise due to spontaneous emission and the recombination of electrons and holes in the semiconductor material (shot noise) and the fluctuation of the amplitude and phase of the emitted light (relative intensity noise RIN). Nonlinear effects in the fiber will increase the noise as well. This subject will be treated in detail in the corresponding chapters of this book. Other sources of noise include reflections on plugs and other devices in the system. Amplifiers inside the transmission link will not only show stimulated emission, but a part of the amplified intensity comes from the spontaneous emission of the amplifier. This part will increase with the amplification and is called amplified spontaneous emission (ASE).

2.5 Amplification

Before optical amplifiers were installed, optical telecommunications was in principle nothing other than the connection between two points with optical fibers. At the end of the fiber the optical signal was transformed into the electrical domain. All functionality, essential for a network, was done with electronic devices. The signal was regenerated and amplified and may in turn switch on another link. Afterwards the signal was transformed back and the procedure repeated.

This picture completely changed with the implementation of the erbium-doped fiber amplifier (EDFA). With this device it was possible, for the first time, to skip the electrical domain and hence, to transmit optical signals

[5] This rule does not hold for a code division multiple access (CDMA) system where each signal bit is multiplied with a code known by the receiver. Therefore, a correlation between the detected noisy signal and the code in the receiver can increase the signal contrary to the noise. This is the so called process gain and depends on the bit rate of the code. On the other hand, due to the multiplication with the temporally short code, for a CDMA-system more bandwidth is required.

over large distances. Together with optical switches and multiplexers it was, in principle, possible to build transparent optical networks.

Contrary to conventional electronic amplifiers, the amplification of an EDFA is independent of the data rate and the modulation format of the data. An EDFA requires no additional fast electronic modules. The amplification bandwidth of the EDFA is sufficient for the C-band (see Fig. 3.4) of optical telecommunications. Under special circumstances it is possible to enhance the bandwidth in a way that simultaneous amplifications in the C- and L-band are possible. Therefore EDFAs are predestined for the WDM systems of today. All these points lead to a strong enhancement of the capacity of optical communication systems, whereas at the same time the cost per bandwidth could be reduced.

With the implementation of optical amplifiers it was possible to transmit optical signals over large distances. On the other hand, this increased the effective length and therefore the impact of nonlinear effects as well (see Sect. 4.9).

The first EDFA was introduced in 1987 [3, 4]. It consists mainly of an optical fiber doped with erbium-ions. Usually semiconductor lasers with wavelengths of 980 or 1480 nm are used as pump sources for the EDFA. The injection of the pump wave is possible from the input and the output side of the fiber.

If the EDFA is pumped with a wavelength of 1480 nm, the erbium ions in the fiber are excited into a higher energy level, resulting in a high inversion state. The erbium ion energy levels are separated into various sublevels but first, only one ground and excited state is assumed. An optical wave with the right wavelength, propagating in the fiber, will be amplified when the Erbium ions fall back onto their ground state. In this case, a photon with the energy $\hbar\omega$ will be emitted. This energy, and therefore the frequency, corresponds to the energy difference between ground and excited state of the ion. Due to the separation of the energy diagram into sublevels, the pump and emitted frequencies are different. Because the decay was initiated by an optical wave, the emitted photons have the same wavelength and phase. Hence, the superposition with the original wave leads to its amplification. The EDFA can thus be seen as a laser without resonator.

If a pump wavelength of 980 nm is used, a third energy level will be involved; in this case the pump excites the ions first into a metastable inter state. From the third level they can cross emissionless to the second level. The ions will stay for around 10 ms in this second level before they decaying spontaneously to the ground level. Due to the slow dynamics, intersymbol interference and cross talk between the channels is negligible in the EDFA. For a pump wavelength of 980 nm a complete inversion is possible. This is not the case for a pump wavelength of 1480 nm, but the quantum efficiency will be higher at 1480 nm [5].

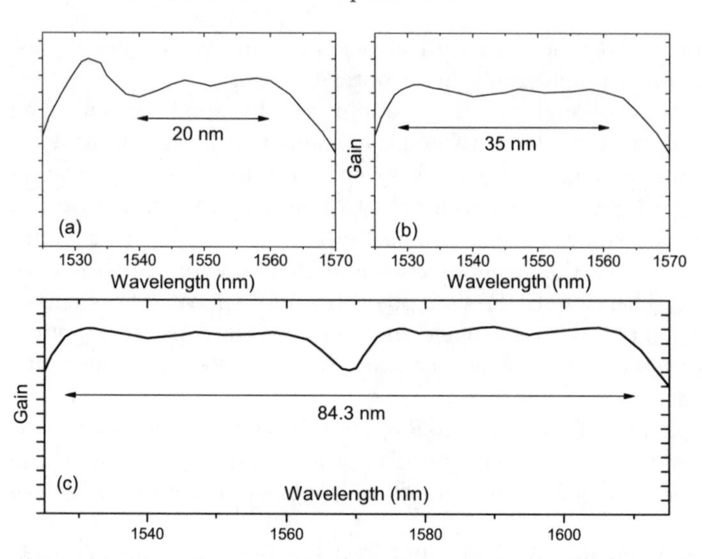

Fig. 2.5. Logarithmic amplification of an EDFA against the wavelength. **a** Amplification for an inversion of 40%-60%. **b** Amplification profile if an additional filter is used. **c** Amplification for the C- and L-band with two EDFA's with different inversions and lengths [5]

The separation of the erbium ion energy level into sublevels has the advantage that a frequency or wavelength band can be amplified. If all wavelength channels of a WDM system should be amplified simultaneously, a flat amplification gain over the whole bandwidth of the amplifier is required. The amplification gain of a pure EDFA is only flat in the range between 1540 and 1560 nm, as shown in Fig. 2.5a. Hence, without additional devices the bandwidth of an EDFA is only 20 nm. In order to enhance this range to 35 nm, special filters are required to reduce the peak of the amplification profile at 1530 nm (Fig. 2.5b).

At a much lower inversion level an amplification in the L-band is likewise possible with an EDFA, but with a much smaller gain than in the C-band. An EDFA with a high inversion level and gain in the C-band and a second one with a much lower inversion level and gain in the L-band can be used to built an amplifier for the C- and L- band together (84 nm) (Fig. 2.5c). For a flat amplification gain over the whole bandwidth the difference in the gain is compensated by different lengths of the doped fibers [5].

The enhancement of the amplification bandwidth of an EDFA to 84 nm is associated with a rather high effort. Due to the fact that the amplification of an EDFA is based on the resonances of erbium ions, a further enhancement of its bandwidth seems to be impossible. On the other hand, the transparency range of modern optical fibers is more than 300 nm. Therefore, in the last few years other amplification mechanisms, especially the Raman and the parametric amplifier came again into consideration. The Raman amplifier,

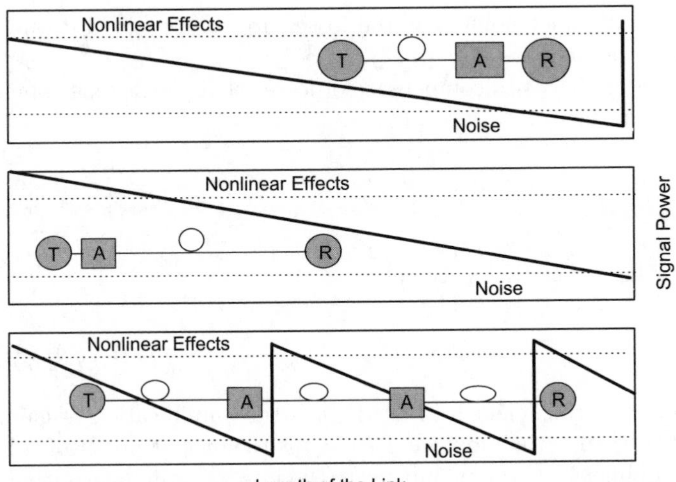

Fig. 2.6. Signal power and possibilities for an amplification (T=Transmitter, R=Receiver, A=Amplifier)

for instance, is only dependent on the pump wavelength and can therefore be used in the whole transparency range of the fiber. With parametric amplifiers an amplification bandwidth of around 200 nm is possible. The exploitation of nonlinear optical effects for amplification is discussed in depth in Chap. 13.

Conventional amplifiers can, in principle, be used in three distinct ways for optical communication systems, as shown in Fig. 2.6 and as follows:

- As input amplifiers in the receiver to increase the signal power before detection. In this case the noise is amplified as well and the signal to noise ratio remains constant.
- As power boosters in the transmitter to amplify the input power before it is transmitted over the system. This set-up can alter the signal to noise ratio and therefore the bit error rate of the system. If the amplification of the amplifier is to high, the risk for signal degradation due to nonlinear effects increases, whereas if it is to low, the signal to noise ratio in the receiver decreases and hence the bit error rate will increase.
- As amplifiers inside the transmission system to compensate for losses in the link. The origin of these losses are for instance fiber losses, or the losses in passive components like switches, cross connects and multiplexers. In this case the cost and complexity of the system will increase. Furthermore, if the number of intermediate amplifiers is increased, due to the amplified spontaneous emission ASE of each amplifier, the noise will increase as well.

Nonlinear amplifiers, based on the Raman effect for example, offer the possibility for a distributed amplification of the signal. In this case the signal is not amplified at a distinct point, but the losses are compensated during the transmission in the fiber (see Chap. 13).

The task of an amplifier is to compensate of losses that the signals experience at a distance L, hence:

$$\frac{\mathrm{d}P(z)}{\mathrm{d}z} = gP(z). \tag{2.49}$$

If the amplifier has a length L_A, then

$$P(L_\mathrm{A}) = P(0)\mathrm{e}^{gL_\mathrm{A}}. \tag{2.50}$$

Like the attenuation coefficient α in (2.41), g is the amplification or gain coefficient in (2.49), $P(0)$ is the input, and $P(L_\mathrm{A})$ is the output power of the amplifier. If the amplifier is used to compensate the losses, so that the output power of the device is equal to the power at the fiber input, a comparison between the equations (2.50) and (2.44) shows that the gain coefficient is

$$g = \alpha \frac{L}{L_\mathrm{A}}. \tag{2.51}$$

The gain of an amplifier (G) is defined as the ratio between input and output power of the device as follows:

$$G = \frac{P(L_\mathrm{A})}{P(0)} = \mathrm{e}^{gL_\mathrm{A}}. \tag{2.52}$$

2.6 The Refractive Index

In linear optics, the refractive index is responsible for an alteration of the phase of an electromagnetic wave propagating in a material. As shown in Sect. 2.4, the imaginary part of the complex refractive index leads to an attenuation of the wave in the medium. We will discuss in this section the real part of the refractive index.

The real part is responsible for phenomena like reflection, refraction and scattering. For optical telecommunications especially an effect called dispersion is very important. Dispersion is a result of the wavelength dependence of the refractive index which leads to different velocities for the components of the signal and, hence, a distortion.

2.6.1 The Sellmeier Equation

The solution of the wave equation in insulators can be described with (2.39) if the term responsible for attenuation is added to the complex amplitude, as follows:

$$\boldsymbol{E}(z,t) = \frac{1}{2}\left(\hat{E}\mathrm{e}^{\mathrm{j}((n-j\kappa)k_0 z-\omega t)} + \mathrm{c.c.}\right)\boldsymbol{e}_i = \frac{1}{2}\left(\hat{E}(z)\mathrm{e}^{\mathrm{j}(nk_0 z-\omega t)} + \mathrm{c.c.}\right)\boldsymbol{e}_i, \tag{2.53}$$

with

$$\hat{E}(z) = \hat{E}\mathrm{e}^{-\alpha/2z} \tag{2.54}$$

as a complex amplitude dependent on z. The phase of this wave is the argument of the exponential function:

$$\Phi = nk_0 z - \omega t. \tag{2.55}$$

If this result is compared with the phase of a wave in vacuum, it follows that the real part of the refractive index is responsible for an alteration of the phase of the wave. Since the real part of the refractive index is dependent on the frequency (2.37), the phase of the wave is frequency-dependent as well. According to (2.37) the real part of the refractive index is

$$n^2(\omega) \approx 1 + \omega_\mathrm{p}^2 \sum_i \frac{f_i\left(\omega_i^2 - \omega^2\right)}{\left(\omega_i^2 - \omega^2\right)^2 + \left(\gamma_{\mathrm{m}i}\omega\right)^2}. \tag{2.56}$$

For the calculation of the refractive index, information about the oscillators in the material is essential. The origin of the refractive index is the shift of the charges in the atoms or molecules of an insulator due to the external electric field. This effect is called polarization. If the frequency of the external field is to high, only the relatively light valence electrons can follow. Hence, this kind of polarization is called an electronic polarization. But if the frequency is low enough the ions of the crystal lattice can follow it as well. In this case an ionic polarization takes place.

Depending on the different masses and bond energies, the resonance frequencies of the electronic polarization are in the UV-range of the electromagnetic spectrum (for silica glass for instance at ≈ 68 and 116 nm). On the other hand, the ionic polarization has resonance frequencies in the infrared (for silica glass at 9.89 μm).

Equation (2.56) is an approximation that only holds if the imaginary part of the refractive index is much smaller than the real part. This condition, is only fulfilled for frequencies of the external field that are far enough from resonances of the material.

If the attenuation of the oscillators is neglected ($\gamma_{\mathrm{m}i} \approx 0$), (2.56) can be written as

$$n^2(\omega) = 1 + \omega_\mathrm{p}^2 \sum_{i=1}^{N} \frac{f_i}{\omega_i^2 - \omega^2} \tag{2.57}$$

with $f_i = \omega_{\mathrm{p}i}^2/\omega_\mathrm{p}^2$ this leads to the Sellmeier equation [6]:

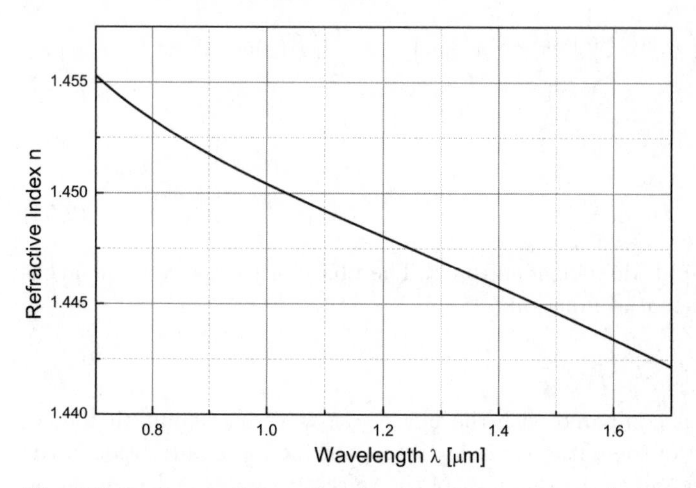

Fig. 2.7. Real part of the refractive index for pure silica glass versus the wavelength, according to the Sellmeier equation

$$n^2(\lambda) = 1 + \sum_{i=1}^{N} \frac{S_i \lambda^2}{\lambda^2 - \lambda_i^2} \tag{2.58}$$

where $\omega = 2\pi f$, and $c = \lambda f$ was used. The summation in (2.58) will be carried out over all material resonances λ_i that are important for the considered frequency range. In most cases, for the calculation of the refractive index two electronic and one ionic resonance will be sufficient. The parameters $S_i = \lambda_i^2/\lambda_{\mathrm{p}i}^2$ and λ_i in (2.58) can be found experimentally. For pure silica glass Table 2.1 shows these parameters. With these values and (2.58) the refractive index against the frequency or wavelength can be calculated. The result for pure silica glass is shown in Fig. 2.7.

Table 2.1. Values for the Sellmeier parameters for pure silica [7]

$S_1 = 0.6961663$	$\lambda_1 = 0.0684043 \mu m$
$S_2 = 0.4079426$	$\lambda_2 = 0.1162414 \mu m$
$S_3 = 0.8974794$	$\lambda_3 = 9.896161 \mu m$

A comparison between the inset of Fig. 2.3 and Fig. 2.7 shows the same behavior of the real part of the refractive index. If the frequency of the optical field is far enough from any resonances in the material, the refractive index will decrease with wavelength or increase with the frequency. For the field of optical telecommunications, the behavior around a wavelength of 1.27 μm is especially important. For pure silica the refractive index shows an inflection at this location.

2.6.2 Phase Velocity, Group Velocity, and Dispersion

In this section, we will discuss in detail the different velocities of an electromagnetic wave and their influences on optical telecommunications. Photons have only one velocity. This velocity is determined by the velocity of light, $c \approx 3 \times 10^8$ m/s. In a material the photons move with the velocity of light in the free space between the atoms. If they hit an atom they will be absorbed and can be re-emitted by the atom. In the field of linear optics, this re-emmission takes place with the same frequency, but at a phase shift with respect to the absorbed photon. According to (2.55), the phase of the wave in the material is:

$$\Phi(z,t,\omega) = n(\omega)k_0 z - \omega t. \tag{2.59}$$

If the waveguide properties of the wave number k are neglected, it can be written as $k = n(\omega) \times k_0 = n(\omega) \times \omega/c$. Hence, the wave number is frequency-dependent $k(\omega)$. This fact has a large impact on the propagation of wave packets in media and is referred to as the dispersion relation.

Let's assume an observer is moving with the wave and watches its phase. If the observer has the velocity $v = \mathrm{d}z/\mathrm{d}t$, then he will see a phase alteration of

$$\frac{\partial \Phi}{\partial t} = k(\omega)\frac{\partial z}{\partial t} - \omega = k(\omega)v - \omega. \tag{2.60}$$

If the observer has the velocity $v = \omega/k(\omega)$ and the wave is monochromatic, then $\omega =$ const and, therefore, $k(\omega) =$ const. Hence, the phase alteration that the observer watches is:

$$\frac{\partial \Phi}{\partial t} = k\left(\omega\right)\frac{\omega}{k\left(\omega\right)} - \omega = 0 \tag{2.61}$$

Therefore, the phase remains constant for the observer i.e., the observer moves with a velocity equal to the velocity with which each point of constant phase (for instance the maximum of the wave) will move. The velocity,

$$v_{\mathrm{ph}} = \frac{\omega}{k(\omega)}, \tag{2.62}$$

is called the phase velocity of the wave. In a vacuum, the wave number is $k_0 = \omega/c$ and the phase velocity is independent of the wavelength and equal to the velocity of light (c). In a medium, the phase velocity contains the frequency-dependent refractive index. Hence, the phase velocity in a medium is a function of the frequency or wavelength:

$$v_{\mathrm{ph}} = \frac{\omega}{k(\omega)} = \frac{c\omega}{n(\omega)\omega} = \frac{c}{n(\omega)}. \tag{2.63}$$

According to (2.63) and the refractive index of Fig. 2.7 the phase velocity against the wavelength for silica glass can be calculated as shown in Fig. 2.8.

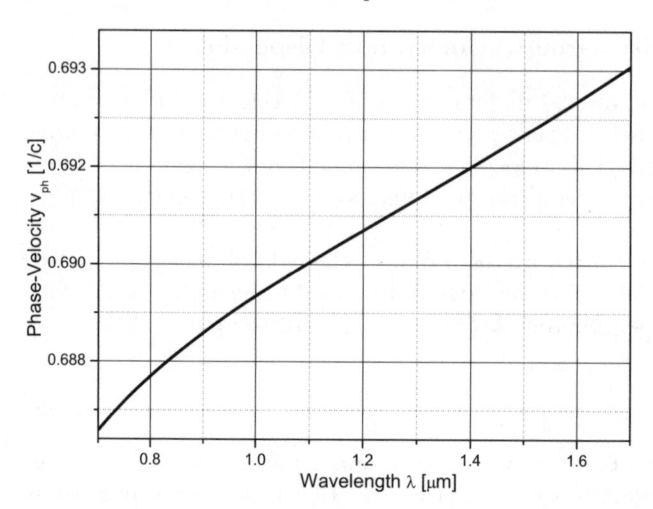

Fig. 2.8. Phase velocity in relation to the velocity of light for pure silica glass against the wavelength

According to Fig. 2.8 the phase velocity of an electromagnetic wave with a wavelength in the telecommunications range in pure silica glass is around 2/3 of the velocity of light in free space and the phase velocity increases with wavelength.

If a mixture of waves with different wavelength instead of a monochromatic wave is injected into the medium, the different spectral components of this mixture will move with different phase velocities through the material. If, for instance, white light is used as the mixture, the different colors of the light will arrive at the output of the material at different times. This phenomenon is the origin of the separation of white light into all colors of the rainbow at glass edges, crystal glasses or diamonds.

Most materials show, as does silica glass, a so called normal dispersion, i.e., as the wavelength increases the refractive index decreases and the phase velocity increases. Anomalous dispersion, the converse behavior, is only achievable if the frequency is near a resonance frequency of the material. But in this case the attenuation is very large, as can be seen in Fig. 2.3.

Until now, we have been discussing waves that are either monochromatic or a mixture of monochromatic waves injected into the material. But the transmission of information is only possible if the wave is modulated. This means that a parameter of a carrier wave (amplitude, frequency or phase) is altered in dependence on the signal that has to be transmitted. According to the theorems of Fourier, the time-dependent alteration of parameters of a monochromatic carrier results in new frequency components. Hence, every modulated wave consists of different frequency components. A very short amplitude in time results, for instance, in a very broad spectrum and vice

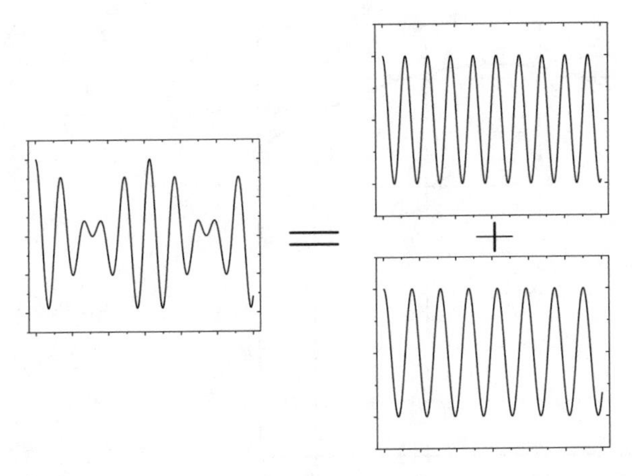

Fig. 2.9. Fading between two waves with different frequencies. The amplitude modualted wave on the *left* side consists of the two monochromatic waves on the *right*

versa. According to this, a monochromatic wave is infinitely long in the time domain.

Additionally, not just one channel should be transmitted over one waveguide, but – in order to save costs – as many as possible. If the optical division multiple access (OTDMA) scheme is used, the channels are transmitted in timeslots. If the number of channels should be increased, the duration of the timeslots has to decrease. Hence, the pulses of high bit rate OTDMA systems have a rather broad spectrum.

The left side of Fig. 2.9 shows an amplitude-modulated wave. This wave consists of the two frequencies ω_1 and ω_2 and can be seen as the superposition of the two waves on the right side.

At all points where the two waves are in phase they will interfere constructively, whereas a destructive interference occurs for the points where this is not fulfilled. Both waves move with their phase velocity through the medium. If the formerly mentioned observer were to follow the envelope of the amplitude modulated wave, he has to move with a velocity that corresponds to the phase difference between the two waves. If this phase difference remains constant, the interference and therefore the phase of the resulting wave will remain constant as well. The two waves have the frequencies ω_1 and ω_2 and the wave vectors k_1 and k_2. Therefore, the velocity of the observer (v) can be calculated:

$$k_1 v - \omega_1 = k_2 v - \omega_2$$

$$v = \frac{\omega_1 - \omega_2}{k_1 - k_2}. \tag{2.64}$$

The bits for optical telecommunications do not only consist of two frequencies. If only one optical pulse is assumed, according to the Fourier trans-

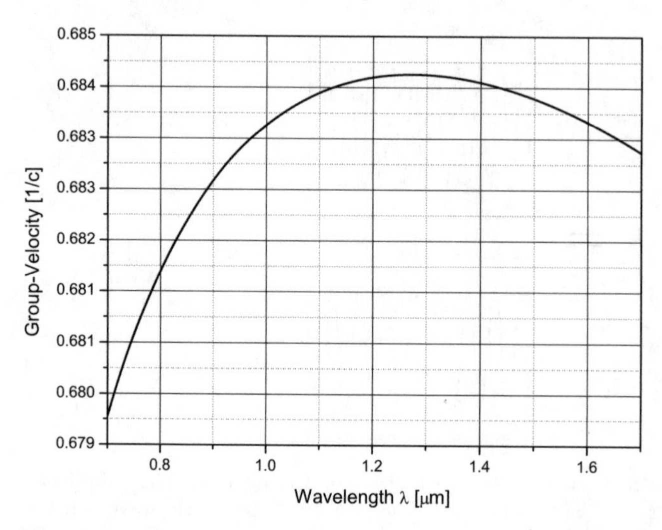

Fig. 2.10. Group velocity for pure silica glass as a function of the wavelength

formation [8]

$$B(t) = \frac{1}{\sqrt{2\pi}} \int_{-\infty}^{+\infty} \tilde{B}(\omega) \mathrm{e}^{-\mathrm{j}\omega t} \mathrm{d}\omega, \tag{2.65}$$

the spectrum is theoretically infinite. The spectrum of the pulse is shifted by the optical carrier frequency. Due to (2.65), there are infinite sidebands on both sides of the carrier, but the higher sidebands decay as faster as longer the pulse duration is. If the pulse duration is infinite it theoretically consists of only one single wavelength (continuous wave or cw). According to the above discussion, the optical pulse transmitting information along the fiber consists of a group of waves with different frequencies. From (2.64), the velocity of this wave group is

$$v_\mathrm{g} = \frac{\mathrm{d}\omega}{\mathrm{d}k}. \tag{2.66}$$

This is the so called group velocity of the pulse (v_g), the velocity with which the maximum of the pulse propagates through the waveguide. Using (2.63), the group velocity (2.66) can be written as

$$v_\mathrm{g} = \frac{c}{n + \omega \frac{\mathrm{d}n}{\mathrm{d}\omega}}. \tag{2.67}$$

The group velocity against the wavelength for pure silica glass is shown in Fig. 2.10. According to (2.67), the group velocity depends on the slope of the refractive index. Hence, at the inflection point of the refractive index curve (around 1.27 μm, see Fig. 2.7) a maximum of the group velocity occurs.

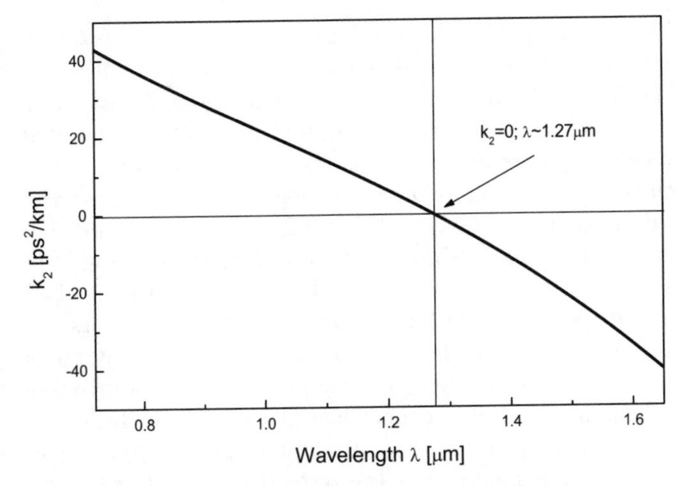

Fig. 2.11. Group velocity dispersion for pure silica glass

As mentioned earlier, if the frequency components of the pulse are in phase, they will interfere constructively. In this case the pulse is narrow and has a high amplitude. Assuming such a pulse is injected into a waveguide consisting of pure silica glass and the pulse has a wavelength of 1.6 μm, the different frequency components propagate with different velocities in the material, as shown in Fig. 2.10. Hence, they will become out of phase. The duration of the pulse is thus broadened and the amplitude decreases.

The alteration of the propagation time of the group, $\tau_g = L/v_g$ per distance L, is the dispersion of the group velocity, or the group velocity dispersion (GVD, k_2). Using the product rule (2.67) yields

$$k_2 = \frac{1}{L}\frac{\mathrm{d}\tau_g}{\mathrm{d}\omega} = \frac{\mathrm{d}}{\mathrm{d}\omega}\left(v_g^{-1}\right) = \frac{1}{c}\left(2\frac{\mathrm{d}n}{\mathrm{d}\omega} + \omega\frac{\mathrm{d}^2 n}{\mathrm{d}\omega^2}\right). \tag{2.68}$$

The group velocity dispersion for pure silica glass is shown in Fig. 2.11. The group velocity dispersion is zero where the group velocity has its maximum. According to the behavior of the group velocity, the dispersion is positive when the group velocity increases and it is negative when it decreases. Since (2.68) involves the second derivation of the refractive index, the curve is zero at the inflection point (at 1.27 μm) of the refractive index. This behavior depends on the optical properties of the material and is therefore often referred as material dispersion (k_M).

According to the normal and anomalous dispersion already mentioned in the case of the phase velocity, this concept can be introduced for the group velocity as well. If the group velocity increases with wavelength – or k_2 is positive – the behavior is called normal, otherwise it is called anomalous.

Until now we have only considered pure silica glass as a bulk material, but in the field of optical telecommunications very thin waveguides are used for the transmission of information. The construction of the waveguides obviously has an influence on the dispersion. The effective refractive index is, due to the waveguide properties, a little smaller than in the bulk material. Based on its origin, this behavior is called waveguide dispersion (k_W). Both the material and the waveguide dispersion are dependent on the wavelength. Hence, they are summarized under the name chromatic dispersion. While the material dispersion changes its sign – between $k_W = 20$ ps^2/km and $k_W = -28$ ps^2/km for wavelengths in the range from 1.015 μm to 1.55 μm, as shown in Fig. 2.11 – the waveguide dispersion is always positive and has a value of $k_M \approx 1.02$ ps^2/km [54] in standard single mode fibers. Hence, in standard fibers out of the range of the zero-crossing of the material dispersion, the waveguide dispersion is negligible. But near the zero-crossing point it can be involved in the compensation of the material dispersion. Since the waveguide dispersion is positive it shifts the zero-crossing of the chromatic dispersion to a higher wavelength. In standard single mode fibers the zero-crossing is at ≈ 1.3 μm.

The waveguide dispersion depends on constructive parameters of the waveguide such as the core diameter and the refractive index difference between core and cladding. Hence, the waveguide dispersion can be altered by constructive means and the zero-crossing point can be shifted further to higher wavelengths. From a technical point of view, the minimum of the attenuation of the fiber at 1.5 μm (see Fig. 2.4) is especially interesting. These fibers, with a zero-crossing of the chromatic dispersion in the range of the minimum of the attenuation, are called dispersion-shifted fibers (DSF). In the following chapters the term dispersion always means the chromatic dispersion of the waveguide.

In optical telecommunications light pulses propagate in a waveguide. Hence, the wave equation can not be solved with the ansatz for monochromatic waves according to (3.4). The solution of the wave equation is discussed in Sect. 3.2 and only the consequences for the wave vector k will be considered here.

According to (2.65) optical pulses consist of different frequencies centered around a central frequency ω_0. Hence, the resulting wave vector of these pulses, k, depends on different frequencies as well. In order to involve this mathematically let's approximate k by a Taylor series around the central frequency ω_0:

$$k = k_0 + k_1(\omega - \omega_0) + \frac{1}{2}k_2\,(\omega - \omega_0)^2 + ..., \tag{2.69}$$

where the different k_n with $n = 0, 1, 2, ..$ are

$$k_n = \left[\frac{\mathrm{d}^n k}{\mathrm{d}\omega^n}\right]_{\omega=\omega_0} \tag{2.70}$$

Hence, k_1 in the second term of (2.69) is $k_1 = \mathrm{d}k/\mathrm{d}\omega$ and as can be seen from (2.66), this is the reciprocal group velocity $v_g^{-1} = \mathrm{d}k/\mathrm{d}\omega$. Accordingly, k_2 in the third term of (2.69) is the group velocity dispersion $k_2 = \mathrm{d}^2k/\mathrm{d}\omega^2 = \mathrm{d}/\mathrm{d}\omega\left(v_g^{-1}\right)$ which leads again to (2.68).

In optical telecommunications the dispersion parameter D is often used instead of k_2 for the characterization of the dispersion properties of waveguides. While the GVD parameter, k_2, describes the frequency dependence of the propagation time, D describes its wavelength dependence. Hence, it can be written as

$$D = \frac{1}{L}\frac{\mathrm{d}}{\mathrm{d}\lambda}\left(\tau_g\right) = \frac{\mathrm{d}}{\mathrm{d}\lambda}\left(v_g^{-1}\right). \tag{2.71}$$

Using (2.67) and $\omega \mathrm{d}n/\mathrm{d}\omega = -\lambda \mathrm{d}n/\mathrm{d}\lambda$, it follows that

$$D = 1/c\left[\mathrm{d}n/\mathrm{d}\lambda - \left(\mathrm{d}n/\mathrm{d}\lambda + \lambda \mathrm{d}^2n/\mathrm{d}\lambda^2\right)\right] = -\frac{\lambda}{c}\frac{\mathrm{d}^2n}{\mathrm{d}\lambda^2}. \tag{2.72}$$

The broadening of the pulse in the fiber can be calculated with the dispersion or the GVD parameter. With $\Delta\omega$ as the spectral width of the pulse it can be written

$$\Delta\tau = Lk_2\Delta\omega = LD\Delta\lambda. \tag{2.73}$$

Therefore, the dispersion parameter D behaves in the same way for a wavelength alteration as the GVD parameter does for a frequency change, it follows $D\mathrm{d}\lambda = k_2\mathrm{d}\omega$. With $\omega = 2\pi c/\lambda$ the relation between dispersion parameter and GVD is

$$D = -\frac{\lambda}{c}\frac{\mathrm{d}^2n}{\mathrm{d}\lambda^2} = -\frac{2\pi c}{\lambda^2}k_2. \tag{2.74}$$

Figure 2.12 shows the wavelength dependence of the dispersion parameter D for pure silica glass. As shown in the inset of Fig. 2.12, the dispersion shows an approximately linear dependence on the wavelength around the crossing point. This behavior is especially important for the dispersion compensation in optical transmission systems that will be discussed in Sect. 5.4.

In pure silica glass, the dispersion parameter D in the C-band (at 1550 nm) is 21.9 ps/(nm $\times$ km) and the dispersion in waveguides depends on constructive parameters. In a standard single mode fiber, the dispersion parameter is $D = 17$ ps/(nm $\times$ km) whereas nonzero dispersion-shifted fibers have a dispersion of $D = 2 - 5$ ps/(nm $\times$ km). Fibers are designed to have a fixed value of dispersion for a distinct wavelength, but the dispersion can vary along the fiber length over a considerable range because of unavoidable variations in the core diameter [55, 56, 57]. These variations are static; dynamic variations of the dispersion are possible due to environmental changes such as temperature.

Dispersion fluctuations are small and will only rarely impact low bit rate systems, but for systems with bit rates of 40 Gbit/s and above they have a considerable impact that must be included in the system design [58].

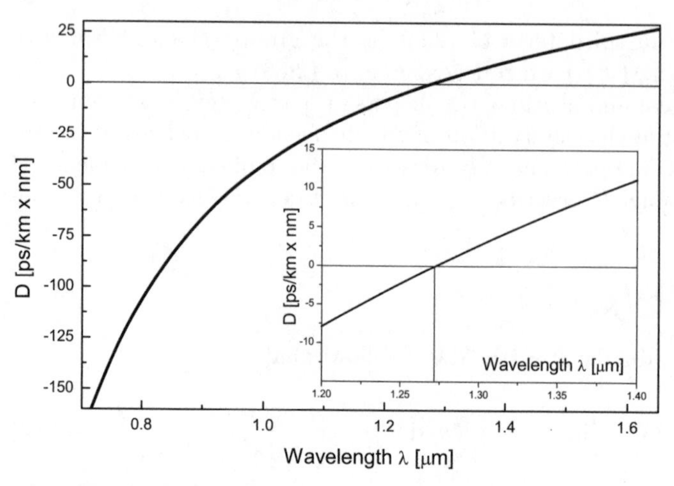

Fig. 2.12. Dispersion parameter D for pure silica glass versus wavelength

2.7 Birefringence

Optical fibers consist of a cylindrical core and a cladding surrounding it. The refractive index of the core is not completely isotropic, as assumed until now, but it shows a slight dependence on the polarization of the propagating field. As a result, if a wave propagates in the fiber their vector components have different phase velocities. Section 4.10 and Fig. 4.19 show how fast and slow axes can be defined.

Without this birefringence, it would be theoretically possible for two waves with an orthogonal polarization state to independently propagate in the fiber. On the other hand, the polarization-dependent refractive index leads to a mixing of the orthogonal components. Furthermore, if only one wave propagates in the fiber with a polarization state that shows an angle with respect to these two axes, then its polarization will be changed during the propagation. This happens because the vector components of the wave experience slightly different refractive indices ($n_{\mathrm{x}} \neq n_y$). Hence, the wave numbers are different ($k_{\mathrm{x}} \neq k_y$) and they propagate with slightly different phase velocities. The component parallel to the fast axis is faster then the component parallel to the slow axis.

Both components are phase shifted against each other and the polarization changes with propagation distance. If the phase shift is in the range between 0 and $\pi/2$ the wave shows an elliptical polarization; for a phase shift of exactly $\pi/2$ the wave is circularly polarized; after that again an elliptical polarization occurs. For a phase shift of π between the two components the wave is again linearly polarized, but now perpendicular to the original wave. For a shift of $3/2\pi$ the wave is again circularly polarized, but into the contrary direction. If the phase shift attains 2π the wave has again its original polarization. The

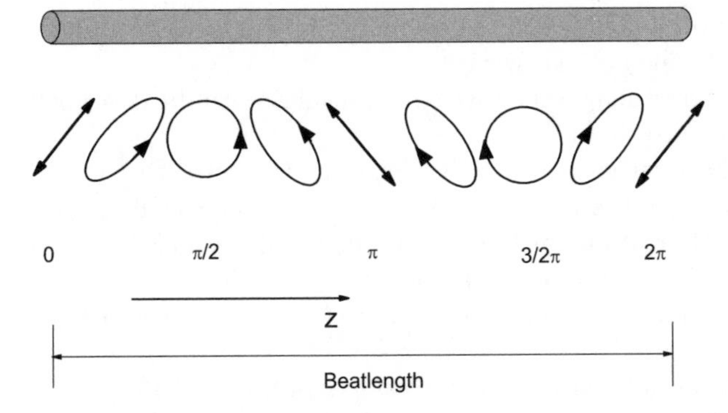

Fig. 2.13. Phase alteration along a fiber as a result of the birefringence in the fiber core. If the phase change is 2π the original polarization is reached again, with the propagation distance corresponding to the beatlength

change of the polarization of a wave propagating in a birefringent fiber is shown in Fig. 2.13.

The distance in the fiber after which the original polarization is reached again is called the "beatlength" (L_B). According to the discussion above it can be calculated as [51]:

$$L_{\mathrm{B}} = \frac{2\pi}{|k_x - k_y|} = \frac{\lambda}{|n_x - n_y|} = \frac{\lambda}{\Delta n}. \tag{2.75}$$

The origins of the birefringence in the fiber are disturbances of the geometry, for instance, due to the fabrication process. Other sources include the environmental temperature and external stress due to a bending or a torsion of the fiber. Hence, the birefringence is not constant. Consequently, the polarization state of the waves in a communication system is completely random.

Typical values for the birefringence in a standard fiber are between $\Delta n \approx 10^{-6}$-10^{-5} [139], [140]. Therefore, at a wavelength of 1.55 μm, the beatlength ranges between 15 cm and 1.5 m.

Summary

- The basis of the mathematical description of linear optical effects are the Maxwell equations.
- An external field leads to an internal charge displacement in the molecules of an insulator. This effect is called polarization.
- The polarizability of a material can be described by the susceptibility χ. The susceptibility is defined as the relative permittivity minus 1.
- In an insulator the propagating wave can be seen as a superposition between the original injected wave and a secondary wave. This secondary

wave has its origin in the polarization of the material and leads to a phase shift in comparison to the original wave.

- The propagation of electromagnetic waves in an insulator can be described by the harmonic oscillator model.
- The real part of the relative permittivity leads to the refractive index, responsible for a phase shift of the wave. On the other hand, the imaginary part results in the attenuation constant of the material, responsible for a decrease of the intensity during propagation.
- Both the refractive index and the attenuation constant are wavelength dependent.
- The attenuation is especially large near resonances of the oscillator.
- The losses in the fiber due to attenuation can be compensated by optical amplifiers. Each optical amplifier adds noise to the signal.
- The refractive index outside the range of resonances can be approximated by the Sellmeier equation.
- The velocity with which each point of constant phase will move is the phase velocity of the wave.
- The wavelength dependence of the phase velocity is called dispersion, far away from resonances it will always increase with wavelength. This is called normal dispersion.
- Every pulse consists of a group of frequencies. The group velocity is the velocity with which the maximum of the pulse propagates through the waveguide.
- The alteration of the propagation time of the group per distance L is the dispersion of the group velocity. In optical telecommunications this group velocity dispersion (GVD) is especially important. If the group velocity increases with wavelength it shows a normal or positive GVD, otherwise it is negative or anomalous.
- In optical fibers the dispersion parameter D is used for the calculation of the dispersion behavior. D describes the wavelength dependence of the reciprocal group velocity. In standard single mode fibers the dispersion parameter is zero at 1.3 μm. In dispersion-shifted fibers this zero-crossing is shifted to the minimum of attenuation. If D is positive the GVD is negative or it is anomalous and vice versa.
- Figure 2.14 shows the refractive index, the reciprocal group velocity (k_1), the group velocity dispersion (k_2) and the dispersion parameter (D) for pure silica glass against the wavelength.
- Standard fibers show birefringence, i.e., the refractive index a wave experiences depends on its polarization state.
- Due to the birefringence the polarization state of signals in standard fibers is completely random.

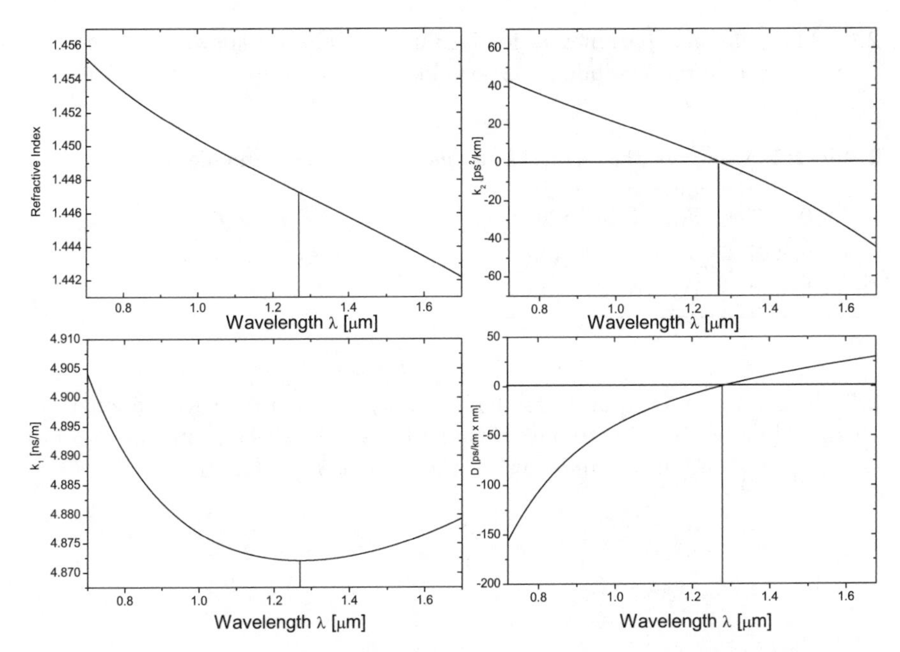

Fig. 2.14. Refractive index, reciprocal group velocity (k_1) , group velocity dispersion (k_2), and dispersion parameter (D) against the wavelength for pure silica glass

Exercises

2.1 Assume two waves with different frequencies in the mobile communications range (0.95 GHz, 0.96 GHz) will be amplified by a nonlinear amplifier before they are beamed by an antenna. The outgoing intensity can be approximated by a power series up to the fourth order. Calculate the frequencies that will be transmitted to the antenna of this system.

2.2 The effective core area of a highly nonlinear fiber HNLF at 1550 nm is 13.9 μm^2 [60]. Calculate the intensity in the core for 100 DWDM channels with a power of 3 mW each.

2.3. Derive from the Maxwell equations the wave equation for the magnetically part of the electromagnetic wave in vacuum and its solution for a plane monochromatic wave.

2.4. Show that equation (2.18) is in fact the solution of the differential equation (2.17).

2.5. The attenuation of a HNLF at 1550 nm is ≈ 0.7 dB/km. Calculate the power and intensity for Exc. 2.2 after a propagation distance of 2 km in the fiber. Compare this with the case for a standard single mode fiber with an effective core area of 50 μm^2 and $\alpha \approx 0.2$ dB/km.

2.6. The Sellmeier parameters for barium fluoride are shown in Table 2.2. Compute the refractive index, phase velocity, and group velocity at 500 nm.

Table 2.2. Values for the Sellmeier parameters of barium fluoride [7]

$S_1 = 0.643356$	$\lambda_1 = 0.057789 \mu m$
$S_2 = 0.506762$	$\lambda_2 = 0.10968 \mu m$
$S_3 = 3.8261$	$\lambda_3 = 46.3864 \mu m$

2.7. Estimate the temporal broadening of a pulse in a fiber with a length of 10 km. The fiber is made of pure silica and the pulse has a center wavelength of 1.55 μm. The initial bandwidth of the pulse is $\Delta\lambda = 0.5$ nm.

3. Optical Telecommunications

Optical telecommunications is, in principle, based on the technology of optical waveguides. The possibility of guiding light at a low rate of loss on defined paths over extremely large distances has led to a kind of revolution in the telecommunications world. A broad acceptance of applications like the Internet depends on the money every subscriber has to pay. Only the high bandwidth at rather low costs, provided by optical waveguides in the backbone of the world wide telecommunications network, opens the possibility of offering high data rates economically.

The first waveguides were commercially available at the end of the seventies, today they have a very broad range of applications. Optical fibers are deployed for speech and data transmission and for cable TV, but they are also used for the guidance of high power laser fields for material processing like welding and cutting and to guide images of diseases from inside a stomach.

The transmission over large distances is almost exclusively carried out by fibers today. In data networks fibers are used at all levels (in local area networks (LAN)- metropolitan area networks (MAN)- and wide area networks (WAN)-area networks (AN)). The distances that are bridged by optical waveguides range from a few hundred meters (less than 300 m for a LAN) to a few thousand kilometers (for intercontinental connections). The data rates vary from a few Mbit/s to several Tbit/s.

For the guidance of the optical signals thin cylindrical fibers made of silica glass are used. These waveguides show a refractive index alteration along their radius. The refractive index of the core is higher than the refractive index of the cladding. Hence, in principle, the waves are guided by total internal reflection. A completely different guidance mechanism is shown by photonic crystal fibers. With this kind of waveguides very large nonlinear coefficients are possible, while at the same time their dispersion behavior can be tailored. Hence, photonic crystal fibers are very interesting for nonlinear applications.

In order to transmit a signal, for instance a music video, over an optical fiber the baseband information has to be modulated onto an optical carrier. Furthermore, together with cross connects, add-drop multiplexers, and other switching elements, the fibers build up a network. So optical telecommunications is not restricted to the transmission of modulated light.

In this chapter the basic principles of optical fibers are first treated. After the description of single mode fiber types, Sect. 3.3 deals exclusively with highly nonlinear and photonic crystal fibers. In Sect. 3.4 the basic types of modulation formats with their corresponding spectra are shown. Finally, the last Section gives a short overview of optical transport systems.

3.1 Fiber Types

Typically, the optical range is the part of the electromagnetic spectrum that can be detected by the human eye. Hence, it is in principle restricted to the range between ≈ 400 and ≈ 800 nm. On the other hand, in optical telecommunications, wavelengths in the infrared part of the spectrum between 1.2 and 1.6 μm are used, due to the frequencies emitted by semiconductor lasers and the transmission range offered by optical waveguides made from silica glass.

Essential to almost all optical waveguides, except for photonic crystal fibers, is the difference between the refractive index of the core and cladding of a cylinder type element called optical fiber. These optical fibers are drawn from a preform, which determines the refractive index progression along the diameter. Most of the commonly used fibers consist of silica (SiO_2) as the host material in the central region, with GeO_2 added to increase the refractive index and hence, form the waveguide. Figure 3.1 shows three fundamental fiber types with the corresponding refractive index progression.

All three fiber types in Fig. 3.1 are radially symmetrical and the basic material is silica glass. The main difference between them is the refractive index progression and the core diameter. The fibers (a) and (b) have, for

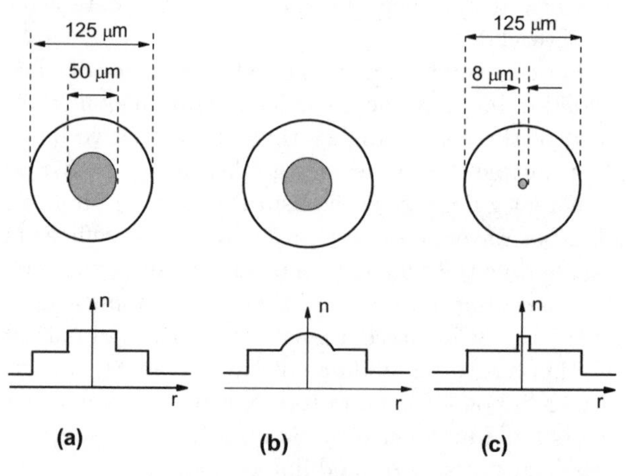

Fig. 3.1. Typical refractive index progression for a step-index (**a**), graded-index (**b**), and single mode fiber (**c**) (sketch approximately true to scale)

instance, a rather large core diameter of 50 μm whereas the core diameter of the third fiber is much smaller and is only about 8 to 10 μm.

In the following sections the different fiber types are discussed in detail, with special emphasis on single mode fibers because this is the most important fiber type for optical communications today. Photonic fibers, which exploit a completely different physical waveguide mechanism and have the possibility to be the most important fibers of tomorrow, are treated in Sect. 3.3.

3.1.1 Step-Index and Graded-Index Fibers

In a multimode fiber (Fig. 3.1a) the refractive index between core and cladding shows a step. Hence, these fibers can be called step-index fibers. Step-index fibers were the first optical waveguides used in optical transmission systems. The wavelength of these first optical communication systems was 0.85 μm. The core diameter of step-index fibers is ≈ 50 μm and hence, nearly 60 times the wavelength. If the core diameter is much bigger than the wavelength, as in the case of step-index fibers, it is justified to approximate the guidance of the optical waves by a ray tracing model. Therefore, the rules of geometrical optics can be used for the description of the wave guiding mechanism in step-index fibers.

In principle the guidance of light in optical fibers is based on the effect of total internal reflection. The refractive index of the core region is higher than that of the cladding. According to law of refraction, if the angle of incidence is higher than a particular value then the whole wave is refracted back into the medium with a higher refractive index. The limiting angle for this behavior is.

$$\sin \alpha_l = \frac{n_2}{n_1} \tag{3.1}$$

With typical refractive indices of core, $n_1 = 1.4516$, and cladding, $n_2 = 1.4473$, in a standard fiber, the angle of total internal reflection in the fiber is $\alpha_l \approx 86°$. Hence, all rays with incident angles greater than 86° can be propagated in the fiber.

Figure 3.2 (top) shows the ways two different rays travel in the optical fiber are shown. Since both rays see the same refractive index they will have the same velocity in the waveguide. But, as can be seen from the figure, the paths of the rays have different lengths. Hence, they will arrive at the output at different times. If a pulse with a particular duration is injected into the fiber input, then different copies of the pulse will arrive at the fiber output at different times. These different copies will interfere with each other at the photo-diode of the receiver, leading to broadening of the pulse at the fiber output. The amount of the pulse broadening depends on the difference in propagation time between the copies. The different copies are called modes, whereas the broadening of the pulse due to the multipath components is called

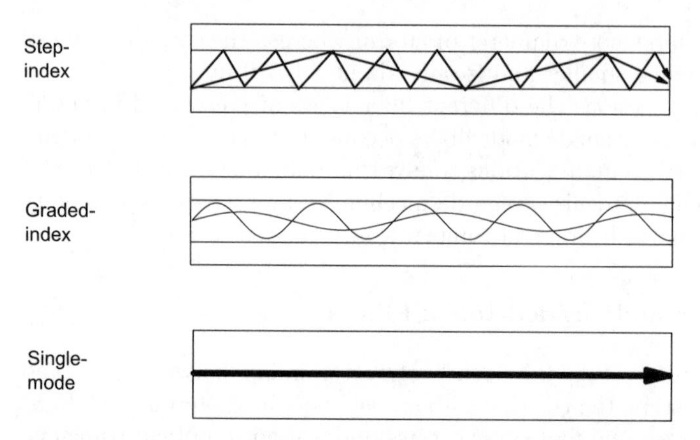

Fig. 3.2. Guidance of electromagnetic waves in step-index (*top*), graded- index (*middle*) and single mode fibers (*bottom*)

mode dispersion. This effect limits the capacity of multimode or step-index fibers (see Exc. 3.1).

In graded-index fibers (Fig. 3.1 b) the refractive index shows a continuous change from the center of the waveguide to the beginning of the cladding. Assuming, for convenience, a monochromatic wave with frequency ω, then the phase velocity of this wave is $v_{\mathrm{ph}} = c/n(\omega)$, as shown in (2.63). Waves propagating near the center of the fiber have a shorter way, but at the same time they will experience a higher refractive index. Hence, due to the refractive index dependence of the phase velocity, they are rather slow. Waves propagating near the cladding have a longer way, but they are faster because the refractive index in this region of the fiber is low. Therefore, the different copies or modes will arrive at the output of a graded-index fiber at nearly the same time. It is thus possible to significantly reduce the mode dispersion in these fibers.

Until the mid-1980s, graded-index fibers were used for optical communication systems. But today they are nearly displaced. Today's optical telecommunication systems use single mode fibers almost exclusively.

3.2 Single Mode Fibers

If the diameter of a step-index fiber is reduced drastically, only one mode will be able to propagate in the waveguide. Hence, this kind of fibers is called mono or single mode fiber. In a first approach it is possible to assume that the core diameter is so small that all modes will propagate the same way and hence, unite to only one residual mode. But the core radius for this case is about 4 to 5 μm, the range of the wavelength of the propagating waves. Hence, the wave character is no longer negligible in this case. Consequently, the wave will not only be guided in the core of the fiber, but a significant part of the intensity will propagate in the boundary region between core and cladding.

If an electromagnetic wave propagates in a waveguide, the amplitude depends on the coordinates perpendicular to the propagation direction. For a cylindrical waveguide, for instance, it depends on the radius (r) and the angle (φ). Hence, the wave equation for plane waves, (2.17) derived in Chap. 2.2, cannot be used. With the Laplace operator in cylindrical coordinates (see Appendix), (2.13) can be written as.

$$\frac{\partial^2 \boldsymbol{E}}{\partial r^2} + \frac{1}{r}\frac{\partial \boldsymbol{E}}{\partial r} + \frac{1}{r^2}\frac{\partial^2 \boldsymbol{E}}{\partial \varphi^2} + \frac{\partial^2 \boldsymbol{E}}{\partial z^2} - \frac{n^2}{c^2}\frac{\partial^2 \boldsymbol{E}}{\partial t^2} = 0. \tag{3.2}$$

If all derivations perpendicular to the propagation direction are summarized in the transverse Laplace operator (Δ_T) [9], it results in.

$$\Delta_\mathrm{T}\boldsymbol{E} + \frac{\partial^2 \boldsymbol{E}}{\partial z^2} = \frac{n^2}{c^2}\frac{\partial^2 \boldsymbol{E}}{\partial t^2} \tag{3.3}$$

For the solution of (3.3), an amplitude dependent on r and φ is required:

$$\boldsymbol{E}(z,t,r,\varphi) = \left(\hat{E}(r,\varphi)\mathrm{e}^{\mathrm{j}(n_W k_0 z - \omega t)} + \mathrm{c.c.}\right)\boldsymbol{e}_i, \tag{3.4}$$

where n_W is an effective refractive index that includes the waveguide properties of the fiber and $\boldsymbol{e}_i$ the unity vector in an arbitrary direction i. If (3.4) is introduced into (3.3) it follows that

$$\Delta_\mathrm{T}\hat{E}(r,\varphi) - n_\mathrm{W}^2 k_0^2 \hat{E}(r,\varphi) = -\frac{n^2}{c^2}\omega^2 \hat{E}(r,\varphi). \tag{3.5}$$

Using $n_\mathrm{W}k_0 = k$ and $\frac{n}{c}\omega = \kappa$, it can be written as

$$\Delta_\mathrm{T}\hat{E}(r,\varphi) + (\kappa^2 - k^2)\hat{E}(r,\varphi) = 0. \tag{3.6}$$

The wave vector κ considers only the material- and frequency-dependent refractive index, whereas k includes the material as well as the waveguide properties.

Waveguides are only able to guide a limited number of modes. The electric field strength perpendicular to the propagation direction has to fulfil certain conditions. The spatial distribution of the modes is a solution of (3.6) when the appropriate boundary conditions are considered. The exact solution should not be of further interest here. But in principle there are two kinds of fiber modes, the HE_{mn} and the EH_{mn}-modes. The intensity distribution of the first nine modes is shown in Fig. 3.3.

For $m = 0$ the axial component vanishes and the modes correspond to the TE and TM modes in planar waveguides. The number of possible modes in the waveguide depends on the frequency, or wavelength of the wave, and its

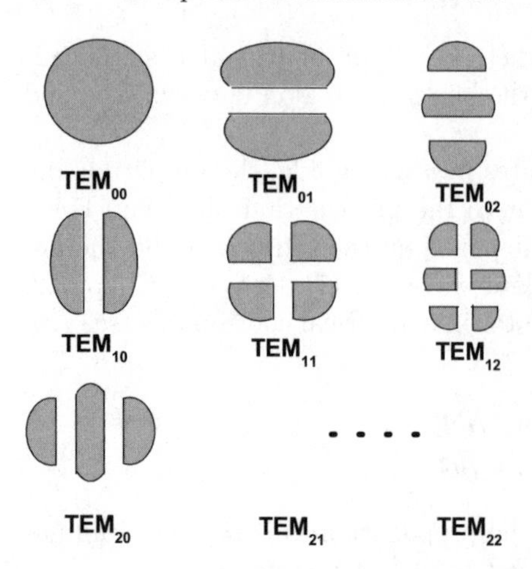

Fig. 3.3. Modes TE_{mn} in a cylindrical optical waveguide

construction, especially on the core radius a and the difference of the refractive indices between core and cladding $(n_1 - n_2)$. The normalized frequency V is thus

$$V = k_0 a \sqrt{n_1^2 - n_2^2}. \quad (3.7)$$

If $V < 2.405$, only one single mode can propagate in the fiber. In other words, the fiber is single mode. The intensity distribution is that of the TEM_{00} mode in the upper left corner of Fig. 3.3. For the fiber in Fig. 3.1(c) with a core diameter of 8.5μm using the usual values for the refractive index of core and cladding at 1.3μm ($n_1 = 1.4516$, $n_2 = 1.4473$) and $k_0 = 2\pi/\lambda$, the limiting wavelength is $\lambda_{co} \approx 1.24$ μm. This means that if the wavelength of the injected light is smaller than λ_{co} the light will propagate in several modes in the waveguide. Hence, a single mode fiber with the above values is only single mode if the wavelength of the injected light is greater than 1.24 μm. This value is the so called cut-off wavelength.

3.2.1 Single Mode Fiber Types

Due to the advantages offered by single mode fibers (high capacity, low losses, no mode dispersion), single mode fibers are nearly exclusively used for the transmission of signals over larger distances today. Since optical waveguides can be used for many different fields, there are many single mode fibers offered. These fibers can differ in their attenuation, dispersion, or nonlinear behavior.

Due to the superposition of the different attenuation mechanisms in a fiber, there are several optical windows in which the fiber has only small losses. These optical windows are shown in Fig. 3.4.

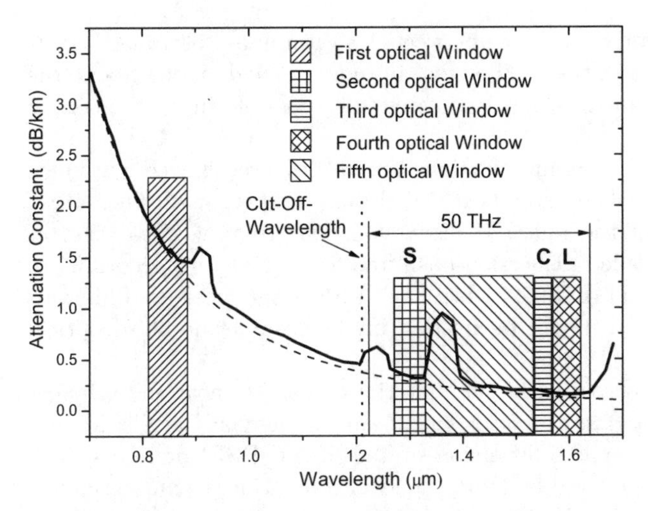

Fig. 3.4. Attenuation regime in a fiber of silica glass and resulting optical windows [62]

The first generation of optical fibers were step-index fibers bustled at a wavelength of 0.85 μm. Single mode fibers are, according to (3.7), only single mode if the wavelength is greater than ≈ 1.2 μm. Above 1.6 μm, strong losses due to the absorption of the basic material will occur. Hence, for single mode fibers only the ≈ 50-THz broad range between ≈ 1.2 and ≈ 1.6 μm is usable.

The first single mode fibers were used in the so called short (S)-band around 1.3 μm historically, this was the second optical window opened for telecommunications. As can be seen from Fig. 3.4 the attenuation in this range is much lower than for the first optical window ≈ 0.35 dB/km. Furthermore, the material dispersions zero-crossing of a standard fiber is in this window.

The attenuation minimum of a silica fiber is at ≈ 1.55 μm (≈ 0.2 dB/km), however. This is the so called third optical window, or conventional (C)-band from 1.53 μm to 1.565 μm. Lasers and receivers were available for this wavelength range in the late 1980s. With special constructive conditions it is possible to shift the zero-crossing of the material dispersion into the third optical window. Amplifiers (EDFAs) working in this band were commercially available by the beginning of the 1990s.

Current wavelength division multiplexing (WDM) systems are working in the C-band with a standardized channel spacing of 100 or 50 GHz (≈ 0.8 or 0.4 nm). In order to increase the bandwidth in standard single mode fibers the long (L)-band, or fourth optical window from 1.57 μm to 1.62 μm, is used for the guidance of optical signals as well.

With special procedures in the manufacturing process of optical fibers, it was possible over the last few years to avoid the remaining OH^-ions in the fiber. Hence, the attenuation peak at 1.365 μm is not present in these fibers and the fifth optical window between 1.35 μm and 1.53 μm is usable for optical telecommunications. Therefore in modern single mode fibers, the entire 50 THz bandwidth, which is theoretically possible in silica glass, is available.

The International Telecommunications Union (ITU) recognized four kinds of single mode fibers. The "standard" single mode fiber, used since 1983, has a zero-crossing of the material dispersion near that of silica glass at 1.31 μm, hence it is called dispersion-unshifted fiber (USF) or, according to the standard number, ITU G.652. Until now, USF is the most used fiber for telecommunications and it will be bustled in the second and third optical window.

Fibers with a zero-crossing of the material dispersion near the attenuation minimum of silica glass at 1.5 μm have been commercially available since 1985. These fibers are called dispersion-shifted fibers (DSF) or ITU G.653. Due to the advantages offered by DSF, the low attenuation, low material dispersion and because of the EDFA's that became available later, these fibers were assumed to be ideal for the third optical window. But a better understanding of nonlinear optical effects changed this point of view [62]. DSFs are ideal for long haul optical time domain multiplexing OTDM transmission systems with a high bit rate at one carrier wavelength, but not for WDM systems. In this case the low material dispersion leads to very good phase matching and hence to the efficient generation of four-wave-mixing products (FWM) (see Sect. 7.3).

Optical fibers with a particularly small attenuation were developed (ITU G.654) especially for submarine systems. With these fibers it is possible to bridge large distances without active components. The attenuation constant in these fibers is smaller than 0.18 dB/km.

Nonzero-dispersion fibers (NZDF) are available for DWDM systems in the third optical window. These fibers have minimal- and maximal-specified dispersion in the range of 1.55 μm. The dispersion is high enough for phase mismatching to suppress FWM products between the channels, but at the same time small enough to enable bit rates of 10 Gb/s per WDM-channel over a distance of 250 km without dispersion compensation [62].

Due to the wavelength dependence of the refractive index, the different frequency components of an optical pulse will propagate with slightly different velocities. This effect leads to a temporal broadening of the pulse and is called dispersion (see Sect. 2.6.2). In a standard single mode fiber, the dispersion parameter (D) has a positive slope. In order to compensate this temporal broadening (see Sect. 5.4), special optical fibers with a negative slope of the dispersion parameter were developed over the last few years. These fibers are called dispersion-compensating fibers (DCF).

In telecommunications literature, additional notations related to brand names of manufacturers for optical fibers are used, such as SMF-28 and LEAF from Corning Inc., or AllWave and TrueWave from Lucent Technolo-

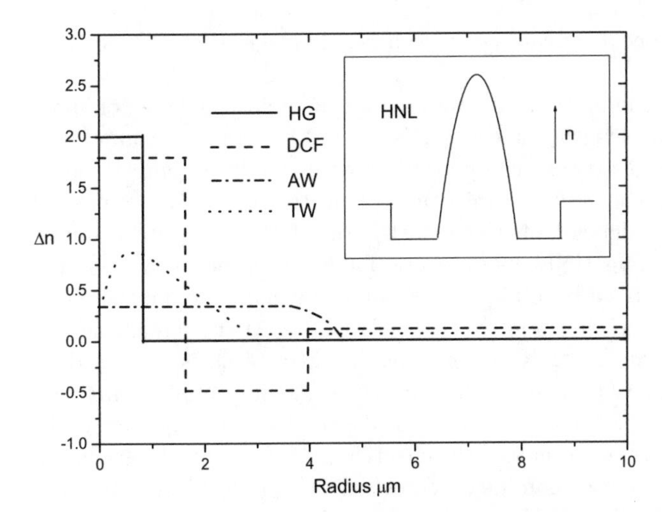

Fig. 3.5. Refractive index regimes for four different types of single mode fibers [267]. The inset shows a W-shaped refractive index profile of a HNL fiber, according to [11]. (AW=AllWave, TW=TrueWave, DCF=dispersion-compensating fiber, HG=highly germanium-doped fiber, HNL=highly nonlinear fiber)

gies. These fibers are different in the refractive index regime, endowment and other constructive parameters, and offer different advantages for optical telecommunications. In Fig. 3.5, for instance, the refractive index regime for an AllWave, a TrueWave, a DCF, and an HG (highly germanium-doped) fiber is shown.

3.3 Highly Nonlinear and Photonic Fibers

Nonlinear processes depend on the intensity of the field hence, it is possible to enhance their efficiency if the field is localized in a very small area. This is the basic idea of highly nonlinear (HNL) as well as photonic fibers. Both can have exceptional high nonlinearities. Due to the effectiveness of nonlinear processes in these fibers, they offer the possibility of designing compact devices that rely on nonlinear effects. In both kinds of fibers, the high nonlinearities are mainly caused by very small effective areas (see Sect. 4.8). Hence, contrary to semiconductor optical amplifiers (SOAs) (see Sect. 12.2.3), no resonant physical effects are exploited and the speed of signal processing in these devices is extremely high.

The easiest way to decrease the effective area in a fiber is used by so called tapered fibers. Tapering refers to a process where the diameter of the fiber is reduced simply by heating and stretching. The drawing of single mode fibers with a standard size of the diameter of 125 μm over a flame can generate tapered fibers with waist diameters of $\approx 1 - 3$ μm and waist length up to

90 mm [10]. The special advantage of tapered fibers is that they can be home-made.

Commercially available HNL fibers were especially designed for nonlinear applications like wavelength converters (Sect. 12.2), parametric amplifiers (Sect. 13.4) for the distortion compensation with nonlinear optical phase conjugation (Chap. 14), and several others. Due to the fact that most of the nonlinear effects depend on the polarization of the interacting waves, polarization-maintaining HNL fibers are available. In order to increase the intensity in the core area a HNL fiber has a small core diameter and therefore, a small effective area, as shown in the inset of Fig. 3.5. At the same time, the nonlinear refractive index, or Kerr constant (see Sect. 4.7), is increased by increasing the amount of germanium dopant. Hence, nonlinear effects are very effective in these fibers. A measure of the strength of the nonlinearity is the ratio between the Kerr constant and the effective area. For a fixed wavelength this is proportional to the nonlinearity coefficient (γ) (5.29). In standard single mode fibers this coefficient has a value of only $\gamma \approx 1.3\ \mathrm{W}^{-1}\mathrm{km}^{-1}$, whereas for HNL fibers it ranges from $12.5\ \mathrm{W}^{-1}\mathrm{km}^{-1}$ to $25\ \mathrm{W}^{-1}\mathrm{km}^{-1}$[11]. For low attenuation, low splicing and bending losses special designs similar to that in the inset of Fig. 3.5 but with a triple cladding structure are possible [12].

Parametric amplifiers, wavelength converters and optical demultiplexers that are based on the effect of four-wave-mixing (Chap. 7) depend not only on the strength of the nonlinearity but the phase matching as well (Sect. 7.3). Therefore, the dispersion and dispersion slope are very important for the effectiveness of nonlinear interactions and for the usable bandwidth. Consequently, HNL fibers show beside their high nonlinearity, a slow dispersion slope in the 1550 nm range (0.013 - 0.016 $\mathrm{ps}/(\mathrm{nm}^2 \times \mathrm{km})$). Furthermore, to reduce the space occupied by the fibers in a nonlinear device, HNL fibers with reduced cladding and coating diameter (90 and 145 µm, respectively [11], contrary to 125 and 250 µm for standard single mode fibers (SSMFs)) were designed as well.

A completely new development for the guidance of optical waves is the use of photonic fibers, sometimes as well referred as microstructured or holey fibers. Photonic fibers are based on the so called photonic crystals. These crystals were reported theoretically for the first time in 1987 by Eli Yablonovitch and Sajeev John [13, 14]. Photonic crystals show a behavior for lightwaves similar to what semiconductors show for electron waves. Semiconductors have, due to their crystal lattice structure, an electronic band gap. The electrons in a pure semiconductor crystal cannot have energies within this band gap. In other words, electrons with particular energies are unable to propagate in the semiconductor. But if the lattice of the semiconductor is doped with other atoms, the band structure of the crystal is altered and it will be possible to change the conduction properties of the semiconductor.

In photonic crystals the conduction properties for lightwaves can be altered in a specific way as well. In order to do this, a transparent dielectric material is periodically structured. The structures can simply be periodically ordered holes in the host medium. The holes are filled with air and the periodic structures have dimensions in the range of the wavelength of light. Hence, the refractive index is periodically altered, with an refractive index of ≈ 1 in the holes, and a specific refractive index in the host material. This periodic modulation of the refractive index corresponds to the lattice structure of a semiconductor crystal. On the other hand, the grating constant of a semiconductor is around 6000 times smaller than that of a photonic crystal designed for wavelengths in the telecommunications range.

Electrons behave – quantum mechanically – like waves. The wavelength of an electron wave depends on its velocity or energy. If the wave propagates in a semiconductor, it will be scattered by atoms of the lattice. Hence, a part of the wave is thrown back. The atoms of the lattice are ordered in planes so on the next plane the wave will be scattered again and so on. Whether the backscattered waves can interfere constructively or not depends on the relative phases between the scattered parts of the wave and hence, on the distance between the planes or the grating period of the lattice. If all backscattered waves interfere constructively, the electron wave cannot propagate in the crystal, but is completely reflected at the grating. If such a total reflection takes place over a range of wavelengths for the electron waves, this range is called a band gap. Electrons with energies that fall inside the band gap are forbidden in the semiconductor.

Photonic crystals will show the same behavior for light waves. Assuming a lightwave propagating in a photonic crystal, a part of the lightwave will be reflected on every border plane between the air in a hole and the material of the dielectric. If the waves reflected on different planes of the periodic structure can interfere constructively, light with this particular wavelength cannot propagate in the crystal. If this condition is fulfilled not just for one particular wavelength, but for a band of frequencies, then the material shows a photonic band gap.

The simplest "photonic crystals", although not real crystals, are one dimensional periodic structures. These 1-D structures are, for example used for mirrors with a high reflectivity. In nature, such structures can be found as well in the wings of a butterfly, for instance, or in the stings of a sea mouse (a 20 cm long marine worm living on the coast of Australia at a depth of 2000 m [16]). The structure is formed by a periodic arrangement of chitin layers and air, for the butterfly or sea water for the sea mouse.

The opalstone is, in some respect, a photonic crystal as well. The periodic modulation of the refractive index is caused by hard silica balls embedded into a matrix of softer silica in a periodic manner [17]. Due to its periodic structures, the wings of a butterfly or an opal will exhibit iridescence or opalescence in different colors. An interesting review about photonic structures in biology is given in [18].

A real photonic crystal is three dimensional. This means the crystal has a band gap independent of the direction the wave will propagate. Whether the backscattered waves can interfere constructively or not depends on the grating constant of the periodic structure, the refractive index difference, and on the wavelength of the wave. Hence, the construction of the first real photonic crystals was rather difficult. The first photonic crystal was built 4 years after its theoretical conception. Since it was easier to build periodic structures for longer wavelength, the crystal did not work in the visible range of the electromagnetic spectrum. But they have found applications in the microwave area as perfectly reflecting substrates in the design of antenna elements [19, 20]. The first crystal was built by the mechanical drilling of holes [21]. Another possibility is offered by a stack of silica or alumina rods [22], lithographic etching, or the principle of self organization after a liquid evaporates. Photonic crystals thus have the potential for large scale integration of optical components and devices [1].

Light with a wavelength inside the forbidden range (determined by the band gap) cannot propagate in a photonic crystal. But, if the periodic structure is broken on a particular location of the crystal, then the whole field is trapped at this point. If the periodicity-breaking region is continued in a particular direction, then the light can propagate in this direction. This is the basis of photonic fibers. The first photonic fibers were built at the University of Bath [23], they are now commercially available.

Many kinds of photonic fibers are possible, although not all of them require a photonic band gap effect. Therefore, the notation microstructured fiber is more generally. Two principle arrangements are shown in Fig. 3.6. The cladding region of the fiber is built by a periodic structure, where in the core region this periodicity is broken for both arrangements. In Fig. 3.6b the central periodicity breaking region is built by a hole, whereas in Fig. 3.6a it is built by the absence of a hole. Both kinds differ significantly from each other. The guidance of light in fiber (a) is based on the mechanism of total internal reflection, as in the case of normal fibers. The central region of the fiber shows a higher refractive index than the cladding. On the other hand the second type (b) offers a completely different physical mechanism of waveguiding than conventional fibers. Here the central region has the refractive index of air where the light is confined by the bandgap of the periodical glass-air structure of the cladding. Other possibilities of photonic crystal structures include different periodicities of the periodic region, as in the case of hollow core concentric ring structures [24], or other materials where the holes can be filled with a high index liquid for example. Another very interesting development is the incorporation of compound glasses with a high nonlinearity [25, 32].

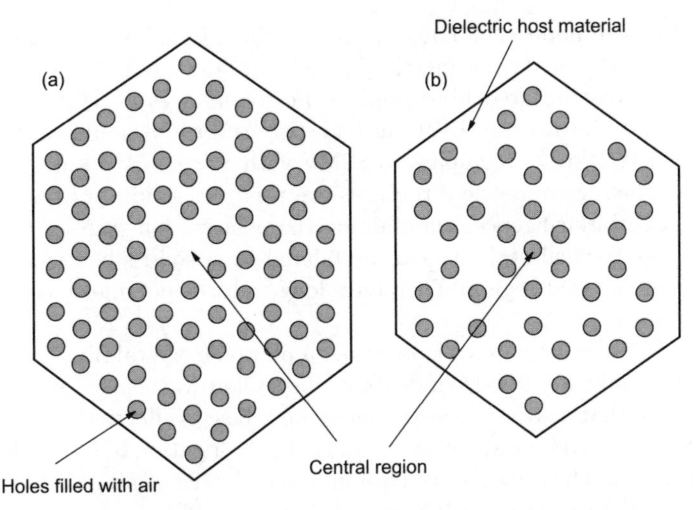

Fig. 3.6. Two dimensional structure of a photonic crystal fiber. The interstitial holes are not shown

Photonic crystal fibers require a much higher refractive index contrast than conventional fibers. Therefore, a completely different technique has to be incorporated in the manufacturing process. The tube stacking technique [26] is widely used where a preform on a macroscopic scale is built by stacking rods and tubes of silica by hand. A typical preform is a meter long and 20 mm in diameter and contains several hundreds of capillaries [27]. In order to build the periodicity breaking region, a tube is left out or replaced by a rod in the center of the preform. The completed preform will then be heated and drawn down on a fiber drawing tower to the dimensions of a fiber. In order to form other structures and use materials other than silica, different fabrication processes, such extrusion through a die, are possible as well [28].

Photonic fibers with a random distribution of the air holes in the cladding region are much easier to built. Although these fibers show not all properties that can be expected from ordered cladding structures, the power is confined in the central region of these fibers. The hole structure is produced by bubbles that were generated by a mixture of silica powder and a gas producing powder in the cladding of the fiber. By the fiber drawing the bubbles are drawn into tubes of random size, location and length [29].

Although the propagation of the lightwave in the fiber in Fig. 3.6a is based on total reflection, as in conventional fibers, these fibers show some interesting advantages. This kind of fibers is often referred to as holey or microstructured fibers. Since the whole field will only propagate in the central region of the fiber where the periodicity is broken, it barely extends beyond the first row of air holes. Hence, with a small core diameter, it is possible to concentrate the field in a small area. Due to the resulting high intensity, these kinds of fibers can show a highly nonlinear behavior. This nonlinear behavior depends on the fiber design and mode size. The effective nonlinearity of the fiber can be altered by increasing or decreasing the intensity inside the fiber [30]. A highly nonlinear microstructured fiber used by Petropoulos et al. [32] with an effective area of $\approx 2.8\ \mu m^2$ at 1550 nm had, for instance, a nonlinearity constant of $\gamma \approx 35\ W^{-1}km^{-1}$ (compare to SSMF with $\gamma \approx 1.2\ W^{-1}km^{-1}$).

Furthermore, when asymmetric structures are used, these kind of fibers can offer a very high birefringence, maintaining the polarization state of the launched waves. Mechanical stability can be offered by a solid silica jacket around the microstructured region. Therefore, long, robust, polymer-coated fibers can be produced [246].

The authors of [31] investigated theoretically a photonic crystal fiber with an effective area of only $1 - 2\ \mu m^2$ and a zero dispersion wavelength below 500 nm. Contrary to that a standard single mode fiber has an effective area of $\approx 80\ \mu m^2$ (see Sect. 4.8). These structures require the fabrication of interhole distances near 600 nm. The authors used a honeycomb lattice similar to that in Fig. 3.6b, but in the center of each honeycomb circular glass regions were doped with germanium to increase the refractive index. For the guiding core defect they used the absence of a doped region.

One of the most important properties of photonic fibers is their ability to show a single mode behavior for a large range of wavelengths. In conventional fibers the number of guided modes depends on the normalized frequency parameter V (3.7) which depends on the frequency:

$$V = \frac{\omega}{c} a \sqrt{n_1^2 - n_2^2}. \tag{3.8}$$

If the frequency is increased, or if the wavelength is decreased, V is no longer smaller than 2.405 and the fiber will no longer be single mode. In a conventional single mode fiber, the single mode behavior is only fulfilled if the wavelength is in the infrared region. In the visible range, the waveguide acts like a multimode fiber. Contrary to that, in a photonic crystal fiber the refractive index of the cladding depends on the frequency. For high frequencies the refractive index of the cladding approaches that of the core. Hence, if the frequency increases, the square root in (3.8) will decrease. Therefore, a stationary effective V value, independent of the frequency, is possible in photonic band gap fibers. This fiber will show a single mode behavior over a wide wavelength range. For example Birks et al. [23] built a single mode fiber in the frequency range between 337 and 1550 nm. With such a fiber it would be possible to extend the wavelength range of optical communications to the visible spectrum.

Another advantage of photonic crystal fibers is the possibility to alter the dispersion behavior by the ratio between the air hole diameter (d) and the pitch between the holes (Λ). Figure 3.7 shows the dispersion (D-parameter) for four different values of the ratio d/Λ, with $\Lambda = 2.3\ \mu m$.

If the air hole diameter is small, the influence of the periodic structure on the dispersion is limited. As can be seen from Fig. 3.7, the dispersion

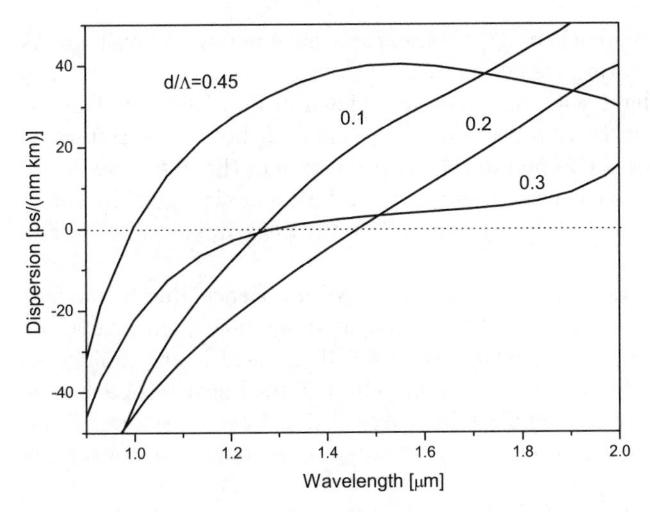

Fig. 3.7. Dispersion profile of photonic band gap fibers for various ratios of d/Λ [61]

curve is close to the material dispersion of silica for $d/\Lambda = 0.1$. The dispersion increases with wavelength and the zero-crossing is between 1.2 μm and 1.3 μm. On the other hand, for a large air hole diameter the waveguide part of the dispersion is rather strong. Hence, at a ratio of 0.45 the zero-crossing is shifted to 1 μm.

These fibers are suitable for the efficient generation of a white light continuum. With photonic fibers it is possible to shift the range of anomalous dispersion to the visible range, therefore the generation and propagation of solitons in the visible will be possible with such fibers [35, 36].

Another interesting aspect of photonic fibers is the possibility to shift the zero-crossing of the dispersion down to the range of visible wavelengths. With designs such as the one described in [31] and mentioned above, zero dispersion below 500 nm was possible. In the visible region, very efficient laser sources for the generation of ultrashort pulses are available (e.g. Titan Sapphire Laser at 800 nm). These sources together with photonic fibers allow the construction of supercontinuum sources that will deliver wavelengths from the UV to the IR [35].

For a ratio of $d/\Lambda = 0.3$ a very low dispersion with a flat profile between 1.2 μm and 1.8 μm can be achieved. With a flat dispersion curve it is possible to yield phase matching over a wide wavelength range. Hence, these fibers are very interesting for all nonlinear applications depending on phase matching such as the generation of higher harmonics, parametric amplifiers and lasers, and wavelength converters.

For other ratios of d/Λ, it is possible to produce photonic crystal fibers with a large and flat negative dispersion where dispersion values below

-100 ps/km/nm can be reached [33]. These fibers have potential applications as dispersion-compensating elements [37].

The guidance of light with the structure shown in Fig. 3.6b is not based on the effect of total internal reflection. The central air hole has a refractive index of one. The light is trapped in the central region of the fiber because at this point, the periodicity is broken due to the additional air hole. The entire field is localized around the central region, but the peak intensity is outside the air hole [33].

Nonlinear effects can only take place in a medium. Hence, due to the fact that a significant part of the intensity is propagated in a hole filled with air or vacuum, nonlinear effects are strongly reduced. If the field is only guided in air it exhibits negligible dispersion as well [34]. The nonlinearity determines the ultimate limits of telecommunications over optical fibers (see Sect. 4.11). With photonic fibers it may therefore be possible to solve current limitations [33].

A very important factor for optical telecommunications is the loss during propagation. In conventional fibers the losses are a result of scattering and absorption, with a minimum at 1.55 μm, and of the confinement of the field and the bending of the fiber. With solid core photonic fibers losses of a fraction of a dB have been reported [38]. But losses significantly lower than in conventional fibers are not expected from this kind of waveguides [27]. Fibers that guide the light in a capillary have losses of more than 100 dB/km in the visible range. The lowest reported loss in a hollow core fiber is 13 dB/km [39]. But the fundamental loss limits are not yet established. The losses decrease drastically with increasing wavelength due to the wavelength dependence of Rayleigh scattering in silica. Furthermore, in hollow core fibers the majority of the light propagates in air in which scattering and absorption is very low. Therefore, photonic crystal fibers have an immense potential in telecommunications and it might be possible to reduce the losses below that of conventional fibers [27, 40].

3.4 Modulation

In telecommunications the signal cannot be propagated in the baseband, a carrier is required. The carrier serves, in principle, two purposes. It enables the propagation in the medium and it offers the possibility to exploit the entire bandwidth of the medium via the multiplex technique.

For the modulation of a carrier with the information, many different ways are possible. Modulation means that certain parameters of the carrier are changed in their dependence on the baseband signal. Two basically different ways can be distinguished. If the information has a direct influence on the carrier parameters it is called an analog modulation. Whereas if, before the modulation, the information is transformed into a digital signal it is a digital modulation. For a digital modulation of the carrier, several bits of the transformed signal can be collected to a symbol with the length of n-bit. These n-bits can be transmitted simultaneously, leading to an increase of the bandwidth efficiency (bit rate per 1Hz bandwidth). For the simultaneous transmission of a n-bit symbol, one of $m = 2^n$ possible carrier signals is required. Due to this the corresponding modulations are called m-ary (e.g. binary for a 1-bit symbol with $n = 1$ and $m = 2$). In the following only digital binary modulations are considered. Therefore, two carrier signals are required ($m = 2^1$) for the transmission of one bit in two possible states ("0" and "1").

Equation (3.9) describes an optical carrier signal as a monochromatic, electro-magnetic wave

$$\boldsymbol{E}(z,t) = \frac{1}{2}\left(\hat{E}\mathrm{e}^{\mathrm{j}(\hat{n}k_0 z - \omega t)} + \mathrm{c.c.}\right)\boldsymbol{e}_i. \tag{3.9}$$

For the modulation of the carrier signal with the information only four different possibilities exist in principle. These are the frequency, phase, amplitude or polarization which can be changed in dependence on the signal.

- Frequency ω: The frequency of the carrier is switched between two possible states. This kind of modulation is called frequency shift keying, or FSK for short.
- Phase $\hat{n}k_0 z - \omega t$: For this modulation the phase of the carrier is switched between the states 0 and π, for instance. A digital phase modulation is called PSK (phase shift keying).
- Amplitude $\hat{E}$: In dependence on the information the amplitude or intensity, respectively of the carrier is switched. For optical telecommunications, the carrier is usually switched on and off. Such a modulation is called ASK (amplitude shift keying).
- Polarization $\boldsymbol{e}_i$: The two bits 0 and 1 are transmitted by waves with different polarization states. But due to the mentioned birefringence in standard single mode fibers (Sect. 2.7), the two polarization states are not independent of each other. The result is a mutual influence.

Therefore, three basic kinds of possible modulations are left for the transmission of signals in optical fibers, ASK, FSK, and PSK. Of course combinations between the basic modulation formats are possible as well. The three basic modulations with their corresponding spectra for a fixed bit pattern and non-return-to-zero (NRZ[1]) pulses are shown in Fig. 3.8. It is assumed here that the temporal width T of the logical "1" is the same as that of the logical "0".

Since the carrier is modulated by a fixed unlimited bit pattern, based on the theorems of Fourier transformation, the spectrum of the modulated

[1] In non return-to-zero pulses, the amplitude of the carrier is not switched to zero for the duration of a bit. Contrary to RZ pulses these pulses offer no easy way for a synchronisation between the transmitter and receiver, but on the other hand, they have a shorter spectrum and require less bandwidth.

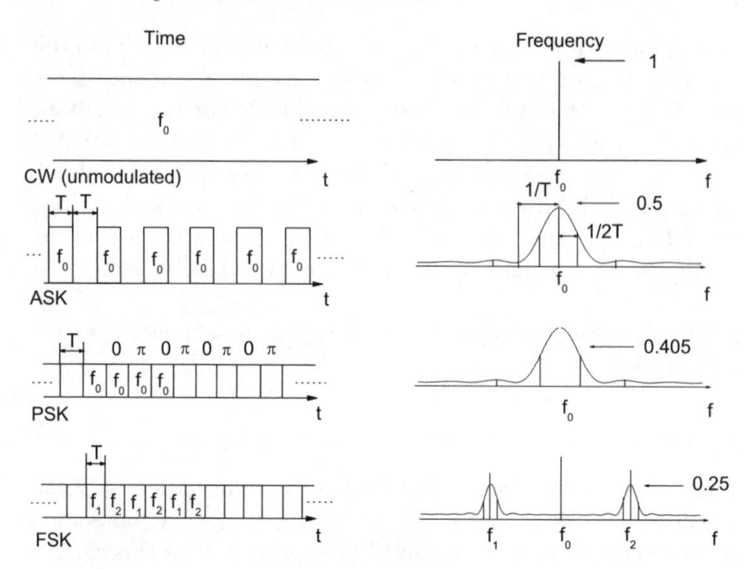

Fig. 3.8. Basic digital binary modulations in the time and frequency domain [271]

carrier would consist of distinct frequencies spaced by $1/2T$. Therefore, the fundamental modulation frequency $f_\mathrm{m} = 1/2T$ is half as large as the bit rate B. The envelope of the spectra follows a $\sin(x)/x$ function with zeros at $f_0 \pm n/T$ for ASK and PSK and at $f_{1,2} \pm n1/T$ for FSK. Hence, the modulated waves can be expressed as a sum of their spectral components. The distinct components are different in their amplitude $E_{\mathrm{m}n}(z)$, their frequency $\omega_{\mathrm{m}n}$ and wave number $k_{\mathrm{m}n}$. Thus for the modulated wave it follows [271]

$$E_\mathrm{m}(z,t) = \frac{1}{2} \sum_n E_{\mathrm{m}n}(z)\, e^{\mathrm{j}(\omega_{\mathrm{m}n} t - k_{\mathrm{m}n} z)} + \mathrm{c.c.}. \tag{3.10}$$

If the bit pattern is not fixed, the analysis is more complicated because the spectrum cannot be expressed as a sum of distinct, equally-spaced frequency components. In the case of one single pulse, the summation in (3.10) becomes an integral and the spectrum contains all frequencies in the $\sin(x)/x$ shape as shown in Fig. 3.9.

In optical telecommunications the pulses do not have sharp transitions such as those shown in the Figs. 3.8 and 3.9. The actual pulse shape has, of course, an influence on its spectrum as well. If the pulse shows a Gaussian shape its spectrum is Gaussian as well. Furthermore, due to the modulation, an additional frequency modulation of the pulse can occur (chirp)[2] and nonlinear effects can add new frequency components to the spectrum.

[2] A modulation of the laser diode control current results in the intended intensity alteration but the frequency of the laser will be changed simultaneously as well. This additional frequency modulation is called chirp. If the control current is increased it is followed by an increase of the temperature and the charge density in the laser

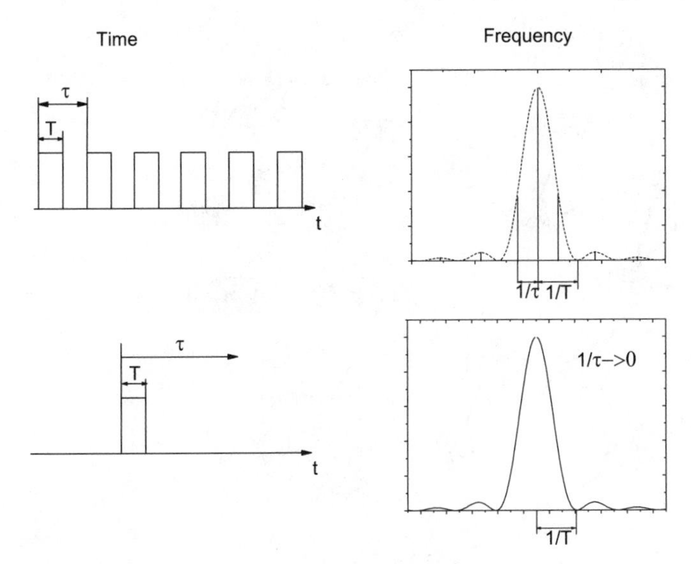

Fig. 3.9. Spectrum for an ASK modulated fixed bit pattern and for the modulation with one single pulse

3.5 Optical Transport Technologies

Early telecommunications networks were telephone networks, so the connection through the network was carried out in a manner that was called "circuit switched". This meant a call from one subscriber to another always built up a virtual connection between themselves all over the network. This connection lasted as long as the conversation was continued. Today, the former analog networks have been replaced by digital ones. In the access points of a digital network the analog speech signal is converted into a digital data stream. The conversion is ruled by the sampling theorem of Shannon, so the sample rate must be at least twice the highest frequency of the signal. Since filters in telephones restrict the highest frequency in speech signals to 3.4 kHz, the sample rate must be at least 6.8 kHz, but in practice 8 kHz are used. Then the amplitudes of the samples are quantized into only 256 distinct values coded by an 8 bit word. According to the sample rate, the time between every two 8-bit words is always $1/(8 \times 10^3\ \text{s}^{-1}) = 125\ \mu\text{s}$. During this time period, other channels can be transmitted over the network. According to this calculation, the basic frame duration in digital networks is 125 μs. Figure 3.10 shows the hierarchical structure of a digital transport network with its different layers.

Over the last few years the early plesiochrone digital hierarchy (PDH) technology was replaced by synchronous optical networks (SONET), adopted by the American national standards institute (ANSI), and by the interna-

diode. Both effects cause, beside other things, a change of the refractive index and therefore an alteration of the output frequency of the laser.

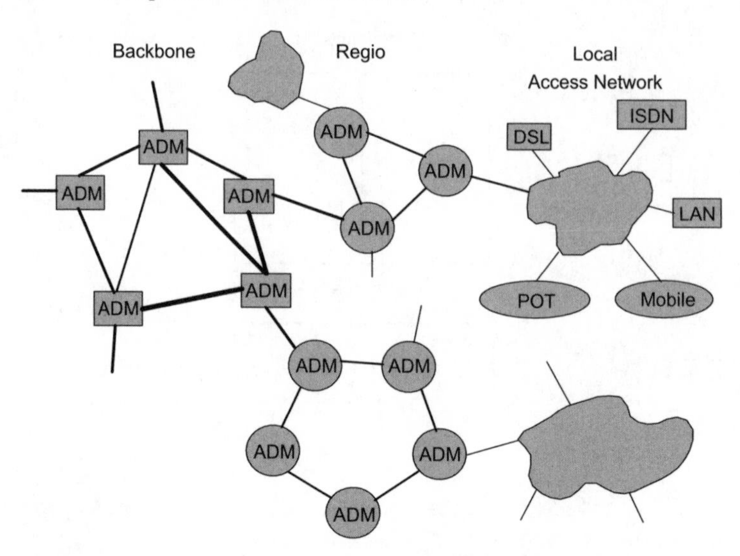

Fig. 3.10. The hierarchical structure of a transport network. (ADM=add-drop multiplexer, LAN=local area network, POT=plain old telephone, ISDN=integrated service digital network, DSL=digital subscriber line)

tional telecommunications unions (ITU-T's) synchronous digital hierarchy (SDH) technology. Although SDH is slightly different from SONET, the basic principle is the same. The rate hierarchy of SONET ranges from 51 Mb/s (OC-1) up to 9.8 Gb/s (OC-192).

Today, the internet is responsible for an ever increasing demand for bandwidth in telecommunications networks. The internet uses the internet protocol (IP) which is packet-oriented in nature for computer communication around the world. This means the data is transferred in autonomous packets, each of which has a destination and source address. Each packet can be transmitted on another route in the network.

The constant bit rate of SONET frames is a serious problem for the bursty like traffic of packet data signals. In order to evade the issue there are several interfaces allowing IP protocols to run over SONET networks. One of the most discussed possibilities to overcome the differences between circuit-switched and packet-oriented systems is the multi protocol label switching (MPLS) protocol in which the labels of the first packets were used to build up a constant route in the network over which all other packets are transmitted.

If the different channels are interlaced in the time domain, this technique is called optical time division multiplexing (OTDM). As mentioned in Sect. 3.2.1, the usable bandwidth of an optical fiber in the S, C and L-band is ≈ 90 nm. SONET uses only a small fraction of this bandwidth. For a multiplication of the capacity of the world's installed fiber network, it is possible to transmit the signals with different wavelengths or frequencies as carriers. Therefore, the SONET networks are successively replaced by (dense)

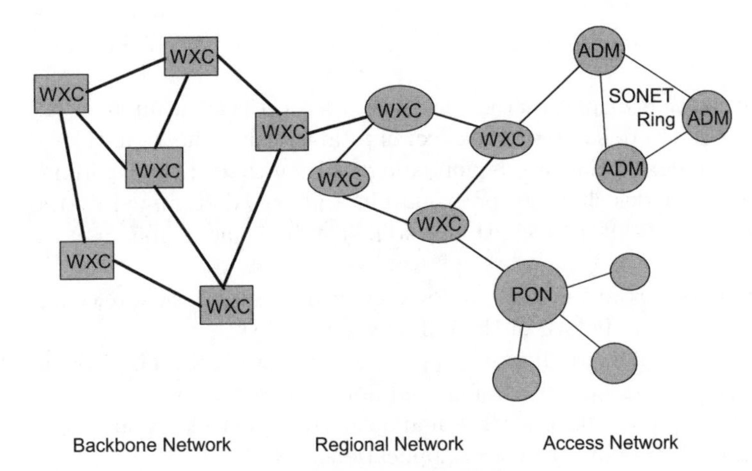

Fig. 3.11. Hierarchical structure of a WDM network. (WXC=wavelength cross-connect, ADM=add-drop multiplexer, PON=passive optical network) [318]

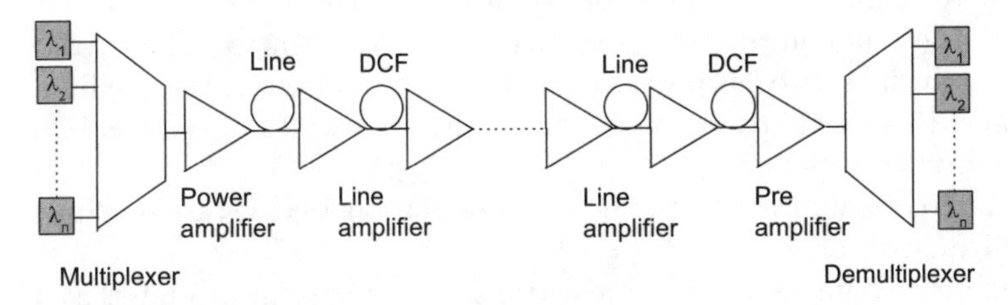

Fig. 3.12. Link over a WDM network (DCF=dispersion-compensating fiber)

wavelength division multiplexed ((D)WDM) networks in the backbones. The structure of a WDM-network is shown in Fig. 3.11. It has in principle the same hierarchical structure as the SONET network in Fig. 3.10. The main The main difference between the two are the multiplexers and demultiplexers at the access points and the wavelength cross-connect nodes in the network. Both devices are essential to build up dynamic WDM networks.

Figure 3.12 shows a typical WDM-link. The channels are dedicated to different wavelengths as carriers, which are then multiplexed and boosted together with the first power amplifier. In the network, the attenuation and linear distortion of the signals is compensated by line amplifiers and an intelligent dispersion management. Cross connects and add-drop multiplexers are able to process and switch the channels in the network while wavelength converters can change the carriers. In add/drop multiplexers or demultiplexers at the border of the network the channels are demultiplexed and transferred to their destination.

In a purely optical network all of these different tasks can be carried out in the optical domain, eliminating the necessity to convert the data at each processing point between the optical and electrical domain.

Summary

- With the exception of photonic crystal fibers with a central air hole, all optical waveguides are based on the effect of total internal reflection. Hence, the refractive index of the core region is higher than that of the cladding.
- The number of modes that can propagate in a fiber is determined by the wavelength. The cut-off wavelength of a standard single mode fiber is $\approx$ 1.2 μm.
- In modern fibers, 5 transmission windows are available for optical telecommunications. The bandwidth of these fibers is $\approx$ 50 THz.
- A large number of fibers for different applications are available. These fibers differ in their attenuation, dispersion, and nonlinear behavior.
- Highly nonlinear fibers have a large nonlinear coefficient. They are especially designed for nonlinear optical applications.
- Photonic crystal fibers offer the possibility of drastically reducing the effective area. Hence nonlinear effects are very efficient at the same time, it is possible to tailor the dispersion properties by design.
- In optical telecommunications, an optical carrier is required. The carrier serves, in principle, two purposes. It enables propagation in the medium and it offers the possibility to exploit the entire bandwidth of the medium via the multiplex technique.
- In optical telecommunications three basic modulation schemes are possible: ASK, FSK and PSK.
- If the different channels of an optical transmission system are interlaced in the time domain, this technique is called OTDM; if they are interlaced in the frequency domain it is DWDM. A combination between both is possible as well. If the channels are different in a specific code sequence with which the signal is multiplied, the technique is called code division multiple access (CDMA).
- For optical transport networks router, cross connects and add-drop multiplexers are required.

Exercises

3.1 Assume a monochromatic wave with a wavelength of 1.55 μm propagating in a multimode fiber. The difference in the refractive indices between core and cladding is $\Delta n = 0.004$. Calculate the angle of total internal reflection. (The cladding is made of pure silica, waveguide influences on the dispersion should be neglected.)

3.2. Estimate the maximum bit rate that can be transmitted via 10 km of the fiber in Exc. 3.1.

3.3. Suppose a fixed bit pattern "010101..." is transmitted in a transmission system with a rate of 40 Gbit/s where each individual bit is ideally rectangular and the duty cycle is 1/3. Calculate the spectrum of this fixed pattern and the spectrum if only one single bit of this pattern is transmitted.

3.4. An analog TV signal is to be transmitted in a cable TV system over optical fibers. The TV standard scans every picture line by line (625 lines) and requires 25 full frames per second. Determine the main spectral components of the signal.

3.5. Determine the bit rate of a digital audio signal if each amplitude value is coded with 16 bit and the analog signal has a highest frequency component of 22 kHz.

3.6. Calculate the bit rate of an uncompressed RGB video signal with a resolution of 1024×768 pixel and a repeating frequency of 70 Hz (whole frame).

4. Nonlinear Effects

In the previous chapters it was assumed that the discrete oscillators, which compose the material, would react linearly on the external field. This simplification is only valid if the optical field strength is rather low. Hence, the intensity has to be small. This is the case for noncoherent light sources such as lamps, light emitting diodes, and sunlight. Since the invention of the laser, a light source that delivers coherent light with extremely high intensities is available. The field strength of this light source can be so large that it is in the range of the inner atomic field. In this case, the excursion of the valence electron is no longer linear and hence, the rules of nonlinear optics are valid.

Compared to solid state lasers the semiconductor laser diodes, used in the field of telecommunications, deliver rather low powers in the mW range. Nevertheless, due to the small effective areas, as described in this chapter, the intensity in the core area is very high. Furthermore, the large effective length in optical fibers can lead to very strong nonlinear effects.

In principle two possibilities are feasible for the mathematical description of nonlinear optics. One is the so called semi-classical theory, the other is the theory of quantum-electrodynamics. In the semi-classical theory the external field is described with the Maxwell equations. But the medium by itself is considered as consisting of atoms and molecules and is hence described with the rules of quantum mechanics. In the theory of pure quantum-electrodynamics the external field as well as the medium are described with the equations of quantum mechanics.

Both approaches have particular advantages and disadvantages, but in most cases they deliver the same results and predictions for the nonlinear optical response of the medium. In the following the semi-classical theory is used almost exclusively. Nevertheless, for the phenomenological description of harmonic generation and wave mixing, the quantum-electrodynamical description is preferred here.

This chapter describes the fundamentals of some optical effects whose origin is the nonlinear response of the dipole system to the external optical field. The interaction between light and matter will be restricted to nonresonant effects, which is valid in optical fibers, for example, but not in resonant nonlinear optical devices like semiconductor optical amplifiers. We first start with an extension of the linear oscillator, introduced in Sect. 2.3. From the model the nonlinear susceptibility tensor elements will be derived. As will be seen in Sect. 4.3, a large number of these elements can be responsible for the generation of nonlinear effects. But due to symmetry arguments, only one residual element, considered a constant, is used to describe the nonlinear effects in optical fibers.

The nonlinear susceptibility is used to expand the wave equation derived in Sect. 2.2 to a nonlinear wave equation. With this equation nonlinear effects that depend on the distinct susceptibilities are described. After that the important concepts of effective area and length are introduced and the phase matching condition, essential for wave mixing processes, will be developed. The Chapter is finished with an estimation of the ultimate capacity limit of optical transmission systems. As can be seen from the last Section, contrary to the well-known Shannon condition, the maximum bit rate in an optical fiber is limited by nonlinear effects.

4.1 The Nonlinear Oscillator

In Sect. 2.3 the wave propagation in a medium was described with the model of the harmonic oscillator. This model is qualified for the description of nonlinear effects as well.

Assume only one oscillator, or dipole, excited by a strong external electric field. Due to the excitation, the dipole may oscillate in the x−direction. To simplify the derivation, the vector character of the field may be neglected. With the parameters of a mechanical oscillator – spring constant k, attenuation constant γ and mass m – the equation of motion yields (see (2.21))

$$\frac{\mathrm{d}^2x}{\mathrm{d}t^2} + \frac{\gamma}{m}\frac{\mathrm{d}x}{\mathrm{d}t} + \omega_o^2 x = \frac{q}{m}E(t). \tag{4.1}$$

As in the case of the linear harmonic oscillator for the resonance frequency, the relation $\omega_0 = \sqrt{k/m}$ is introduced into (4.1). The reset force of this oscillator is

$$F_{\mathrm{reset}} = -m\omega_0^2 x. \tag{4.2}$$

Hence, the electron moves in a field with potential energy

$$U = -\int F_{\mathrm{reset}}\mathrm{d}x = 1/2m\omega_0^2 x^2. \tag{4.3}$$

As mentioned in Sect. 2.3, the oscillator can be contemplated as an ensemble consisting of a valence electron and the rest of the atom. For the high frequencies of optical waves only the rather light electron can follow the fast fluctuations of the external field. According to (4.3), which is valid for the linear case, the electron moves in a parabolic potential.

If the excursion is too large, the reset force will no longer be linearly proportional to the x-coordinate, hence (4.2) no longer holds. Assume now that the external field is so intense that the x-coordinate exceeds the linear range. In classical mechanics this is the case if you hit a tuning fork too strong. The tuning fork will not only oscillate on their fundamental frequency, as in the linear case, many additional frequencies can be heard as well. Hence, the sound of the tuning fork will be distorted. The same behavior can be seen in an overdriven amplifier that leaves the linear range, resulting in a distortion of the amplified signal. In the output of the amplifier, new frequency components can be found. This can be done purposely, as in the case of old tube amplifiers for E-guitars. The sound of Jimi Hendrix or the Beatles would be unthinkable without these characteristic distortions. In new amplifier systems, this nonlinear behavior is artificial cloned.

The cohesion between the atom and valence electron (the spring which is responsible for the reset force in the mechanical model) is warranted by the mutual Coulomb attraction. Hence, the electron moves in a Coulomb potential

$$U = \left(U_0 - \frac{Q}{4\pi\varepsilon x} \right) . \tag{4.4}$$

Where Q is a carrier density distribution. In fact, only in the case of very small excursions (4.4) can be approximated by (4.3). For higher excursions, the Coulomb potential differs significantly from the parabolic form. If the excursion is not too large, the potential function can be approximated by a power series, as was be done in the case of a nonlinear element like a diode in a circuit (Sect. 1).

Since the oscillators are embedded in a medium, such as a crystal with particular symmetry conditions, the environment of the single oscillator must be taken into account. Hence, the excursion of the oscillator depends on the symmetry of the material.

If the material has no precedent direction and hence, the crystal shows the same behavior in all directions (inversion symmetry), then the carriers are isotropically distributed and the electron moves symmetrically. Therefore, the potential function, responsible for the description of the motion of the electron, has to fulfil the symmetry condition $U(x) = U(-x)$. Consequently, the function can only contain even multiples in the power exponent of x. Hence, the lowest correction term in the power series must depend on x to the fourth power

$$U = 1/2 m\omega_0^2 x^2 - 1/4 m b x^4 , \tag{4.5}$$

where b is a parameter that describes the strength of the nonlinearity. Due to the fact that $F = -\mathrm{grad} U$, the reset force is then

$$F_{\text{reset}} = -m\omega_0^2 x + m b x^3 . \tag{4.6}$$

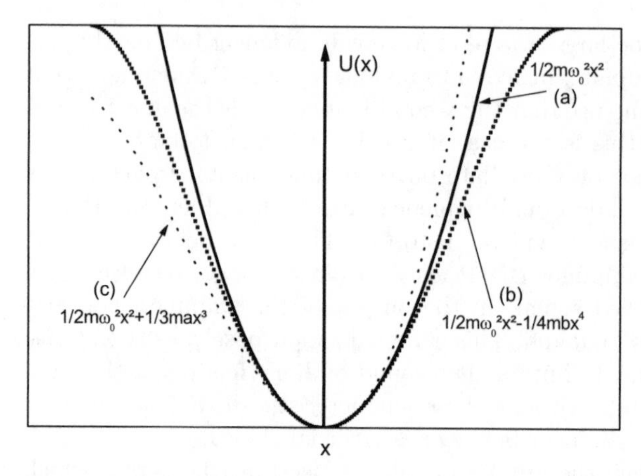

Fig. 4.1. Potential energy function for linear and nonlinear optical media with different symmetry

If the material shows no inversion symmetry then the carriers are not distributed isotropically. In this case, the Coulomb potential is not symmetric and $U(x) \neq U(-x)$. In consequence, odd and even powers are possible in the power series and the first correction term depends cubically on x. The potential energy equation is thus

$$U = 1/2m\omega_0^2 x^2 - 1/3amx^3. \tag{4.7}$$

Where the factor a is again responsible for the description of the strength of the nonlinearity. According to this discussion, the reset force is then

$$F_{\text{reset}} = -\text{grad}U = -m\omega_0^2 x + amx^2. \tag{4.8}$$

Figure 4.1 shows the different described potential energy functions in which the electron can move if its excursion exceeds the linear case.

If the intensity of the optical wave is very high then the external electric field, exciting the oscillator, can be extremely strong. Additional terms are possibly necessary in the potential equation. These extremely intensive fields are available, for instance, with modern amplified ultrashort-pulse laser systems. The pulses of such laser systems can contain more power than the total power of all power stations in the world. If these pulses are focused on a small spot, the highest intensities and electric fields on earth are then available. If they interact with matter it is possible to reproduce in the laboratory the extreme physical conditions that can be found in the center of stars or in the neighborhood of black holes [42].

The external fields generated by these lasers can be so high that they are much stronger than the internal electric field between atom and electron. Hence, the external field is no longer a little disturbance, as in the case before. In the radiated field, a large number of harmonics can be found (up to the

70th harmonic has been observed [43]). Therefore, with high intensity pulses in the infrared, it is possible to generate pulses in the x-ray range. Due to the phase matching condition (Sect. 4.10), this frequency conversion is carried out on the surface of materials. If the pulses are focused on a thin metal foil ions can be accelerated over a distance of 10 μm to energies of 100 MeV [44].

If the field strength is very high, the underlying physics is completely different. If the intensity of the applied field is above 10^{18} W/cm^2, the magnetic part of the electromagnetic wave can no longer be neglected. At these intensities the magnetic force that the electron experiences, is comparable to the electric force. The electrons are accelerated so strongly that the relativity theory must be included in the description of their motion.

In the field of telecommunications, these very high intensities will not be reached. If the higher order terms are neglected, the entire reset force of a non-inversion symmetric medium becomes

$$F_{\text{reset}} = -m\omega_0^2 x + amx^2 + bmx^3. \tag{4.9}$$

Hence, with (4.1) the equation of motion of the oscillator can be expressed as

$$\frac{\mathrm{d}^2 x}{\mathrm{d}t^2} + \frac{\gamma}{m}\frac{\mathrm{d}x}{\mathrm{d}t} + \omega_o^2 x - ax^2 - bx^3 = \frac{q}{m}E(t). \tag{4.10}$$

For the solution of (4.10), it is assumed that the external field is so low that the nonlinear terms ax^2 and bx^3 are much lower than the linear one, $\omega_o^2 x$. If this is the case, (4.10) can be solved with the disturbance theory, where the electric field will be multiplied by an additional parameter, λ, that quantifies the strength of the disturbance. To determine all nonlinear effects and susceptibilities resulting from (4.10), the external electric field has to consist of three distinct frequency components

$$E(t) = \hat{E}_1 \mathrm{e}^{\mathrm{j}\omega_1 t} + \text{c.c.} + \hat{E}_2 \mathrm{e}^{\mathrm{j}\omega_2 t} + \text{c.c.} + \hat{E}_3 \mathrm{e}^{\mathrm{j}\omega_3 t} + \text{c.c.}. \tag{4.11}$$

A complete derivation of all susceptibilities is lengthy and rather complicated. To simplify it, only the susceptibilities responsible for the generation of the second and third harmonic should be considered. Therefore, for the description of the electric field, the equation

$$E(t) = \hat{E}\mathrm{e}^{\mathrm{j}\omega t} \tag{4.12}$$

is sufficient. After multiplying $E(t)$ by λ (4.10) becomes

$$\frac{\mathrm{d}^2 x}{\mathrm{d}t^2} + \frac{\gamma}{m}\frac{\mathrm{d}x}{\mathrm{d}t} + \omega_o^2 x - ax^2 - bx^3 = \lambda\frac{q}{m}E(t). \tag{4.13}$$

If x is developed in a power series of λ, one yields

$$x = x^{(1)}\lambda + x^{(2)}\lambda^2 + x^{(3)}\lambda^3 + ... \tag{4.14}$$

Equation (4.13) contains, beside the first, the second and third power of x as well

$$x^2 = \lambda^2 \left[x^{(1)}\right]^2 + \lambda^4 \left[x^{(2)}\right]^2 + 2\lambda^3 x^{(1)} x^{(2)} + ... \quad (4.15)$$

and

$$x^3 = \lambda^3 \left[x^{(1)}\right]^3 + \lambda^6 \left[x^{(2)}\right]^3 + ... \quad (4.16)$$

Every term that is proportional to λ^n must be a solution of the differential equation (4.13). If the equations (4.15) and (4.16) are introduced in (4.13), it follows after a sorting of the powers of λ [41, 73] that:

$$\frac{\mathrm{d}^2 x^{(1)}}{\mathrm{d}t^2} + \frac{\gamma}{m}\frac{\mathrm{d}x^{(1)}}{\mathrm{d}t} + \omega_o^2 x^{(1)} = \frac{q}{m}E(t), \quad (4.17)$$

$$\frac{\mathrm{d}^2 x^{(2)}}{\mathrm{d}t^2} + \frac{\gamma}{m}\frac{\mathrm{d}x^{(2)}}{\mathrm{d}t} + \omega_o^2 x^{(2)} = a\left[x^{(1)}\right]^2, \quad (4.18)$$

$$\frac{\mathrm{d}^2 x^{(3)}}{\mathrm{d}t^2} + \frac{\gamma}{m}\frac{\mathrm{d}x^{(3)}}{\mathrm{d}t} + \omega_o^2 x^{(3)} = 2a x^{(1)} x^{(2)} + b\left[x^{(1)}\right]^3, \quad (4.19)$$

and

$$\frac{\mathrm{d}^2 x^{(n)}}{\mathrm{d}t^2} + \frac{\gamma}{m}\frac{\mathrm{d}x^{(n)}}{\mathrm{d}t} + \omega_o^2 x^{(n)} = f\left(x^{(1)} + x^{(2)} + ... + x^{(n)}\right). \quad (4.20)$$

Equation 4.17 complies to the equation of motion of a linear oscillator. Hence, the solution derived in Sect. 2.3 can be used. According to (2.22) $x^{(1)}(\omega)$ can be written as

$$x^{(1)}(\omega) = \frac{qE(\omega)}{m}\frac{1}{Z(\omega)} = x_0^1(\omega)\mathrm{e}^{\mathrm{j}\omega t}, \quad (4.21)$$

where $Z(\omega)$ is the complex

$$Z(\omega) = \left(\omega_0^2 - \omega^2\right) + \mathrm{j}\gamma_m\omega. \quad (4.22)$$

Therefore, the amplitude in (4.21) is

$$x_0^{(1)}(\omega) = \frac{q\hat{E}}{m}\frac{1}{Z(\omega)}. \quad (4.23)$$

For a solution of (4.18), the square of (4.21) is required. Using (4.12), it follows as a function that depends on the second harmonic of the fundamental frequency, 2ω

$$\left[x^{(1)}(\omega)\right]^2 = \frac{q^2\hat{E}^2}{m^2}\frac{\mathrm{e}^{\mathrm{j}2\omega t}}{Z^2(\omega)}. \quad (4.24)$$

The required solution of x must be dependent on the second harmonic as well

$$x^{(2)}(t) = x_0^{(2)}(2\omega)\mathrm{e}^{\mathrm{j}2\omega t}. \tag{4.25}$$

If the equations (4.25) and (4.24) are introduced into (4.18), one yields for the amplitude

$$x_0^{(2)}(2\omega) = a\frac{q^2\hat{E}^2}{m^2}\frac{1}{Z^2(\omega)Z(2\omega)}, \tag{4.26}$$

where

$$Z(2\omega) = \left(\omega_0^2 - 4\omega^2\right) + \mathrm{j}2\gamma_m\omega. \tag{4.27}$$

With these solutions (4.19) can be solved. As already mentioned, only the response of the oscillator to higher harmonics is of interest here. The third harmonic can only be generated through the second term on the right side. Hence, the first term can be neglected. With a completely analogous derivation as used for (4.26), the amplitude $x_0^{(3)}(3\omega)$ is then

$$x_0^{(3)}(3\omega) = b\frac{q^3\hat{E}^3}{m^3}\frac{1}{Z^3(\omega)Z(3\omega)}, \tag{4.28}$$

where

$$Z(3\omega) = \left(\omega_0^2 - 6\omega^2\right) + \mathrm{j}3\gamma\omega. \tag{4.29}$$

In Sect. 4.3 the amplitude functions are used to derive the nonlinear susceptibilities, but before that the nonlinear polarization will be introduced.

4.2 Nonlinear Polarization

An external field applied to an atom leads to a shift of the internal carriers. If the intensities are not too high, a pure electronic polarization can be seen as a distortion of the electron cloud. The distortion results in a field-induced dipole moment which in turn is again a source of an electromagnetic radiation. The secondary wave, radiated by the atom, superimposes on the primary one. This is the origin of the propagation of an optical field in a medium on a microscopic level. For consideration on a macroscopic scale, a macroscopic measure is required. This is the polarization, defined as the sum of distinct microscopic, field-induced dipole moments per unit volume in the considered medium.

Assuming a given medium consists of N molecules and the field-induced dipole moment of the ith molecule is $\boldsymbol{p}_i$, the polarization of the medium is then:

$$\boldsymbol{P}(t) = \sum_{i=1}^{N} \boldsymbol{p}_i(t). \tag{4.30}$$

The result of the summation over the distinct dipole moments depends on the macroscopic symmetry of the medium. Since the polarization of the medium is induced by the external electric field, the polarization must be connected to the electric field as well. This connection is determined by (2.10).

In the nonlinear case the electron moves – as already mentioned – in a Coulomb-potential, which cannot be neglected if the applied field is too high. This fact was taken into account on the microscopic level by the expansion of the reset force and the potential in a power series. Hence, the macroscopic polarization can be expanded in a power series as well. If the disturbance is small, the polarization equation $\boldsymbol{P} = \varepsilon_0 \chi \boldsymbol{E}$ is then

$$\boldsymbol{P} = \varepsilon_0 \left(\chi^{(1)} \boldsymbol{E} + \chi^{(2)} \boldsymbol{EE} + \chi^{(3)} \boldsymbol{EEE} + ... \right) = \boldsymbol{P}_{\mathrm{L}} + \boldsymbol{P}_{\mathrm{NL}}^{(2)} + \boldsymbol{P}_{\mathrm{NL}}^{(3)} + ... \tag{4.31}$$

The distinct $\chi^{(n)}$ are the linear and nonlinear susceptibilities of the material, which will be described in the next Section in detail. The term $\boldsymbol{P}_{\mathrm{L}}$ determines the linear part of the polarization, whereas $\boldsymbol{P}_{\mathrm{NL}}^{(2)}$ and $\boldsymbol{P}_{\mathrm{NL}}^{(3)}$ describe the second and third order nonlinear polarization, respectively. According to the corresponding correction term in the reset force, the response of the oscillator shows harmonic distortions. The response of the oscillator reflects the electronic polarization and, due to the fact that every accelerated current is a source of a secondary wave, is responsible for the generation of new frequency components in the electromagnetic field.

Figure 4.2 shows the polarization or the response of a material to an external monochromatic field and the corresponding Fourier transform. If the intensity of the applied field is small the response is linear, as shown in Fig. 4.2a. The electron moves in a parabolic potential, the polarization shows no distortion and in the corresponding Fourier transform, or in the secondary wave, only the fundamental frequency ω_0 is present. Consequently, the forced oscillation of the dipoles is only shifted in phase, resulting in the refractive index of the material.

If the intensities are increased, the response of the material will become nonlinear. The concrete response depends on the material symmetry. In Fig. 4.2b the potential function of an inversion symmetric material is shown. Due to the symmetry center of such materials no preference direction in the material is present. Hence, the potential energy function (Fig. 4.1b) is symmetrical around the zero point. As a result, in the secondary radiated field of the oscillators, a new frequency component – the third harmonic – can be found as well.

If no symmetry center is present in the crystal, the symmetry rule no longer holds and it follows a potential energy function according to Fig. 4.2c. In the radiated field, even ($2\omega_0$) and odd ($3\omega_0$) frequency components are

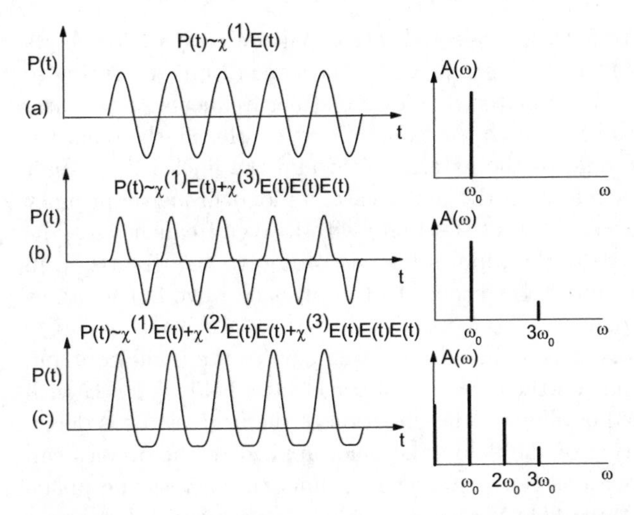

Fig. 4.2. Potential energy function for linear and nonlinear media and the corresponding Fourier transforms

present. As shown in Fig. 4.1c an additional constant polarization of the medium can be found.

If the medium is excited with several monochromatic waves with different frequencies, mixing products between these waves are present in the radiated field as well.

4.3 The Nonlinear Susceptibility

Like the refractive index, the distinct susceptibilities $\chi^{(n)}$ of a real material are as well complex as dependent on the frequency of the field. Hence, the linear polarization is

$$\boldsymbol{P}_{\mathrm{L}}(\omega) = \varepsilon_0 \chi^{(1)}(\omega) \boldsymbol{E}(\omega). \tag{4.32}$$

In principle, this is the relation between the Fourier component of the polarization at frequency ω and the corresponding Fourier component of the electric field at the same frequency. The relation is established over the susceptibility tensor $\chi^{(1)}$.

If the vector component is neglected, (2.25) leads to

$$P_{\mathrm{L}} = lqx^{(1)}(\omega) = \varepsilon_0 \chi^{(1)}(\omega) E(\omega). \tag{4.33}$$

For the first order susceptibility $\chi^{(1)}$, equations (4.21) and (4.23) yield

$$\chi^{(1)}(\omega) = \frac{lqx^{(1)}(\omega)}{\varepsilon_0 E(\omega)} = \frac{lq^2}{m\varepsilon_0} \frac{1}{Z(\omega)} = \chi'^{(1)}(\omega) - \mathrm{j}\chi''^{(1)}(\omega). \tag{4.34}$$

If the oscillatory strength is neglected, a comparison with (2.26) shows (with $Z(\omega) = (\omega_0^2 - \omega^2) + i\gamma\omega)$ – as expected – the same result. According to (4.32), the polarization of the material is for a distinct frequency, ω, linearly dependent on an external field with the same frequency. Hence, the polarization is linearly proportional to the external field and the linearity constant is the susceptibility. Therefore, in the linear case, a monochromatic primary wave can only animate the dipoles to oscillate with its own frequency ω. The secondary wave radiated by the dipoles has, according to the discussion in Sect. 4.1, only a phase shift with respect to the primary wave but contains no new frequency components.

If the vector component is included, the excursion of the oscillators and the susceptibility depends on the direction of the electric field vector $\boldsymbol{E}$ with respect to the embedded oscillator. The direction of the field vector is determined by the polarization of the field. The field, applied to the dipoles can have three distinct directions (for instance x, y, and z), whereas the polarization can consist of three field vectors as well. In consequence, the linear susceptibility $\chi^{(1)}$ is a tensor, consisting of nine elements. In a Cartesian coordinate system with components x, y, and z, (4.32) can be rewritten in matrix form [59, 129]

$$\begin{bmatrix} P_{\mathrm{L}x}(\omega) \\ P_{\mathrm{L}y}(\omega) \\ P_{\mathrm{L}z}(\omega) \end{bmatrix} = \varepsilon_0 \begin{bmatrix} \chi_{xx}^{(1)}(\omega) & \chi_{xy}^{(1)}(\omega) & \chi_{xz}^{(1)}(\omega) \\ \chi_{yx}^{(1)}(\omega) & \chi_{yy}^{(1)}(\omega) & \chi_{yz}^{(1)}(\omega) \\ \chi_{zx}^{(1)}(\omega) & \chi_{zy}^{(1)}(\omega) & \chi_{zz}^{(1)}(\omega) \end{bmatrix} \begin{bmatrix} E_x(\omega) \\ E_y(\omega) \\ E_z(\omega) \end{bmatrix} . \tag{4.35}$$

Equation 4.35 can be expressed as a summation over the distinct components as well

$$P_{\mathrm{L}i}(\omega) = \varepsilon_0 \sum_j \chi_{ij}^{(1)}(\omega) E_j(\omega)$$

where $i, j = x, y, z$.

In the case of the second order nonlinear polarization, the external fields can interact with each other. If both fields have different frequencies, they produce a polarization of the material at the sum- and difference-frequencies, or multiples of the input frequencies. Hence, the secondary field – radiated by the dipoles – consists of the original and the new frequency components. According to this, the nonlinear susceptibility of second order depends on the combination of the input frequencies, the field vectors of the different input fields, and the field vectors of the polarization itself. The second order nonlinear polarization is then

$$\boldsymbol{P}_{\mathrm{NL}}^{(2)}(\omega = \omega_1 + \omega_2) = \varepsilon_0 \chi^{(2)}(\omega_1, \omega_2) \boldsymbol{E}(\omega_1) \boldsymbol{E}(\omega_2). \tag{4.36}$$

For the vectors of the applied field $3^2 = 9$ different combinations, from $E_x E_x$ to $E_z E_z$, exist. The polarization itself can again consist of three field vectors. Hence, due to the different possible combinations of the several field vectors, the susceptibility $\chi^{(2)}$ consists of $3^2 \times 3 = 27$ elements and is a tensor of third rank. The matrix form of (4.36) is

$$\begin{bmatrix} P^{(2)}_{\mathrm{NL}x}(\omega) \\ P^{(2)}_{\mathrm{NL}y}(\omega) \\ P^{(2)}_{\mathrm{NL}z}(\omega) \end{bmatrix} = \varepsilon_0 \begin{bmatrix} \chi^{(2)}_{xxx} & \chi^{(2)}_{xxy} & \cdots & \chi^{(2)}_{xzz} \\ \chi^{(2)}_{yxx} & \chi^{(2)}_{yxy} & \cdots & \chi^{(2)}_{yzz} \\ \chi^{(2)}_{zxx} & \chi^{(2)}_{zxy} & \cdots & \chi^{(2)}_{zzz} \end{bmatrix} \begin{bmatrix} E_x(\omega_1)E_x(\omega_2) \\ E_x(\omega_1)E_y(\omega_2) \\ E_x(\omega_1)E_z(\omega_2) \\ E_y(\omega_1)E_x(\omega_2) \\ E_y(\omega_1)E_y(\omega_2) \\ E_y(\omega_1)E_z(\omega_2) \\ E_z(\omega_1)E_x(\omega_2) \\ E_z(\omega_1)E_y(\omega_2) \\ E_z(\omega_1)E_z(\omega_2) \end{bmatrix} . \tag{4.37}$$

In analogy to the derivation of the linear polarization (4.37) can be written as a summation as well

$$P^{(2)}_{\mathrm{NL}i}(\omega = \omega_1 + \omega_2) = \varepsilon_0 \sum_{jk} \chi^{(2)}_{ijk}(\omega_1,\omega_2) E_j(\omega_1) E_k(\omega_2) \tag{4.38}$$

where $i, j, k = x, y, z$. If the vector component is neglected, the polarization responsible for the second harmonic in (4.31) becomes

$$P^{(2)}_{\mathrm{NL}} = \hat{P}^{(2)}_{\mathrm{NL}} \mathrm{e}^{\mathrm{j}2\omega t} = lqx^{(2)}_0 \mathrm{e}^{\mathrm{j}2\omega t} = \varepsilon_0 \chi^{(2)}(2\omega, \omega, \omega) \hat{E}^2 \mathrm{e}^{\mathrm{j}2\omega t} . \tag{4.39}$$

Using (4.26), the second order susceptibility is expressed as

$$\chi^{(2)}(2\omega, \omega, \omega) = \frac{lqx^{(2)}(2\omega)}{\varepsilon_0 \hat{E}^2 \mathrm{e}^{\mathrm{j}2\omega t}} = \frac{lq^3}{m^2 \varepsilon_0} \frac{a}{Z^2(\omega) Z(2\omega)} \tag{4.40}$$

where $Z(\omega)$ is defined according to (4.22) and $Z(2\omega)$ according to (4.27), respectively.

The nonlinear polarization of third order connects three fields with each other. The dipole oscillates again on the sum, difference, and multiples of the input frequencies. The third order nonlinear polarization is

$$P^{(3)}_{\mathrm{NL}}(\omega = \omega_1 + \omega_2 + \omega_3) = \varepsilon_0 \chi^{(3)}(\omega_1, \omega_2, \omega_3) \boldsymbol{E}(\omega_1) \boldsymbol{E}(\omega_2) \boldsymbol{E}(\omega_3) . \tag{4.41}$$

Now three fields are involved, hence $3^3 = 27$ combinations of these fields are possible. With the three possible directions of the nonlinear polarization it follows that the nonlinear susceptibility $\chi^{(3)}$ can consist of $3^3 \times 3 = 81$ elements and is a tensor of fourth rank. In matrix form, (4.41) is represented by

$$\begin{bmatrix} P^{(3)}_{\mathrm{NL}x}(\omega) \\ P^{(3)}_{\mathrm{NL}y}(\omega) \\ P^{(3)}_{\mathrm{NL}z}(\omega) \end{bmatrix} = \varepsilon_0 \begin{bmatrix} \chi^{(3)}_{xxxx} & \chi^{(3)}_{xxxy} & \cdots & \chi^{(3)}_{xzzz} \\ \chi^{(3)}_{yxxx} & \chi^{(3)}_{yxxy} & \cdots & \chi^{(3)}_{yzzz} \\ \chi^{(3)}_{zxxx} & \chi^{(3)}_{zxxy} & \cdots & \chi^{(3)}_{zzzz} \end{bmatrix} \begin{bmatrix} E_x(\omega_1)E_x(\omega_2)E_x(\omega_3) \\ E_x(\omega_1)E_x(\omega_2)E_y(\omega_3) \\ \cdots \\ \cdots \\ \cdots \\ E_z(\omega_1)E_z(\omega_2)E_y(\omega_3) \\ E_z(\omega_1)E_z(\omega_2)E_z(\omega_3) \end{bmatrix} . \tag{4.42}$$

In summation form the above equation can be expressed as

$$P_{\mathrm{NL}i}^{(3)}(\omega) = \varepsilon_0 \sum_{jkl} \chi_{ijkl}^{(3)}(\omega_1,\omega_2,\omega_3) E_j(\omega_1) E_k(\omega_2) E_l(\omega_3) \tag{4.43}$$

where $i, j, k, l = x, y, z$. Without the vector component the susceptibility responsible for the generation of the third harmonic from (4.28) is then

$$\chi^{(3)}(3\omega,\omega,\omega,\omega) = \frac{lqx^{(2)}(3\omega)}{\varepsilon_0 \hat{E}^3 \mathrm{e}^{\mathrm{j}3\omega t}} = \frac{lq^4}{m^3\varepsilon_0} \frac{b}{Z^3(\omega)Z(3\omega)}. \tag{4.44}$$

According to the above derivations, the nonlinear susceptibilities are rather complicated. They consist of a large number of elements that depend on the polarization of the electric fields involved and the direction of the generated polarization. Furthermore, they depend on the frequencies of the fields and are complex, i.e., they consist of a real and an imaginary part. For most applications in the field of optical telecommunications, this complexity as well as most of the tensor elements can be neglected and hence, the analyses can be simplified considerably.

According to Sect. 2.4, the imaginary part of the linear susceptibility leads to an attenuation of the wave in the medium. This attenuation is especially large if the frequency of the applied field is in the range of material resonances. In the case of nonlinear susceptibilities, the imaginary part has to be involved if linear combinations of the frequencies of the applied fields coincide with resonances of the material. The imaginary part of the second order susceptibility $\chi^{(2)}$ is, for instance, very strong if the sum of the input frequencies coincide with the two photon absorption in the material. In this case the real part is also enhanced.

On the other hand, if the frequencies of the exciting fields and linear combinations of them are far enough from these resonances, the imaginary parts of the linear and nonlinear susceptibilities can be neglected. For optical telecommunications, the usable wavelengths in optical fibers are far away from linear and nonlinear absorptions, hence the imaginary part plays no significant role. But in the case of semiconductor optical amplifiers, used for nonlinear optical signal processing, resonant effects are exploited (Sect. 12.2.3). Hence, the description of nonlinear effects in such devices is rather complicated.

Furthermore, for most materials of interest for technical applications, the number of tensor elements can be reduced drastically. This follows from symmetry arguments. According to the symmetry of the considered crystals, the corresponding tensor elements can be zero or dependent on other elements (see for instance, [2]). For an isotropic material, for example, the first order susceptibility $\chi^{(1)}$ contains only one single element. This element depends on the frequency and leads to the refractive index, as shown in Sect. 2.6.

For materials with a symmetry center, such as the silica glass of optical waveguides, all 27 elements of the second order susceptibility $\chi^{(2)}$ are zero and hence do not have to be considered (see Sect. 4.1). According to this, in a material with inversion symmetry, the generation of the second harmonic as well as the generation of sum and difference frequencies is impossible, as already explained in Sect. 4.2. This rule only holds if a pure local interaction of the field is assumed, i.e., every dipole or oscillator is considered separately. If neighboring dipoles are included, non-local electric quadrupole and magnetic dipole interactions will take place [59, 63]. But in most cases they are negligibly small.

On the surface of an inversion symmetric crystal, the symmetry partners are lost and the symmetry is broken. Hence, on the surface a second order susceptibility – responsible for the generation of a second harmonic wave – is present. This only holds for the upper most atomic layers, hence the surface is limited to a few Ångstroms. A linear interaction with the surface is reflection where the penetration depth of the wave corresponds to the wavelength of the light. Due to this, reflected light in the visible range comes from areas some thousand times deeper than in the case of a nonlinear interaction. This fact has made surface second harmonic generation (SHG) a powerful, widely used, and commonly recognized tool for the characterization of surface and interface properties. As shown in [64], it is possible to investigate the symmetry of the surface layer in materials with a very weak nonlinearity like transparent insulators, e.g., barium fluoride. In a pump-probe arrangement in combination with ultrashort laser pulses, it has been possible to study ultrafast surface dynamics like the lifetime of electron excitation in metals [65, 66] and semiconductors [67], laser-induced order/disorder transitions [67, 68] coherent surface optical phonon oscillations [69], and several other effects. The pump pulse excites a physical process at the surface which changes the nonlinear optical properties of the material for the variably delayed probe pulse. Hence, the alteration of the probe pulse as a function of the delay is a measure for the lifetime of the surface excitation.

The second order electronic polarization follows directly from the oscillations of the electrons. Hence, the second harmonic generated at the surface depends on the degrees of freedom of the electron motions on the surface. The second harmonic thus reflects the surface geometry. If the crystal is turned around the surface normal, one yields an alteration of the intensity of the second harmonic generated at the surface. This means the intensity of the generated signal depends on the rotation angle and is anisotropic. According to the anisotropy, direct conclusions on the surface symmetry of the crystal and the values of the corresponding susceptibility $\chi^{(2)}$ can be drawn.

Figure 4.3 shows the second harmonic (SHG) measured in reflection from the surface of a barium fluoride crystal against the rotation angle.

As can be seen from the sketch of the atom layers, the signal clearly reflects the $C_{3\nu}$-Symmetry of an unreconstructed [111] surface [70, 71].

Photonic fibers used in telecommunications consist of silica glass which has an inversion center. Hence, the second order susceptibility vanishes.

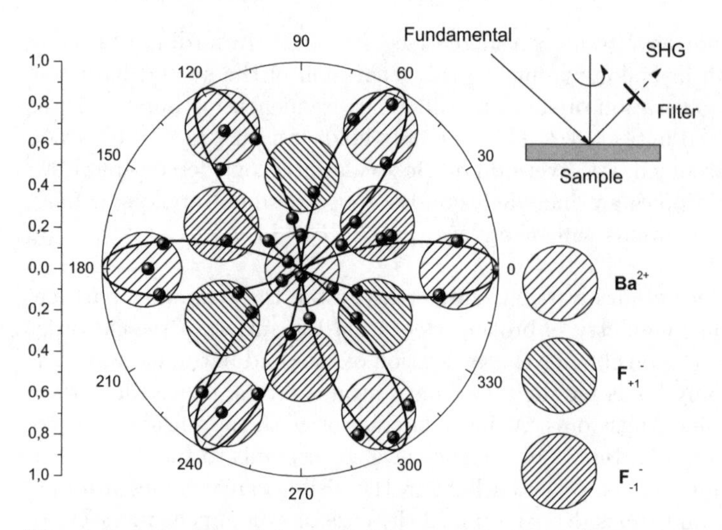

Fig. 4.3. Polar plot of the second harmonic intensity from the surface of a barium fluoride [111] crystal against the rotation angle and symmetry of the surface. The inset shows the experimental set-up. The surface was excited with a 120-fs pulse from a titanium sapphire laser (800 nm). F^-_{-1} denotes the lower layer of the fluorine atoms, whereas F^-_{+1} denotes the upper layer. Ba^{2+} denotes the barium atom layer

Therefore, for nonlinear optical effects in the telecommunications range, the third order susceptibility $\chi^{(3)}$ is of particular importance. As mentioned, for materials without a symmetry, $\chi^{(3)}$ consists of 81 elements. Due to the symmetry of silica glass, only 21 of these are not zero. These 21 residual elements are not independent of each other and the tensor $\chi^{(3)}$ for optical fibers can be reduced to only 3 elements. If the dependence of the third order susceptibility on the involved frequency components can be neglected and if there is no Raman contribution to the susceptibility, then Kleinmans symmetry condition is valid [72]. In consequence, the third order susceptibility can be reduced to only one element and $\chi^{(3)}$ can be used as a constant.

4.3.1 Estimation of the Order of Magnitude of the Susceptibilities

In this Section, the order of magnitude of the distinct susceptibilities will be estimated. To simplify this estimation, the dependence of the susceptibility on the field polarization and the wavelength will be neglected and each tensor is assumed to be a constant.

The photon energy of the wavelength range and linear combinations of them used for optical telecommunications is far away from any material resonances in the optical fiber, as can be seen from Sect. 2.6.1. Hence, the distinct photons were not absorbed by the material system and the origin of the nonlinearity of the material is a pure electronic one. In consequence, the electron cloud is distorted due to the external field. For small intensities of the applied field, the Coulomb force can be approximated to be linearly dependent on the field strength. This is the range of linear optics. Hence, for the nonlinear case, the electric field has to be rather strong, but not too strong because then the approximation of the polarization into a power series is no longer valid, and the magnetic part of the field plays a significant role as well.

Let's assume that the electric field amplitude of the external field is – in order of magnitude – close to the inner atomic field strength, responsible for the Coulomb binding of the ground state electron in a hydrogen atom E_{at} (Exc. 4.2). This is represented by

$$E_{\mathrm{at}} = \frac{e}{4\pi\varepsilon_0 a_0^2} \tag{4.45}$$

where e is the electron charge and a_0 is the Bohr radius. The correction term of lowest order, $P^{(2)} = \varepsilon_0\chi^{(2)}E^2$ in (4.31), is in the range of the linear polarization $P^{(1)} = \varepsilon_0\chi^{(1)}E$ for a pure electronic polarization:

$$\chi^{(2)}E_{\mathrm{at}}^2 \approx \chi^{(1)}E_{\mathrm{at}}. \tag{4.46}$$

Therefore, the second order polarization is

$$\chi^{(2)} \approx \frac{\chi^{(1)}}{E_{\mathrm{at}}}. \tag{4.47}$$

In analogy to this consideration, the higher order susceptibilities are

$$\chi^{(3)} \approx \frac{\chi^{(1)}}{E_{\mathrm{at}}^2}$$

and

$$\chi^{(4)} \approx \frac{\chi^{(1)}}{E_{\mathrm{at}}^3}.$$

For a refractive index of 1.5, (4.87) yields a value of 1.25 for $\chi^{(1)}$. Hence we can assume that the value of $\chi^{(1)}$ for solid materials is in the range of 1. The charge of the electron e is 1.60218×10^{-19} C, the permittivity of vacuum ε_0 is 8.8542×10^{-12} C/Vm, and the Bohr radius a_0 is 5.29177×10^{-11} m. With these values, (4.45) yields $E_{\mathrm{at}} \approx 5.14\times10^{11}$ V/m. Table 4.1 lists the estimated values of the distinct susceptibilities. The ratio between two successive susceptibilities is

$$\frac{\chi^{(n)}}{\chi^{(n-1)}} \approx \frac{1}{E_{\mathrm{at}}} \approx 2\times10^{-12}. \tag{4.48}$$

According to this estimation, the next higher susceptibility is 12 orders of magnitude smaller than the previous one. With the standard value for

Table 4.1. Estimation of the order of magnitude for the susceptibilities of a solid material

$\chi^{(1)}$	1
$\chi^{(2)}$	2×10^{-12} m/V
$\chi^{(3)}$	3.8×10^{-24} m^2/V^2
$\chi^{(4)}$	7.4×10^{-36} m^3/V^3

n_2 in optical fibers ($n_2 = 2.35 \times 10^{-20}$ m^2/W [30]) and the equation for the nonlinear refractive index (4.90) that will be derived in Sect. 4.7, one yields $\chi^{(3)} \approx 1.72 \times 10^{-22}$ m^2/V^2 for the effective third order susceptibility in optical fibers at a wavelength of 1.5 µm. As can be seen, this value is larger than the estimated one according to Table 4.1.

In the literature, the susceptibility is sometimes referred to in electrostatic units (esu). This unit is based on the Gaussian system, in which all field quantities have the same unit. Material polarization and electric field strength in the Gaussian system have the unit statvolt/cm. Contrary to (4.31), here the polarization of the medium $\boldsymbol{P}$ depends on the electric field $\boldsymbol{E}$ via:

$$\boldsymbol{P} = \chi^{(1)}\boldsymbol{E} + \chi^{(2)}\boldsymbol{EE} + \chi^{(3)}\boldsymbol{EEE} + \ldots \tag{4.49}$$

Hence, the units for the different susceptibilities are dimensionless for $\chi^{(1)}$, cm/statvolt for $\chi^{(2)}$, cm^2/statvolt^2 for $\chi^{(3)}$, and so on, in most cases this is not explicitly indicated, but the unit esu is used. A conversion between the units can be carried out with the equations (4.50) to (4.52) [73]:

$$\chi^{(1)} = 4\pi\chi^{(1)}(\mathrm{esu}), \tag{4.50}$$

$$\chi^{(2)}(\mathrm{m/V}) = 4.189 \times 10^{-4}\chi^{(2)}(\mathrm{esu}), \tag{4.51}$$

and

$$\chi^{(3)}(\mathrm{m^2/V^2}) = 1.4 \times 10^{-8}\chi^{(3)}(\mathrm{esu}). \tag{4.52}$$

According to the estimation, nonlinearities with a pure electronic origin are rather small. Due to the symmetry of silica glass, only nonlinear effects that depend on odd susceptibilities are possible. The first odd susceptibility is $\chi^{(3)}$. Hence, the ratio between the linear polarization, responsible for the refractive index, and the nonlinear one, for instance responsible for the nonlinear refractive index, is

$$\frac{P_\mathrm{L}}{P_\mathrm{NL}} = \frac{\chi^{(1)}}{\chi^{(3)}EE} \approx \frac{1}{1.72 \times 10^{-22}\mathrm{m^2/V^2}\ EE}. \tag{4.53}$$

Assuming that the intensity of the sun, the solar constant, is 1.37 kW/m^2, then (4.88) with $n_0 = 1$ yields an electric field strength of $E \approx 1016$ V/m.

Table 4.2. Estimation of the order of magnitude for the susceptibilities of a solid material [73]

Mechanism	$\chi^{(3)}$ (esu)
Electronic Polarization	10^{-14}
Molecule Orientation	10^{-12}
Elektrostriction	10^{-12}
Thermal Effects	10^{-4}

The ratio between the linear and third order nonlinear polarization is in this case, according to (4.53), $\frac{P_\mathrm{L}}{P_\mathrm{NL}} \approx 5.6 \times 10^{15}$. Hence, the linear response preponderates by far and the nonlinearity of the interaction can be completely neglected.

On the other hand, let's consider the example from the introduction of this book: a DWDM system with 100 WDM channels and an input power of 3 mW each, propagating in a single mode fiber with an effective core area of 80 μm^2 and an attenuation constant of $\alpha = 0.2$ dB/km. The intensity in the core of the fiber at the input is 3.75×10^9 W/m^2 and has a respectable electric field strength of $E \approx 1.7 \times 10^6$ V/m. Therefore, the ratio between the polarizations is $\frac{P_\mathrm{L}}{P_\mathrm{NL}} \approx 2 \times 10^9$. This is still nothing compared to the intensities used for nonlinear optical studies with solid state laser systems ($10^{12} - 10^{14}$ W/m^2), but these intensities are only reached if the beam is focused and hence – due to the diffraction of light – only for a few μm. In the mentioned example, after a distance of 5 km the intensity will still be $\approx 3 \times 10^9$ W/m^2. Hence, the nonlinearity cannot be neglected. These two points, the rather high intensities – due to the small core area – and the large interaction lengths – due to the slight attenuation – are the major reasons why nonlinear effects are so important in optical telecommunications.

If the electronic polarization is not the only effect involved, considerably larger susceptibilities can be found. This is the case in semiconductor optical amplifiers where resonant optical effects are exploited. Table 4.2 shows the corresponding value of $\chi^{(3)}$ for several mechanisms that cause nonlinear susceptibilities.

4.4 The Nonlinear Wave Equation

The nonlinear response of an oscillator, or dipole, to an intense optical field was considered by the expansion of the polarization equation in a power series (4.31). Hence, to obtain the nonlinear wave equation, only this power series has to be introduced into the wave equation for insulators (2.12). The result is

$$\Delta \boldsymbol{E} = \frac{1}{c^2}\frac{\partial^2 \boldsymbol{E}}{\partial t^2} + \frac{1}{c^2}\chi^{(1)}\frac{\partial^2 \boldsymbol{E}}{\partial t^2} + \frac{1}{c^2}\chi^{(2)}\frac{\partial^2 \boldsymbol{EE}}{\partial t^2} + \frac{1}{c^2}\chi^{(3)}\frac{\partial^2 \boldsymbol{EEE}}{\partial t^2} + ... \quad (4.54)$$

Using $n = \sqrt{\varepsilon_{\mathrm{r}}} = \sqrt{1+\chi^{(1)}}$ it follows that

$$\Delta \boldsymbol{E} - \frac{n^2}{c^2}\frac{\partial^2 \boldsymbol{E}}{\partial t^2} = \frac{1}{c^2}\frac{\partial^2}{\partial t^2}\left(\chi^{(2)}\boldsymbol{EE} + \chi^{(3)}\boldsymbol{EEE} + ...\right). \quad (4.55)$$

Equation (4.55) is the wave equation for nonlinear media. With this equation, all nonlinear effects of second and third order can be described. The first term on the right hand side is responsible for the generation of the second harmonic and the sum and difference frequency generation. As described in Sect. 4.3, these effects and $\chi^{(2)}$ depend on the material symmetry. Therefore, the first term will be zero in materials with inversion symmetry.

The second term on the right hand side is responsible for the intensity-dependent refractive index, the self-phase modulation (SPM), cross-phase modulation (XPM), four-wave-mixing (FWM), and several other effects, all treated in this book in detail. All these effects are of importance to optical telecommunications because they limit the capacity of optical transmission systems. On the other hand, they offer the possibility of many interesting applications, like pure optical signal processing, optical amplification, and distortion compensation.

The following sections give a rather simple overview of nonlinear effects based on the responsible polarization term. A detailed treaty, especially under consideration of the waveguide properties, follows in special sections of this book.

For the description, only the response of the dipole on the external field is of interest. The oscillator is assumed to be fixed at a particular point in the host medium and the z dependence of the wave is included in the complex amplitude. Therefore, the external monochromatic plane wave exciting the oscillator can be written as

$$\boldsymbol{E}(z,t) = \frac{1}{2}\left(\hat{\boldsymbol{E}}(z)\mathrm{e}^{-\mathrm{j}\omega t} + \mathrm{c.c.}\right) \quad (4.56)$$

and

$$\hat{\boldsymbol{E}}(z) = \hat{E}_0 \mathrm{e}^{\mathrm{j}kz}\boldsymbol{e}_i, \quad (4.57)$$

with $\boldsymbol{e}_i$ as the unit vector in an arbitrary i–direction (x or y in this case) and $\hat{E}_0$ as the complex amplitude. In order to simplify the derivation further it is assumed that the polarization of the wave has no influence on the response of the oscillator, hence, the vector character can be neglected.

4.5 Second Order Nonlinear Phenomena

The origin of all second order nonlinear phenomena is the first nonlinear term in (4.31). For a pure electronic polarization, the second order susceptibility

$\chi^{(2)}$ is 12 orders of magnitude higher than $\chi^{(3)}$, as was shown in (4.48). Hence, rather low field strengths can also lead to strong nonlinear effects of second order. This only holds, as mentioned in Sect. 4.3, for materials without inversion symmetry.

According to (4.31), the second order nonlinear polarization depends quadratically on the electric field E:

$$P_{\mathrm{NL}}^{(2)} = \varepsilon_0 \chi^{(2)} E^2 . \tag{4.58}$$

If (4.56) is – neglecting the vector component – introduced into (4.58), the nonlinear polarization becomes

$$\begin{aligned} P_{\mathrm{NL}}^{(2)} &= \frac{1}{4}\varepsilon_0 \chi^{(2)} \left(\hat{E}\mathrm{e}^{-\mathrm{j}\omega t} + \hat{E}^* \mathrm{e}^{\mathrm{j}\omega t}\right)\left(\hat{E}\mathrm{e}^{-\mathrm{j}\omega t} + \hat{E}^* \mathrm{e}^{\mathrm{j}\omega t}\right) \\ &= \frac{1}{4}\varepsilon_0 \left(2\chi^{(2)} \left|\hat{E}\right|^2 + \chi^{(2)} \hat{E}^2 \mathrm{e}^{-\mathrm{j}2\omega t} + \mathrm{c.c.}\right) , \end{aligned} \tag{4.59}$$

where the relation $\hat{E}\hat{E}^* = \left|\hat{E}\right|^2$ is used. As can be seen from (4.59), a constant part dependent on the field intensity ($I \sim \left|\hat{E}\right|^2$) can be found in the polarization of the medium. This part leads to a static electric field in the crystal (optical rectification). The second term in (4.59) describes a part of the material polarization oscillating with twice the frequency of the fundamental:

$$P_{\mathrm{NL}}^{(2)}(2\omega, 2k) = \frac{1}{4}\varepsilon_0 \chi^{(2)} \left(\hat{E}^2 \mathrm{e}^{-\mathrm{j}2\omega t} + \mathrm{c.c.}\right) . \tag{4.60}$$

Hence, the material polarization has twice the fundamental frequency 2ω. The wave number of this polarization (k_2) results from $\hat{E}^2$ and (4.57). Therefore, the phase velocity of the polarization propagating through the material is $v_{\mathrm{ph}} = 2\omega/k_2$, where $k_2 = k(2\omega)$ is the wave number at twice the fundamental frequency ω. The polarization itself is generated by the product of the fundamental waves moving through the crystal, which has a wave number of $2k$. The phase velocity of this wave is $v_{\mathrm{ph}} = 2\omega/2k$. Since the wave numbers are dependent on the refractive index of the material at the given frequency, in most cases $k_2 \neq 2k$. Hence, due to the dispersion of the material, the polarization responsible for the generation of the new wave and the new wave itself come out of phase. For an effective generation of the second harmonic the phases between both must be matched, this is the so called phase matching condition which is considered in detail in Sect. 4.10.

From a quantum mechanical point of view, two photons of the primary wave are annihilated simultaneously in the material as a new photon with twice the fundamental frequency is generated. This behavior is shown in Fig. 4.4. In the first step the atom or molecule of the medium is excited into a virtual inter state. In the second step it comes back to the ground state. If the interaction between the photons and the atom is finished, the

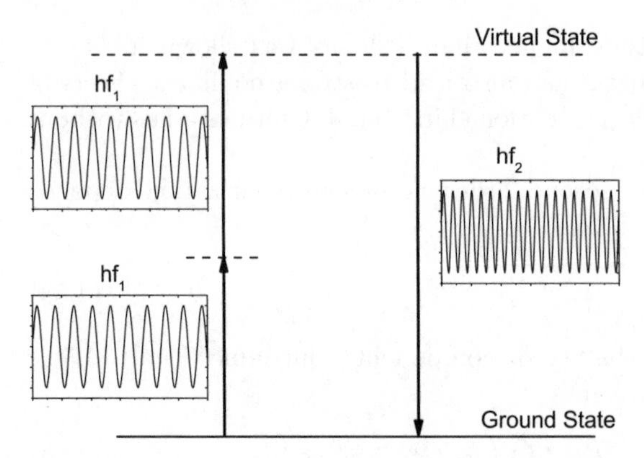

Fig. 4.4. Energy diagram for second harmonic generation

atom experiences no long lasting change, i.e., it is not different from the atom before the interaction. Hence, the rules of energy and momentum conservation are valid between the generating and the generated photons and the atom is not involved.

Since two photons with the same energy are annihilated simultaneously, the new photon must show twice the energy of the original ones. With the quantum mechanical formulation for the photon energy $E = \hbar\omega$ where $\hbar$ as the Planck constant divided by 2π, the rule of conservation of energy yields

$$\begin{aligned} \hbar 2\omega &= \hbar\omega + \hbar\omega \\ 2\omega &= \omega + \omega . \end{aligned} \tag{4.61}$$

Hence, the generation of the second harmonic can be seen as a direct consequence of the energy conservation during the interaction. At the same time, the momentum has to be exchanged between the photons in a manner where the total momentum of the interaction will remain conserved. With $\boldsymbol{p} = \hbar\boldsymbol{k}$ the rule of conservation of momentum gives

$$\begin{aligned} \hbar\boldsymbol{k}_2 &= \hbar\boldsymbol{k} + \hbar\boldsymbol{k} \\ \boldsymbol{k}_2 &= \boldsymbol{k} + \boldsymbol{k} . \end{aligned} \tag{4.62}$$

The generation of the second harmonic is of technical importance, especially for the frequency doubling of coherent laser light. It offers the possibility to increase the available spectrum of lasers. For example the fundamental frequency of a Nd:YAG -laser is in the invisible infrared range at 1064 nm. With the second harmonic generation in nonlinear crystals it is then possible to generate intense green light at 532 nm.

If the laser delivers short pulses with a particular spectral width, each frequency component of the pulse is translated into the second harmonic. Figure 4.5 shows the second harmonic of a Titanium Sapphire laser. The central wavelength of the fundamental is at $\approx$ 790 nm and the temporal duration of the pulse is $\approx$ 80 fs.

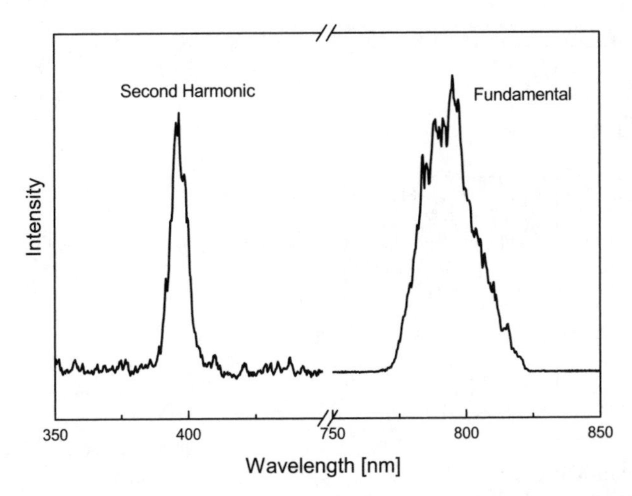

Fig. 4.5. Fundamental and second harmonic of a femtosecond pulse from a titanium sapphire laser

If both photons are different in their frequency (or energy), the energy of the new generated photon is the sum (or difference) of the original energies. As in the former case, the incident photons are annihilated simultaneously and the process is executed via a virtual inter state of the atom:

$$\omega_3 = \omega_1 + \omega_2 \tag{4.63}$$

and

$$\boldsymbol{k}_3 = \boldsymbol{k}_1 + \boldsymbol{k}_2 . \tag{4.64}$$

The quantum mechanical picture of sum frequency generation is shown in Fig. 4.6.

Until now, we have considered two incident photons generating a new one, but one incident photon is annihilated whereas two new photons are simultaneously generated. The sum of the energy of the new photons corresponds to the energy of the original one. In the moment in which the incident photon will be annihilated, the atom transits to a virtual interstate. When it returns back to its ground state it radiates a photon with frequency ω_1 and simultaneously one with the difference frequency between ω_1 and the frequency of the incident photon. The energy diagram of this process is shown in Fig. 4.7.

Therefore, this nonlinear mechanism offers the possibility to generate two new waves or photons from one incident photon. If the transparency range of the crystal is not exceeded and the conservation of momentum or phase matching is fulfilled, the generation of any frequency combination is possible as long as the sum of the frequencies of the two new photons corresponds to the energy of the original one. The incident wave is in this case called pump, whereas the two new waves are called signal and idler. Due to the fact

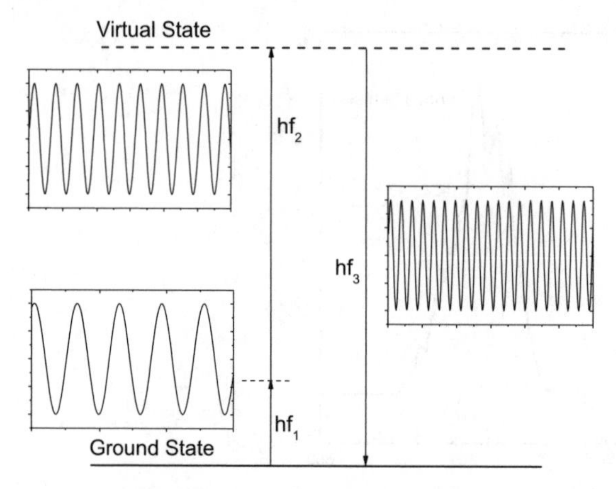

Fig. 4.6. Quantum mechanical picture of the sum frequency generation in a nonlinear crystal without inversion symmetry

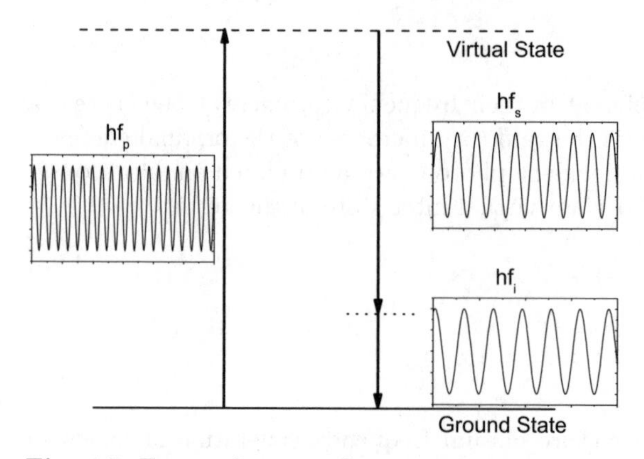

Fig. 4.7. Energy diagram of an optical parametric amplifier or oscillator

that the quantum mechanical state of the atom remains unchanged after the interaction it follows from the rules of conservation of energy and momentum between the photons that

$$\omega_{\mathrm{p}} = \omega_{\mathrm{s}} + \omega_{\mathrm{i}} \tag{4.65}$$

and

$$\boldsymbol{k}_{\mathrm{p}} = \boldsymbol{k}_{\mathrm{s}} + \boldsymbol{k}_{\mathrm{i}}. \tag{4.66}$$

This nonlinear mechanism is of great importance for the generation of coherent light in a large wavelength range. The nonlinear crystal, for instance beta barium borate (BBO), is arranged in a resonator, like in a laser. To tune the arrangement the size of the resonator is adjusted. The initial break down of the pump photon into the signal and idler photon is random. Only some atoms emit photons that correspond to the wavelength of the resonator. The resonator causes a feedback, the photons hit the crystal again and are amplified. The amplification increases with the number of round trips in the resonator until a stationary electromagnetic wave is built up. This can be coupled out of the resonator with a semi-transparent mirror.

The tuning of the arrangement is carried out via a change of the phase matching according to (4.66). At the same time, the size of the oscillator has to be matched to the new wavelength. Such a system is called an optical parametric oscillator (OPO). The notion "Parametric" comes from the fact that a parameter of the material – the polarization – is changed.

If the signal wave is not randomly generated in the crystal, but introduced from outside the device, the signal wave is amplified by the source. Hence, such an arrangement is called parametric amplifier. The parametric amplifiers in Sect. 13.4 of this book rely on a different physical mechanism, namely a third order parametric process.

4.6 Third Order Nonlinear Phenomena

If materials with an inversion center are considered, the first nonlinear term in (4.31) is that of third order:

$$P_{\mathrm{NL}}^{(3)} = \varepsilon_0 \chi^{(3)} EEE. \tag{4.67}$$

Introducing (4.56) into (4.67) yields

$$\begin{aligned} P_{\mathrm{NL}}^{(3)} &= \frac{1}{8}\varepsilon_0 \chi^{(3)} \left(\hat{E}\mathrm{e}^{-\mathrm{j}\omega t} + \hat{E}^*\mathrm{e}^{\mathrm{j}\omega t}\right)^3 \qquad (4.68) \\ &= \frac{1}{8}\varepsilon_0 \chi^{(3)} \left(\hat{E}^3\mathrm{e}^{-\mathrm{j}3\omega t} + 3\hat{E}^2\hat{E}^*\mathrm{e}^{-\mathrm{j}\omega t} + 3\hat{E}\hat{E}^{2*}\mathrm{e}^{\mathrm{j}\omega t} + \hat{E}^{3*}\mathrm{e}^{\mathrm{j}3\omega t}\right) \\ &= \frac{1}{8}\varepsilon_0 \chi^{(3)} \left((\hat{E}^3\mathrm{e}^{-\mathrm{j}3\omega t} + \mathrm{c.c.}) + 3\left|\hat{E}\right|^2 (\hat{E}\mathrm{e}^{-\mathrm{j}\omega t} + \mathrm{c.c.})\right). \end{aligned}$$

The first term in (4.68) leads to a material polarization with three times the fundamental frequency:

$$P_{\mathrm{NL}}^{(3)}(3\omega,3k) = \frac{1}{8}\varepsilon_0 \chi^{(3)} \left(\hat{E}^3\mathrm{e}^{-\mathrm{j}3\omega t} + \mathrm{c.c.}\right). \tag{4.69}$$

The wave number of this polarization, k_3, can be derived – as in the case of the second order phenomena – from $\hat{E}^3$ and (4.57). With $k_3 = k(3\omega)$ as the wave number at three times the fundamental frequency, the material polarization moves through the crystal with a phase velocity of $v_{\mathrm{ph}} = 3\omega/k_3$. Whereas it is generated by the product of three times the fundamental wave, this product moves with a wave number of $3k$ and hence its phase velocity is $v_{\mathrm{ph}} = 3\omega/k_3$. Since the difference between the frequencies of fundamental and

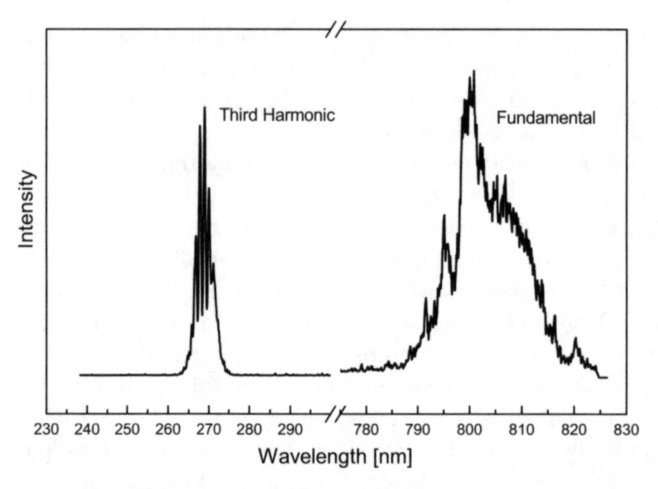

Fig. 4.8. Fundamental and third harmonic of a pulse from a titanium sapphire laser (temporal duration=80 fs, central wavelength=805 nm)

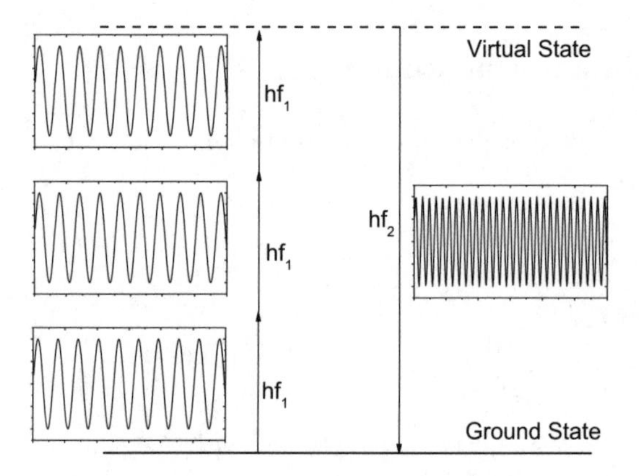

Fig. 4.9. Energy diagram for third harmonic generation

third harmonic is greater than in the case of second harmonic generation, the difference between the wave numbers is larger as well, i.e., $k_3 \neq 3k$. Therefore, the polarization and the new generated wave will grow out of phase faster than in the case of second harmonic generation. The third harmonic of a short femtosecond pulse is shown in Fig. 4.8.

From a quantum mechanic point of view, the atom is excited into a virtual inter state by the simultaneous annihilation of three photons. If the atom returns back to the ground state it re-emmits a photon, as shown in Fig. 4.9.

For a pure electronic polarization, the excitation of the atom corresponds to the distortion of the electron cloud. Hence, the dwell time of the atom in the excited state is very short, in principle only for the duration when

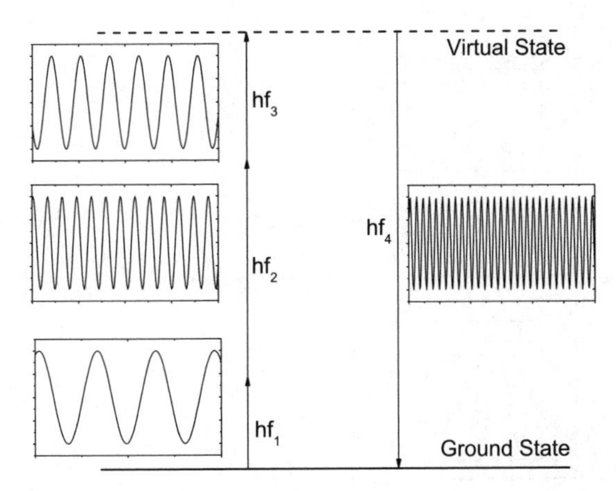

Fig. 4.10. Energy diagram for the nonlinear effect of sum frequency generation

the external field is present ($\approx 10^{-16}$ s). Annihilation and generation of the photons is therefore approximately instantaneous.

As in the case of second order phenomena, the atom remains unchanged after the interaction, hence the rules of conservation of energy and momentum have to be applied only to the involved photons:

$$3\omega = \omega + \omega + \omega \tag{4.70}$$

$$\boldsymbol{k}_3 = \boldsymbol{k} + \boldsymbol{k} + \boldsymbol{k}. \tag{4.71}$$

If the three annihilated photons have different energies, the energy of the generated photon corresponds to the sum of these energies such that

$$\omega_4 = \omega_1 + \omega_2 + \omega_3 \tag{4.72}$$

and

$$\boldsymbol{k}_4 = \boldsymbol{k}_1 + \boldsymbol{k}_2 + \boldsymbol{k}_3. \tag{4.73}$$

Figure 4.10 shows the quantum mechanical picture of this process. This effect corresponds to the sum frequency generation, already treated in the last section.

Like the case of second order effects, the converse effect to the sum frequency generation is possible as well; an incident photon of high energy generates three new photons with less energy. The sum of the energies of the new photons corresponds to the energy of the original one:

$$\omega_1 = \omega_2 + \omega_3 + \omega_4 \tag{4.74}$$

and

$$\boldsymbol{k}_1 = \boldsymbol{k}_2 + \boldsymbol{k}_3 + \boldsymbol{k}_4. \tag{4.75}$$

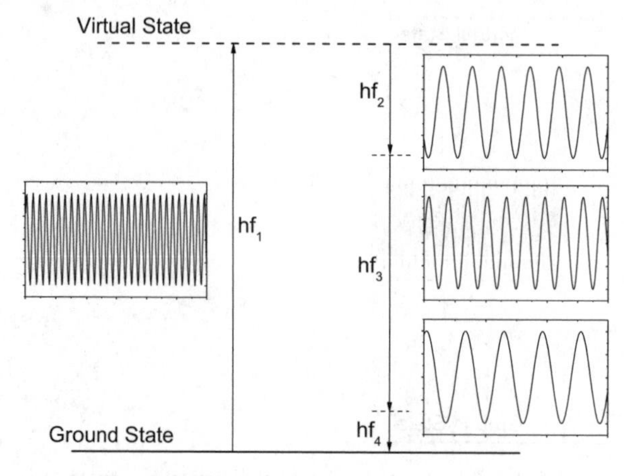

Fig. 4.11. Energy diagram of an optic parametric process

The quantum mechanical picture of this process is shown in Fig. 4.11. For third order nonlinear effects it is also possible for two photons to be annihilated by the atom, generating two new photons simultaneously. In this case, the rules of conservation of energy and momentum are again valid:

$$\omega_1 + \omega_2 = \omega_3 + \omega_4$$

and

$$\boldsymbol{k}_1 + \boldsymbol{k}_2 = \boldsymbol{k}_3 + \boldsymbol{k}_4. \tag{4.76}$$

This third order phenomenon is called four-wave-mixing (FWM) and is of special interest for the range of optical telecommunications. Since FWM involves a mixing between adjacent channels in a WDM system, the FWM degrades the system to noise ratio and hence the capacity of optical transmission systems (Sect. 7.6). On the other hand, the effect of FWM is exploited in all-optical signal processing (Sect. 12.2.1), optical amplification (Sect. 13.4), and distortion compensation (Chap. 14). The energy diagram of FWM is shown in Fig. 4.12.

All mentioned wave generation and mixing effects in this Section combine the annihilation and generation of four distinct photons. Hence, these effects are called four photon interaction.

The second term in (4.68) describes a nonlinear part of the polarization at the frequency of the applied optical field. The polarization leads to an additional, intensity-dependent contribution to the refractive index of the material. This effect is called nonlinear refractive index and is treated in the next section in detail.

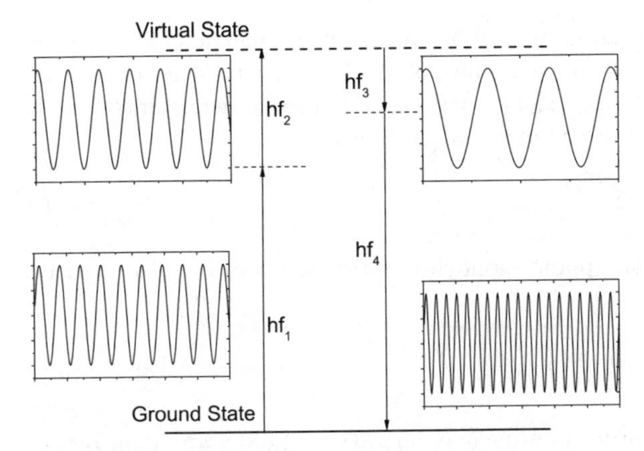

Fig. 4.12. Energy diagram of four-wave-mixing

4.7 The Nonlinear Refractive Index

The nonlinear refractive index is one of the most important nonlinear effects based on the third order susceptibility. It is responsible for the self-phase modulation (SPM) of waves (Sect. 6.1), the cross-phase modulation (XPM) between different waves (Sect. 6.2), the self focussing of a beam in a crystal, the soliton (Chap. 9), and some other effects.

As shown in Sect. 2.6.1, the refractive index depends on the frequency for small intensities. In the nonlinear case, i.e., for high intensities, it becomes intensity-dependent as well, as will be shown here. The derivation of the nonlinear refractive index starts with the linear wave equation. The electric fields are assumed to be plane monochromatic waves propagating in the same direction so that the vector component can be neglected. Hence, the linear wave equation is

$$\Delta E = \frac{1}{c^2}\frac{\partial^2 E}{\partial t^2} + \frac{1}{\varepsilon_0 c^2}\frac{\partial^2 P}{\partial t^2}. \tag{4.77}$$

With the linear polarization, $P = \varepsilon_0 \chi E$, it becomes

$$\Delta E = \frac{1}{c^2}\frac{\partial^2 E}{\partial t^2}(1+\chi). \tag{4.78}$$

Using the equation for the linear refractive index $n = \sqrt{1+\chi} = \sqrt{\varepsilon_{\mathrm{r}}}$ the equation can be rewritten as

$$\Delta E = \frac{n^2}{c^2}\frac{\partial^2 E}{\partial t^2}. \tag{4.79}$$

According to this, the polarization is the origin of the linear, frequency-dependent, refractive index. Assume now that the intensities are so high that the response of the material system of an inversion-symmetric crystal to the external applied field is nonlinear. The polarization can then be developed in a power series, as in (4.31). Hence, the second order susceptibility $\chi^{(2)}$ is zero and the polarization of the material becomes

$$P = \varepsilon_0 \left(\chi^{(1)} E + \chi^{(3)} EEE + ... \right). \tag{4.80}$$

The electric field of the plane monochromatic wave can be described by (4.56) such that

$$E(z,t) = \frac{1}{2} \left(\hat{E}(z) \mathrm{e}^{-\mathrm{j}\omega t} + \mathrm{c.c.} \right). \tag{4.81}$$

Therefore, the third order nonlinear polarization (4.68), can again be derived. The first term in (4.68) is responsible for an interaction between four photons. But this interaction can lead to significant results only if the phase matching or conservation of momentum is fulfilled, as mentioned in Sect. 4.6. Here it is assumed that phase matching for this interaction is not given. Hence, the term can be neglected and the third order nonlinear polarization becomes

$$\begin{aligned} P_{\mathrm{NL}}^{(3)} &= \frac{3}{8} \varepsilon_0 \chi^{(3)} \left| \hat{E} \right|^2 (\hat{E} \mathrm{e}^{-\mathrm{i}\omega t} + \mathrm{c.c.}) \\ &= \frac{3}{4} \varepsilon_0 \chi^{(3)} \left| \hat{E} \right|^2 E. \end{aligned} \tag{4.82}$$

Introducing (4.82) into (4.77) yields

$$\begin{aligned} \Delta E &= \frac{1}{c^2} \frac{\partial^2 E}{\partial t^2} + \frac{1}{\varepsilon_0 c^2} \frac{\partial^2}{\partial t^2} \left[\varepsilon_0 \chi^{(1)} E + \frac{3}{4} \varepsilon_0 \chi^{(3)} \left| \hat{E} \right|^2 E \right] \\ &= \frac{1}{c^2} \frac{\partial^2 E}{\partial t^2} \left[1 + \chi^{(1)} + \frac{3}{4} \chi^{(3)} \left| \hat{E} \right|^2 \right]. \end{aligned} \tag{4.83}$$

With the new refractive index of the material,

$$n_{\mathrm{ges}} = \sqrt{1 + \chi^{(1)} + \frac{3}{4} \chi^{(3)} \left| \hat{E} \right|^2}, \tag{4.84}$$

(4.83) becomes

$$\Delta E = \frac{n_{\mathrm{ges}}^2}{c^2} \frac{\partial^2 E}{\partial t^2}. \tag{4.85}$$

As can be seen from (4.85), the nonlinear refractive index leads to an additional phase shift of the wave in the material. Assuming that the nonlinear part of the refractive index is relatively small compared to the linear component, then (4.84) can be developed in a Taylor series around $\frac{3}{4}\chi^{(3)} \left| \hat{E} \right|^2$. Hence, the total refractive index in the medium is

$$n_{\text{ges}} = n + \frac{3}{8n}\chi^{(3)}\left|\hat{E}\right|^2 \tag{4.86}$$

with the linear component

$$n = \sqrt{1+\chi^{(1)}}. \tag{4.87}$$

The intensity of the electro-magnetic field is given by

$$I = \frac{1}{2}\varepsilon_0 cn\left|\hat{E}\right|^2. \tag{4.88}$$

Hence, the square of the absolute value of the amplitude of the electric field is linearly proportional to the intensity. Therefore, (4.86) can be interpreted as

$$n_{\text{ges}} = n + \frac{3}{8n}\chi^{(3)}\left|\hat{E}\right|^2 = n + n_2 I. \tag{4.89}$$

The material dependent parameter, n_2, gives the relation between the intensity and the nonlinear part of the refractive index. The total refractive index in the material thus consists of a linear part – dependent on the frequency – and a nonlinear part. This nonlinear part depends on the intensity and the frequency of the applied optical field.

In the case of the traditional electro-optic Kerr effect, the change in refractive index of a crystal is proportional to the square of the strength of an applied static electric field. In analogy to this behavior, the intensity dependence of the refractive index is called optical or nonlinear Kerr effect. Hence, n_2 is the material-dependent Kerr constant.

Using (4.88) a comparison of both sides of (4.89) gives for the Kerr constant n_2

$$n_2 = \frac{3\chi^{(3)}}{4\varepsilon_0 cn_0^2}. \tag{4.90}$$

According to (4.90), the nonlinear refractive index depends, among other things, on the third order susceptibility $\left(\chi^{(3)}\right)$ and the linear refractive index (n). Both values are not constant and are dependent on the frequency. Hence, n_2 is not a real constant and is frequency-dependent as well. The standard value for n_2 for optical fibers in the telecommunications range is $n_2 = 2.35 \times 10^{-20}$ m^2/W [30]. The nonlinear refractive index can be increased with the germanium concentration in the core therefore, highly nonlinear fibers can have a much higher n_2, between 4.56×10^{-20} m^2/W and 6.0×10^{-20} m^2/W [11]. In glasses other than fused silica, the nonlinear refractive index can be very high. Arsenic chalcogenide $As_{40}Se_{60}$ has, for instance, a Kerr constant 930 times higher than that of fused silica at a wavelength of 1550 nm ($n_2 = 2.3 \times 10^{-17}$ m^2/W) [186].

For the derivation of n_2, the nonlinear susceptibility $\chi^{(3)}$ as well as the linear refractive index n are used as scalars, but normally both depend on the

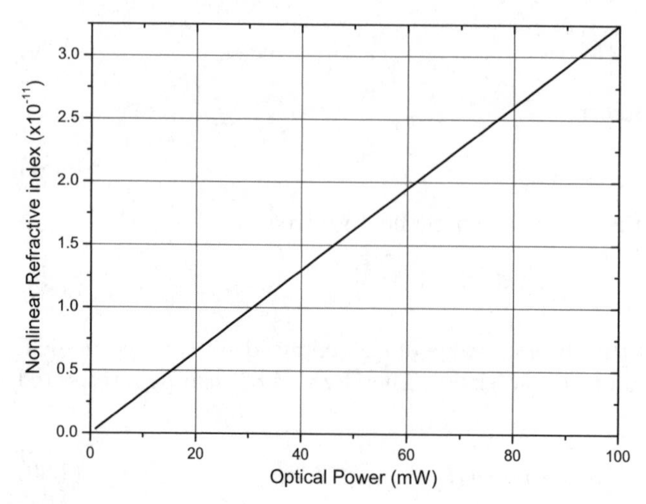

Fig. 4.13. Nonlinear refractive index for fibers made of silica glass ($n_2 = 2.6 \times 10^{-20}$ m^2/W, $A_{\text{eff}} = 80$ μm^2)

polarization state of the waves. Therefore, the above values of n_2 can be seen as averaged over all possible polarization states, which is only appropriate for sufficiently long fibers.

In optical fibers the nonlinear Kerr constant is relatively small, hence the nonlinear refractive index is significant only for very high intensities. In single mode waveguides, due to the small core area the field is concentrated on, the intensities are very high. With the effective area (Sect. 4.8) A_{eff}, (4.89) yields

$$n_{\text{ges}} = n + n_2 \frac{P}{A_{\text{eff}}}. \tag{4.91}$$

Figure 4.13 shows the nonlinear refractive index against the optical power for a fiber of silica glass with an effective area of 80 μm^2.

Even though the alteration of the refractive index is rather small, it has a great impact nevertheless on optical transmission systems. This is due to the large interaction length (Sect. 4.9). Contrary to the wave mixing phenomena, the nonlinear refractive index does not suffer from the phase matching condition. Hence, the nonlinear effects aligned with it can accumulate over the propagation distance.

If material resonances are involved, the nonlinear refractive index can increase drastically. Such material resonances include one and two photon transitions of the molecules or atoms, a Raman transition, and similar effects. Table 4.3 gives an overview of the nonlinear refractive indices of different physical processes.

The behavior of the nonlinear refractive index near a one photon transition is shown in Fig. 4.14, where near the material resonance, both the absolute value and the sign of the nonlinear refractive index change. Hence,

Table 4.3. Nonlinear refractive indices of various physical effects [73]

Mechanism	$n_2(\mathrm{cm}^2/\mathrm{W})$
Electronic polarization	10^{-16}
Molecular orientation	10^{-14}
Electrostriction	10^{-14}
Thermal effects	10^{-6}

under particular circumstances, it will be possible to compensate a positive dispersion with the nonlinear refractive index.

The nonlinear refractive index a wave experiences depends not only on its own intensity and frequency, but also on the intensity and frequency of all other waves propagating at the same time in the material. This can be exploited for a resonance enhancement of the refractive index for a wave with any frequency, ω_1, in the case of a two photon transition. The adaptation to the resonance frequency ω_{R} of the transition can be done by tuning the frequency of a second wave, ω_2, with $\omega_{\mathrm{R}} = \omega_1 + \omega_2$.

The dependence of the refractive index on the optical field offers a possibility for the manipulation of light by light, an important effect for optical switching which is treated in detail in Sect. 12.3.

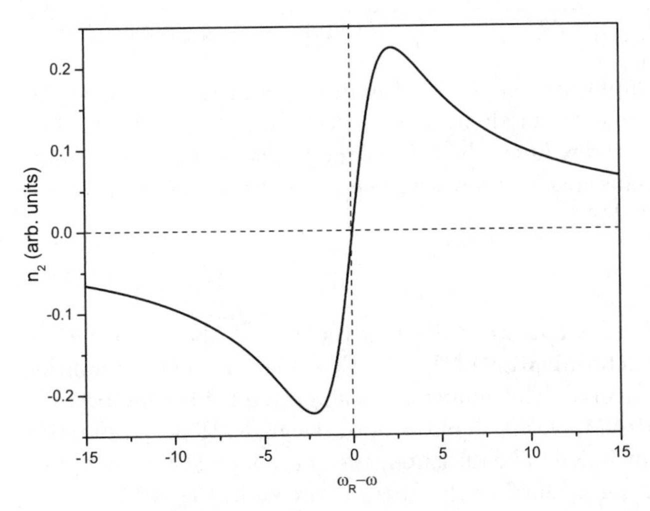

Fig. 4.14. Nonlinear refractive index near a one photon transition versus the difference between resonance frequency (ω_{R}) and the frequency of the optical wave (ω) [129]

4.8 Effective Area and Nonlinear Coefficient

As can be seen from (4.55), all nonlinear effects depend on the field strength of the electromagnetic wave or, due to the fact that the intensity is proportional to the amplitude squared, the intensity of the field. Since the measurable quantity in telecommunications is the power of the input and output signal of an optical fiber, a relation between the intensity and the power is required to compare the theoretical and experimental results. This relation is given over the effective area A_{eff}. The intensity depends on the area the power is concentrated on, as follows:

$$I = \frac{P}{A_{\mathrm{eff}}}. \tag{4.92}$$

In optical fibers the wave is guided mostly in the core area. Hence, it could be assumed that the effective area corresponds to the core area of the fiber, but the field distribution is not equable. The field strength near the axis is higher than that in the range between core and cladding. Furthermore, a non vanishing part of the field propagates in the cladding region of the fiber. The effective area depends on the modal field distribution and can be calculated with overlapping integrals [84], and is defined as [80, 83]

$$A_{\mathrm{eff}} = \frac{2\pi\left(\int_0^\infty |E_{\mathrm{a}}(r)|^2\, r\mathrm{d}r\right)^2}{\int_0^\infty |E_{\mathrm{a}}(r)|^4\, r\mathrm{d}r} = \frac{2\pi\left(\int_0^\infty I(r) r\mathrm{d}r\right)^2}{\int_0^\infty I^2(r) r\mathrm{d}r}, \tag{4.93}$$

where $E_{\mathrm{a}}(r)$ is the amplitude and $I(r)$ is the intensity of the near field distribution of the fundamental mode at a radius r from the axis of the fiber. In a single mode, step-index fiber, the field of the fundamental mode can be approximated by a Gaussian function with the beam radius[1] w. In this case the effective area is [80]

$$A_{\mathrm{eff}} = \pi w^2(\lambda). \tag{4.94}$$

$2w(\lambda)$ is the mode field diameter (MFD) of the fiber at the wavelength λ. For the experimental determination of the effective area, the field distribution outside the fiber is measured. The diameter of a single mode fiber lies between 5 μm and 10 μm near the wavelength of the optical signals. Hence, diffraction phenomena must be included. The measurement of the intensity distribution can be carried out in the far field of the fiber, as shown in Fig. 4.15.

Due to the diffraction at the fiber endface, the relation between the near and far fields is given by the diffraction integral, where the near field is within the region w^2/λ. According to [80, 81], the MFD is thus

[1]The parameter w lies in a plane perpendicular to the propagation direction in the fiber. It determines the distance between the axis of the fiber, or the point of highest intensity, and the position where the intensity is decreased to $1/e^2$.

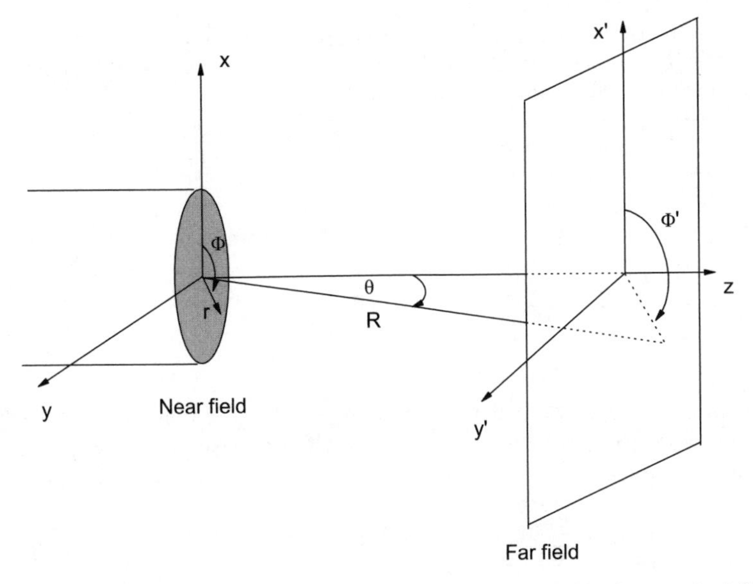

Fig. 4.15. Coordinate system for the near and far field of an optical fiber [80]

$$\mathrm{MFD_{FF}} = \frac{\sqrt{2}\lambda}{\pi}\left[\frac{\int_0^{\pi/2}|F(\theta)|^2\tan(\theta)\,\mathrm{d}\theta}{\int_0^{\pi/2}|F(\theta)|^2\tan(\theta)\sin^2(\theta)\,\mathrm{d}\theta}\right]^{1/2}, \tag{4.95}$$

where $F(\theta)$ is the observable far field amplitude, as a function of the angle θ. Dispersion-shifted, dispersion-compensating, and some other fibers have no step-index geometry, hence for these fibers the relation (4.94) no longer holds. To calculate the effective area in these fibers, correction terms can be included [82]. Table 4.4 lists the effective areas for different fiber types, as quoted in the references.

Table 4.4. Effective areas at 1550 nm for different fiber types (SSMF=standard single mode fiber, DSF=dispersion-shifted fiber, DCF=dispersion-compensating fiber, HNL=highly nonlinear, MSF=micro structured fiber)

Fiber type	A_{eff} (μm^2)	Reference
SSMF	80	[85]
DSF	54	[30]
DCF	20	[85]
HNL	9 – 15	[11]
MSF	2.8	[32]
MSF	1 – 2	[31]

Table 4.5. Nonlinear coefficients at a wavelength of 1550 nm for different fiber types (SSMF=standard single mode fiber, DSF=dispersion-shifted fiber, HNL=highly nonlinear, MSF=micro structured fiber)

Fiber type	γ $(\mathrm{W}^{-1}\mathrm{km}^{-1})$	Reference
SSMF	1.2	
DSF	1.76	
HNL	12.5-25	[11]
MSF	35	[32]

An important value for the calculation of the strength of nonlinear effects is the ratio between the Kerr constant (n_2) and the effective area for a given frequency of the optical field. This ratio is called nonlinear coefficient and is

$$\gamma = k_0 \frac{n_2}{A_{\text{eff}}} = \frac{\omega n_2}{c A_{\text{eff}}} = \frac{2\pi n_2}{\lambda A_{\text{eff}}}. \tag{4.96}$$

In standard single mode fibers with an effective area of $80 \mu\mathrm{m}^2$ and a nonlinear Kerr constant of $n_2 = 2.35 \times 10^{-20}\ \mathrm{m}^2/\mathrm{W}$, the nonlinearity coefficient is $\gamma = 1.2\ \mathrm{W}^{-1}\mathrm{km}^{-1}$ for a wavelength of $\lambda = 1.55\ \mu\mathrm{m}$. On the other hand, dispersion-shifted fibers with smaller effective areas have a higher nonlinear coefficient of $\gamma = 1.76\ \mathrm{W}^{-1}\mathrm{km}^{-1}$. Therefore, nonlinear effects are stronger in these fibers. The highest nonlinear coefficients can be found in specially designed nonlinear fibers like HNL and MSF. The values for different fiber types, according to the literature, are listed in Table 4.5.

4.9 Effective Length

Nonlinear effects do not only depend on the intensity but also on the interaction length at which the power density remains sufficiently high. Due to diffraction effects, a focused laser beam falls apart after a distance of some wavelength of the light. Hence, after a few µm the intensity will decrease drastically. Whereas, the field is guided in an optical fiber, the intensity is only decreased by the very low attenuation. Therefore, the interaction length can be several kilometers.

A measure for the relevance of nonlinear effects in optical fibers is given by the product of the intensity and effective length, as follows:

$$I L_{\text{eff}} = \frac{P}{A_{\text{eff}}} L_{\text{eff}} = \int_0^L \frac{P(0)}{A_{\text{eff}}} \mathrm{e}^{-\alpha_{\mathrm{km}^{-1}} z}\, \mathrm{d}z = \frac{P(0)}{A_{\text{eff}}} \frac{1 - \mathrm{e}^{-\alpha_{\mathrm{km}^{-1}} z}}{\alpha_{\mathrm{km}^{-1}}}. \tag{4.97}$$

Hence, the effective length of interaction is

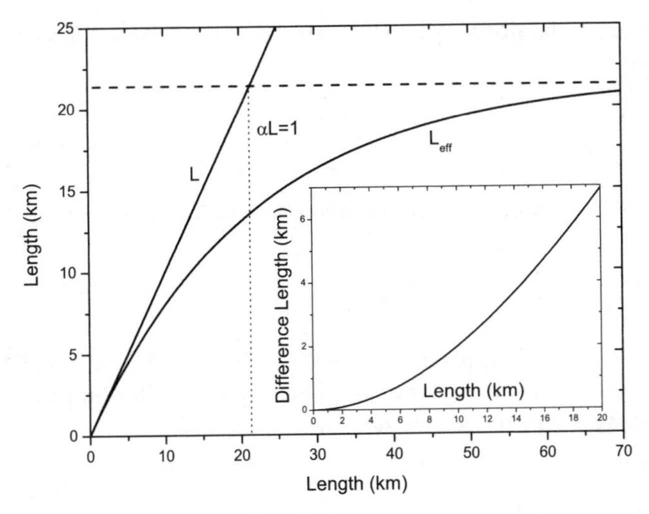

Fig. 4.16. Effective interaction length against the length of the optical waveguide, for a fiber with an attenuation of 0.2 dB/km. The inset shows the difference between the effective and fiber lengths against the length of the fiber

$$L_{\mathrm{eff}} = \frac{1 - \mathrm{e}^{-\alpha_{\mathrm{km}^{-1}} z}}{\alpha_{\mathrm{km}^{-1}}}. \tag{4.98}$$

The effective length against the real fiber length is shown in Fig. 4.16 for a fiber with an attenuation of 0.2 dB/km. As can be seen from the inset, if the length of the waveguide is rather small ($\alpha L \ll 1$), the difference between effective and real length is barely significant $L_{\mathrm{eff}} \approx L$. Whereas, if the length is high enough ($\alpha L \gg 1$), the effective length approaches a limiting value due to the attenuation:

$$\lim_{z \longrightarrow \infty} L_{\mathrm{eff}}^{G} = \frac{1}{\alpha_{\mathrm{km}^{-1}}}. \tag{4.99}$$

According to (4.99), the efficiency of nonlinear effects in optical waveguides does not increase linearly with the fiber length. The limiting value in the above example is, for instance, $L_{\mathrm{eff}}^{G} = 21.72$ km. In earlier transmission systems without optical amplifiers and a fiber attenuation of 0.2 dB/km, this was the maximum interaction length at which nonlinear effects can happen. But, in the fiber spans of today the attenuation is compensated in every optical amplifier. Hence, the intensity at the output of the amplifier is the same as it was at the fiber input. Therefore, the effective length in amplified systems is multiplied by the number of amplifiers.

If a transmission system with the length L consists of n fibers with the same length L_{A} ($n = L/L_{\mathrm{A}}$), where L_{A} determines the distance between the amplifiers, then $n - 1$ is the number of additional amplifiers in the system. If, furthermore, every amplifier has the same gain and the power after each amplifier is the same as at the fiber input $P(0)$, the product between power and effective length is

$$n\frac{P}{A_{\mathrm{eff}}}L_{\mathrm{eff}} = n\int_0^{L_{\mathrm{A}}} \frac{P(0)}{A_{\mathrm{eff}}}\mathrm{e}^{-\alpha_{\mathrm{km}^{-1}}z}\,\mathrm{d}z = n\frac{P(0)}{A_{\mathrm{eff}}}\frac{1-\mathrm{e}^{-\alpha_{\mathrm{km}^{-1}}L_{\mathrm{A}}}}{\alpha_{\mathrm{km}^{-1}}} \tag{4.100}$$

The effective interaction length for an amplified system L_{eff} is therefore

$$L_{\mathrm{eff}} = n\left(\frac{1-e^{-\alpha_{\mathrm{km}^{-1}}L_{\mathrm{A}}}}{\alpha_{\mathrm{km}^{-1}}}\right). \tag{4.101}$$

Hence, the effective length in a transmission system with $n-1$ additional amplifiers is n times greater than in a nonamplified system. The limiting value of the effective length can be determined by:

$$\lim_{z\longrightarrow\infty} L_{\mathrm{eff}}^{G_v} = \frac{n}{\alpha_{\mathrm{km}^{-1}}}. \tag{4.102}$$

4.10 Phase Matching

In principle the wave mixing and harmonic generation effects can happen spontaneously in the atoms somewhere in a crystal, if this is not restricted by the symmetry of the material. But if a macroscopic effect should occur – i.e., if the new generated wave should be measurable behind the crystal – the conservation of momentum must be fulfilled during the interaction. The conservation of momentum depends on the wave numbers of the involved waves and requires the matching of the phases between the generating and generated waves.

To simplify the consideration, phase matching will be described here using the example of higher harmonic generation. The case of third harmonic generation will be treated in depth because the generation of the second harmonic is forbidden in inversion symmetric crystals. The derived predicates can be generalized to the sum and difference frequency generation and to four-wave-mixing (FWM). Nevertheless, the phase matching of FWM is treated in detail again in Sect. 7.3.

Until now, we have only considered one single oscillator or dipole consistently integrated in a host medium. If the response of this dipole is nonlinear, the radiated field will contain higher harmonics. If the polarization is purely electronical in nature, the nonlinear susceptibilities are very low. Hence, the intensity of the higher harmonic is very weak. A real medium consists of a large number of such embedded dipoles. The generation of higher harmonics is a measurable effect only if the responses of a large number of them interfere constructively.

Let us now assume a dipole dl located at the point (x, y, z) that radiates an electromagnetic field containing the third harmonic of the fundamental frequency. A second dipole $d2$, radiating as well the third harmonic, is located at $(x + \Delta x,\ y + \Delta y,\ z + \Delta z)$. If the third harmonic of both should interfere constructively, then the phase of the third harmonic generated by $d1$ after the distance $\Delta r = (\Delta x,\ \Delta y,\ \Delta z)$ must be the same as the phase of the third harmonic radiated by $d2$. The origin of the third harmonic is the third order polarization generated by the product of three times the fundamental wave during its propagation through the medium. The wave number of the polarization is determined by k_3. Hence, the phase velocity of the third order polarization depends on k_3, whereas the phase velocity of the product generating the polarization depends on three times the fundamental wave number $3k_1$. The third harmonic generated by $d1$ can only be in phase with the third harmonic generated by $d2$ if the wave number of the third order polarization coincides with the wave number of three times the fundamental, i.e., $k_3 = 3k_1$. In a material that shows dispersion, the wave number and therefore the phase velocity depend on the frequency of the waves. Hence, this condition is in most cases not fulfilled, as shown in Fig. 2.8. If only these two oscillators are considered and both phases are out of phase by π then the field radiated by $d2$ destroys the field radiated by $d1$ and no effect can be measured.

Assuming the fundamental is a plane monochromatic wave, then its field can be described by

$$\boldsymbol{E} = \frac{1}{2}\left(\hat{\boldsymbol{E}}\mathrm{e}^{\mathrm{j}(\boldsymbol{k}_1\cdot\boldsymbol{r}-\omega t)} + \text{c.c.}\right). \tag{4.103}$$

Ideally, all waves propagate in the same direction and the wave vector has the absolute value $|\boldsymbol{k}_1| = n(\omega)\,\omega/c$, with ω as the angular frequency of the fundamental. The third order polarization is responsible for the generation of the third harmonic:

$$P_{\mathrm{NL}}^{(3)} = \varepsilon_0\chi^{(3)}EEE.$$

As described in Sect. 4.1, it is by itself a wave propagating through the medium as well. If only the part leading to the generation of the third harmonic $P_{\mathrm{NL}}^{(3)}(3\omega)$ is considered, the polarization wave can be written as

$$P_{\mathrm{NL}}^{(3)}(3\omega) = \frac{1}{2}\left(\hat{P}_{\mathrm{NL}}(3\omega)\mathrm{e}^{\mathrm{j}(\boldsymbol{k}_3\cdot\boldsymbol{r}-3\omega t)} + \text{c.c.}\right). \tag{4.104}$$

The term $\hat{P}_{\mathrm{NL}}(3\omega)$ describes the complex amplitude of the polarization. The wave vector of the third harmonic has the absolute value $|\boldsymbol{k}_3| = n(3\omega)\,3\omega/c$. The origin of the polarization is the fundamental wave propagating through the medium. Under consideration of the vector character, a comparison with the first term in (4.68) yields

$$P_{\mathrm{NL}}^{(3)}(3\omega) = \frac{1}{2}\left(\hat{P}_{\mathrm{NL}}(3\omega)\mathrm{e}^{\mathrm{j}(\boldsymbol{k}_3\cdot\boldsymbol{r}-3\omega t)} + \text{c.c.}\right) = \frac{1}{8}\left(\varepsilon_0\chi^{(3)}\hat{E}^3\mathrm{e}^{\mathrm{j}(3\boldsymbol{k}_1\cdot\boldsymbol{r}-3\omega t)} + \text{c.c.}\right). \tag{4.105}$$

The phases of both waves in (4.105) match if the arguments of the exponential functions coincide:

$$\boldsymbol{k}_3 \cdot \boldsymbol{r} - 3\omega t = 3\boldsymbol{k}_1 \cdot \boldsymbol{r} - 3\omega t. \tag{4.106}$$

This leads to the condition $\boldsymbol{k}_3 = 3\boldsymbol{k}_1$. If the wave vector is multiplied by the Planck constant $\hbar$, one yields the momentum of the photon, $\boldsymbol{p} = \hbar\boldsymbol{k}$. A multiplication of the frequency with the Planck constant yields to the energy $E = \hbar\omega$. Hence, from a quantum physics point of view, the fact that three photons with the energy $E_1 = \hbar\omega_1$ generate one new photon with the energy $E_2 = \hbar 3\omega_1$ describes the conservation of energy $3\hbar\omega_1 = \hbar 3\omega_1$. At the same time, the phase matching provides the conservation of momentum during the process $\hbar\boldsymbol{k}_3 = 3\hbar\boldsymbol{k}_1$.

Since the polarization and fundamental propagate in the same direction in collinear plane waves, the vector character can be neglected. Hence, a phase mismatching can be calculated from the refractive index difference of the medium:

$$\Delta k = k_3 - 3k_1 = \frac{3\omega}{c}\left(n(3\omega) - n(\omega)\right). \tag{4.107}$$

Let us then assume a monochromatic plane wave is propagating through a crystal in the z-direction and generating a polarization with three times the fundamental frequency. We will consider planes of the crystal perpendicular to the propagation direction z. At $z = 0$, all dipoles in the plane oscillate with the same phase and hence, the third harmonic generated by the dipoles of this plane interfere constructively. If the wave propagates further the polarization and the generating waves will grow out of phase. But as long as the phase difference between the polarization and three times the fundamental is smaller than π, the contributions from the dipoles on the distinct planes add to the intensity of the third harmonic. If the phase difference is π, the dipoles on this plane oscillate with the opposite phase. They cannot contribute to the third harmonic and hence the intensity reaches its maximum. This is the case if

$$\pi = \Delta k z \tag{4.108}$$

is valid. If the phase difference exceeds π, the contributions from the dipoles on the planes interfere destructively with the third harmonic wave generated, the intensity decreases. At a distance of 2π, the number of planes that interfere constructively equals the number of planes with destructive interference, hence the intensity of the third harmonic will be zero.

The distance in the crystal at which the generated wave reaches its maximum is called the coherence length. According to (4.108) it is

$$z = L_{\mathrm{coh}} = \frac{\pi}{\Delta k}. \tag{4.109}$$

The coherence length for the generation of the third harmonic is derived as below

$$\Delta k = k_3 - 3k_1 \tag{4.110}$$

where

$$k_i = \frac{n_i \omega_i}{c} = \frac{2\pi n_i}{\lambda_i}$$

and

$$i = 1, 3; \lambda_3 = 1/3\lambda_1$$

Therefore,

$$L_{\mathrm{coh}} = \frac{\lambda_1}{6\,(n_3 - n_1)}$$

where λ_1 and λ_3 as the wavelengths of the fundamental and the third harmonic, respectively, and n_1, n_3 are the corresponding refractive indices. For a fundamental wave with a wavelength of 1.6 μm, the third harmonic has a wavelength of 0.53 μm. Hence, according to (2.58) and Fig. 2.7, the refractive index difference is $\approx 1.7 \times 10^{-2}$. Therefore, according to (4.110) the coherence length is merely $L_{\mathrm{coh}} \approx 15.7$ μm. This means for a propagation distance of only15.7 μm, the third harmonic reaches its maximum, after which the intensity decreases. If the propagation distance is 31.4 μm, the intensity is zero and will increase again and so on.

The intensity of the third harmonic can be described [74] (Sect. 7.3) as

$$I_{\mathrm{TH}} = const \left|\chi^{(3)}\right|^2 I^3 l^2 \left(\frac{\sin \Delta k l/2}{\Delta k l/2}\right)^2 \tag{4.111}$$

with l as the length of the medium, Δk as the phase mismatching, and I as the intensity of the applied field. Thus if the phase mismatch is zero, the intensity of the third harmonic increases with the propagation distance in the material. Since the number of dipoles which contribute in phase to the field on the distinct planes increases quadratically with the propagation distance, the increase of the third harmonic intensity is quadratic as well. For a constant phase mismatch, the intensity of the third harmonic against the length is shown in Fig. 4.17 where the intensity oscillates along the crystal with a periodicity of two times the coherence length.

Figure 4.18 shows the intensity of the third harmonic against the phase mismatch for a constant propagation distance. The intensity decreases very fast if the phase matching condition is not fulfilled. Hence, if the nonlinear effect should be exploited for an application, it requires the matching of the phases.

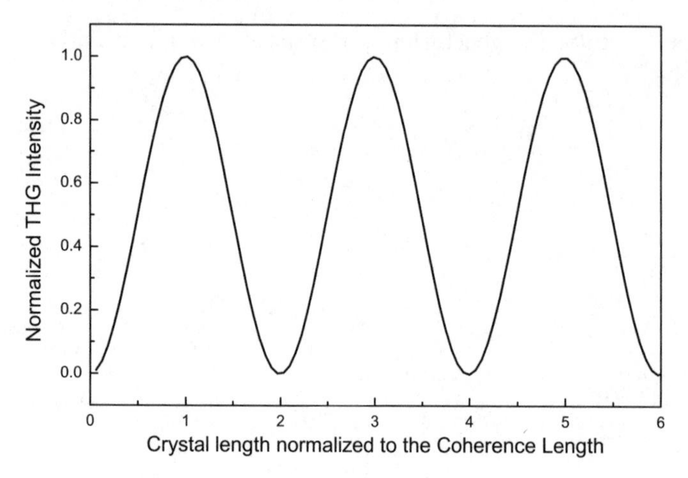

Fig. 4.17. Normalized intensity of the third harmonic for a constant phase mismatch versus the coherence length

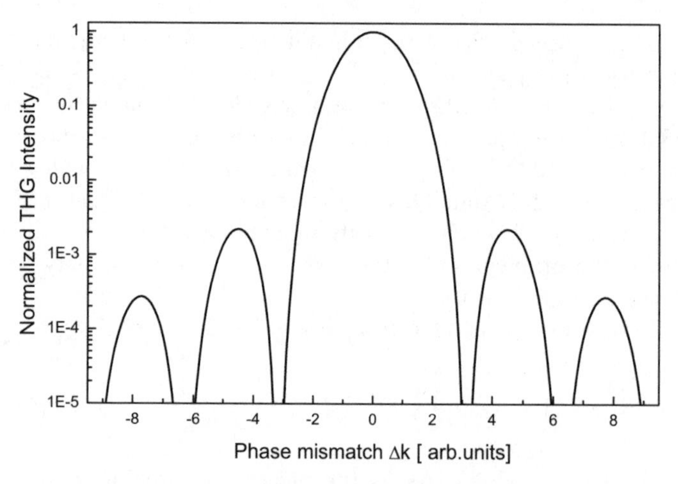

Fig. 4.18. Normalized intensity of the third harmonic against the phase mismatch. Note the logarithmic scale of the intensity axes

To achieve phase matching in spite of dispersion, there exist a number of different possibilities. In principle the refractive indices of the fundamental and the higher harmonic have to be adapted in a defined way or the discrete wave vectors must be set in a manner such that $\Delta \boldsymbol{k}$ will be as small as possible.

A frequently used method for phase matching, for the case of second harmonic generation in particular, is the exploitation of the birefringence of some crystals. Birefringence means that the refractive index of the crystal depends on the polarization direction of the light wave. This is the case in anisotropic crystals like calcite. As carried out in Sect. 4.3, the linear susceptibility in an anisotropic crystal is a tensor. Hence, the tensor elements, and therefore the refractive index, depend on the field polarization and the direction of the material polarization. Furthermore, anisotropic crystals do not have inversion centers. Hence, the generation of the second harmonic in the bulk of the material is possible and the second order susceptibility is partly very strong. This is the case in kalium titanyl phosphate (KTP) or beta barium borate (BBO).

The anisotropy is a consequence of the structure of the crystal. If the structure has a symmetry axes, the crystal is called uniaxial and the axes are denoted as the optical axes. If the lightwave propagates parallel to this axes, its electronic E field component is perpendicular to the optical axes and the crystal behaves like an isotropic material. In this case the wave is called an ordinary one and the refractive index the wave experiences is n_o. Whereas, if the wave propagates perpendicular to the optical axes, with an angle between the field vector and the axes, the vector can be split into a parallel component and one perpendicular to it. The crystal shows different bond forces in these directions and hence, different electronic polarizations. These differences result, according to Sect. 2.6.1, in a different refractive index and hence, according to (2.63) a different phase velocity of the wave. Therefore, the axes are called fast and slow axes.

Assume in such a crystal a wave with arbitrary polarization (beside perpendicular to the optical axes) that starts from a point inside the crystal. In the case of an isotropic crystal, the wave would propagate with a spherical wavefront. Whereas in the birefringent crystal the wavefront is an ellipsoid because the components can propagate faster in direction of the optical axes, (see Fig. 4.19a). This wave is called extraordinary. If the wave, generated at the point, is polarized perpendicular to the optical axes as well, a spherical wavefront will be formed.

If, in the case of higher harmonic generation, the fundamental propagates in the crystal as an ordinary wave, it propagates with a spherical wavefront. If the generated second harmonic is an extraordinary wave it will propagate with an ellipsoidal wavefront. At particular locations, the sphere of the fundamental with frequency ω crosses the ellipsoid of the second harmonic with frequency 2ω (Fig. 4.19a). At these points of intersection ordinary and extraordinary wave are phase-matched. This is the so called type I phase matching:

$$2k_1^{(\mathrm{o})} = k_2^{(\mathrm{e})} \tag{4.112}$$

where o denotes ordinary wave and e denotes extraordinary wave. For type II phase matching, the fundamental will be split in ordinary wave and extraordinary parts. The second harmonic is then an extra ordinary wave:

$$k_1^{(\mathrm{o})} + k_1^{(\mathrm{e})} = k_2^{(\mathrm{e})}. \tag{4.113}$$

The described phase matching method depends very critically on the angle between the field vector and the optical axes. Hence, a small change – for

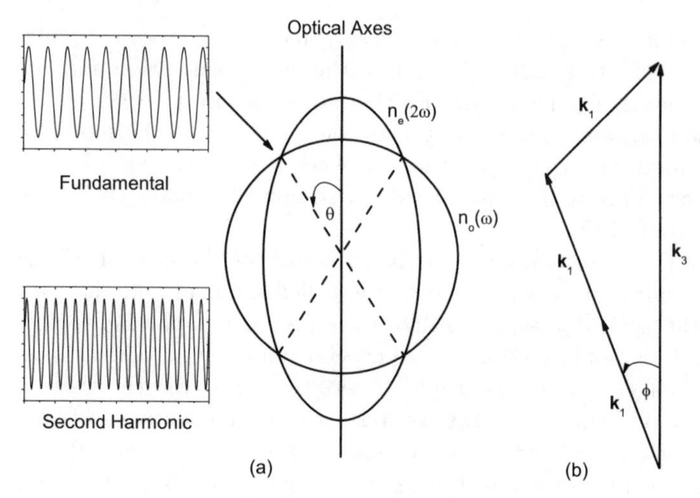

Fig. 4.19. Birefringent phase matching for second harmonic generation (**a**) and geometrical phase matching for third harmonic generation (**b**)

instance due to heating of the set-up – results in a drastic change of the conversion efficiency. Furthermore, the frequency range for which conversion is possible depends on the refractive indices of fast and slow axes, and on the transparency range of the material.

An alternative method is the quasi phase matching (QPM), where the phase mismatching of the phases between a fundamental and a higher harmonic wave is relocated at every coherence length. This will be done by a periodic modulation of the nonlinear susceptibility of the medium. The periodicity of this modulation is twice the coherence length, i.e., $\Lambda = 2L_{\text{coh}}$. An advantage of QPM is that phase matching can be achieved over the entire transparency range of the medium. This holds as well for frequencies for which birefringent adaptation would be impossible. Due to the fact that anisotropic crystals are required for the conversion, the tensor of the nonlinear susceptibility consists of different values. Hence, with QPM it will be possible to choose the polarization of the waves that conversion will take place via the largest tensor element $\chi^{(2)}$. The periodic grating can contain several periodic structures at the same time in order to adapt to several frequencies in phase.

As material for QPM devices, lithium niobate ($LiNbO_3$) is used in most cases. The device is than called periodically poled lithium niobate (PPLN). Lithium niobate is a ferro-electric material. These materials behave in an electric field as ferromagnetic materials in a magnetic field, i. e., an applied electric field is able to align the elementary dipoles of the material. If the electric field is strong enough, this alignment will remain after the field is switched off. Hence, the material conserves the periodical poling of the dipoles. Due to the periodical poling of the ferro-electric domains, the sign of the nonlinear coefficient changes at every domain. This change of the sign is responsible for the relocation of the phase mismatch.

A PPLN is used for the frequency conversion of femtosecond pulses delivered from an erbium-doped fiber laser, for example. The frequency of the fundamental is in the range between 1530 and 1610 nm, hence the converted frequency range is 765–805 nm. The conversion efficiency is 25 % and the device is commercially available [75].

Phase matching of the third harmonic is much more critical than that of the second harmonic, because the refractive index difference for the fundamental and the third harmonic is much higher in dispersive materials. One possibility of phase matching is offered by anomalous dispersion (where the refractive index increases with increasing wavelength). In this case the wave vector of the third harmonic is smaller than that of three times the fundamental. Hence, it is possible to achieve phase matching via the angles between the wave vectors, as shown in Fig. 4.19b.

An anomalous dispersion is shown by some dyes and gases [76] near material resonances. If gases with normal and anomalous dispersion are mixed in a defined way, phase matching for a collinear geometry is possible [77].

If the intensities of the participating waves are high enough, phase matching can be reached via the nonlinear refractive index. Tamaki et al. [78] used a hollow fiber filled with air and obtained phase matching for very high intensities of $\approx 10^{14}$ W/cm^2.

Self-phase matching can be obtained using a transient ultrafast refractive index grating, created by femtosecond pulses in a transparent dielectric. Phase matching takes place through a momentum exchange via the grating [79].

Four-wave-mixing (FWM) is dependent on phase matching as well. If the frequency difference between the distinct waves is rather low, several possibilities of phase matching exist. One method is to exploit the different phase velocities of the modes in a multi mode fiber [84]. Due to the fact that FWM has possibly many important applications in the field of telecommunications, phase matching for FWM is considered in detail in Sect. 7.3.

4.11 Capacity Limit of Optical Fibers

The performance of a practical optical transmission system depends on a large number of factors, such as the choice of the modulation format, the detection technique and so on. A more fundamental statement can be made with the concept of the channel capacity. The capacity (C) of a transmission channel is the maximum possible bit rate at which an information can be transmitted over a noisy channel without an error. The capacity is determined by the usable bandwidth of the channel (B) and the spectral efficiency (E). In the classical case the spectral efficiency depends only on the ratio between the signal and the noise power, i.e., the signal to noise ratio (S/N). For a linear transmission system with additive white Gaussian noise, the Shennon formula [45] yields

$$C = BE = B \log_2 (1 + S/N) . \tag{4.114}$$

The usable bandwidth of single mode fibers is – as derived in Sect. 3.2.1 – determined by the cut off wavelength and the absorption in the material. In a standard single mode fiber only the S-, C- and L-bands are used, hence it follows a wavelength range of ≈ 90 nm for a central wavelength of 1.57 nm. The frequency bandwidth can be calculated with

$$\Delta f \approx \frac{\Delta\lambda \, c}{\lambda^2} \approx 11\text{THz}. \tag{4.115}$$

For a signal to noise ratio of 25 dB, the capacity of the channel, according to (4.114) is $C \approx 91$ Tbit/s. In optical fibers without the OH^-absorption peaks, the usable spectral range is much higher, between ≈ 1.2 and 1.6 μm (see Fig. 3.4). Hence, the usable bandwidth is ≈ 50 THz and the capacity $C \approx 415$ Tbit/s.

In classical information theory, the channel is considered as linear where the noise is additive and independent of the signal power. Hence, according to (4.114), the channel capacity increases with the power of the signal. On the contrary, optical fibers are nonlinear. If the signal power is increased, the intensity and therefore the efficiency of nonlinear effects will be increased as well. Hence, signal power and noise are not independent of each other in optical fibers. An increase of the signal power involves an increase of the noise. Furthermore, if the input power exceeds a particular threshold, no additional optical power reaches the fiber output. The total power exceeding the threshold will be scattered back. Therefore, (4.114) cannot be used to determine the capacity limit.

In a nonlinear channel, the capacity cannot increase up to infinity with the signal power as in (4.114), but an optimum of the signal power must exist at which the channel capacity reaches its maximum. If the intensity is increased further, the stronger nonlinear effects lead to a higher noise power and hence, to a reduction of the channel capacity.

Nonlinear effects in optical fibers can be divided into two groups; the origin of the first one is the nonlinear refractive index, the origin of the second are nonlinear scattering effects. Scattering effects are, first of all, the Raman (Chap. 10) and the Brillouin scattering (Chap. 11). If the input power of the signals is under a particular threshold – ≈ 1 W for standard single mode fibers (Sect. 10.3.2) – the Raman scattering is only spontaneous. The Raman scattering is stimulated and strong enough to have a large influence on the channel only if the input power exceeds this threshold. This only holds if the entire available bandwidth is modulated at once. But in WDM systems the bandwidth is divided into sub bands, where each of which is modulated separately. In this case, even when the power is well below threshold, the Raman scattering reduces the signal to noise ratio (SNR) for the lower wavelength channels and therefore the capacity of the system (Sect. 10.4). On the other hand, the influence of Raman scattering can be compensated by mid-span spectral inversion (see Sect. 14.5).

The threshold of the stimulated Brillouin scattering is much lower than the Raman threshold. The value is approximately 7 mW in standard single mode fibers (Sect. 11.4). This threshold determines the maximum input power for the transmission system and all power exceeding this threshold is scattered back in the fiber. But the threshold can be drastically increased with a PSK modulation of the input signals (Fig. 11.12). According to this discussion, the scattering effects can be neglected for the estimation of the capacity limit.

The effects related to the nonlinear alteration of the refractive index can be estimated with the third order nonlinear polarization (4.67):

$$P_{\mathrm{NL}} = \varepsilon_0 \chi^{(3)} ABC \tag{4.116}$$

where A, B, and C are the three distinct electric fields that can be coupled with each other by the nonlinear susceptibility $\chi^{(3)}$. As result a large number of terms are obtained, the relevance of each depends on the nonlinear effect under consideration and special additional conditions. An example is a WDM system. Hence, the optical bandwidth is broken up into sub-channels, each modulated in intensity.

- In the case of self-phase modulation (SPM, Sect. 6.1), all three electric fields A, B, and C in (4.116) come from the same channel and act back on this channel. For the SPM, the phase of the pulse is changed by its own intensity. This phase alteration leads to a frequency change (chirp) and can affect – together with the dispersion of the fiber – the temporal width of the pulse. Hence, the SPM determines the minimal guard time between adjacent time channels. But if the pulses are injected with an opposite chirp, the temporal broadening due to SPM can be compensated.
- In the case of four-wave-mixing (FWM, Chap. 7) all three fields A, B, and C are channels of the WDM system acting on a different channel D. In channel D, mixing products are generated. In the receiver, a coherent superposition between the signal in the channel and the new mixing products takes place. The dependence on the relative phase between both the current intensity at the output of the receiver fades. Hence, the FWM leads to additional noise and thus a reduction of the capacity. The FWM depends on the phase matching condition between the involved channels and can be effectively suppressed by a high local dispersion (see Dispersion Management, Sect. 5.4).
- In the case of cross-phase modulation (XPM, Sect. 6.2), two fields of another channel in the system (for instance B and C in (4.116)) act on channel A. The intensity of the other channels alters the phase of the pulses in channel A. As in the case of SPM this can lead to a temporal broadening of the pulses and to additional noise in the system, and hence a reduction

of the capacity. Contrary to SPM, this effect cannot be suppressed because the information in the other channels is independent of the information in channel A.

Hence all mentioned nonlinear effects, except the XPM, have either a very small, neglectable influence on the capacity of the fiber or can be suppressed by an intelligent design of the transmission system.

Due to the nonlinear behavior of optical fibers, the capacity could not be determined over a long period. It was not until the year 2001 that Mitra and Stark [46] were able to give a lower bound of the capacity and hence the transmissible bit rate. Shortly after that Narimanov and Mitra [47] presented an exact solution based on perturbation theory. In the estimation of the lower bound, the nonlinear channel was led back to a linear one with a multiplicative noise. We will consider this calculation in the following section.

If the third order nonlinear polarization – if the conditions in a waveguide and the intensity modulation are taken into account – is introduced into the wave equation, one yields the nonlinear Schrödinger equation (Sect. 5, (5.36)). If only XPM is considered for all channels in the system, the nonlinear Schrödinger equation is

$$\mathrm{j}\frac{\partial B_i}{\partial z} = k_2 \frac{1}{2}\frac{\partial^2 B_i}{\partial T^2} - 2\gamma \sum_{j \neq i} \left|B_j\left(z,t\right)\right|^2 B_i \tag{4.117}$$

with B_i as the complex amplitude of the considered channel and B_j as the complex amplitude of all other channels in the WDM system. Due to the fact that the distinct pieces of information in the WDM channels are independent of each other, it is possible to consider $-2\gamma \sum_{j \neq i} \left|B_j\left(z,t\right)\right|^2 = V(z,t)$ as an arbitrary noise term. In this case, the nonlinear Schrödinger equation is reduced to a linear one with a random potential. Hence, the nonlinear channel is led back to a linear one with multiplicative noise:

$$\mathrm{j}\frac{\partial B_i}{\partial z} = k_2 \frac{1}{2}\frac{\partial^2 B_i}{\partial T^2} + V(z,t)B_i . \tag{4.118}$$

The advantage of this simplification is that it is possible to give an analytical solution from (4.118) for the lower bound of the channel capacity. The solution is [46]

$$C = n_{\mathrm{c}} B \log_2 \left(1 + \frac{\mathrm{e}^{-(I/I_0)^2} I}{I_{\mathrm{n}} + \left(1 - \mathrm{e}^{-(I/I_0)^2}\right) I}\right), \tag{4.119}$$

with

$$I_0 = \sqrt{\frac{BD\Delta\lambda}{2\gamma^2 \ln\left(n_{\mathrm{c}}/2\right) L_{\mathrm{eff}}}} . \tag{4.120}$$

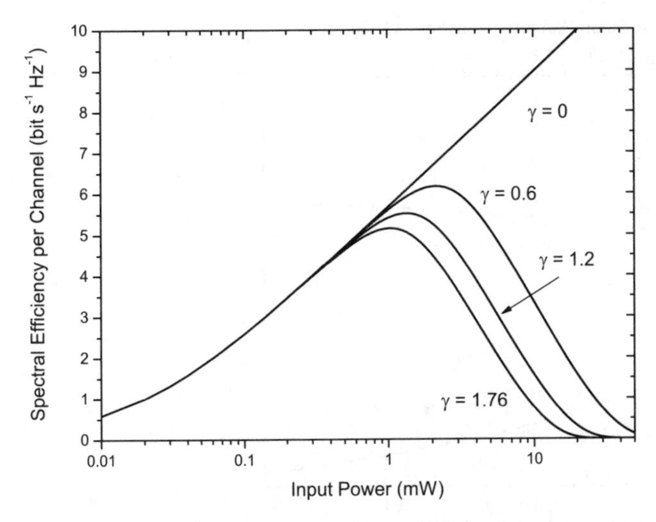

Fig. 4.20. Spectral efficiency of a nonlinear channel against the input power. The channel parameters are mentioned in the text

Here, B is the occupied bandwidth per channel in the WDM system, I is the input intensity of each channel, D is the dispersion parameter of the fiber, $\Delta\lambda$ is the channel spacing between the WDM-channels and the number of channels in the system is n_{c}, γ is the nonlinearity coefficient, α is the attenuation constant and L_{eff} is the effective length of interaction in the considered fiber.

In a transmission system, the absorption losses of each fiber span are compensated by optical amplifiers. Each of these optical amplifiers adds spontaneous noise to the system. This noise depends on the amplifier gain, hence the noise increases with the length of the transmission system. The additional noise due to the amplifiers can be calculated using

$$I_{\mathrm{n}} = aGh\nu Bn_{\mathrm{s}} \tag{4.121}$$

where all amplifiers are assumed to be equal, G is the gain of the amplifiers, h is the Planck constant, ν is the central frequency, a is a constant, and n_{s} determines the number of amplifiers. With $B = 15$ GHz, $D = 20$ ps$\times$nm^{-1}km^{-1}, $\Delta\lambda = 1$ nm, $\gamma = 1.2$ W^{-1}km^{-1}, $n_{\mathrm{c}} = 100$, $n_{\mathrm{s}} = 5$, $\alpha = 0.2$ dB/km, and $L_{\mathrm{eff}} \approx n_{\mathrm{s}}/\alpha = 108$ km, a value of 15.7 mW for I_0 follows. Using (4.121), with $a = 2$, $G = 1000$, $h = 6.6 \times 10^{-34}$ W s^2, and $\nu = 200$ THz a calculation for I_{n} yields 0.02 mW.

The lower bound of the spectral efficiency of a nonlinear channel with the above values against the input power for four distinct values of the nonlinearity coefficient γ is shown in Fig. 4.20:

If the channel is linear ($\gamma = 0$), (4.119) leads to (4.114). Hence, the spectral efficiency increases with the signal power. Whereas, in the nonlinear case

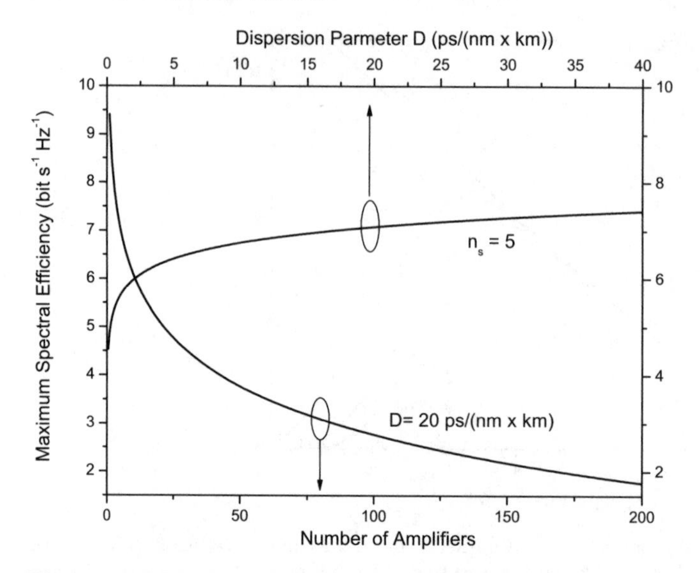

Fig. 4.21. Maximum spectral efficiency per channel versus the dispersion parameter (*top*) and against the number of amplifiers of the transmission system (*bottom*). The used parameters for the calculation are described in the text

a maximum of the spectral efficiency will be reached. The input power at which the capacity has its maximum can be calculated with

$$I_{\mathrm{max}} \approx \left(I_0^2 \frac{I_{\mathrm{n}}}{2}\right)^{\frac{1}{3}} . \tag{4.122}$$

With the above values the maximum input power for a nonlinearity coefficient of $\gamma = 1.2\ \mathrm{W}^{-1}\mathrm{km}^{-1}$can be calculated as $I_{\mathrm{max}}(\gamma = 1.2) \approx 1.05$ mW. The maximum channel capacity is

$$C_{\mathrm{max}} \approx \frac{2}{3} n_{\mathrm{c}} B \log_2 \left(2I_0/I_{\mathrm{n}}\right) . \tag{4.123}$$

The capacity of the distinct WDM channel increases logarithmically with the distance to other channels and the dispersion, as can be seen from (4.120) and Fig. 4.21. As will be described in Chap. 6.2.2, there is no residual phase shift and hence no additional noise remains if the pulses propagate completely through each other. An influence on the capacity due to XPM only exists if the interaction between the pulses is incomplete. For a high dispersion and a large channel spacing, the difference in the group velocity of the pulses is very high. Hence most interactions are complete and no additional noise is added to the system.

If the length of the transmission system is increased, the number of amplifiers will increase as well. Every amplifier adds noise to the transmission system, hence I_{n} is increased (4.121). At the same time, the effective length of interaction of the nonlinearity is increased ($L_{\text{eff}} \approx n_s/\alpha$) and I_0 is decreased (4.120). Both effects lead to a reduction of the capacity of the transmission system if the length is increased, as shown in Fig. 4.21.

The above calculations assumed that only the nonlinear effect of cross-phase modulation has an impact on the system capacity. The noise generated by XPM depends on the intensity change of the pulses in the other channels. Hence, one can assume that for modulation techniques that modulate the phase or frequency instead of the intensity of the optical carrier, the nonlinearity has no influence on the system. These modulation techniques can show a nominally constant intensity. Therefore, the phase distortion of the other channels is not time variant and hence introduces no noise to the other channels. The capacity of these systems should increase with the signal power as in the linear case. That is not the case was shown in [48], whose authors found that modulation techniques with constant intensity can in fact improve the spectral efficiency of WDM systems. But for constant intensity modulation, the effects of four-wave-mixing (FWM) cannot be neglected [49]. Hence, the nonlinearity of the channel still has a limiting influence on the system capacity.

The capacity of a transmission channel is only a theoretical value and says nothing about how to approach it. Hence, from a practical point of view, it is important to come as close as possible to this limit for as minimal costs. The practical, achievable spectral efficiency depends on the modulation format and the detection technique. In this sense it is possible to achieve a higher spectral efficiency with a non-binary, constant-intensity modulation and coherent detection [50].

Summary

- Linear and nonlinear effects can be described with the model of the harmonic oscillator.
- If the intensity of the electrical force is increased, the motion of the oscillator will become nonlinear. In the radiated field, harmonics of the fundamental wave as well as sum and difference frequencies will be found.
- The nonlinear susceptibilities of the material, describing the nonlinear interaction between applied optical fields and the medium, are complex tensors. If the frequencies of the applied fields, as well as linear combinations of them, are far away from material resonances, the imaginary part can be neglected. According to symmetry arguments, the number of tensor elements can be reduced significantly in most materials of practical interest.
- In the nonlinear regime, the refractive index of optical waveguides becomes intensity-dependent; $n_{\text{ges}} = n_0 + n_2 I$. The nonlinear refractive index of standard single mode fibers is $n_2 = 2.35 \times 10^{-20}\ \text{m}^2/\text{W}$.
- The length at which the power of the signal is concentrated in the fiber, and hence responsible for nonlinear interaction, corresponds to the effec-

tive length (L_{eff}). For a rather small fiber length and a weak absorption ($\alpha L \ll 1$), the effective length corresponds to the fiber length ($L_{\text{eff}} \approx L$). Whereas in the opposite limit ($\alpha L \gg 1$), the fiber length comes to a limiting value of $L_{\text{eff}} = \frac{1}{\alpha}$. The effective length for standard single mode fibers with an attenuation constant of $\alpha = 0.2$ dB/km is $L_{\text{eff}} \approx 22$ km. It starts again after every optical amplification.

- The power concentration in the fiber is determined by the effective area (A_{eff}). Standard single mode fibers have an effective area of $\approx 80\mu\text{m}^2$, whereas the effective area of photonic fibers can be as small as 1-3 μm^2.
- Nonlinear effects are very weak, but in fibers the high intensity due to the small core area and the very long interaction length, together with the small attenuation, leads to an efficient nonlinear interaction.
- For the generation of new frequencies the phase matching condition must be fulfilled. It corresponds to the rule of conservation of momentum in the quantum mechanical picture and means that the wave vectors of the interacting waves must be matched. If the phase mismatch is zero, the intensity of the new wave increases quadratically with the propagation distance in the material. If the phases are not completely matched, the intensity of the new wave oscillates with a periodicity of two times the coherence length.
- The capacity limit of optical fibers is not only determined by the signal to noise ratio and the bandwidth, as in the linear case, but by the nonlinear behavior of the channel as well. Hence, the capacity that is dependent on the signal power has a maximum.

Exercises

4.1. Determine the number of tensor elements that the fifth order susceptibility $\chi^{(5)}$ can consist of.

Which index has the matrix element in the second row and fifth column of the third order susceptibility $\chi^{(3)}$?

4.2. Determine the intensity at which the applied field reaches the Coulomb field, binding the ground state electron in the hydrogen atom.

4.3. Assume a DWDM system with 50 channels and an input power of 5 mW, each in a standard single mode fiber with an $A_{\text{eff}} = 80\ \mu\text{m}^2$ and $n_2 = 2.6 \times 10^{-20}\ \text{m}^2/\text{W}$. Estimate the total refractive index at the input of the fiber for a wavelength of 1.55 μm, assuming that it is made of pure silica.

4.4. Determine the effective length of a 15-km fiber with an attenuation constant of $\alpha = 0.35$ dB/km. How large is the limiting value of the effective length in this fiber?

4.5. Estimate the coherence length for second harmonic generation in pure silica if the waves propagate collinearly and the fundamental wavelength is 1.6 μm!

4.6 Derive the relation between the wavelength and frequency bandwidth (4.115).

4.7. Determine the maximum bit rate which can be transmitted free of error over a noisy linear channel if the signal to noise ratio is 20 dB and only the C-band is used.

4.8. Determine the maximum bit rate that can be transmitted over a non-linear channel according to the lower bound given by [46] (4.123). The usable bandwidth of 400 nm ($1.2 - 1.6$ μm) in the fiber is divided into 300 channels with a channel spacing of 1 nm and a bandwidth of 15 GHz each. The transmission system consists of 20 identical optical amplifiers, the gain of each amplifier is 30 dB and it is assumed that it only amplifies the bandwidth of the channels. The centre wavelength is 1.55 μm and the constant a is assumed to be 2. The attenuation constant of the fiber spans is 0.2 dB/km; the fiber has a nonlinearity coefficient of 1.2 $\mathrm{W}^{-1}\mathrm{km}^{-1}$ and a dispersion parameter of 20 $\mathrm{ps}\times\mathrm{nm}^{-1}\mathrm{km}^{-1}$.

Part II

Nonlinear Effects in Optical Waveguides

5. The Nonlinear Schrödinger Equation

The nonlinear Schrödinger equation (NSE) is of particular importance in the description of nonlinear effects in optical fibers. It is named after the similarity that it shows to the well known equation derived by Erwin Schrödinger. This equation is one of the basics of quantum theory and describes the behavior of a particle in an arbitrary potential. Contrary to that, the NSE is based on the classical Maxwell Theory and it describes the behavior of wave packets in a nonlinear environment. But, as can be seen from the soliton section, these classical wave packets behave – under particular circumstances – like particles as well.

In this section, the NSE is derived from the nonlinear wave equation and the two terms are described independently. The linear term leads to a temporal broadening of the wave packets in the waveguide and results in the already-mentioned dispersion. This dispersion limits the capacity of transmission systems, but can be circumvented with a technique called "Dispersion Management". This technique will be described in Sect. 5.4.

The nonlinear term of the NSE leads to a spectral broadening of the pulses which can interact with the dispersion in different ways. The basics will be described here, but the details of this behavior are the subjects of the next few chapters.

5.1 Derivation of the Nonlinear Schrödinger Equation

Since the waves propagating in a nonlinear material are assumed to be classical, the framework used to describe the nonlinear effects will be delivered by the Maxwell equations. Hence, the starting point for the derivation of the NSE is the wave equation for nonlinear, transparent media derived in Sect. 4.4. The nonlinear wave equation is given by (4.55):

$$\Delta \boldsymbol{E} - \frac{n^2}{c^2}\frac{\partial^2 \boldsymbol{E}}{\partial t^2} = \frac{1}{c^2}\frac{\partial^2}{\partial t^2}\left(\chi^{(2)}\boldsymbol{E}\boldsymbol{E} + \chi^{(3)}\boldsymbol{E}\boldsymbol{E}\boldsymbol{E} + ...\right). \tag{5.1}$$

Here, the solution in cylindrical waveguides is of particular interest. For the simplification of the derivation, the following assumptions will be made:

- Only one mode is present in the waveguide. Hence, a single mode fiber will be considered.

- The material is perfectly transparent and the wavelength – as well as harmonics of the wavelength – of the applied field is far away from any material resonances. Thus absorption and nonlinear resonances can be neglected.
- All scattering effects in the waveguide are neglected. Therefore, Rayleigh, Rayleigh-wing, Raman, and Brillouin scattering play no significant role.
- The amplitude of the considered wave packet changes very slowly with respect to its carrier. This is the so called slowly varying envelope approximation.
- The field strength of the applied field is, compared to the inner atomic field, relatively small. Hence, the nonlinearity can be seen as a small disturbance of the linear behavior.
- The fields are linearly polarized in the same direction and the polarization status remains the same during the propagation. This means that the birefringence of the fiber and the vector nature of the waves are neglected.
- The nonlinearity has no influence on the field components perpendicular to the propagation direction.

The fourth assumption implies that the alteration of the carrier wave happens much faster than the alteration of the envelope of the pulse. If the spectral width of the pulse ($\Delta\omega$) to the carrier frequency (ω_0) is much smaller than 1, i.e.,

$$\frac{\Delta\omega}{\omega_0} \ll 1, \tag{5.2}$$

then it is possible to assume nearly monochromatic waves [51]. The spectral width of a pulse is determined by the Fourier transformation. For carrier frequencies of $\omega_0 > 1.5 \times 10^{14}$ Hz, which corresponds to a wavelength $\lambda_0 <$ 2 μm, the inequality is fulfilled if the duration of the pulses are $\tau_0 \geq 0.7$ ps and $\tau_{\mathrm{FWHM}} \geq 1.2$ ps. Here it is assumed that the pulses have a Gaussian shape and that the carrier frequency is 100 times the spectral pulse width (see Appendix).

Under these simplifications the quasi monochromatic wave propagating in the z- direction can be expressed as

$$E = \frac{1}{2}\left(\hat{E}(\boldsymbol{r})\mathrm{e}^{\mathrm{j}(kz-\omega t)} + \mathrm{c.c.}\right), \tag{5.3}$$

where k is the wave number, including the refractive index and the properties of the waveguide. The slowly varying envelope, $\hat{E}(\boldsymbol{r})$, consists of two parts: one in the propagation direction (longitudinal, $A(z)$) and the other perpendicular to the propagation direction (transverse, $F(r,\varphi)$). Hence, it is

$$\hat{E}(\boldsymbol{r}) = F(r,\varphi)A(z). \tag{5.4}$$

Here, a cylindrical coordinate system is used with r and φ as the perpendicular variables radius and azimuth, and z as the longitudinal component.

According to [9] and [84], it is more convenient for the derivation to describe the wave in terms of its power. The temporal mean value of the intensity depends on the amplitude of the wave

$$I = \frac{1}{2}\varepsilon_0 cn \left|\hat{E}\right|^2 . \tag{5.5}$$

The power of the wave is the intensity multiplied by the area it propagates through. For fibers this is of course a cylindrical area, with the intensity decreasing quadratically with an increase of the radius r. Hence, the power is

$$P = \int\int I\, r\, \mathrm{d}r\mathrm{d}\varphi. \tag{5.6}$$

The intensity distribution of the mode is determined by the transverse part of the amplitude. Therefore, inserting (5.5) and (5.4) into (5.6) yields:

$$P = \frac{1}{2}\varepsilon_0 cn \left|A(z)\right|^2 \int\int \left|F(r,\varphi)\right|^2 r\, \mathrm{d}r\mathrm{d}\varphi. \tag{5.7}$$

For a single mode waveguide the field distribution is independent on φ, it follows $\int\int \left|F(r,\varphi)\right|^2 r\, \mathrm{d}r\mathrm{d}\varphi = 2\pi \int \left|F(r)\right|^2 r\, \mathrm{d}r$. With

$$N^2 = \frac{1}{2}\varepsilon_0 cn 2\pi \int \left|F(r)\right|^2 r\, \mathrm{d}r, \tag{5.8}$$

the equation can be simplified to

$$P = N^2 \left|A(z)\right|^2 . \tag{5.9}$$

Now it is possible to introduce a longitudinal amplitude, $B = N\ A(z)$. The square of the absolute value of this amplitude determines the power of the wave:

$$P = \left|B\right|^2 . \tag{5.10}$$

Hence, the equation that describes the propagating wave in the single mode waveguide can be expressed as

$$E = \frac{1}{2}\left[\frac{B}{N}F(r)\mathrm{e}^{\mathrm{j}(kz-\omega t)} + \mathrm{c.c.}\right]. \tag{5.11}$$

As already discussed in Sect. 3.2, the Laplace operator Δ in (5.1) can be divided into a transverse (Δ_{T}) and a longitudinal component $\left(\partial^2/\partial z^2\right)$. Furthermore, fused silica – optical fibers consist of – is a material with an inversion center. Hence, as mentioned in Sect. 4.4, according to symmetry arguments the second order susceptibility $\left(\chi^{(2)}\right)$ can be neglected. Therefore, the nonlinear wave equation in an optical fiber can be described by

$$\Delta_{\mathrm{T}} E + \frac{\partial^2 E}{\partial z^2} - \frac{n^2}{c^2}\frac{\partial^2 E}{\partial t^2} = \frac{\chi^{(3)}}{c^2}\frac{\partial^2}{\partial t^2} EEE. \tag{5.12}$$

First let's consider the left side of (5.12). Since the amplitude in (5.11) depends on z ($B = B(z)$), the product rule yields the following for the longitudinal part:

$$\frac{\partial^2 E}{\partial z^2} = \frac{1}{2}\left[\frac{\partial^2 U}{\partial z^2} + 2jk\frac{\partial U}{\partial z} - k^2 U\right] F(r)\mathrm{e}^{\mathrm{j}(kz-\omega t)}, \tag{5.13}$$

where

$$U = \frac{B}{N}. \tag{5.14}$$

If the envelope of the wave changes slowly during propagation, then the second derivation is much smaller than the first one and, hence, negligible:

$$\frac{\partial^2 U}{\partial z^2} \ll k\frac{\partial U}{\partial z}. \tag{5.15}$$

The temporal derivation of (5.11) is

$$\frac{n^2}{c^2}\frac{\partial^2 E}{\partial t^2} = -\frac{1}{2}\frac{n^2}{c^2}\omega^2 U\ F(r)\mathrm{e}^{\mathrm{j}(kz-\omega t)}. \tag{5.16}$$

Therefore, equations (5.13) to (5.16) yield the following for the left side of (5.12):

$$\begin{aligned}\Delta_{\mathrm{T}} E + \frac{\partial^2 E}{\partial z^2} - \frac{n^2}{c^2}\frac{\partial^2 E}{\partial t^2} &= \frac{1}{2}U\ \mathrm{e}^{\mathrm{j}(kz-\omega t)}\Delta_{\mathrm{T}} F(r) \\ &\quad + \frac{1}{2}\left[2\mathrm{j}k\frac{\partial U}{\partial z} - k^2 U\right] F(r)\mathrm{e}^{\mathrm{j}(kz-\omega t)} \\ &\quad + \frac{1}{2}\frac{n^2}{c^2}\omega^2 U\ F(r)\mathrm{e}^{\mathrm{j}(kz-\omega t)} \\ = \frac{1}{2}\left[2\mathrm{j}k\frac{\partial U}{\partial z}F(r) + U\left(\Delta_{\mathrm{T}}\ F(r) + \left(\kappa^2 - k^2\right)F(r)\right)\right]&\mathrm{e}^{\mathrm{j}(kz-\omega t)}. \end{aligned} \tag{5.17}$$

Contrary to the wavenumber k, $\kappa = \frac{n}{c}\omega$ does not depend on the waveguide properties. Using (5.3), the right side of (5.12) can be deduced:

$$\frac{1}{c^2}\frac{\partial^2}{\partial t^2}\chi^{(3)} EEE = \frac{\chi^{(3)}}{8c^2}\frac{\partial^2}{\partial t^2}\left[(E + E^*)(E + E^*)(E + E^*)\right]. \tag{5.18}$$

If (5.11) is introduced for the electric field:

$$E = V\mathrm{e}^{-\mathrm{j}\omega t} + \mathrm{c.c.} \tag{5.19}$$

with

$$V = \frac{B}{N} F(r) \mathrm{e}^{\mathrm{j}kz}, \tag{5.20}$$

the right side of the wave equation is then

$$\frac{1}{c^2} \frac{\partial^2}{\partial t^2} \chi^{(3)} EEE = \tag{5.21}$$
$$\frac{\chi^{(3)}}{8c^2} \frac{\partial^2}{\partial t^2} \left[\left(V^3 \mathrm{e}^{-\mathrm{j}3\omega t} + V^{*3} \mathrm{e}^{\mathrm{j}3\omega t} \right) + 3 \left(V^2 V^* \mathrm{e}^{-\mathrm{j}\omega t} + V^{*2} V \mathrm{e}^{\mathrm{j}\omega t} \right) \right].$$

The first term on the right side of (5.21) describes the generation of the third harmonic in the waveguide. This process is determined by the phase matching condition. According to the estimation of the coherence length in Sect. 4.10, the phase mismatch for this process is very high in optical fibers. Therefore, the term can be neglected.

The second term describes an additional contribution to the polarization. This contribution has the frequency of the fundamental wave and hence, does not suffer from any phase matching condition. With

$$VV^* = |V|^2 = \left| \frac{B}{N} \right|^2 |F(r)|^2 \tag{5.22}$$

and the second temporal derivation, the right side of the wave equation (5.12) becomes

$$\frac{1}{c^2} \frac{\partial^2}{\partial t^2} \chi^{(3)} EEE = -\frac{3\omega^2}{8c^2} \chi^{(3)} \left| \frac{B}{N} \right|^2 |F(r)|^2 \frac{B}{N} F(r) \mathrm{e}^{\mathrm{j}(kz - \omega t)}. \tag{5.23}$$

According to the assumptions, the transversal component of the field, $\Delta_{\mathrm{T}} F(r)$, is not influenced by the nonlinearity. Hence, if the left (5.17) and right side (5.23) are joined together, we have

$$\frac{1}{2} \left[2\mathrm{j}k \frac{\partial U}{\partial z} + U \left(\kappa^2 - k^2 \right) \right] = -\frac{3\omega^2}{8c^2} \chi^{(3)} |F(r)|^2 |U|^2 U. \tag{5.24}$$

Due to (5.24), the alteration of the amplitude during its propagation in the waveguide is

$$\frac{\partial U}{\partial z} = \frac{\mathrm{j}}{k} \left\{ \frac{3\omega^2}{8c^2} \chi^{(3)} |F(r)|^2 |U|^2 U + \frac{1}{2} U \left(\kappa^2 - k^2 \right) \right\}. \tag{5.25}$$

With the approximations $k \approx n\omega/c$ and $\left(\kappa^2 - k^2 \right) \approx 2k \{ \kappa - k \}$, it can be written as

$$\frac{\partial U}{\partial z} = \mathrm{j} \left[\frac{3\omega}{8cn} \chi^{(3)} |F(r)|^2 |U|^2 U + U (\kappa - k) \right]. \tag{5.26}$$

If (5.14) is introduced for U and both sides are multiplied with the transverse field distribution, $2\pi \int |F(r)|^2 r dr$ to eliminate the transverse coordinates, (5.26) can be written as

$$2\pi \int |F(r)|^2 r\mathrm{d}r \frac{\partial B}{\partial z} = \mathrm{j}\Big[\frac{3\omega}{8cnN^2}\chi^{(3)} 2\pi \int |F(r)|^4 r\mathrm{d}r\, |B|^2 B \tag{5.27}$$

$$+2\pi \int |F(r)|^2 r\mathrm{d}r B(\kappa - k)\Big].$$

Introducing (5.8) and (4.93) for the effective core area A_{eff} in which the field is concentrated yields

$$\frac{\partial B}{\partial z} = \mathrm{j}\left[\frac{3\omega\chi^{(3)}}{4c^2n^2\varepsilon_0 A_{\mathrm{eff}}} |B|^2 B + B(\kappa - k)\right]. \tag{5.28}$$

With the nonlinear refractive index n_2 (4.90), derived in Sect. 4.7, and the nonlinearity coefficient (see Sect. 4.8),

$$\gamma = \frac{\omega n_2}{cA_{\mathrm{eff}}}, \tag{5.29}$$

the alteration of the envelope during its propagation in the waveguide becomes

$$\frac{\partial B}{\partial z} = \mathrm{j}\left[\gamma |B|^2 B + B(\kappa - k)\right]. \tag{5.30}$$

For a further expansion of (5.30), a transformation into the frequency domain can be used. The Fourier transform of the amplitude is

$$B(t) = \frac{1}{\sqrt{2\pi}} \int_{-\infty}^{+\infty} \tilde{B}(\omega) \mathrm{e}^{-\mathrm{j}\omega t} \mathrm{d}\omega. \tag{5.31}$$

Thus (5.30) can be written as

$$\frac{\partial \tilde{B}}{\partial z} = \mathrm{j}\left[\gamma \left|\tilde{B}\right|^2 \tilde{B} + \tilde{B}(\kappa - k)\right] \tag{5.32}$$

If the wave number k is developed in a Taylor series around the frequency ω_0 (2.69), (5.32) yields

$$\frac{\partial \tilde{B}}{\partial z} = \mathrm{j}\gamma \left|\tilde{B}\right|^2 \tilde{B} - \mathrm{j}\tilde{B}\left[(\omega - \omega_0) k_1 + \frac{1}{2}(\omega - \omega_0)^2 k_2\right] \tag{5.33}$$

Since the wave number $\kappa = n\omega/c$ complies to the first term of the Taylor series in (2.69), both are cancelled in (5.33). For the inverse Fourier transformation of (5.33), $(\omega - \omega_0)$ can be replaced by $-j\partial/\partial t$, whereas $(\omega - \omega_0)^2$ will be replaced by $\partial^2/\partial t^2$. Hence, it follows that

$$\frac{\partial B}{\partial z} = \mathrm{j}\gamma |B|^2 B - k_1 \frac{\partial B}{\partial t} - \mathrm{j}k_2 \frac{1}{2}\frac{\partial^2 B}{\partial t^2}. \tag{5.34}$$

As discussed in Sect. 2.6.2, the pulse maximum moves through the waveguide with the group velocity v_{g}. If instead of the time t in (5.34) the equation

$$T = t - \frac{z}{v_{\mathrm{g}}} \tag{5.35}$$

is introduced, the pulse will remain stationary in this moving frame because the frame propagates with the group velocity as well. According to the fact that k_1 is the reciprocal group velocity (see Sect. 2.6.2), (5.35) can be written as $T = t - k_1 z$ and the second term in (5.34) will vanish. Finally, the nonlinear Schrödinger equation (NSE) can be represented by

$$\mathrm{j}\frac{\partial B}{\partial z} = -\gamma \left|B\right|^2 B + k_2 \frac{1}{2}\frac{\partial^2 B}{\partial T^2}. \tag{5.36}$$

Equation (5.36) describes the propagation of a wave packet with the amplitude envelope (B) in a moving frame (T). The name relates to the similarity of this equation to the Schrödinger equation of quantum mechanics.

The first term on the right side describes the influence of the nonlinearity on the pulse, whereas the second term can be addressed to the linear effect of dispersion in the waveguide. If the absorption in the material is included in the consideration, the NSE takes the form of

$$\mathrm{j}\frac{\partial B}{\partial z} + \mathrm{j}\frac{\alpha}{2}B = -\gamma \left|B\right|^2 B + k_2 \frac{1}{2}\frac{\partial^2 B}{\partial T^2}, \tag{5.37}$$

with α as the attenuation constant. In the following, both terms of the nonlinear Schrödinger equation without absorption will be treated in detail. But first, two important length values of nonlinear optics in fibers will be described: the dispersion and the nonlinear length [51].

5.2 Dispersion and Nonlinear Length

First, assume that the time scale in (5.36) is normalized to the initial pulse width (τ_0) at the fiber input via $\tau = T/\tau_0$. At the same time – according to (5.10) – the absolute value will be separated from the phase and hence, a normalized amplitude A will be defined by

$$B(z,\tau) = \sqrt{P}A(z,\tau), \tag{5.38}$$

where P is the power of the pulse. The initial pulse width (τ_0) is a constant and therefore independent of the derivation. If (5.38) is introduced into the NSE, (5.36) becomes

$$\mathrm{j}\frac{\partial A}{\partial z} = -\gamma P \left|A\right|^2 A + \mathrm{sgn}(k_2)\frac{|k_2|}{2\tau_0^2}\frac{\partial^2 A}{\partial \tau^2}, \tag{5.39}$$

where $\mathrm{sgn}(k_2) = \pm 1$ describes the sign of the dispersion parameter k_2. With (2.74), the dispersion length can be written as

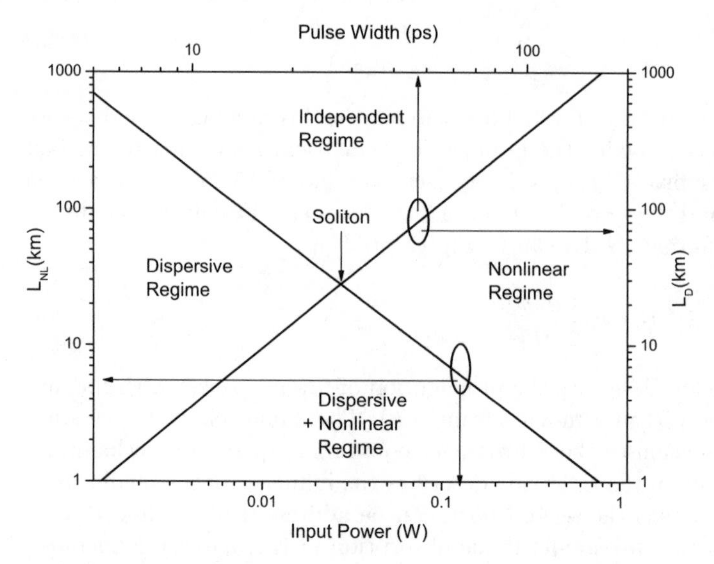

Fig. 5.1. Nonlinear and dispersion length for a standard single mode fiber at a wavelength of 1.55 μm, with a nonlinearity coefficient of $\gamma = 1.3\ \mathrm{W}^{-1}\mathrm{km}^{-1}$ and a dispersion constant for pure silica of $D = 21.89$ ps/(km×nm)

$$L_{\mathrm{D}} = \frac{\tau_0^2}{|k_2|} = \frac{\tau_0^2 2\pi c}{|D|\,\lambda^2}, \tag{5.40}$$

where D is the dispersion parameter introduced in Sect. 2.6.2. The nonlinear length is then

$$L_{\mathrm{NL}} = \frac{1}{\gamma P}. \tag{5.41}$$

With these definitions, (5.39) can be rewritten as

$$\mathrm{j}\frac{\partial A}{\partial z} = -\frac{1}{L_{\mathrm{NL}}}|A|^2 A + \frac{\mathrm{sgn}(k_2)}{2L_{\mathrm{D}}}\frac{\partial^2 A}{\partial \tau^2}. \tag{5.42}$$

For large nonlinear lengths, the first term in (5.42) vanishes. Hence, in this case the fiber behaves linearly. According to (5.41), this happens if the nonlinearity of the waveguide, or the power of the pulses, is rather small. Whereas in the case of a large dispersion length, the second term in (5.42) vanishes and hence, the broadening of the pulses can be neglected. As can be seen from (5.40), this will happen if either the initial pulse width is large or if the dispersion of the fiber is rather small.

Figure 5.1 shows the nonlinear and dispersion length for a standard single mode fiber. Depending on the fiber length in relation to the system parameters four different regimes are depicted. For a small input power and a rather long pulse, the nonlinear as well as the dispersion lengths are very large. For a pulse width of 100 ps, the dispersion length is, for instance, 358 km and the nonlinear length for an input power of 2 mW is 385 km. If the fiber length is much smaller than these values, neither the dispersion nor the nonlinearity has an influence on the transmission of the pulses. Hence, this regime is called independent.

In the dispersive regime, the fiber length is smaller than the dispersion length but larger than the nonlinear length. According to Fig. 5.1, this is the case if the pulse width is small but the power of the pulse is rather low. If the pulse width is 10 ps, for instance, then it can be seen from the figure that the dispersive length is only 3.6 km. If, at the same time, the pulse power is 5 mW, then the nonlinear length is 154 km. Hence, in this regime the dispersion has an influence on the transmission and the nonlinearity can be neglected.

In the nonlinear regime, the dispersion can be neglected in relation to the nonlinear behavior. This is the case if the fiber length is larger than the dispersive but smaller than the nonlinear length. According to Fig. 5.1, this happens if the pulse power is high but the pulse length is rather large. For example, a pulse with a power of 500 mW and a width of 100 ps has a dispersion length of 358 km but a nonlinear length of only 1.54 km.

The last regime is determined by the nonlinear as well as the dispersive effects. In this case, the fiber length is larger than the nonlinear and the dispersion lengths. A pulse with a power of 600 mW and a width of 6 ps has, for instance, a nonlinear and dispersion length of 1.3 km. In this case, both effects play together and it depends on the pulse as well as on the fiber parameters if the pulse is temporally broadened or compressed. If the nonlinear equals the dispersive length the pulse can maintain its width during propagation if its carrier wavelength is in the anomalous dispersion regime of the fiber, the resulting pulse shape is a so called soliton (see Sec. 9.1). For the example above a soliton can be created if the pulse width is 28 ps and the pulse power 28 mW.

5.3 The Linear Term

In this section, only the predictions of the NSE for the pulse propagation in the linear regime will be considered. If the nonlinearity in (5.36) is assumed to be zero ($\gamma = 0$), then only the linear term remains. Hence, the envelope of the wave packet in this linear case is

$$\mathrm{j}\frac{\partial B}{\partial z} = k_2 \frac{1}{2} \frac{\partial^2 B}{\partial T^2}. \tag{5.43}$$

As already discussed in Sect. 2.6.2, k_2 is the alteration of the propagation time of the group $\tau_\mathrm{g} = L/v_\mathrm{g}$ per distance L and hence, the dispersion of the group velocity. According to the behavior of the group velocity, the dispersion is positive when the group velocity increases and it is negative when it decreases. In pure silica, the dispersion is zero at the inflection point of the refractive index (at 1.27 μm) and can be shifted by the waveguide dispersion. The group velocity dispersion leads to an alteration of the pulse shape. Equation (5.43) gives a mathematical description of this alteration.

Differential equations like (5.43) are conveniently solved in the frequency domain. Hence, (5.43) has to be transformed with the rules of the Fourier transformation. According to (5.31), $\partial B(t)/\partial t$ and $-\mathrm{j}\omega \tilde{B}(\omega)$ as well as $\partial^2 B(t)/\partial t^2$ and $-\omega^2 \tilde{B}(\omega)$ are Fourier transform pairs. Thus (5.43) can be written as

$$\frac{\partial \tilde{B}}{\partial z} = \mathrm{j}\frac{1}{2}k_2\omega^2\tilde{B}, \tag{5.44}$$

where $\tilde{B}$ is the envelope of the pulse in the frequency domain. According to (5.44), the envelope has to be a function that is dependent on the distance z, that will remain the same after a spatial derivation. This property is fulfilled if an exponential function is chosen for $\tilde{B}(z,\omega)$. Hence, the complex amplitude of the pulse in the frequency domain is

$$\tilde{B}(z,\omega) = \tilde{B}(0,\omega)\mathrm{e}^{\mathrm{j}1/2k_2\omega^2 z}, \tag{5.45}$$

where $\tilde{B}(0,\omega)$ is the Fourier transform of the pulse shape at the fiber input ($z = 0$). The argument of the exponential function describes the phase. As can be seen from (5.45), the phase of the distinct frequency components of the pulse alters during the propagation in the waveguide. This alteration depends on the frequency (ω) and the propagation distance (z). But it follows from (5.45) as well that no new frequency components will be generated. Hence, due to the alteration of the distinct frequency components, the pulse can merely change its shape.

This alteration of the shape takes place in the time domain. For an investigation of the temporal pulse shape, the inverse transformation of (5.45) is required. The easiest way for an inverse transformation is offered by a Gaussian shape of the input pulse. Hence, the input pulse in the time domain (see Appendix) is

$$B(0,T) = \mathrm{e}^{-\frac{T^2}{2\tau_0^2}}. \tag{5.46}$$

The amplitude of this pulse is 1 and the half-width at an intensity of 1/e is τ_0. The FWHM pulse width of a Gaussian pulse is $\tau_{\mathrm{FWHM}} = 2\sqrt{\ln 2}\,\tau_0 \approx 1.665\,\tau_0$. Therefore, the dispersion length of a Gaussian pulse defined via its FWHM width is

$$L_{\mathrm{D}} = \frac{\tau_{\mathrm{FWHM}}^2}{4\ln 2\,|k_2|} = \frac{\tau_{\mathrm{FWHM}}^2 \pi c}{2\ln 2\,|D|\,\lambda^2}. \tag{5.47}$$

In the frequency domain, the pulse has a Gaussian shape as well and can be expressed as

$$\tilde{B}(0,\omega) = \mathrm{e}^{-\frac{\omega^2}{2\tau_0^2}}. \tag{5.48}$$

For a bandwidth limited pulse, i.e., if the pulse has no additional chirp, the spectral width is $\Delta\omega = \frac{1}{\tau_0} \approx \frac{1.665}{\tau_{\mathrm{FWHM}}}$. If (5.48) is introduced into (5.45) the general solution, according to (5.31) is

$$B(z,T) = \frac{1}{\sqrt{2\pi}} \int \tilde{B}(0,\omega)\mathrm{e}^{\mathrm{j}(1/2k_2\omega^2 z-\omega T)}\mathrm{d}\omega. \tag{5.49}$$

If the integration is carried out, the temporal alteration of the pulse shape is [51]

$$B(z,T) = \frac{\tau_0}{\sqrt{(\tau_0^2 - \mathrm{j}k_2 z)}}\mathrm{e}^{-\frac{T^2}{2(\tau_0^2-\mathrm{j}k_2 z)}}. \tag{5.50}$$

According to (5.50), the pulse retains its Gaussian shape during propagation, but its width will be increased and the phase alters. If (5.50) is separated into an absolute value with a phase component it can be rewritten as

$$B(z,T) = |B(z,T)|\,\mathrm{e}^{\mathrm{j}\varphi(z,T)}. \tag{5.51}$$

Thus the absolute value of the pulse as a function of T at a distance z is [51]

$$|B(z,T)| = \frac{\tau_0}{\sqrt[4]{\tau_0^2 + k_2^2 z^2}}\mathrm{e}^{-\frac{T^2\tau_0^2}{2(\tau_0^4+k_2^2 z^2)}} \tag{5.52}$$

and the phase factor is

$$\varphi(z,T) = \frac{1}{2}\tan^{-1}\frac{k_2 z}{\tau_0^2} - \frac{T^2 k_2 z}{2\left(\tau_0^4 + k_2^2 z^2\right)}. \tag{5.53}$$

The half-width of the pulse during its propagation in the waveguide follows from a comparison between the phases of the equations (5.46) and (5.52)

$$\tau_1 = \tau_0\sqrt{1 + \frac{k_2^2 z^2}{\tau_0^4}}. \tag{5.54}$$

Using the dispersion length in (5.40), already introduced in Sect. 5.2, the width of the pulse is

$$\tau_1 = \tau_0\sqrt{1 + \left(\frac{z}{L_{\mathrm{D}}}\right)^2}. \tag{5.55}$$

The dispersion length determines the propagation distance in the waveguide, after which the original Gaussian pulse is broadened by a factor of $\sqrt{2}$:

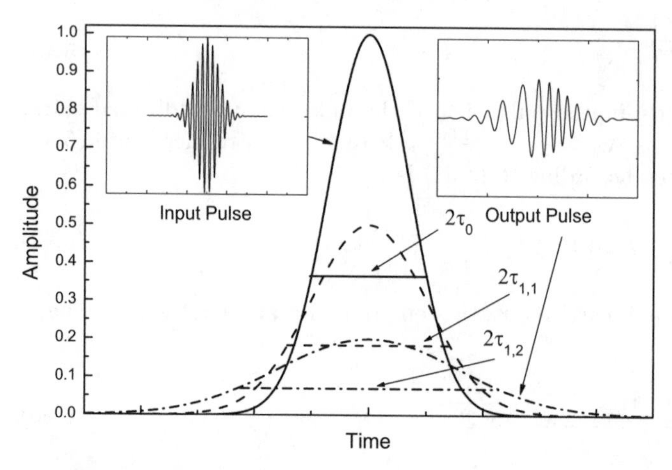

Fig. 5.2. Broadening of a pulse with a Gaussian shape during its propagation in a dispersive waveguide. Note that the attenuation of the pulse in the waveguide is neglected. The *left* inset shows the input pulse, whereas in the *right* inset the output pulse in the case of a positive dispersion is depicted

$$\tau_1 \left(z = L_D\right) = \tau_0 \sqrt{2}. \tag{5.56}$$

Hence, a pulse with a temporal width of 25 ps at the fiber input has a width of 35 ps after a propagation distance of 30.6 km in a standard single mode fiber ($k_2(1.55\ \mu\text{m}) = -20.4\ \text{ps}^2/\text{km}$).

Figure 5.2 shows the principal behavior of the broadening of a pulse with Gaussian shape during its propagation in a dispersive waveguide. As can be seen, the rearrangement of the frequency components in the pulse leads to a decreasing of the amplitude whereas at the same time, the temporal width of the pulse is increased. Note that the effects of absorption are not included in the consideration. Hence, the effect of dispersion has a strong impact on communication systems. Due to the decreasing amplitude, the signal to noise ratio in the transmission system is reduced. At the same time, if the broadening exceeds the guard time between channels, it comes to an interference between adjacent channels. To avoid intersymbol interference, the relation [52]

$$B\sqrt{|k_2|\, L} \leq 1/4 \tag{5.57}$$

has to be fulfilled if the source has a small spectral width, where B is the bit rate of the system and L is the fiber length. With (2.74), the group velocity dispersion (k_2) can be replaced by the dispersion parameter (D). According to (5.57), the total dispersion of the transmission link ($|k_2|\, L$) determines the maximum transmittable bit rate or, with a defined bit rate, the maximal bridgeable distance. In a standard single mode fiber with a dispersion parameter of $D = -17\ \text{ps}/(\text{nm} \times \text{km})$, the transmission distance

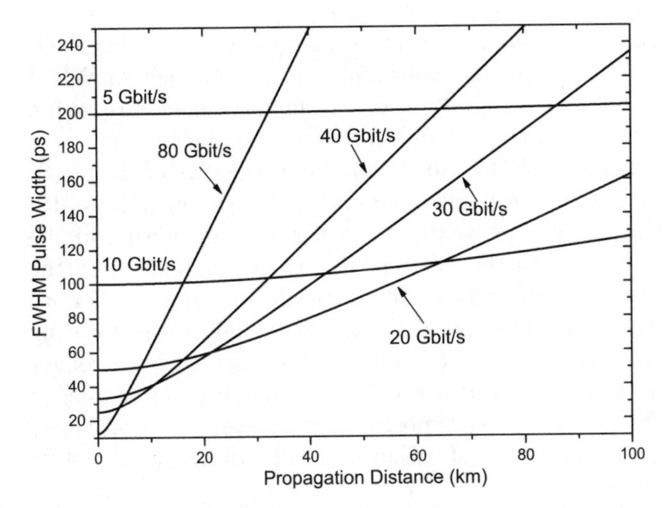

Fig. 5.3. Dispersive pulse broadening against the propagation distance in a standard single mode fiber. For the dispersion parameter, the value for pure silica at a wavelength of 1.55 μm was assumed: $|D| = 21.89$ ps/(km×nm)

for a 10 Gbit/s system in the C-band is therefore $\approx$ 29 km, but it will be reduced to only 1.8 km for a 40 Gbit/s system.

According to (5.40) pulses with a short duration have a short dispersion length hence, its broadening is – due to (5.55) – stronger than the broadening for longer pulses. In a standard single mode fiber ($D = -17$ ps/(nm × km)) a pulse with a duration of 1 ps is therefore already broadened by a factor of $\sqrt{2}$ after a distance of only 46 m.

Figure 5.3 shows the temporal broadening in dependence on the propagation distance for a transmission system with different bit rates. An optical time division multiplex (OTDM) non-return-to-zero system with bandwidth limited pulses and a pulse width (τ_{FWHM}) that is simply determined by 1/bit rate is assumed. As can be seen from the figure, for 5 and 10 Gbit/s systems with a carrier wavelength of 1.55 μm in a standard single mode fiber, the dispersive broadening of the pulses is negligible for a propagation distance of 100 km. The dispersion length for these cases is 517 km and 129 km, respectively. On the other hand, the behavior is completely different at higher bit rates. A transmission system of 80 Gbit/s has, for instance, a dispersion length of only 2 km. Hence the pulses experience very strong broadening only after short distances. After a propagation distance of 32 km, the pulse width of the 80 Gbit/s system exceeds that of a 5 Gbit/s system. Therefore, the propagation of this bit rate is impossible under these conditions.

In principle, three possibilities to transfer high bit rates over optical waveguides exist. The first is to use a short repeater distance, but this approach is very cost intensive. The second exploits the zero dispersion of the group velocity dispersion in fibers. In this case, higher order dispersion terms have to be included in the consideration and the efficiency of nonlinear effects, especially the FWM is very high, thus limiting the approach for WDM systems in particular. The last possibility uses a technique called "dispersion management" that will be treated in detail in the next section.

According to the equations (5.40) and (5.55), the broadening of the pulses is independent of the sign of the dispersion parameter. Hence, the pulse width will be increased for a positive, as well as for a negative, dispersion in the fiber. Only the manner in which the different frequency components of the pulse are altered depends on the sign of the dispersion parameter. Due to (5.45), the phase of the pulse envelope in the frequency domain is given by $\varphi_\omega = 1/2 k_2 \omega^2 z$. Hence, higher frequency components experience a stronger phase shift during their propagation in the fiber than the lower frequency ones. This phase shift is positive if k_2 is positive, otherwise it is negative.

With (5.40) and (5.53) the phase of a Gaussian pulse during its propagation in an optical fiber is

$$\varphi(z,T) = \operatorname{sgn}(k_2)\frac{1}{2}\arctan\left(\frac{z}{L_\mathrm{D}}\right) - \frac{T^2}{2\tau_0^2}\frac{\operatorname{sgn}(k_2)\frac{z}{L_\mathrm{D}}}{\left(1+\left(\frac{z}{L_\mathrm{D}}\right)^2\right)}, \tag{5.58}$$

where $\operatorname{sgn}(k_2)$ determines again the sign of the group velocity dispersion (GVD). For non-dispersion-shifted fibers, the zero-crossing of the dispersion is at a wavelength of ≈ 1.3 µm. Therefore, if the carrier wavelength of the pulses is smaller, then k_2 is positive and vice versa, as shown in Fig. 2.11. According to (5.58), the phase dependence on the time T is a parable. The higher frequency components experience, due to (5.45), a stronger positive phase shift than the lower one if the dispersion is positive. In the case of negative dispersion, this behavior is inverted. The momentary frequency of the pulse is the temporal derivation of the phase. Hence, it follows that

$$\mathrm{d}\omega = -\frac{\mathrm{d}\varphi}{\mathrm{d}T} = \frac{\operatorname{sgn}(k_2)}{\tau_0^2}\frac{\frac{z}{L_\mathrm{D}}}{\left(1+\left(\frac{z}{L_\mathrm{D}}\right)^2\right)}T. \tag{5.59}$$

The negative sign in (5.59) is a result of the negative frequency in (5.3). According to (5.59), the momentary frequency changes linearly along the time T and depends on the propagation distance, the pulse width, and the dispersion length of the fiber. Figure 5.4 shows this behavior for a positive and a negative dispersion.

According to Fig. 5.4, a positive GVD[1] leads to a shifting of the lower frequency components to the pulsefront ($T < 0$), whereas the higher frequency components are shifted to the rear side of the pulse ($T > 0$). Hence, in the case of a positive GVD, the front of the pulse is red-shifted, whereas the rear

[1]Positive GVD means that k_2 is positive, the dispersion parameter D is negative, and the pulse propagates in the range of normal dispersion.

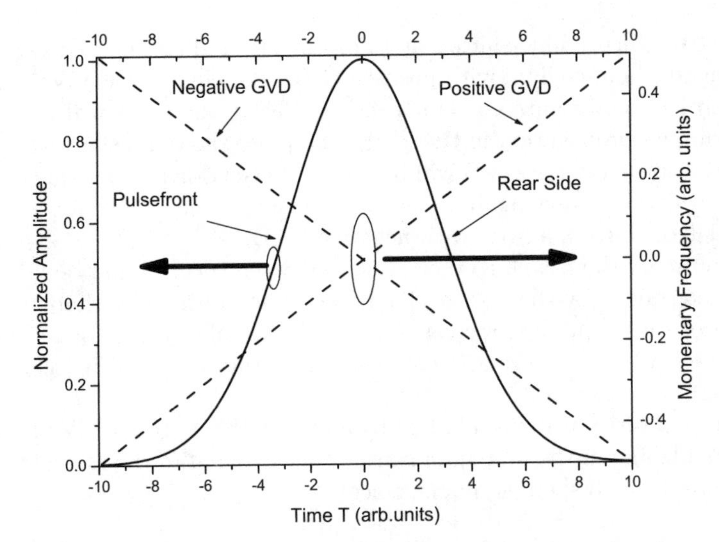

Fig. 5.4. Amplitude and momentary frequency of a pulse with Gaussian shape as it propagates through an optical fiber with positive and negative GVD. A negative time T describes the pulse components that will arrive before the maximum. A negative momentary frequency describes a frequency lower than the mean frequency of the pulse

side will be blue-shifted, as can be seen from the inset of Fig. 5.3. On the other hand, if the carrier frequency of the pulse lies in the range of negative GVD, the pulsefront will be blue- and the rear side red-shifted. The frequency modulation of the pulse is called a "chirp".

Due to the dispersion, the different spectral components of the pulse have different velocities in the fiber. They incur out of phase and the duration of the pulse will be increased. This behavior is the same for a positive and a negative dispersion, only the direction differs.

Short pulses, for instance, in the time range of a few femtoseconds (10^{-15}s), have a large number of Fourier components. The spectral width can be broadened additionally with nonlinear effects. The spectrum can reach over the entire spectral range of an optical fiber. Due to the GVD of the waveguide, it is possible to shift the different spectral components in time. Hence, a fast modulator at the end of the fiber can modulate the different wavelengths independently. This is a source of a large number of WDM channels with only one laser (see Sect. 12.1).

Until now, it has been assumed that the Fourier components of the pulse are in phase at the fiber input. In practical laser diodes, the emitted frequency depends on the injection current. Hence, the intensity modulated pulses show an additional frequency modulation, a chirp. As a result, the different Fourier components are already out of phase at the fiber input. This chirp can be compensated or increased by the dispersion of the fiber. If a pulse with negative chirp (blue-shifted at the front and red-shifted at the rear side) propagates in a medium with positive GVD, then due to the GVD the different Fourier components are slid together. Hence, the pulse will be compressed during its propagation in the fiber. The pulse shows its shortest width if the initial chirp and the GVD will balance out each other. After that point the pulse width increases again. The same behavior takes place for a pulse with a positive chirp in a fiber with negative GVD.

For optical pulses without a chirp the relation between spectral and temporal width is determined by the Fourier Transformation of the pulse shape. If the spectrum of such a pulse is increased by nonlinear effects, it is possible to exploit the GVD for a dramatic compression of the pulse width (see Sect. 12.1).

If a fiber with a positive GVD is followed by a fiber with a negative GVD, the temporal broadening in the first span can be compensated in the second one. This is the basis of "dispersion management".

5.4 Dispersion Management

In the case of a positive GVD, the front of the pulse is red-shifted whereas the rear side will experience a blue-shift. The Fourier components of the pulse propagate with different velocities and as they incur out of phase, the pulse will be temporally broadened. In the case of a negative GVD, the lower frequency components of the pulse are faster than the red. Hence, the front of a pure Gaussian pulse will be blue-shifted during propagation in a material with negative dispersion, whereas the rear side will be red-shifted. In consequence, the Fourier components will incur out of phase as well and the pulse will be broadened.

On the other hand, if a fiber with positive GVD is followed by a fiber with a negative GVD, it will be possible for both effects to balance out. Therefore, in the fiber with positive GVD, the pulse will be temporally broadened and in the fiber with negative GVD, it will be compressed again. The temporally pulse shape breathes. If both effects are completely equal and all other effects, especially the nonlinear, are neglected, the pulse has again its original shape. After that it will be broadened again. This is the so called "dispersion management" (DM).

In systems consisting of standard single mode fibers, the negative dispersion, required for dispersion compensation, can be delivered by dispersion-compensating modules. They consist of specially designed fibers with high negative dispersion ($D < -70$ ps/(nm$\times$km)) furled in a ring. The fibers are called dispersion-compensating fibers (DCF). Another possibility for dispersion compensation is offered by chirped fiber Bragg gratings. Using a mechanical stretching of the fiber containing the adjustable fiber Bragg grating, it is possible to vary the dispersion between -404.5 ps/nm and $+445.9$ ps/nm

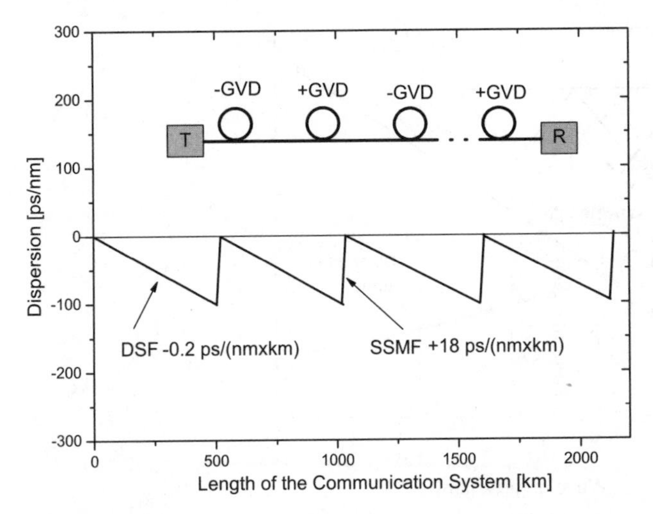

Fig. 5.5. Dispersion management in a transmission system by fibers of different sign of the GVD. The total accumulated dispersion over the propagation distance is zero

[53]. The special advantage of dispersion-compensating modules is that they can be introduced into existing transmission systems.

For long haul transoceanic optical transmission, the first generation of dispersion management was used in 5 Gb/s single channel systems. In these systems a dispersion-shifted fiber with a negative dispersion of -0.2 ps/(nm×km) was used for the main transmission. After several hundreds of kilometers, the accumulated dispersion was periodically compensated by the positive dispersion of a standard single mode fiber (+18 ps/(km×nm)). Figure 5.5 shows the dispersion map against the propagation distance for such a first generation dispersion-managed system.

The figure shows that the total accumulated dispersion along the system length is zero but, due to the fact that the dispersion parameter D depends on the wavelength, this holds for just one carrier wavelength only. Therefore, for WDM systems, the dispersion slope must be taken into consideration. The dispersion parameter for pure silica, together with the dispersion slope, in the range of 1.4 to 1.6 μm (including the C- and L-band) is shown in Fig. 5.6. As can be seen, the dispersion is not perfectly linear. For a WDM system, the dispersion has to be compensated for a rather broad wavelength range. Hence, it depends on the slope of the GVD if the dispersion can be compensated for all wavelength at once. The inset in Fig. 5.6 shows the dispersion in a DM transmission system for several WDM channels. As shown, only the central wavelength will experience a complete compensation of the dispersion, whereas for the longest and shortest wavelengths a positive and a negative dispersion will remain, respectively. This dispersion accumulates with distance.

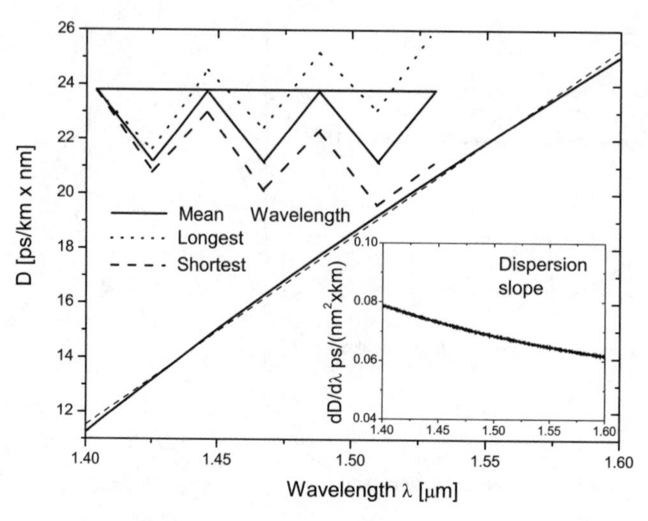

Fig. 5.6. Dispersion parameter (D) for pure silica in the range between 1.4 and 1.6 μm. In the right inset the dispersion slope ($\mathrm{d}D/\mathrm{d}\lambda$) for silica is shown. The left inset shows the accumulation of dispersion for the channels of a WDM system

For pure silica the dispersion slope at a wavelength of 1550 nm is, according to the inset of Fig. 5.6, $\mathrm{d}D/\mathrm{d}\lambda = 0.064$ ps/(nm$^2\times$ km). A WDM channel 10 nm off the systems zero dispersion the channel will then show an accumulated dispersion of more than 6000 ps/nm after a distance of 10 000 km. This is especially true in long transmission systems like undersea cable, or in systems with many WDM-channels – exploiting more than one optical window – where the dispersion slope cannot be neglected.

In order to reduce the degradation of system performance due to the influence of the dispersion slope, nonzero-dispersion-shifted fibers (NZ-DSF) with reduced dispersion slope are introduced into long haul systems. These fibers have higher negative dispersion than DSF (-2 to -3 ps/(nm×km)). For an additional reduction of nonlinear effects, fibers with large effective areas are used together with the NZ-DSFs in a hybrid configuration. The fibers with the large core areas will be positioned after the amplifiers when the signal power is high. These fibers have negative dispersions of -2 to -3 ps/(nm×km) as well but, unfortunately rather high dispersion slopes. The NZ-DSF with a reduced dispersion slope but small effective area is introduced when the signal power is lower. The dispersion compensation is then again carried out with an SSMF which will typically be incorporated after a distance of around 500 km, which corresponds to an accumulated dispersion of -1000 ps/km [100]. Contrary to the first generation of DM systems, the location of the first SSMF is now at half the distance of the dispersion period so that the accumulated dispersion is half of that in Fig. 5.5 and the sign alternates during the compensation period. For long transmission distances, the dispersion at the edges of the WDM band is still rather large. This can be reduced with an additional pre- and post-compensation for each channel.

For an additional increase of the transmittable bandwidth and a decrease of nonlinear effects in long haul systems, fibers with ultralarge effective areas of more than 100 μm^2 (A_{eff} enlarged positive dispersion fiber EE-PDF) were introduced together with slope-compensating, dispersion-compensating fibers (SCDCF) [100, 101]. The EE-PDF has a large positive dispersion of +20 to 22 ps/(nm×km), whereas the SCDCF shows a large negative dispersion between −40 and −60 ps/(nm×km). The enhanced effective area reduces the intensity in the fiber core and, therefore, the strength of nonlinear effects. At the same time, the high local dispersion leads to a strong phase mismatch between the channels (Sect. 4.10) and a high walk-off parameter. As already mentioned, nonlinear effects like four-wave-mixing depend on the matching of the phases between the original and the new generated waves. This phase matching in turn depends on the dispersion in the fiber. If the dispersion is high, the phase mismatch is high as well and, therefore the efficiency of FWM is very low. The high walk-off parameter leads to a reduction of the XPM impact (see Sect. 6.2.2).

A problem of dispersion-managed systems is the additional influence of nonlinear effects to the pulse shape, as will be discussed in Sect. 6.1.1 in detail. Furthermore, a conventional dispersion-compensating fiber (DCF) has a very small effective area on the order of 20 μm^2 or less and, therefore, a higher nonlinearity coefficient γ than standard single mode fibers. Hence, the nonlinear effects are more efficient. Due to the nonlinearity, the positive dispersion does not have to be completely compensated by a negative one. The required compensation values for a given transmission system are rather complicated to calculate. Hence, numerical simulations including all important parameters, and the fine tuning of the real world system is required for good dispersion compensation results [102, 103].

Contrary to DM systems where nonlinearity will be suppressed as good as possible in DM Soliton systems (DMS) the nonlinearity is essential for the propagation of the pulses. In these systems a balance between linear and nonlinear effects is required for pulse propagation. DMS are treated in Sect. 9.6 in detail.

DM systems with a high local dispersion can suppress interchannel nonlinear effects but, for high bit rates, nonlinear effects inside the distinct channels are very effective (intrachannel, see Sect. 8). Especially important for these interactions is the parameter τ_{FWHM}/T, with τ_{FWHM} as the FWHM pulse width and T as the reciprocal bit rate or the time spacing between adjacent pulses. If the pulse width is increased, the interaction between the pulses will increase due to the local dispersion in the spans as well. In order to keep the interactions low the pulses should not overlap in classical DM systems. Therefore, the maximum pulse width the pulses can reach during transmission should be very small. Hence, the maximum allowable dispersion compensation distance ($L_{\mathrm{map}}\tilde{}T^2/|D|$) decreases quadratically with the bit rate.

On the other hand, it was shown that the interaction reaches a maximum if $\tau_{\mathrm{FWHM}}/T \approx 1$ [90], if the pulse width is increased further the effectivity of the intrachannel interaction decreases. Therefore, in the last few years DM transmission systems with a strong overlap of adjacent pulses is widely accepted as a key technology for high bit rate optical systems.

5.5 The Nonlinear Term

The last sections considered the pulse propagation in the linear regime. In principle, it leads to a temporal broadening of the pulses due to the wavelength dependence of the refractive index. In this section, the linear influence should be neglected, hence only the influence of the fiber nonlinearity will be considered. The starting point is again the NSE (5.36). In this case the linear term is zero ($k_2 = 0$), so the equation remains as:

$$\frac{\partial B}{\partial z} = \mathrm{j}\gamma \left|B\right|^2 B. \tag{5.60}$$

This equation is directly solvable in the time domain. The solution is again a function of z that remains the same after a derivation, hence

$$B(z,t) = B(0,t)\mathrm{e}^{\mathrm{j}\gamma|B(0,t)|^2 z}. \tag{5.61}$$

where $B(0,t)$ is the envelope of the pulse at the fiber input ($z = 0$). As can be seen, the phase of the pulse will change during the propagation in the nonlinear waveguide and it depends on $|B(0,t)|^2$. According to (5.10), the intensity of the pulse is proportional to the square of the absolute value of the envelope. Hence, the phase alteration depends on the intensity of the pulse. Since its own intensity is responsible for its phase alteration, this phenomenon is called self-phase modulation (SPM). This will be treated in detail in Chap. 6.

According to (5.61), in spite of the phase alteration, the shape of the pulse remains constant. If again a Gaussian shape is assumed for the envelope of the input pulse $B(0,t)$ (5.46), the Fourier transform of the pulse at a point z in the waveguide is

$$\tilde{B}(z,\omega) = \frac{1}{\sqrt{2\pi}} \int_{-\infty}^{+\infty} B(0,T)\mathrm{e}^{\mathrm{j}\left(\gamma|B(0,t)|^2 z + \omega t\right)} \mathrm{d}T. \tag{5.62}$$

Using (5.46), it can be written as

$$\tilde{B}(z,\omega) = \frac{1}{\sqrt{2\pi}} \int_{-\infty}^{+\infty} \mathrm{e}^{-\frac{T^2}{2\tau^2}} e^{\mathrm{j}\left(\gamma e^{-\frac{T^2}{\tau^2}} z + \omega t\right)} \mathrm{d}T. \tag{5.63}$$

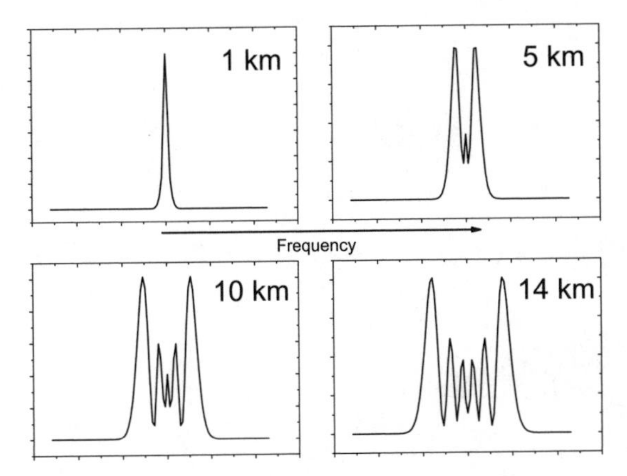

Fig. 5.7. Spectra of a 25 ps pulse at different propagation distances in a standard single mode fiber (nonlinearity coefficient $\gamma = 1.3\ \text{W}^{-1}\text{km}^{-1}$, Power = 1W). Note that the amplitude value of each graph is normalized to its maximum

Hence, in the spectrum of the pulse, additional frequency components will appear. The spectral broadening of the pulse depends on the nonlinearity of the waveguide (γ), the intensity of the pulse ($|B(0,t)|^2$), and the propagation distance (z). The result is a function with several maxima. Figure 5.7 shows the spectra of a pulse with a duration of 25 ps after a propagation distance of $1, 5, 10$, and 14 km in a standard single mode fiber without dispersion.

The linear effect of dispersion led to a broadening of the pulse in the time domain. The different Fourier components of the pulse propagated with different velocities, but no new frequencies were generated in the linear regime. Whereas in the nonlinear regime, the pulses were broadened in the frequency domain and new frequency components were generated via the intensity dependence of the phase of the pulse. If only the nonlinear regime is considered, the shape of the pulse remains constant.

5.6 Numerical Solution of the NSE

For a real investigation of the propagation of pulses in a waveguide, both terms as well as the attenuation have to be included in the NSE. The NSE was solved for the first time in the year 1971 with the inverse scattering theory by Zakharov and Shabat [211]. But an analytical solution is difficult and only possible with strong simplifications.

On the other hand, a variety of simulation programs are commercially available. The core of these programs consists in most cases of a numerical simulation of the NSE. In these programs, the NSE is solved with a method called "split-step Fourier" analysis. The basic idea behind this method is in principle the same as what we have exploited in the last sections. The

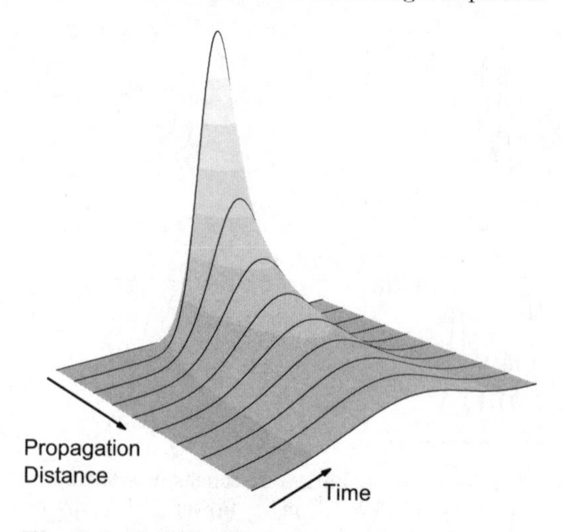

Fig. 5.8. Temporal broadening of the pulse in the time domain due to the dispersion of the waveguide. The nonlinearity is neglected here

waveguide is separated into N equally long subsections, where each subsection has the length l. For each subsection, the nonlinear and dispersive behavior are calculated independently. This simplifies the consideration drastically. According to the fact that the dispersive phenomena can be better solved in the frequency domain, whereas the nonlinear effects are easier to solve in the time domain, the pulse will be transformed with the fast Fourier transform (FFT) into the frequency domain in order to calculate the linear term for the distance l. After that the inverse FFT is used to transform the pulse back into the time domain to calculate its nonlinear behavior along the subsection. As long as the length of the subsections is short enough, this method can deliver very accurate results. On the other hand, for long transmission systems or a large number of WDM channels the simulation can be very time consuming.

The independent linear and nonlinear behavior of a pulse in a dispersive nonlinear waveguide is again shown in the Figs. 5.8 and 5.9.

Summary

- The expansion of the wave equation in a Taylor series leads to the nonlinear Schrödinger equation if only the cubic term will be considered.
- The NSE describes the propagation of the envelope of a pulse in a nonlinear dispersive fiber.
- The linear term of the NSE results in the temporal broadening of the pulses. The different Fourier components of the pulse propagate with different velocities, but no new frequency components are generated.

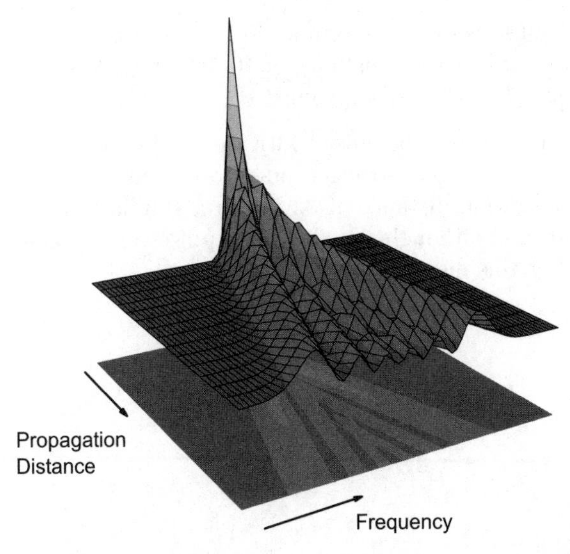

Fig. 5.9. Broadening of a pulse in the spectral domain due to the nonlinearity of the waveguide. The dispersion is neglected here

- The sorting of the Fourier components in the pulse depends on the sign of the dispersion.
- For a positive dispersion, the front of the pulse is red- and the rear side blue-shifted.
- For a negative dispersion, the front is blue- and the rear side red-shifted.
- The temporal broadening due to the positive dispersion can be balanced with a waveguide with negative dispersion. This technique is called "dispersion management".
- The nonlinear term of the NSE leads to an intensity dependence of the phase of the pulse and hence, to a spectral broadening of the pulses.
- If only the nonlinear term is considered the pulse shape remains constant.
- The NSE can be numerically solved with the split step Fourier method.

Exercises

5.1. Determine the nonlinear Kerr constant of an HNL fiber if it has an effective area of $A_{\text{eff}} = 14\ \mu\text{m}^2$ and the nonlinear coefficient in the C-band is $\gamma = 18\ \text{W}^{-1}\ \text{km}^{-1}$.

5.2. You want to use a 2-km long HNL fiber ($\gamma = 18\ \text{W}^{-1}\ \text{km}^{-1}$, $D = 23$ ps/(km×nm)) as a nonlinear wavelength converter for the C-band. Dispersion effects in the device should play a negligible role. Estimate the parameters of the pulses you are able to convert.

5.3. Estimate the FWHM pulse width of a 10-and 40-Gbit/s transmission system in the C-band after a propagation distance of 10, 20, and 100 km in a standard single mode fiber ($|D| = 22$ ps/(nm×km)).

5.4. The pulse broadening depends on the pulse width. According to Fig.5.3 a 80 Gbit/s signal shows a still much stronger pulse broadening then a 5 Gbit/s signal after a propagation distance of 32 km in a standard single mode fiber. Why is this the case, although the 80 Gbit/s pulses are broader then the 5 Gbit/s pulses after this distance?

6. Self- and Cross-Phase Modulation

Self-phase modulation (SPM) and cross-phase modulation (XPM, or CPM) are two of the most important nonlinear effects in optical telecommunications. Both effects lead to a phase alteration of the pulses and are frequently called carrier-induced phase modulation (CIP). The alteration of the phase leads to a change of the pulse spectrum. In the case of SPM, the pulse changes its spectrum due to its own intensity; it will be broadened. Together with the dispersion of the material, this spectral broadening can lead to an alteration of the temporal width of the pulse. The sign of the GVD determines whether the width will be additionally broadened or compressed through the nonlinear effect. The broadening causes, of course, a degradation of system performance, but it can be reduced if the SPM is considered in system design. Dispersion management is especially important in this respect.

On the other hand, the SPM can be exploited for a number of applications as well. If the pulses are very short, the spectral broadening is very strong. This broad spectrum can be used as a source of WDM-channels. The compression of the pulse can generate ultrashort pulses in the femtosecond domain from rather long ps pulses.

The XPM is similar to the SPM but the origin of the spectral broadening of the pulses are other pulses propagating at the same time in the waveguide; they will mutually influence each other. This is especially important for WDM systems where a huge number of pulses with different carrier wavelengths will be transmitted in one fiber. The XPM is, in principle, a deterministic effect, as will be shown in this chapter. Hence, an intelligent system design should be able to cancel its influence on the signals. But, due to the fact that the bit pattern in the other channels is completely random, this will be impossible. The XPM is the fundamental effect that determines the capacity of optical transmission systems in Sect. 4.11.

In this chapter, the SPM will be treated first. The propagation equation will be derived from the NSE and their impact on communications systems will be shown theoretically as well as in experimental results. After that, the XPM will be treated in detail.

6.1 Self-Phase Modulation (SPM)

In the case of SPM no other waves are involved. Hence, for the theoretical description the NSE, derived in Sect. 5.1, is sufficient. Two important simplifications will be made: first, the effect of absorption is neglected; and second, the temporal pulse length is assumed to be rather large – or the dispersion of the fiber is assumed to be rather small – so that the considered fiber length is much less than the dispersion length. Hence, it is possible to consider quasi monochromatic waves and the second term on the right side of (5.37) vanishes. Therefore,

$$\mathrm{j}\frac{\partial B}{\partial z} = -\gamma \left|B\right|^2 B. \tag{6.1}$$

Contrary to Sect. 5.5, it is assumed here that the influence of the actual pulse shape can be neglected. As in the case of a monochromatic wave, according to (5.10), the absolute value of the amplitude squared corresponds to the power of the wave. If absorption is neglected, this is the input power of the pulse. Therefore, (6.1) can be written as

$$\frac{\partial B}{\partial z} = \mathrm{j}\gamma P B. \tag{6.2}$$

The solution of (6.2) is again very simple. For the envelope of the wave at a distance z in the waveguide,

$$B(z) = B(0)\mathrm{e}^{\mathrm{j}\gamma P z}, \tag{6.3}$$

where $B(0)$ is the pulse envelope at the fiber input. Due to (6.3), the slowly varying amplitude $B(z)$ changes its phase during the propagation in the waveguide and the phase alteration is proportional to the propagation distance z. According to (5.11) the total transmitted wave in the fiber is

$$E\left(z,t\right) = \frac{1}{2}\left[\frac{B(z)}{N}F(r)\mathrm{e}^{\mathrm{j}(kz-\omega t)} + \mathrm{c.c.}\right]. \tag{6.4}$$

Using (5.3) and (6.3), it can be written as

$$E\left(z,t\right) = \frac{1}{2}\left[\hat{E}\left(\boldsymbol{r},0\right)\mathrm{e}^{\mathrm{j}([\gamma P+k]z-\omega t)} + \mathrm{c.c.}\right]. \tag{6.5}$$

If the waveguide characteristics are neglected the wave number k is $n_0 k_0$, with $k_0 = \frac{\omega}{c}$ and $n_0 = n(\omega)$ as the linear, frequency dependent refractive index. The nonlinearity coefficient, γ, in (4.96) can be rewritten as:

$$\gamma = k_0 \frac{n_2}{A_{\mathrm{eff}}}. \tag{6.6}$$

Therefore, the wave in the fiber is represented by

$$E(z,t) = \frac{1}{2}\left[\hat{E}(\boldsymbol{r},0)\,\mathrm{e}^{\mathrm{j}\left(\left[\frac{n_2}{A_{\mathrm{eff}}}P+n_0\right]k_0 z-\omega t\right)} + \mathrm{c.c.}\right]. \tag{6.7}$$

The intensity of the wave is its power related to the effective area of the fiber $I = P/A_{\mathrm{eff}}$. Hence, the wave can also be expressed as

$$E(z,t) = \frac{1}{2}\left[\hat{E}(\boldsymbol{r},0)\,\mathrm{e}^{\mathrm{j}([n_0+n_2 I]k_0 z-\omega t)} + \mathrm{c.c.}\right]. \tag{6.8}$$

The phase of the wave is the argument of the exponential function and is represented by

$$\Phi(z) = (n_0 + n_2 I)\,k_0 z - \omega t. \tag{6.9}$$

According to (6.9), the phase alteration Φ of the pulse is proportional to its intensity I. The pulse modulates, due to the intensity-dependent refractive index, its own phase with its intensity. Hence, the name self-phase modulation is used to describe this phenomenon.

With the increasing demand for bandwidth in existing optical transmission systems, the bit rate of the channels will be increased. Therefore, the temporal duration of the pulses are rather short. At the same time, with optical amplification, the interaction length for nonlinear effects and the noise floor is increased dramatically. In order to achieve a particular signal to noise ratio, the intensity of the pulses will be increased. Hence, the short pulses have rather high intensities. Due to (6.9), the intense middle part of the pulse experiences a higher refractive index change than the edges. Therefore, the peak of the pulse undergoes a higher phase shift than the wings.

The upper part of Fig. 6.1 shows the electric field of a Gaussian pulse at the input of a fiber[1]. The bottom of Fig. 6.1 shows the phase shift of the envelope of the pulse due to SPM. As can be seen, the phase shift follows the intensity shape of the envelope.

The temporal derivation of the phase corresponds to the frequency of the wave. Hence, the frequency according to (6.9) and (5.59) is

$$\omega(z) = -\frac{\partial \Phi(z,t)}{\partial t} = \omega_0 - n_2 k_0 \frac{\partial I}{\partial t} z. \tag{6.10}$$

Therefore, along with the temporal change of the pulse intensity, the mean frequency or the mean wavelength will be changed. As described in (6.10) it will decrease if the pulse intensity increases, reach the original frequency in the pulse maximum, and will increase for the decreasing wing. In Fig. 6.2, this is shown for the spectral spreading, i.e., the difference between the momentary and the mean frequency of the pulse.

---

[1]For the slowly varying envelope approximation, required for the derivation of the NSE, the spectral width of the pulse had to be much shorter than its carrier frequency. Hence, the number of oscillations of the field is much higher than shown in Fig. 6.1.

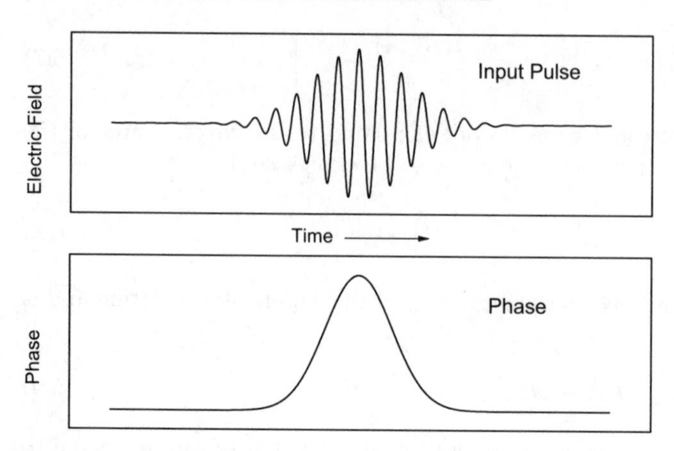

Fig. 6.1. Electric field of a Gaussian pulse at the fiber input (*top*) and phase shift due to SPM (*bottom*)

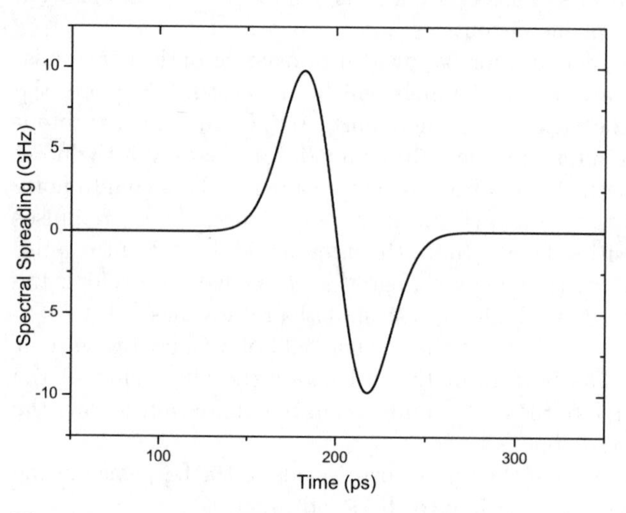

Fig. 6.2. Frequency alteration due to the self-phase modulation of the pulse. The pulse has a temporal width of 25 ps and a peak power of 10 mW. The effective length of the fiber is 22 km, and the use of a standard single mode fiber is assumed here ($\alpha = 0.2$ dB/km, $\gamma = 1.3$ W^{-1}km^{-1})

Figure 6.3 shows the electric field of the input and the spectrally widened pulse due to SPM. Note the similarity to the effect of GVD, mentioned in Sect. 5.3. In the case of SPM, new frequency components are generated whereas for the GVD, only a rearrangement of the existing frequency components takes place.

Due to SPM, the leading edge is stretched. The peak maintains its original frequency and the tailing edge is compressed. For the frequency, this means

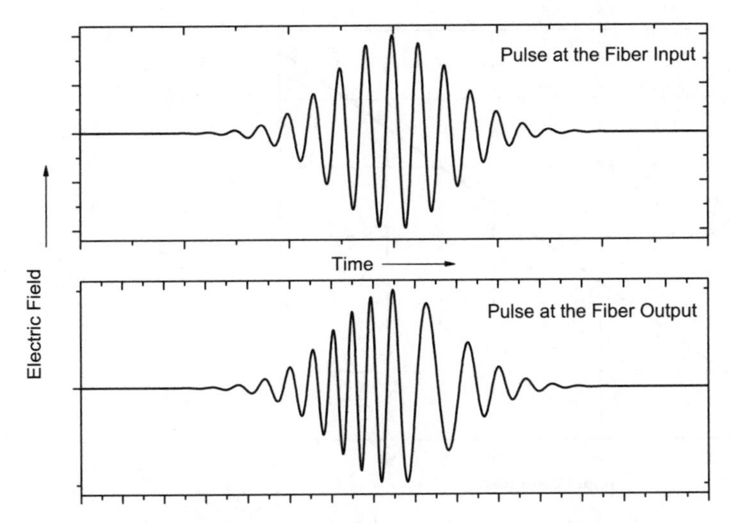

Fig. 6.3. Electric field of a pulse at the fiber input and output. Due to the SPM in the fiber the frequency of the pulse is changed. Note that the pulse here propagates from the left to the right

the leading edge is shifted to the red whereas the tailing edge experiences a blue-shift.

6.1.1 SPM's Impact on Communication Systems

The nonlinear part of the phase alteration due to SPM is, according to (6.9)

$$\Phi_{\mathrm{NL}} = n_2 I k_0 z. \tag{6.11}$$

Introducing A_{eff} into (6.11) and using (6.6), the phase change at the output of a fiber with length L is

$$\Phi_{\mathrm{NL}} = \gamma P L. \tag{6.12}$$

Until now the influence of attenuation has been neglected. The attenuation leads to a reduction of the intensity and hence, a decreasing strength of nonlinear effects along the fiber span. Therefore, (6.12) can be written as

$$\Phi_{\mathrm{NL}} = \gamma P L_{\mathrm{eff}} \tag{6.13}$$

where L_{eff} is the effective length according to (4.98). For an input power of 1 W, one yields a phase shift of $\pi/2$ after a distance of $\approx$ 1.2 km in a standard single mode fiber (nonlinearity coefficient $\gamma = 1.3\ \mathrm{W}^{-1}\mathrm{km}^{-1}$, attenuation $\alpha = 0.2$ dB/km). Whereas, for the same phase shift, an effective length of 1200 km is required in systems with an input power of 1 mW. This is much more than the limiting value of the effective length in single mode

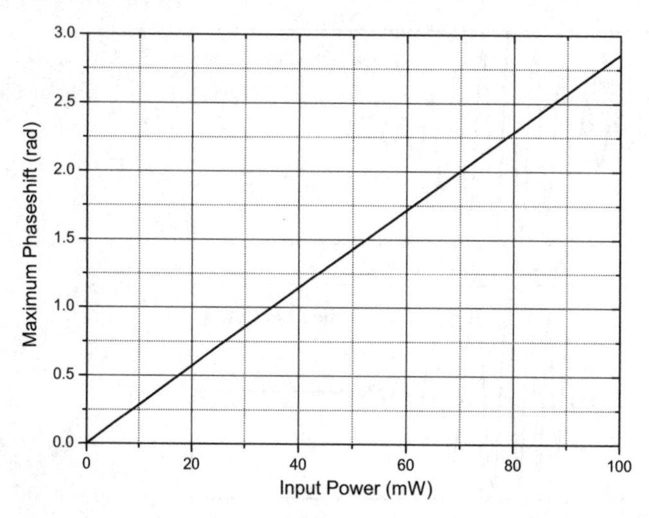

Fig. 6.4. Maximum phase shift for an unamplified system versus input power. (Attenuation constant $\alpha = 0.2$ dB/km, nonlinearity coefficient $\gamma = 1.3\ \mathrm{W}^{-1}\mathrm{km}^{-1}$)

fibers (≈ 22 km, Sect. 4.9). Hence, if no optical amplifiers are present in the system it follows a phase shift of at most ≈ 0.03 rad for a signal power of 1 mW. The maximum phase shift for transmission systems without optical amplifiers in dependence on the input power is shown in Fig. 6.4.

If optical amplifiers are present in the system, the relations are completely different. In this case, the effective length depends on the number of amplifiers. Hence, if the system is long enough small input powers can cause a strong phase shift.

The influence of the phase shift on the transmission system depends on the modulation format of the carrier. In phase shift keying (PSK) systems the information lies in the change of the carrier phase. In binary (B) PSK systems, the phase alters between $+\pi/2$ and $-\pi/2$, corresponding to a logical "1" and "0". Phase noise leads to a degradation of the eye pattern and hence, directly to a reduction of the signal to noise ratio. On the other hand, SPM requires a change of the intensity in order to produce phase noise. Strong intensity alterations can happen if semiconductor lasers are directly phase-modulated. For an external modulation the intensity fluctuations are very slight, hence, the influence of SPM can be neglected.

Large intensity fluctuations on very short time scales can be found in intensity-modulated systems. An alteration of the phase causes an alteration of the pulse frequency as well. According to (6.13), the spectral broadening ΔB of the pulse at the fiber output is

$$\Delta B = \frac{\partial \Phi_{\mathrm{NL}}}{\partial t} = \gamma L_{\mathrm{eff}} \frac{\partial P}{\partial t}. \tag{6.14}$$

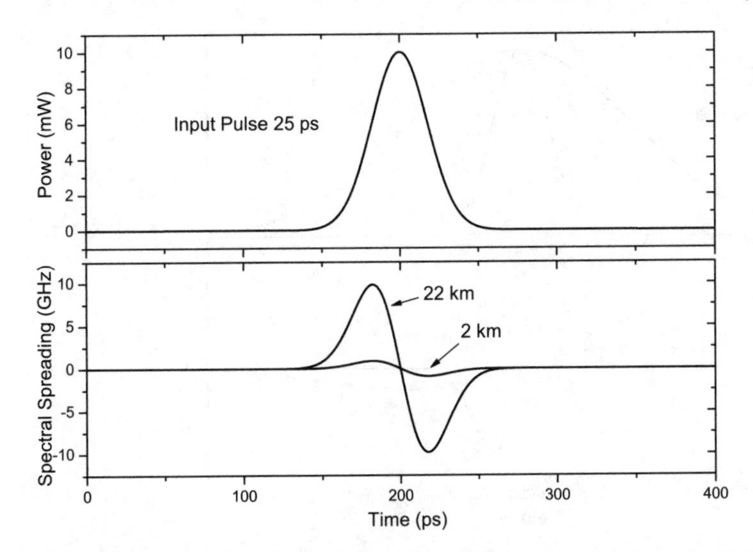

Fig. 6.5. Spectral broadening of a pulse with temporal durations of 25 ps after a propagation distance of 2 and 22 km (effective length) in a standard single mode fiber ($\alpha = 0.2$ dB/km, $\gamma = 1.3\ \mathrm{W}^{-1}\mathrm{km}^{-1}$)

Figure 6.5 shows the spectral broadening of an intensity-modulated pulse with a temporal width of 25 ps after a propagation distance of 2 and 22 km (effective length) in a standard single mode fiber.

As can be seen from (6.14) the spectral broadening depends not only on the effective length but on the temporal alteration of the power of the input pulse as well. Short pulses are much more affected than long pulses due to SPM. In Fig. 6.6, the spectral broadening of two Gaussian pulses with durations of 25 and 100 ps in a fiber with maximum effective length (22 km) are shown. According to the figure, the long pulse has a maximum broadening of $\Delta B \approx 5$ GHz, whereas the shorter pulse shows an additional frequency broadening of $\Delta B \approx 20$ GHz.

Until now the linear effect of dispersion has been neglected in the description of SPM. As discussed in Sect. 5.3, it leads to different velocities of the different spectral components contained in the pulse and hence, a temporal broadening of the pulse in the fiber. As can be seen from Fig. 5.4, the red Fourier components propagate faster than the blue ones in the case of normal dispersion. Therefore, the pulse shows the red components at the leading edge whereas the blue components will travel at the tailing wing. Due to SPM, the leading edge is also shifted to lower frequencies, while the tailing wing will be shifted to higher frequencies. Hence, both effects – nonlinearity and dispersion – act in the same direction and lead to a stronger temporal broadening of the pulse than the dispersion alone. Therefore, in the case of normal dispersion, SPM, together with, GVD, leads to a reduction of the transmittable bit rate.

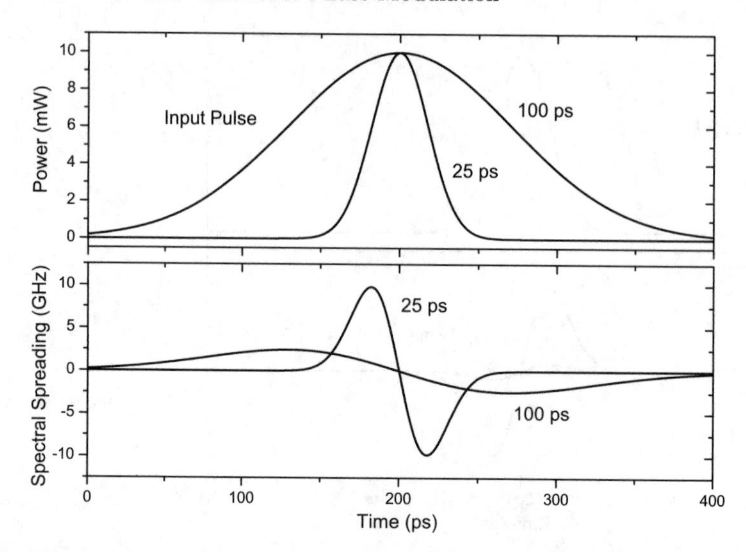

Fig. 6.6. Spectral broadening of two pulses with temporal durations of 25 and 100 ps. The propagation distance in a standard single mode fiber is 22 km (effective length). The other parameters are: $\alpha = 0.2$ dB/km, $\gamma = 1.3$ $\mathrm{W}^{-1}\mathrm{km}^{-1}$)

If the pulses, spectrally-broadened by SPM, are transmitted in the range of an anomalous dispersion instead the behavior is completely different. The red-shifted leading edge of the pulse is slowed down by the dispersion and propagates in the direction of the pulse center. At the same time, the blue-shifted tailing wing speeds up due to the GVD and travels in the direction of the pulse center as well. Therefore, GVD and SPM act in different directions, resulting in a compression of the pulse.

In the range of anomalous dispersion it is possible to find an amplitude or intensity value of the wave where the linear effect of dispersion and the nonlinear effect of SPM can cancel each other completely. These are the so called solitons treated in detail in Chap. 9.

A common method to suppress the fiber dispersion is dispersion management, discussed in Sect. 5.4. The dispersion-managed system consists of alternate spans with positive and negative GVD. If only the linear case is included in the calculation of the transmission system, an overcompensation of the pulses results. The SPM leads to an additional compression in the fiber spans with a negative compensation while an additional broadening takes place in the fiber spans with positive GVD. But both effects will not cancel each other. Furthermore, the nonlinearity coefficient is higher in DCF than in standard single mode fibers. Due to this behavior, the pulses at the fiber output are broader than at the fiber input even though the linear dispersion is completely compensated. The effect accumulates over distance and is shown in Fig. 6.7.

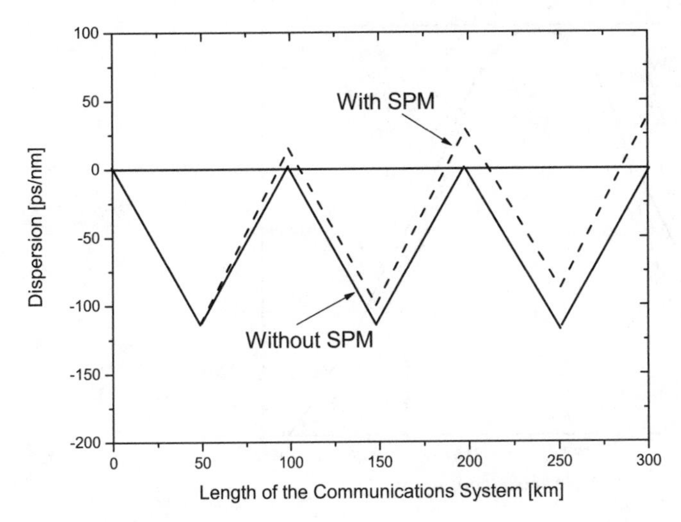

Fig. 6.7. Dispersion management in a transmission system. The SPM in anomalous dispersive fibers leads to an additional contribution to the GVD

The total net dispersion of the system depends on the SPM, and hence the intensity of the pulses. For optimum results, the normal dispersion may not be completely compensated by the anomalous dispersion. The calculation requires computer simulations, including the nonlinearity of the fiber and the fine tuning of the existing system.

6.1.2 Experimental Results

The graphs in this section are based on experimental data from the diploma thesis of Matthias Wolf [141]. The experiments were carried out at Siemens AG in Munich, Germany. The experimental set-up is shown in Fig. 6.8.

In a standard transmitter card for optical transmission systems (OCR 192), low noise NRZ[2] pulses with a data rate of 10 Gbit/s were generated.

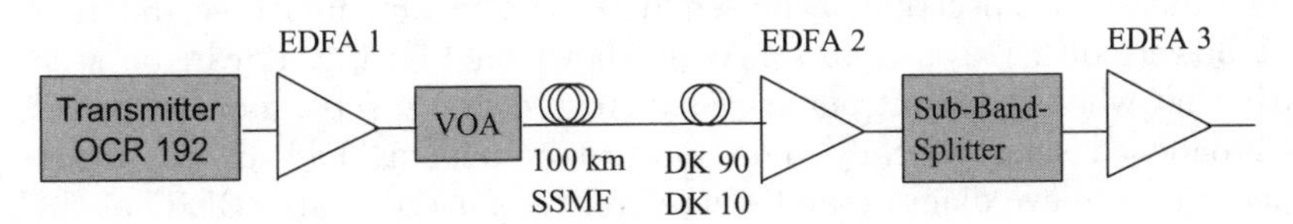

Fig. 6.8. Experimental set-up for the investigation of the influence of SPM on transmission systems (VOA = variable optical attenuator)

[2]NRZ = non-return-to-zero. The pulses do not go back to zero in the time interval of one bit. On the contrary, return-to-zero (RZ) pulses will fall back during the pulse time. An NRZ signal has a lower required bandwidth compared to an RZ signal but the clock recovery is more complicated. With RZ signals, the clock is directly transmitted.

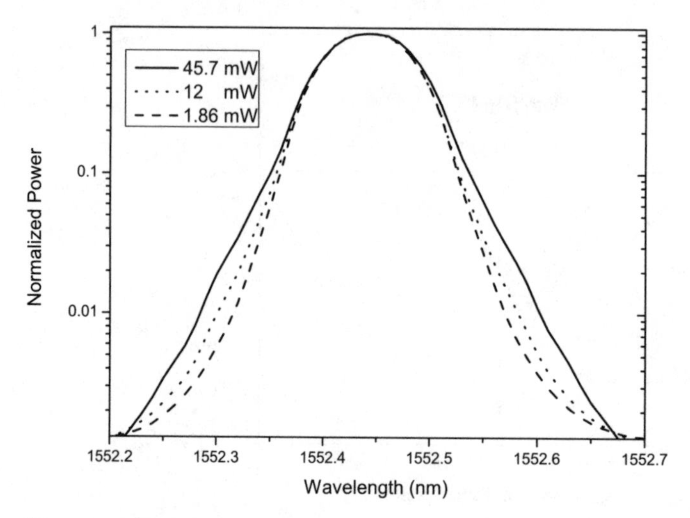

Fig. 6.9. Normalized, measured, output spectrum of three pulses of different bit streams with variable input power

In order to do this the output of a semiconductor laser, with a constant wavelength of 1552.52 nm, was modulated externally by a Mach-Zehnder modulator.[3] The modulated bit stream was then amplified in an erbium-doped fiber amplifier (EDFA1). In order to investigate the SPM dependence on the input power, it was varied by a variable optic attenuator (VOA). The transmission link consisted of a 100 km long standard single mode fiber (SSMF) with an attenuation constant of 0.22 dB/km.

For the dispersion compensation, two commercially available modules (DK 90 and DK 10) were deployed. Both consisted of dispersion-compensating fibers. The first compensated the GVD in 90 km of the SSMF, whereas the second compensated the residual 10 km at a wavelength of 1550 nm. At the fiber output the signals were amplified by a second optical amplifier (EDFA2). In order to minimize the noise of the signal, the amplification was divided into two stages, with a sub-band-splitter in between.

The measured spectrum, achieved with this set-up, for three different input powers of 2, 12 and 45.7 mW is shown in Fig. 6.9. It can be seen clearly that when the input power was increased, the spectrum of the pulses was broadened symmetrically. The spectral broadening had an important influence on the eye diagram and, therefore, the bit error rate (BER) of the

[3]In a Mach-Zehnder modulator the input signal is broken down into two branches. The phase of the wave in one of the branches can be varied, by the Pockels effect, for example. After that, both waves interfere at the output of the device. If the phase shift between both is π, the output signal vanishes, if it is 0 they interfere constructively.

With modulators based on the electro-optic effect, bandwidths of several 10 GHz are possible.

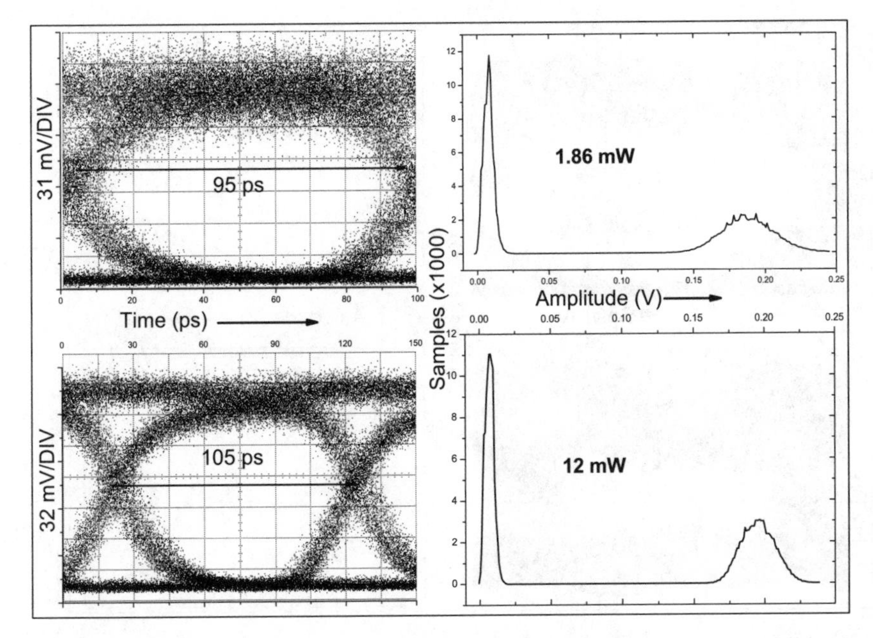

Fig. 6.10. Eye diagram and corresponding amplitude sampling values measured at the output of the transmission system for input powers of 1.86 and 12 mW. Note the different time scale of the abscissae

transmission system. The measured eye diagrams[4] and the corresponding amplitude sampling values at the output of the transmission system for input powers of 1.86 and 12 mW are shown in Fig. 6.10.

As can be seen from a comparison of both eye diagrams, SPM increased the temporal broadening. This happened because the dispersion compensation of the transmission system was optimized without consideration of nonlinear effects. Hence, there was a residual net dispersion which depended on the pulse intensity.

On the other hand, the signal with the higher power showed a much better signal to noise ratio. But when the input power was increased further it led to a distortion of the pulses and, hence, a drastic deterioration of the eye diagram, as can be seen in Fig. 6.11. In transmission systems, this would be directly connected with an increase of the bit error rate (BER) and, therefore, a decrease of the system performance.

As can be seen from the amplitude histograms, the noise influence was not stochastic. In particular, the sampling values of a logical "1" showed a distribution shifted to lower values and hence, there was no Gaussian shape.

[4] An eye diagram of a digital signal will be generated if all possible signal changes ($0 \rightarrow 0, 0 \rightarrow 1, 1 \rightarrow 1, 1 \rightarrow 0$) are superimposed. This can be done with a memory oscilloscope for example.

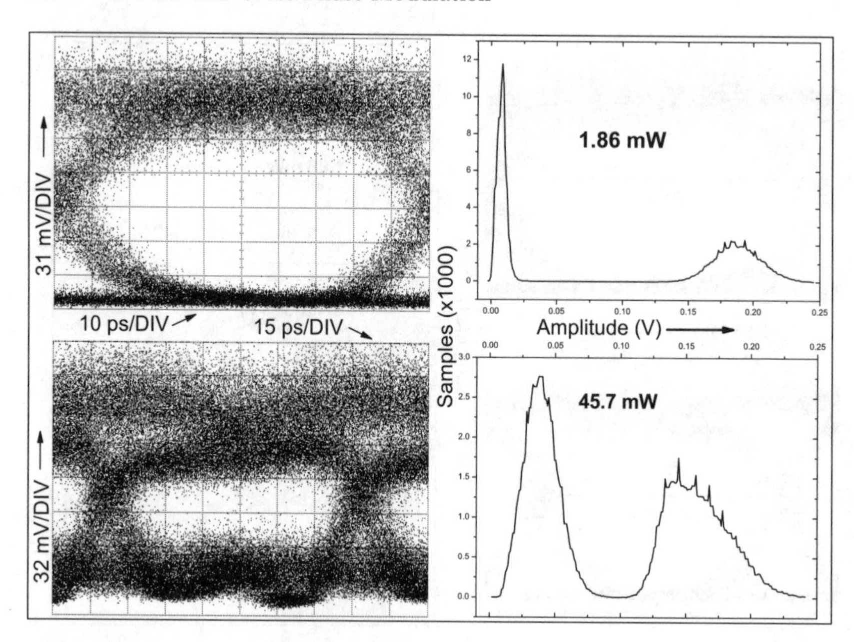

Fig. 6.11. Eye diagram and corresponding amplitude sampling values for input powers of 1.86 and 45.7 mW

6.2 Cross-Phase Modulation (XPM)

Cross-phase modulation (XPM, or sometimes, CPM) is similar to SPM, but contrary to SPM, different channels can interact with each other via the alteration of the intensity-dependent refractive index. The refractive index that a wave experiences can be altered by the intensities of all other waves propagating in the fiber. Hence, a pulse at one wavelength has an influence on a pulse at another wavelength, i.e., light is modulated by light. This effect can be exploited for optical switching and signal processing (see Chap. 12), but it can also lead to a degradation of system performance. Since signals in the other channels are naturally stochastic, the effect of XPM cannot be compensated by intelligent system design, as was the case for SPM. The XPM is the crucial effect that limits the capacity of fiber transmission systems in Sect. 4.11.

Although all waves of a transmission system can interact with each other in the case of XPM, for simplicity, only two channels propagating at the same time in the fiber will be considered. As in the case of SPM, we assume that the pulse duration is long or the dispersion in the fiber is low, so that it is possible to neglect the linear term in (5.37) and to assume quasi monochromatic waves. The attenuation in the fiber is neglected as well. Hence, the NSE turns into

$$\mathrm{j}\frac{\partial B}{\partial z} = -\gamma \left|B\right|^2 B. \tag{6.15}$$

Contrary to the case of SPM, the entire field consists of two parts with different amplitudes, frequencies, and wave numbers. If this fact is considered, then

$$B = B_1 e^{j(k_1 z - \omega_1 t)} + B_2 e^{j(k_2 z - \omega_2 t)}. \tag{6.16}$$

For the right side of (6.15) it can be written as (see 5.22)

$$j\gamma |B|^2 B = j\gamma B B^* B. \tag{6.17}$$

with B^* as the complex conjugate of B:

$$B^* = B_1^* e^{-j(k_1 z - \omega_1 t)} + B_2^* e^{-j(k_2 z - \omega_2 t)}. \tag{6.18}$$

If the multiplication of (6.17) is carried out, a large number of terms will follow. If (6.16) is introduced into (6.15) these terms can be separated by the arguments of the exponential functions. Hence, a coupled differential equation system for the two components B consists of

$$\frac{\partial B_1}{\partial z} = j\gamma \left[\left(|B_1|^2 + 2|B_2|^2 \right) B_1 + B_1^2 B_2^* e^{j(\Delta k z - \Delta \omega t)} \right], \tag{6.19}$$

and

$$\frac{\partial B_2}{\partial z} = j\gamma \left[\left(|B_2|^2 + 2|B_1|^2 \right) B_2 + B_2^2 B_1^* e^{j(-\Delta k z + \Delta \omega t)} \right], \tag{6.20}$$

with $\Delta k = k_1 - k_2$ and $\Delta\omega = \omega_1 - \omega_2$ as the difference in frequencies and wave numbers between the two waves. The nonlinearity constant (γ) contains now the mean frequency ($\bar{\omega}$) of the involved frequencies, as follows:

$$\gamma = \frac{\bar{\omega} n_2}{c A_{\text{eff}}}. \tag{6.21}$$

The last terms in (6.19) and (6.20) describe the generation of mixing frequencies but, as mentioned in Sect. 4.10. The efficiency of their generation depends on the matching of the phases. Since we are only interested in the effect of XPM here, we will assume that the fiber shows dispersion and therefore, the phase matching condition is not fulfilled and the generation of mixing frequencies is suppressed. In consequence, the last term can be neglected in both equations, leading to the following differential equation system:

$$\frac{\partial B_1}{\partial z} = j\gamma \left(|B_1|^2 + 2|B_2|^2 \right) B_1 \tag{6.22}$$

and

$$\frac{\partial B_2}{\partial z} = j\gamma \left(|B_2|^2 + 2|B_1|^2 \right) B_2. \tag{6.23}$$

The equations (6.22) and (6.23) represent a coupled differential equation system, i.e., the solution of one equation depends on the solution of the other. Hence, in principle, they are only solvable together. But, for the sake of simplicity, it should be assumed that everything written in the brackets is constant for a distance z. Then with (5.10), the first equation is now

$$\frac{\partial B_1}{\partial z} = \mathrm{j}\gamma \left[P_1 + 2P_2\right] B_1. \tag{6.24}$$

with P_1 and P_2 as the powers of the first and the second wave, respectively. If the powers are assumed as constant the solution of this equation is relatively simple:

$$B_1(z) = B_1(0)\mathrm{e}^{\mathrm{j}\gamma[P_1+2P_2]z}. \tag{6.25}$$

Hence, according to (6.25), the phase of the first wave is altered due to its own power – this is the SPM effect– but at the same time the power of the other wave has an influence on the phase of wave 1 as well. Furthermore, this influence is twice that of SPM. The total wave propagating in the fiber can be written as

$$E_1\left(z,t\right) = \frac{1}{2}\left[\frac{B_1(z)}{N_1}F(r)\mathrm{e}^{\mathrm{j}(k_1 z - \omega_1 t)} + \text{c.c.}\right]. \tag{6.26}$$

Using (6.25) and (5.3) we can obtain the following

$$\begin{aligned} E_1(z,t) &= \frac{1}{2}\left[\hat{E}_1\left(\boldsymbol{r},0\right)\mathrm{e}^{\mathrm{j}\{(\gamma[P_1+2P_2]+k_1)z-\omega_1 t\}} + \text{c.c.}\right] \\ &= \frac{1}{2}\left[\hat{E}_1\left(\boldsymbol{r},0\right)\mathrm{e}^{\mathrm{j}\left\{\left[\frac{n_2}{A_{\text{eff}}}(P_1+2P_2)+n_0\right]k_{01}z-\omega_1 t\right\}} + \text{c.c.}\right] \\ &= \frac{1}{2}\left[\hat{E}_1\left(\boldsymbol{r},0\right)\mathrm{e}^{\mathrm{j}\{[n_2 I_1+2n_2 I_2+n_0]k_{01}z-\omega_1 t\}} + \text{c.c.}\right] \end{aligned} \tag{6.27}$$

where $\hat{E}_1\left(\boldsymbol{r},0\right)$ is the wave amplitude of the first wave at the fiber input. For the wave $E_2(z,t)$, a similar equation holds where only the indices are different. Due to (6.27) in the case of two pulses in the fiber, the nonlinear refractive index consists of two parts. The total refractive index in the fiber is

$$n_{\text{ges}} = n_0 + n_2 I_1 + 2n_2 I_2. \tag{6.28}$$

The second wave alters the refractive index experienced by the other wave via its intensity and hence, changes the phase of the other wave. The phase of the wave is

$$\Phi_1 = \left[\frac{n_2}{A_{\text{eff}}}\left(P_1 + 2P_2\right) + n_0\right] k_{01} z - \omega_1 t. \tag{6.29}$$

The nonlinear part caused by XPM can be written as

$$\Phi_{1\text{XPM}} = 2\frac{n_2}{A_{\text{eff}}}P_2 k_{01} z = 2\gamma P_2 z. \tag{6.30}$$

Using γ according to (6.21), the frequency change due to XPM at the output of a fiber with length L is therefore

$$\Delta B_{\text{XPM}} = 2\gamma L \frac{\partial P_2}{\partial t}. \tag{6.31}$$

If the attenuation is included into the consideration, (6.31) can be rewritten as

$$\Delta B_{\text{XPM}} = 2\gamma L_{\text{eff}} \frac{\partial P_2}{\partial t}, \tag{6.32}$$

where L_{eff} is the effective length according to (4.98). A comparison between (6.32) and (6.14) shows that the spectral broadening due to XPM is twice that of the spectral broadening caused by SPM. This holds if both pulses are at the same place at the same time. Due to the fiber dispersion, this condition is normally fulfilled only for a rather small propagation distance. This fact limits the influence of the XPM to communication systems and the impact will be treated in detail in Sect. 6.2.2.

6.2.1 Polarization Dependence of XPM

We have been assuming that the polarization of all waves interacting in the fiber is the same. But, as already mentioned in Sect. 2.7, due to the birefringence, the polarization of the waves is completely random after a small propagation distance. The relative polarization between the waves has of course an influence on the effectivity of the XPM. In a first approximation, one could imagine that two perpendicular polarized waves are completely independent of each other, due to the fiber nonlinearity. This is not the case, however.

With two perpendicular vector components, each of which perpendicular to the propagation direction, all possible polarization states of the wave in the fiber can be expressed. Hence, an arbitrary polarization can be written as

$$\boldsymbol{B} = B_x \mathrm{e}^{\mathrm{j}(k_x z - \omega t)} + B_y \mathrm{e}^{\mathrm{j}(k_y z - \omega t)}, \tag{6.33}$$

with k_x and k_y as the wave numbers for the x and y direction, respectively. Due to the fact that the refractive indices for this two directions are different and random (see Sect. 2.7), the polarization of the wave changes randomly during propagation. For the consideration of the polarization dependence, the pulse envelope is expressed as a sum of two vector components with different frequencies:

$$\begin{aligned}\boldsymbol{B} = & B_{1x}\mathrm{e}^{\mathrm{j}(k_{1x}z-\omega_1 t)} + B_{1y}\mathrm{e}^{\mathrm{j}(k_{1y}z-\omega_1 t)} \\ & + B_{2x}\mathrm{e}^{\mathrm{j}(k_{2x}z-\omega_2 t)} + B_{2y}\mathrm{e}^{\mathrm{j}(k_{2y}z-\omega_2 t)} .\end{aligned} \tag{6.34}$$

The polarization of the waves was neglected in the derivation of the NSE. Hence, it cannot be used in the form of (5.37) to investigate the polarization dependence of XPM. In the Appendix, the differential equation system responsible for FWM and XPM is derived from the nonlinear wave equation under consideration of the polarization state of the waves. As shown in (A.46), the result for XPM is

$$\frac{\partial B_{1x}}{\partial z} = 2\mathrm{j}\gamma \left[\left(\frac{1}{2} |B_{1x}|^2 + |B_{2x}|^2 + \frac{1}{3} \sum_i |B_{iy}|^2 \right) B_{1x} \right] , \tag{6.35}$$

$$\frac{\partial B_{2x}}{\partial z} = 2\mathrm{j}\gamma \left[\left(\frac{1}{2} |B_{2x}|^2 + |B_{1x}|^2 + \frac{1}{3} \sum_i |B_{iy}|^2 \right) B_{2x} \right] , \tag{6.36}$$

$$\frac{\partial B_{1y}}{\partial z} = 2\mathrm{j}\gamma \left[\left(\frac{1}{2} |B_{1y}|^2 + |B_{2y}|^2 + \frac{1}{3} \sum_i |B_{ix}|^2 \right) B_{1y} \right] , \tag{6.37}$$

$$\frac{\partial B_{2y}}{\partial z} = 2\mathrm{j}\gamma \left[\left(\frac{1}{2} |B_{2y}|^2 + |B_{1y}|^2 + \frac{1}{3} \sum_i |B_{ix}|^2 \right) B_{2y} \right] , \tag{6.38}$$

where $i = 1, 2$ and all phase-dependent terms were neglected. If both waves have the same polarization, for instance in the x-direction, they can be written as

$$B_{1x} = 1, \quad B_{1y} = 0, \tag{6.39}$$

and

$$B_{2x} = 1, \quad B_{2y} = 0. \tag{6.40}$$

Substituting (6.39) and (6.40) into the differential equations (6.35) to (6.38), we have

$$\frac{\partial B_{1x}}{\partial z} = 2\mathrm{j}\gamma \left[\left(\frac{1}{2} |B_{1x}|^2 + |B_{2x}|^2 \right) B_{1x} \right] , \tag{6.41}$$

$$\frac{\partial B_{2x}}{\partial z} = 2\mathrm{j}\gamma \left[\left(\frac{1}{2} |B_{2x}|^2 + |B_{1x}|^2 \right) B_{2x} \right] ,$$

$$\frac{\partial B_{1y}}{\partial z} = 0,$$

and

$$\frac{\partial B_{2y}}{\partial z} = 0.$$

This is of course the same result as that already obtained from equations (6.22) and (6.23), i.e., XPM is twice as strong as SPM. But if both waves are polarized perpendicular to each other, the equations become:

$$B_{1x} = 1, \;\; B_{1y} = 0 \tag{6.42}$$

and

$$B_{2x} = 0, \;\; B_{2y} = 1. \tag{6.43}$$

Introducing (6.42) and (6.43) into the equations (6.35) to (6.38) then yields

$$\frac{\partial B_{1x}}{\partial z} = 2\mathrm{j}\gamma \left[\left(\frac{1}{2} |B_{1x}|^2 + \frac{1}{3} |B_{2y}|^2 \right) B_{1x} \right], \tag{6.44}$$

$$\frac{\partial B_{2x}}{\partial z} = 0,$$

$$\frac{\partial B_{1y}}{\partial z} = 0,$$

and

$$\frac{\partial B_{2y}}{\partial z} = 2\mathrm{j}\gamma \left[\left(\frac{1}{2} |B_{2y}|^2 + \frac{1}{3} |B_{1x}|^2 \right) B_{2y} \right].$$

Hence, if both waves are perpendicularly polarized, they are not independent of each other, as would be the case for linear behavior of the waveguide. As can be seen from (6.44), there is still a mutual influence between the waves. In the perpendicular case the XPM is weaker (1/3) than the XPM in the parallel case and is weaker (2/3) than the SPM. Similar to the derivation of the spectral broadening for SPM and XPM in the above sections, the spectral broadening under consideration of the polarization is

$$\Delta B_{\mathrm{XPM}\perp} = \frac{2}{3} \gamma L_{\mathrm{eff}} \frac{\partial P_2}{\partial t}. \tag{6.45}$$

Equation (6.45) assumes a perpendicular polarization of the waves which remains constant during propagation. This is true for polarization-maintaining fibers only. The spectral broadening for two interacting pulses with a temporal duration of 25 ps each is shown in Fig. 6.12. In the figure, the effects of SPM and XPM are considered independent of each other and the pulses overlap completely.

6.2.2 XPM's Impact on Communication Systems

For the derivation of the differential equations describing the XPM effect, only two channels were considered. On the other hand, modern WDM systems consist of a huge number of channels at different carrier wavelengths. Each

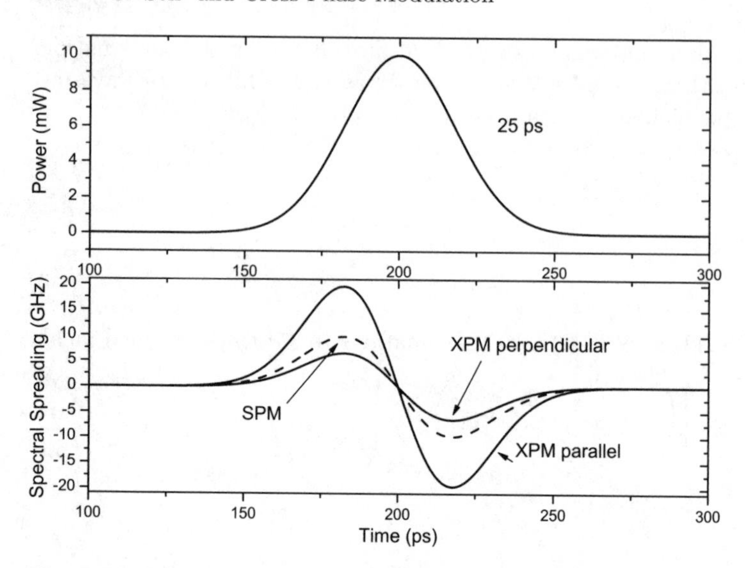

Fig. 6.12. Spectral broadening of a pulse due to SPM and XPM in a standard single mode fiber ($\alpha = 0.2$ dB/km, $\gamma = 1.3$ W^{-1}km^{-1}) after a propagation distance of 22 km (effective length). Both effects are considered independently. The second pulse required for XPM also has a temporal duration of 25 ps and overlaps completely with the first pulse

wavelength delivers a contribution to the nonlinear refractive index that acts back on the distinct channels. Hence, all channels can influence each other via their intensities. In principle, for a mathematical description of the interaction among M WDM channels, M differential equations similar to the equations (6.22) and (6.23) are required. If only SPM and XPM are considered, the equations for M channels are

$$\frac{\partial B_1}{\partial z} = \mathrm{j}\gamma \left(|B_1|^2 + 2\sum_{i=2}^{M} |B_i|^2 \right) B_1, \tag{6.46}$$

$$\vdots$$

and

$$\frac{\partial B_M}{\partial z} = \mathrm{j}\gamma \left(|B_M|^2 + 2\sum_{i=1}^{M-1} |B_i|^2 \right) B_M.$$

Hence, if all channels coincide temporally, the total refractive index in the first channel of a WDM system with M channels experiences is

$$n_{\text{ges}} = n_0 + n_2 I_1 + 2\sum_{i=2}^{M} n_2 I_i. \tag{6.47}$$

Therefore, the phase of wave 1 is

$$\Phi_1 = \left[\frac{n_2}{A_{\text{eff}}}\left(P_1 + 2\sum_{i=2}^{M} P_i\right) + n_0\right] k_0 z - \omega t. \tag{6.48}$$

As can be seen from (6.48), the influence of XPM is much more important than the influence of SPM in WDM systems. This dominance can be relatively weaker if the effects of dispersion are included into the consideration. But, as mentioned in Sect. 4.11, XPM is the most important factor that determines the transmission capacity of optical fibers.

The XPM leads, as does the SPM, to an alteration of the phase of the waves. Hence, the influence of XPM should be considered in phase-modulated systems in particular because an arbitrary phase alteration leads directly to a deterioration of the signal to noise ratio. On the other hand, only an intensity alteration can be translated via the XPM into a phase change. Strong intensity alterations can happen if semiconductor lasers are directly phase-modulated. In this case, a residual amplitude modulation of up to 20% is possible [130]. If in a system with N channels, the penalty of each channel is less than 1 dB, the condition

$$P(\text{mW}) < \frac{21}{N} \tag{6.49}$$

has to be fulfilled [248]. On the other hand, if the systems are intensity-modulated, the XPM has no influence on the system performance if the dispersion is neglected because only a phase modulation takes place. But the phase alteration is associated with a frequency alteration and the frequency change depends on the intensity slope. As in the case of SPM, the dispersion can lead to an additional temporal broadening or compression of the spectral broadened pulses, which has indeed an influence on the system performance. In soliton systems the cross-phase modulation causes the carrier frequencies of the solitons to shift. This shift results in a timing jitter which is the dominant nonlinear penalty in DWDM systems with dispersion-managed solitons [131].

As can be seen from Fig. 6.12, if the two pulses overlap completely, the leading edge of the second pulse is red-shifted whereas the tailing wing will be blue-shifted due to XPM. This is because the leading edge of the other pulse with increasing intensity coincides with the leading edge of the pulse under consideration. On the other hand, the two pulses have different carrier wavelengths. Hence, the chromatic dispersion of the fiber leads to different group velocities of the pulses. If they really overlap, they only do so for a rather short distance.

The time difference related to the propagation distance that is caused by the different group velocities is determined by the "walk-off" parameter (d_{12}):

$$d_{12} = k_1(\lambda_1) - k_1(\lambda_2), \tag{6.50}$$

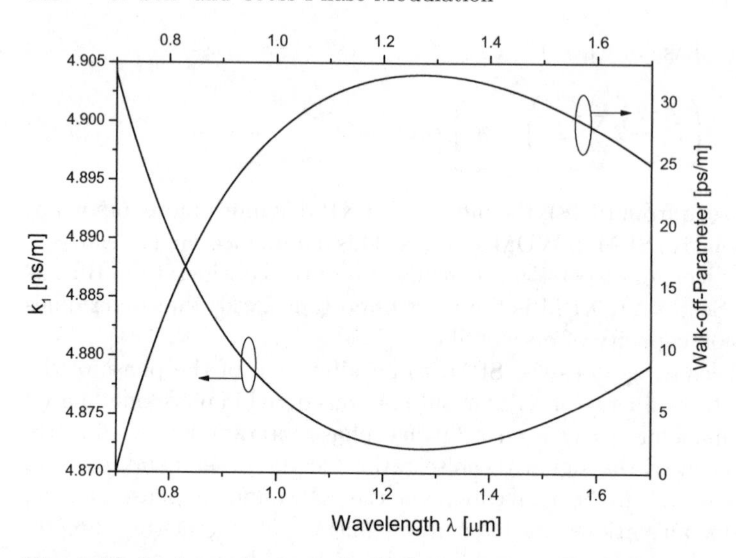

Fig. 6.13. Reciprocal group velocity for pure silica glass and corresponding walk-off parameter against the wavelength of the second pulse. The first pulse has a carrier wavelength of 700 nm

where $k_1(\lambda_{1,2})$ is the reciprocal group velocity v_g^{-1}, according to (2.70):

$$k_1 = \frac{1}{c}\left[n + \omega\frac{\mathrm{d}n}{\mathrm{d}\omega}\right] = \frac{1}{v_g}. \tag{6.51}$$

Hence, the walk-off parameter determines the difference between the reciprocal group velocities of the two pulses:

$$d_{12} = v_g^{-1}(\lambda_1) - v_g^{-1}(\lambda_2). \tag{6.52}$$

Figure 6.13 shows the reciprocal group velocity for pure silica glass and the corresponding walk-off parameter versus the wavelength for a pulse with a carrier wavelength of $\lambda_1 = 0.7$ μm. According to the behavior of the refractive index, the reciprocal group velocity has a minimum in the range of the zero dispersion of the material.

The maximum walk-off parameter appears if the second pulse has a wavelength of 1.27 μm. The time difference between both pulses is 34 ps for every meter they will propagate in the fiber. If the center wavelength of the second pulse increases, the walk-off parameter decreases. For the two pulses at the beginning and end of the L-band ($\lambda_1 = 1.57$ and $\lambda_2 = 1.62$ μm), the walk-off parameter is -1.24 ps/m. This value is clearly smaller for neighboring channels. If the wavelength difference is 1 nm, for instance the walk-off parameter is -0.02 ps/m ($\lambda_1 = 1.577$ and $\lambda_2 = 1.578$ μm).

Around the dispersion minimum, the walk-off between the pulses is very small. Hence, if the channels are arranged symmetrically around the dispersion minimum of a dispersion-shifted fiber, the walk-off can be zero. The XPM

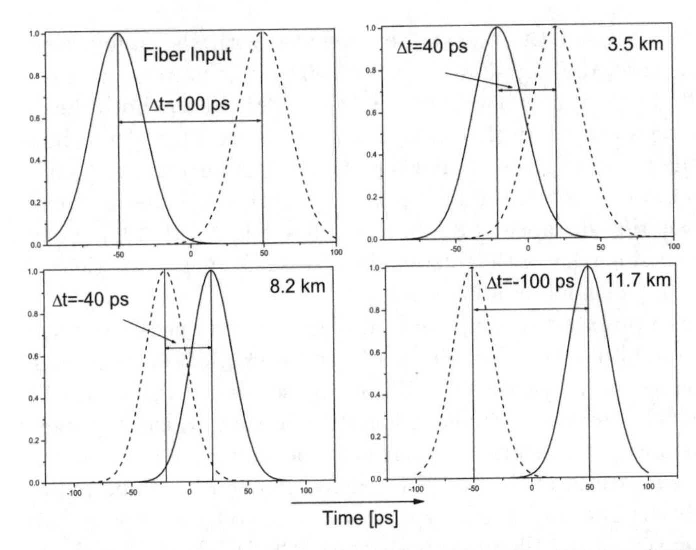

Fig. 6.14. Gaussian pulses with a temporal width of 41.6 ps (FWHM) and a carrier wavelength of 1.577 and 1.578 μm for different propagation distances in a standard single mode fiber $D = 17$ ps/(nm × km)

thus has a much stronger influence on the spectral broadening of the pulses than the SPM. But due to the small dispersion in this range, the spectral broadening is not translated into a temporal broadening of the pulses.

Figure 6.14 shows two pulses with carrier wavelengths of $\lambda_1 = 1.577$ μm (blue), and $\lambda_2 = 1.578$ μm (red) and a temporal width of $\tau_0 = 25$ ps in pure silica. The red pulse was injected into the fiber 100 ps before the blue one. For convenience, the influence of dispersion and nonlinearity to the pulse width should be neglected.

In a standard single mode fiber, the propagation of pulses with a carrier wavelength above 1.3 μm is determined by the anomalous dispersion, as shown in Sect. 2.6.2. The GVD parameter, and therefore the group velocity, decreases with increasing wavelength. Hence, the blue pulse injected into the fiber after the red one moves faster. After a distance of 3.5 km, both pulses overlap halfway and overlap completely after 5.9 km. If the pulses propagate further, they walk apart and after a distance of 12 km, they are again completely separated. But now the blue pulse is in front of the red one.

From the standpoint of XPM, a mutual influence can only happen if both pulses overlap at least partly. Hence, for the above example, it occurred only for a propagation distance between 3.5 and 8 km. According to (6.31), the spectral broadening of one pulse depends on the temporal alteration of the intensity of the other pulse. As can be seen from Fig. 6.14 after a propagation of 3.5 km, the rising edge of the blue pulse hits the falling edge of the red pulse. The temporal derivation $\partial P_2/\partial t$ is positive for the rising edge and hence, the falling edge of the red pulse is red-shifted. After a propagation distance of

5.9 km, the falling edge of the blue pulse coincides with the falling edge of the red. In this case, the temporal derivation $\partial P_2/\partial t$ is negative and the formerly red-shifted edge is now blue-shifted by the XPM. Hence, both effects will compensate each other. At the same time, the rising edge of the blue pulse stretches the rising edge of the red one, thus experiencing a red-shift. If the pulses propagate further, the blue pulse moves through the red pulse and after a propagation distance of 8 km, the falling edge of the blue pulse coincides with the rising edge of the red one. Hence, the formerly experienced red-shift is now compensated by a blue-shift.

If the blue pulse propagates completely through the red one, no residual phase shift of the red pulse remains. But if the pulses can not walk through each other completely a net phase shift of both pulses due to XPM will be conserved. This can be the case in transmission systems that bridge only short distances, for instance. It can also be seen in dispersion-managed systems if the dispersion compensation is incorporated too frequently. If the pulses overlap in time at the fiber input the result is a partial collision as well, this can be avoided by staggering the pulse positions of the WDM channels.

If the interaction length between the pulses is defined as the distance between the beginning and the end of the overlap at half the maximum power, then the time for the interaction is $2\tau_{\mathrm{FWHM}}$, with τ_{FWHM} as the FWHM pulse width. According to (2.73) it follows for the interaction length

$$L_{\mathrm{I}} = \frac{2\tau_{\mathrm{FWHM}}}{D\Delta\lambda}, \tag{6.53}$$

where D is the dispersion parameter and $\Delta\lambda$ is the difference between the carrier wavelengths of the pulses. Accordingly the interaction length for two adjacent pulses of the C-band (channel spacing = 50 GHz, $\lambda_1 = 1544.92$ nm, $\lambda_2 = 1544.53$ nm) with a FWHM width of 25 ps in a standard single mode fiber ($D = 17$ ps/(nm×km)) is $L_{\mathrm{I}} = 7.54$ km. For a nonzero dispersion-shifted fiber with $D = 2$ ps/(nm×km) it follows an interaction length of $L_{\mathrm{I}} = 64.1$ km.

In the example in Fig. 6.14, only two pulses injected at different times were considered. For the case of a real bit pattern, the case is more complicated where the bits of the blue channel move through the bits of the red one.

Summary

- SPM leads to a phase alteration of the wave due to its own intensity.
- Phase alteration causes a spectral broadening of the pulses. The red-shifted frequencies are at the leading edge, whereas the blue components will travel at the tailing wing of the pulse.
- Together with fiber dispersion, spectral broadening can be translated into an alteration of the temporal width of the pulse. In the case of normal

dispersion the pulse is additionally stretched, whereas it will be compressed for an anomalous dispersion.

- Due to the interplay between GVD and SPM, the effects of SPM must be included into the system design for dispersion-managed communication links.
- The XPM leads to a phase alteration of one wave as reaction to the intensity of another and causes a frequency broadening.
- The frequency broadening as a result of XPM is twice that of SPM. If the waves are perpendicular polarized, it is reduced to 2/3 of SPM.
- Due to XPM, two perpendicular polarized waves are not independent of each other in the fiber.
- As a result of the walk-off in dispersive waveguides, the mutual influence between two pulses due to XPM can be cancelled if the pulses can walk through each other completely.

Exercises

6.1. The radiation characteristics of an array of antenna elements can be steered with the phase of the current of each element. These devices are called intelligent or smart antennas. The phase alteration can be done in the optical domain while the radio signals are propagated in optical fibers (radio-over-fiber). You have planned to use the effect of XPM in a 1-km highly nonlinear polarization-maintaining fiber ($\gamma = 18\ \mathrm{W}^{-1}\ \mathrm{km}^{-1}$, $\alpha = 0.3$ dB/km) for a phase alteration of the radio signals.

Determine the required pump power for a phase alteration of $\pi/2$ if the signal power is low so that the SPM can be neglected.

Determine the required pump power if the signal power amounts to 20 mW.

6.2. Derive the differential equation system for self- and cross-phase modulation from the NSE. All phase-dependent terms as well as dispersion and absorption should be neglected. The field consists of three parts:

$$B = B_1 \mathrm{e}^{\mathrm{j}(k_1 z - \omega_1 t)} + B_2 \mathrm{e}^{\mathrm{j}(k_2 z - \omega_2 t)} + B_3 \mathrm{e}^{\mathrm{j}(k_3 z - \omega_3 t)}. \tag{6.54}$$

6.3. Suppose that two perpendicularly polarized waves propagate through a polarization-maintaining fiber. Determine the equation for the electric field of one of the waves and the total refractive index the wave will experience. The effects of birefringence can be neglected.

6.4. Assume two pulses in WDM channels at the begin and end of the C-band ($\lambda_1 = 1528.77$ nm and $\lambda_2 = 1560.61$ nm) and in two neighboring channels with a temporal spacing of 100 GHz. For the transmission, a standard single mode fiber $D = 17$ ps/(nm×km) and Gaussian pulses with a FWHM width of 30 ps are used. At the fiber input, all pulses are introduced simultaneously. Determine the propagation distance after which the pulses are walked-off for both cases.

7. Four-Wave-Mixing (FWM)

Four-wave-mixing (FWM), or sometimes four-photon-mixing (FPM), describes a nonlinear optical effect at which four waves or photons interact with each other due to the third order nonlinearity of the material. As a result, new waves with sum and difference frequencies are generated during the propagation in the waveguide. FWM is comparable to the so called intermodulation in electrical communication systems. This effect was already mentioned in the introduction of this book. The intermodulation, as well as the FWM, leads to noise in the neighboring channels which degrades the system performance. For WDM systems in dispersion-shifted fibers, FWM is the most important nonlinear effect.

On the other hand, FWM can be used for a number of applications in optical telecommunications, such as wavelength conversion and optical switching. Of particular interest is the generation of a phase-conjugated wave due to the FWM effect which can be used to cancel any kind of signal distortions in the system. The applications of FWM and other nonlinear effects are treated in the third part of this book in detail.

In the quantum mechanical model, two photons are annihilated by the atom as it generates two new photons simultaneously (see Fig. 4.12). Only virtual states of the atom are involved and hence, the rules of conservation of energy and momentum have to be fulfilled during the process. The conservation of momentum leads to the phase matching condition. The efficiency of FWM depends very strongly on this matching of the phases of the waves involved. Therefore, the FWM efficiency is a function of the fiber dispersion. In transmission systems using the zero dispersion of the material FWM is a severe problem. On the other hand, if the transmission system shows a high local dispersion, which is the case in dispersion-managed systems, the FWM can be effectively suppressed.

In this chapter, FWM is described mathematically with the help of the NSE. The significance of the effect to WDM systems as well as possibilities for an effective suppression are discussed in detail.

7.1 Mixing between WDM Channels

If three optical waves with frequencies f_i, f_j and f_k are propagating in the fiber, they can interact via the third order susceptibility of the material and new waves with the frequencies

$$f_{i,j,k} = f_i + f_j - f_k \tag{7.1}$$

can be generated by the FWM process, where i, j and k can have the values 1, 2, and 3. Three elements arranged in three classes can lead to 27 possible variations. But if the third frequency (f_k) in (7.1) equals the first, or second (f_i, f_j) frequency no new frequency is generated, resulting in f_i or f_j, respectively ($f_i + f_j - f_i = f_j$). Furthermore, if the first two frequencies in (7.1) are changing their places as well, no new frequencies are generated ($f_i + f_j - f_k = f_j + f_i - f_k$). Therefore, a side condition for (7.1) is $k \neq i, j$.

If the side condition is introduced, 9 residual combinations are left from the 27 variations. Hence, three waves with different frequencies are able to generate 9 new waves which have mixing frequencies [133]. The first column of Table 7.1 shows the possible combinations for the generation of new waves if the original waves are three WDM channels out of the L-band with an equidistant channel spacing of 1 nm ($\lambda_1 = 1578$ nm, $\lambda_2 = 1579$ nm, $\lambda_3 = 1580$ nm).

According to (4.115), the frequency spacing depends on the wavelength. Therefore, if the channels have an equidistant wavelength spacing, their frequency differences are not completely equal. For the example above, the frequencies are $f_1 = 189.98255$ THz, $f_2 = 189.86223$ THz, and $f_3 = 189.74206$ THz. Therefore, the frequency spacings are $\Delta f_{1,2} = 0.12032$ THz

Table 7.1. Possible combinations for the generation of new waves due to the FWM process for channels with equidistant wavelength (*left column*) and frequency spacing (*right column*).

i, j, k	$\lambda_{i,j,k}$/nm ($\Delta\lambda = 1$ nm)	$f_{i,j,k}$/THz ($\Delta f = 120$ GHz)
112	1577.001	189.99
113	1576.005	189.87
123	1577.002	189.99
132	1578.999	190.23
221	1580.001	190.35
223	1578.001	190.11
231	1581.003	190.47
331	1582.005	190.59
332	1581.001	190.47

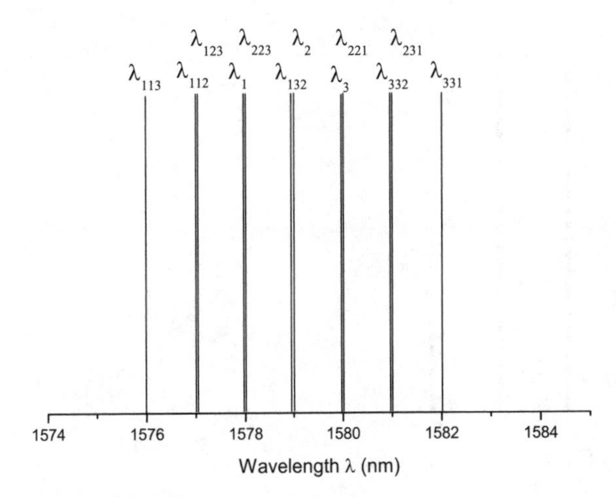

Fig. 7.1. Mixing frequencies due to FWM for equidistant wavelength spacing

and $\Delta f_{2,3} = 0.12017$ THz. Due to the slightly different frequency differences, the mixing products do not fall exactly into one of the original channels. In fact, the three input channels generate 9 new waves with different frequencies. The generated frequencies for the example above are shown in Fig. 7.1.

On the other hand, if the original channels have an equal frequency spacing, the new generated waves will have the same spacing and a large number of them falls exactly into the original channels.[1] The generated waves show the same frequency comb as the original waves did. As an example a frequency spacing of 120 GHz. The distinct channels have the frequencies $f_1 = 190.11$ THz ($\lambda_1 = 1576.942$ nm), $f_2 = 190.23$ THz ($\lambda_2 = 1575.947$ nm), and $f_3 = 190.35$ THz ($\lambda_3 = 1574.954$ nm). The generated mixing products are shown in the second column of Table 7.1 and in Fig. 7.2

According to Fig. 7.2, only 4 new frequencies are generated while the other 5 mixing products coincide with the original channels or with each other. For N original channels, the number of possible mixing products [85] is

$$M = \frac{1}{2}\left(N^3 - N^2\right). \tag{7.2}$$

If the intensities of the new generated waves are strong enough, they can interact again with each other or with the original channels and will produce additional mixing products. Therefore, if the intensity is sufficient, from the 9 wavelength channels of the first example according to (7.2), in a second cycle 324 mixing products could be generated. On the other hand, only 147 mixing products would be possible in a second cycle for the second example with equidistant frequency spacing. The number of possible mixing products versus the number of channels of a WDM system is shown in Fig. 7.3.

[1]Note that in the ITU-T Grid [ITU-T G692], used for WDM systems, the channels have an equidistant frequency spacing of 50 or 100 GHz.

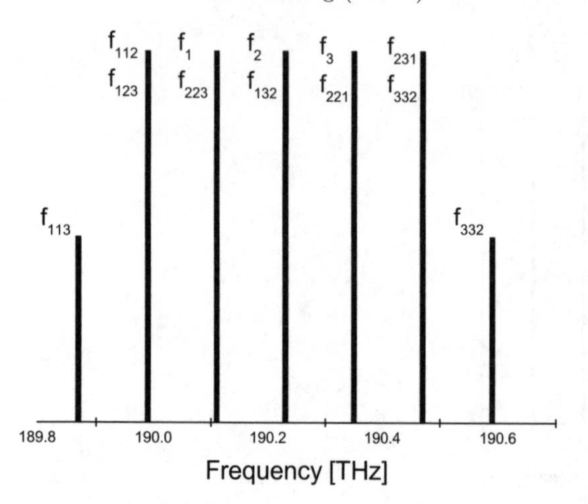

Fig. 7.2. Mixing frequencies due to FWM for an equal frequency spacing between the original channels

The effectiveness of the generation of new mixing products depends strongly on the phase matching condition between the distinct waves. In a dispersive material, a larger frequency spacing results in a higher refractive index difference, and therefore, a higher phase mismatch between the channels. Hence, the FWM efficiency decreases for channels that are farther away (see Fig. 7.13). As a result only adjacent channels generate mixing products effectively in dispersive fibers, leading to a decrease of the signal to noise ra-

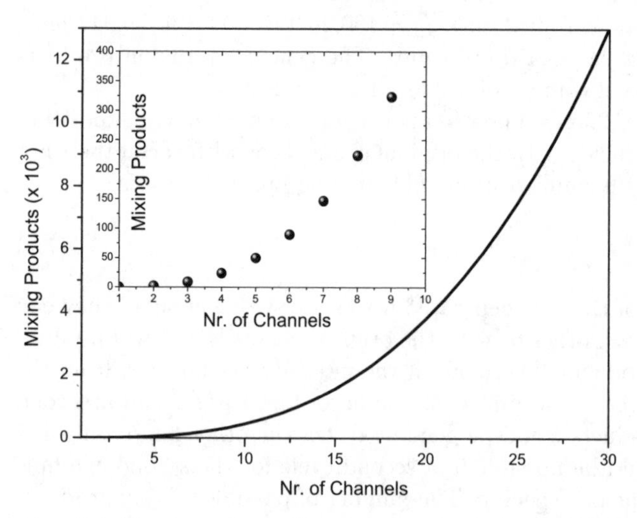

Fig. 7.3. Number of possible mixing products due to FWM versus the number of channels

tio in these channels. The intensity of the mixing products between channels farther away is negligibly small.

The disturbance is a cross talk between neighboring channels resulting in a fading between the original wave and the new generated wave that depends on the relative phase. If the relative phase between both is π, the original wave is weakened. If both waves are in phase, it is reinforced. This leads to an increase of the noise in the system and, therefore, a decrease of the signal to noise ratio (SNR). In WDM systems, especially in dispersion-shifted fibers, the FWM is the most important effect degrading the system performance. Therefore, many proposals have been made to suppress the FWM.

7.2 Mathematical Description of FWM

The starting point for the derivation of the differential equation system responsible for the interaction between the four waves is again the nonlinear Schrödinger equation (NSE). As in the case of SPM and XPM, it is assumed that the pulse duration is long or the dispersion in the fiber is low, so that it is possible to neglect the linear term in (5.37) and to assume quasi monochromatic waves. If the attenuation in the fiber is neglected as well, the NSE becomes

$$\mathrm{j}\frac{\partial B}{\partial z} = -\gamma \left|B\right|^2 B. \tag{7.3}$$

In the case of FWM, the superposition propagating through the fiber consists of four waves with different amplitudes, frequencies, and wave numbers, as represented by

$$B = B_1 \mathrm{e}^{\mathrm{j}(k_1 z - \omega_1 t)} + B_2 \mathrm{e}^{\mathrm{j}(k_2 z - \omega_2 t)} + B_3 \mathrm{e}^{\mathrm{j}(k_3 z - \omega_3 t)} + B_4 \mathrm{e}^{\mathrm{j}(k_4 z - \omega_4 t)}. \tag{7.4}$$

If (7.4) is introduced into (7.3), the NSE can be separated into four coupled differential equations, each of which is responsible for one distinct wave in the fiber. As mentioned in Sect. 7.1 three waves with different frequencies are able to generate nine new waves. Here, for convenience, the consideration is restricted to the case

$$\omega_4 = \omega_1 + \omega_2 - \omega_3. \tag{7.5}$$

The waves with the frequencies $\omega_1 = 2\pi f_1$ and $\omega_2 = 2\pi f_2$ are called pump waves, whereas the wave with frequency $\omega_3 = 2\pi f_3$ is the signal and $\omega_4 = 2\pi f_4$ the so called idler wave. In analogy to SRS (see Chap. 10), the lower sideband can be seen as the Stokes and the higher generated sideband as the anti-Stokes wave. Figure 7.4 shows the position and notation of the four distinct frequencies in the frequency and wavelength domain. If the signal wave is already present in the fiber, the FWM process can be seen as an up conversion of its frequency.

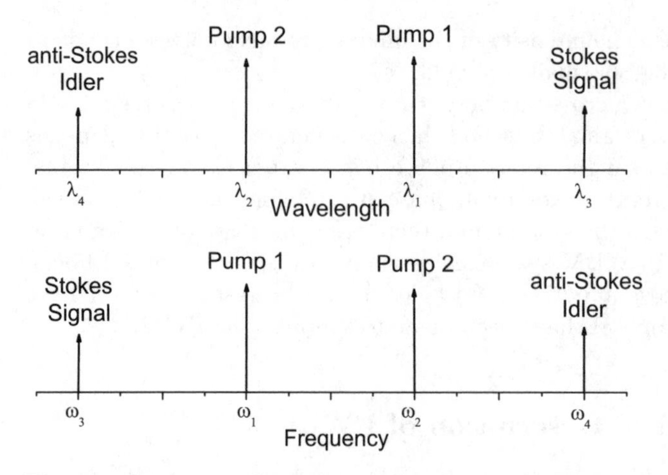

Fig. 7.4. Position and notation of the distinct frequencies for $\omega_4 = \omega_1 + \omega_2 - \omega_3$. Note the different arrangement of the frequencies in comparison to Fig. 7.2

If all three original waves have the same frequency $\omega_1 = \omega_2 = \omega_3$, the process is called a degenerate FWM where the new generated mixing product has the same frequency ω. The degenerate FWM is important if the waves propagate into different directions – the new generated wave has in fact the same frequency as the original ones but can be distinguished by its propagation direction. This is the case for the self diffraction in Sect. 12.3.4.

If only two of the three waves are equal ($\omega_1 = \omega_2 \neq \omega_3$), the process is called partly degenerate. For a partly degenerate FWM the frequency of the mixing product is lower than the pump frequencies by $\Delta\omega$ where $\Delta\omega$ is the frequency separation between the pump waves and the idler. This process is important for applications like the FWM wavelength converter (Sect. 12.2.1) and the parametric amplifier (Sect. 13.4).

If (7.4) is introduced into (7.3), a large number of terms will be obtained. Terms of the form $|B_i|^2 B_i$, with $i = 1, 2, 3, 4$ are responsible for the self-phase modulation of the waves, whereas terms with $|B_i|^2 B_j$ $(i \neq j)$ denote the cross-phase modulation. All other terms are responsible for FWM but an effective generation of new waves can only happen if the phase matching condition is fulfilled. The distinct FWM terms can be separated according to their frequency with (7.5)

$$\omega_1 + \omega_2 = \omega_3 + \omega_4,$$

or, with $\omega_i = 2\pi f_i$

$$f_1 + f_2 = f_3 + f_4. \tag{7.6}$$

Therefore, the frequency f_1 is equal to $f_3 + f_4 - f_2$. If the difference between all three frequencies is equal, then f_1 can be obtained from $2f_2 - f_4$ and

$f_2 + f_3 - f_1$ as well (see Fig. 7.4). Hence, one yields the following from the NSE for the first pump wave (f_1) in the fiber:

$$\frac{\partial B_1}{\partial z} = \mathrm{j}\gamma \left[\left(|B_1|^2 + 2\sum_{i \neq 1} |B_i|^2 \right) B_1 + \mathrm{FWM} \right], \tag{7.7}$$

where $i = 1, 2, 3, 4$, with

$$\begin{aligned} \mathrm{FWM} = {} & 2B_3 B_4 B_2^* \mathrm{e}^{\mathrm{j}\Delta k_{3,4,-2,-1} z} + 2B_2 B_3 B_1^* \mathrm{e}^{\mathrm{j}\Delta k_{2,3,-1,-1} z} \\ & + B_2^2 B_4^* \mathrm{e}^{\mathrm{j}\Delta k_{2,2,-4,-1} z} \end{aligned} \tag{7.8}$$

The wave numbers in (7.8) are

$$\Delta k_{3,4,-2,-1} = k_3 + k_4 - k_2 - k_1, \tag{7.9}$$

$$\Delta k_{2,3,-1,-1} = k_2 + k_3 - k_1 - k_1, \tag{7.10}$$

and

$$\Delta k_{2,2,-4,-1} = 2k_2 - k_4 - k_1. \tag{7.11}$$

For a simplification, we assume that the frequency difference between the pump waves f_1 and f_2 is not equal to the frequency difference between the pump wave f_2 and the signal wave f_3. Hence, only one FWM term remains. Furthermore, the distinct amplitudes are normalized to the input power of the pump wave P_1:

$$B_m = \sqrt{P_1} A_m. \tag{7.12}$$

Therefore, (7.7) is now

$$\frac{\partial A_1}{\partial z} = \mathrm{j}\gamma P_1 \left[\left(|A_1|^2 + 2\sum_{i \neq 1} |A_i|^2 \right) A_1 + 2A_3 A_4 A_2^* \mathrm{e}^{\mathrm{j}\Delta k_{3,4,-2,-1} z} \right]. \tag{7.13}$$

In analogy the relations for the other three differential equations can be separated from the large number of terms obtained through (7.4) and (7.3):

$$\frac{\partial A_2}{\partial z} = \mathrm{j}\gamma P_1 \left[\left(|A_2|^2 + 2\sum_{i \neq 2} |A_i|^2 \right) A_2 + 2A_3 A_4 A_1^* \mathrm{e}^{\mathrm{j}\Delta k_{3,4,-1,-2} z} \right], \tag{7.14}$$

$$\frac{\partial A_3}{\partial z} = \mathrm{j}\gamma P_1 \left[\left(|A_3|^2 + 2\sum_{i \neq 3} |A_i|^2 \right) A_3 + 2A_1 A_2 A_4^* \mathrm{e}^{\mathrm{j}\Delta k_{1,2,-4,-3} z} \right], \tag{7.15}$$

and

$$\frac{\partial A_4}{\partial z} = \mathrm{j}\gamma P_1 \left[\left(|A_4|^2 + 2\sum_{i \neq 4} |A_i|^2 \right) A_4 + 2A_1 A_2 A_3^* \mathrm{e}^{\mathrm{j}\Delta k_{1,2,-3,-4} z} \right], \quad (7.16)$$

where $i = 1, 2, 3, 4$.

The combination of the wave number indices determines their composition, as in (7.9) to (7.11). A comparison between the coupled differential equation system (7.13) to (7.16) and the equations for the XPM (6.22) and (6.23) shows that the first term inside the brackets of each equation is responsible for SPM, whereas the second term describes the effect of XPM. These terms are responsible for a phase alteration and a corresponding frequency broadening of the pulses, but new frequency components can only be generated via the FWM terms. First we assume that only the two pump waves (1 and 2) are present at the fiber input. Since everything written between the brackets in equations (7.13)–(7.16) can only alter the phases, the last terms are especiall- important. As can be seen from (7.15), for the generation of a signal wave, an idler is required, whereas an idler wave can only be generated if a signal is present in the fiber. But, a wide band noise floor is always present in communication systems. The noise power is sufficient for the generation of the signal and idler wave. If the phase matching requirement is fulfilled both waves rise out of the noise during propagation. The optical power is translated from the two pump waves (1 and 2) to the new generated waves (3 and 4). the fiber input, this signal wave is continuously amplified if the phase matching condition is fulfilled, whereas at the same time an idler wave is generated in the fiber. This is the basis of wavelength conversion and parametric amplification. Figure 7.5 shows the conditions for the four waves in an optical fiber if the phases between the distinct waves are matched.

If the intensity of the pump waves increases the behavior differs from that in Fig. 7.5. For a power of 1 W, the simulation shows a periodic function along the fiber in Fig. 7.6 a, under the conditions denoted in the Figure. If the power further increases the cycle duration decreases, as shown in Fig. 7.6b whereas for an input power of 70 W the behavior is chaotic, as shown in Fig. 7.6c.

In the range of optical telecommunications, relatively small input powers propagate in the fibers. Therefore, only the periodic behavior will be seen. The periodicity of the idler as a function of the fiber length depends on the phase matching condition which is treated in detail in the next Section.

7.3 Phase Matching

Contrary to the effects of SPM and XPM where only the phases of the waves are changed, FWM generates waves with new frequencies. In the quantum mechanical model these new waves have different energies and during the interaction, the rules of conservation of energy and momentum must be fulfilled. The conservation of momentum leads to the phase matching condition.

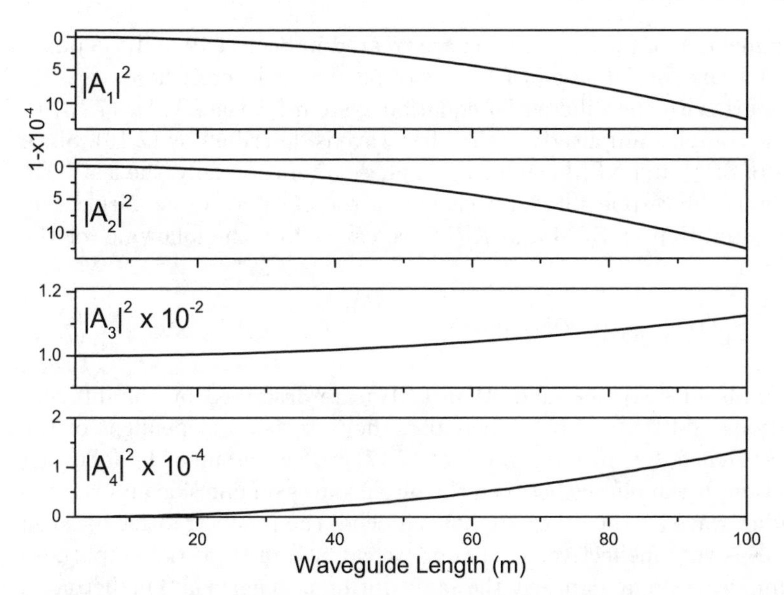

Fig. 7.5. Power exchange between pump ($|A_1|^2$, $|A_2|^2$), signal ($|A_3|^2$), and idler waves ($|A_4|^2$) during the propagation in the fiber (fiber length L = 100 m, γ = 0.0178 $\mathrm{m}^{-1}\mathrm{W}^{-1}$, k_2 = 0.07 ps^2/km, Δf = 10 GHz, P_1 = 0.1W, $A_1 = A_2 = 1$, $A_3 = 0.01$)

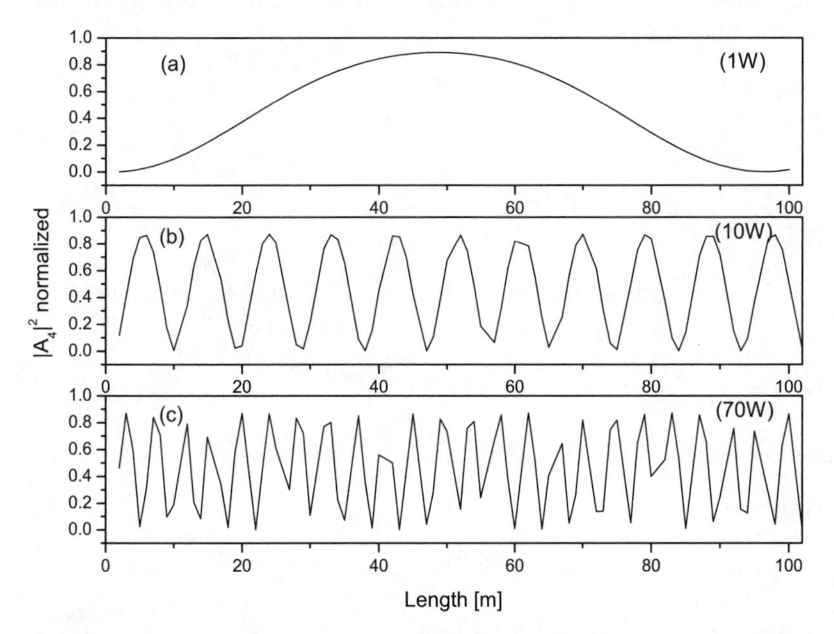

Fig. 7.6. Mixing product versus fiber length for different input powers P_1(γ = 0.0176 $\mathrm{m}^{-1}\mathrm{W}^{-1}$, Δf = 100 GHz, k_2 = 0.07 ps^2/km, $A_1 = A_2 = 1$, $A_3 = 0.1$, $\alpha = 0$)

The fundamentals of phase matching are treated in Sect. 4.10 and the phase matching for the special case of FWM is of particular interest here.

Let's start with the differential equation system for the FWM, (7.13) to (7.16). The complex amplitude of the idler wave is described by (7.16). Since the terms of SPM and XPM can lead to a phase alteration only, the last term in (7.16) must be responsible for a change of the absolute value. Neglecting the phase alteration by SPM and XPM, we can obtain the following for the amplitude A_4:

$$\frac{\partial A_4}{\partial z} = 2\mathrm{j}\gamma P_1 A_1 A_2 A_3^* \mathrm{e}^{\mathrm{j}\Delta k_{1,2,-3,-4} z}. \tag{7.17}$$

The amplitudes A_1, A_2, and A_3^* in (7.17) are described by the differential equations (7.13) to (7.15). Therefore, they are not independent of the distance z. Hence, for an integration of (7.17) we will assume the following simplification; If the phases between the pump waves on one side and the signal and idler wave on the other are not matched, the power transfer between these waves is very ineffective. Therefore, we can assume that the amplitudes of the pump waves remain nearly the same during propagation. Furthermore, the signal wave is not generated from the noise floor of the fiber, but has a rather high amplitude at the fiber input already. Hence, the additional increase due to FWM can be neglected as well. Over all we can assume that the amplitudes of the waves 1 to 3 remain constant and correspond to their value at the fiber input:

$$A_1(z) \approx A_1(0),$$

$$A_2(z) \approx A_2(0), \tag{7.18}$$

and

$$A_3^*(z) \approx A_3^*(0).$$

If (7.17) is integrated from $z = 0$ to $z = l$, the amplitude change of the idler is

$$A_4(z) = \frac{2\gamma P_1}{\Delta k_{1,2,-3,-4}} A_1(0) A_2(0) A_3^*(0) \mathrm{e}^{\mathrm{j}\Delta k_{1,2,-3,-4} z} \Big|_0^l \,. \tag{7.19}$$

At the fiber input ($z = 0$), the amplitude of the idler is zero:

$$A_4(0) \approx 0. \tag{7.20}$$

According to this boundary condition, (7.19) can be written as

$$A_4 = \frac{2\gamma P_1}{\Delta k_{1,2,-3,-4}} A_1(0) A_2(0) A_3^*(0) \left[\mathrm{e}^{\mathrm{j}\Delta k_{1,2,-3,-4} l} - 1 \right] \tag{7.21}$$

Therefore, the intensity change of the idler depends on the propagation distance in the fiber:

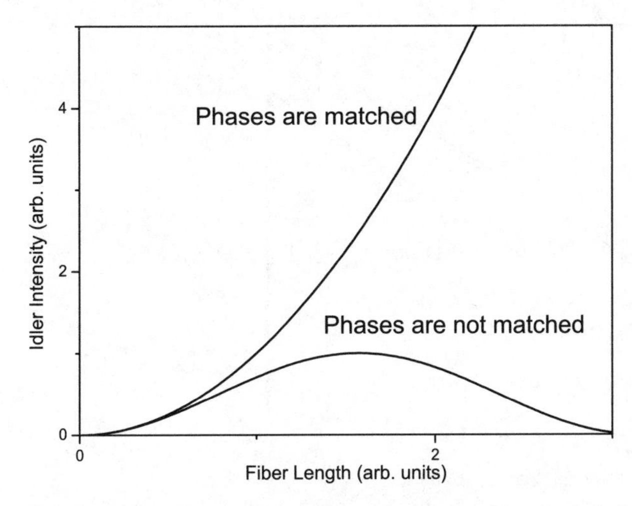

Fig. 7.7. Intensity of the idler wave versus fiber length for phase matching and a phase mismatch between the waves

$$I_4 \sim A_4 A_4^* \sim \text{const} I_1 I_2 I_3 l^2 \left(\frac{\sin \Delta k_{1,2,-3,-4}\, l/2}{\Delta k_{1,2,-3,-4}\, l/2} \right)^2 , \tag{7.22}$$

where the theorem $\sin(x/2) = \sqrt{1/2\,(1 - \cos x)}$ is used, I_n is the intensity of the waves, const describes a number that summarizes all involved constants, l is the fiber length, and $\Delta k_{1,2,-3,-4}$ is the phase mismatch between the waves.

If the phases of the waves are matched $\Delta k_{1,2,-3,-4} = 0$ then according to (7.22), the intensity of the idler increases quadratically with the length of the waveguide. On the other hand, the assumption required for the derivation of (7.22) no longer holds because as A_4 and A_3^* grow, A_1 and A_2 will loose power (Fig. 7.5). But this is neglected here. If the phases are not matched the intensity shows a periodic function along the fiber, as can be seen in Fig. 7.7. The phase matching condition for the idler is

$$\begin{aligned} \Delta k_{1,2,-3,-4} &= k_1 + k_2 - k_3 - k_4 \\ &= \frac{1}{c} \left(n_1 \omega_1 + n_2 \omega_2 - n_3 \omega_3 - n_4 \omega_4 \right) \\ &= 2\pi \left(\frac{n_1}{\lambda_1} + \frac{n_2}{\lambda_2} - \frac{n_3}{\lambda_3} - \frac{n_4}{\lambda_4} \right) . \end{aligned} \tag{7.23}$$

Connected to the phase matching condition is the coherence length (Sect. 4.10). From (4.109), the coherence length for FWM is

$$L_{\text{coh}} = \frac{\pi}{|\Delta k_{1,2,-3,-4}|} . \tag{7.24}$$

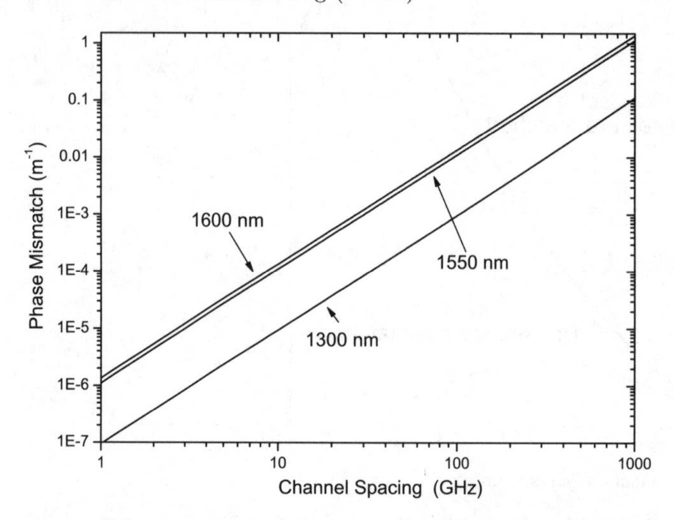

Fig. 7.8. Phase mismatch against the channel spacing in fused silica for an average wavelength of 1600 nm (L-band), 1550 nm (C-band) and 1300 nm (S-band). The values for dispersion and dispersion slope were computed with the Sellmeier equation and Sellmeier parameters (Table 2.1)

The origin of the phase mismatch lies in the frequency dependence of the refractive index and the dispersion. Out of the range of the dispersion minimum, the difference between the refractive indices increases with increasing frequency spacing. Therefore, the phase mismatch depends on the frequency spacing and the average frequency of the waves. Figure 7.8 shows the phase mismatch in fused silica versus the channel spacing for three different wavelengths.

As can be seen from the Figure, the refractive index difference and therefore the phase mismatch is weak for a small channel spacing. Small refractive index differences for large channel spacings can also be found near the zero-crossing of the material dispersion (for fused silica at ≈ 1.27 µm). Hence, large coherence lengths for rather large channel spacings are possible in the range of the fibers zero dispersion. The coherence length for three average wavelengths in the C, S and L-band is shown in Table 7.2. As can be seen, the influence of the FWM on the transmission performance is very effective near the zero-crossing of the dispersion (for pure silica in the S-band). Channels spaced far away from each other can interact and produce mixing products as well. These mixing products increase the noise in the original channels and decrease the performance of the system.

On the other hand, a small dispersion means that the temporal broadening of the pulses due to GVD is small and hence, the transmission of high bit rates is possible. Due to this advantage many thousand kilometers of dispersion-shifted fibers working in the range of zero dispersion – for DSF in the C-band – are introduced in the global transmission systems. The implementation

Table 7.2. Coherence length in fused silica for different average wavelengths and channel spacings

Δf (GHz)	$L_{\text{coh}1.3\mu\text{m}}$ (m)	$L_{\text{coh}1.55\mu\text{m}}$ (m)	$L_{\text{coh}1.6\mu\text{m}}$ (m)
1000	28	2.8	2.3
100	3293	286	233
50	13304	1144	935
1	33587972	2863923	2338930

of WDM is a severe problem in these fibers. Different possibilities for the suppression of FWM in standard and dispersion-shifted fibers are treated in Sect. 7.5 in detail.

The intensity of the generated idler versus the fiber length for pure silica with an average wavelength of 1.3, 1.55 and 1.6 μm and a channel spacing of 50 GHz is shown in Fig. 7.9. The idler shows a periodic intensity fluctuation along the fiber. The periodicity of the fluctuation depends on the phase mismatch and hence on the coherence length which is given in Table 7.2. The intensity in the fiber increases until it reaches a maximum at the coherence length. After the maximum it decreases until the fiber length corresponds to two times the coherence length. If the intensity reaches a minimum the idler will increase again and so on. The maximum idler intensity for 1.55 μm is higher because the coherence length for an average wavelength of 1.55 μm is

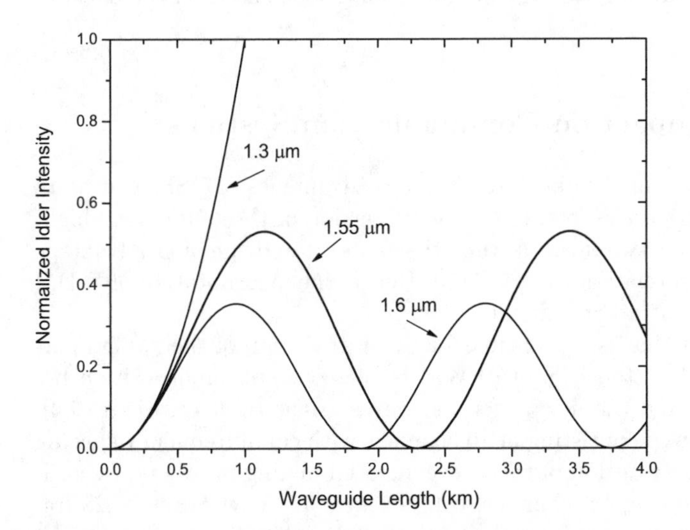

Fig. 7.9. Normalized idler intensities for a 4 km fiber when the attenuation is neglected. (channel spacing = 50 GHz, average wavelengths = 1.3, 1.55 and 1.6 μm, for the dispersion calculation pure silica was assumed)

longer than for a wavelength of 1.6 μm. For an average wavelength of 1.3 μm the coherence length is more than 13 km and therefore, the idler generation is very effective.

The FWM not only leads to a degradation of the system performance, it can be exploited for a number of applications as well. In this case, the phase mismatch has to be compensated. The easiest method for a compensation is using only small channel spacings but this method is limited due to the spectral width of the distinct channels.

In Sect. 4.10, we discussed birefringence as a possibility for phase matching in the case of second harmonic generation. Optical fibers are birefringent as well, as shown in Sect. 7.7.1, but the refractive index difference in standard optical fibers is too small to compensate for phase mismatch. Furthermore, the birefringence is arbitrary. On the other hand, the birefringence is constant in polarization-maintaining fibers and the refractive index difference is much higher than in standard single mode fibers. Hence, it is possible to exploit the birefringence for a phase matching in these fibers especially in the range of positive GVD. For carrier wavelength in negative GVD, the positive nonlinear refractive index n_2 can be used to compensate the phase mismatch [104].

The most often used method for phase mismatch compensation is the introduction of devices with a very small dispersion in the required wavelength range. It is possible to produce fibers with a high nonlinear coefficient and a zero-crossing of the dispersion in the C-band (HNL-DSF), for example. In microstructured fibers the dispersion behavior can be tailored according to the required needs. Furthermore, these fibers show a high nonlinear coefficient. Hence, they are very interesting for applications based on the effect of FWM.

7.4 FWM's Impact on Communication Systems

As already discussed in the above section, the advantages of DSF relating to their dispersion properties lead to a broad insertion of these fibers for high-bit-rate transmission systems. On the other hand, the dispersion advantage is a disadvantage in relation to the FWM. Hence, the incorporation of WDM into these systems is rather difficult.

The main effect that is responsible for the degradation of the system performance in WDM systems due to FWM is the coherent superposition between the original and the new generated waves at the receiver. Figure 7.10 shows a WDM system consisting of 10 channels with equal frequency spacing and the spectral distribution of the new generated mixing products. According to (7.2), the 10 original channels can generate 450 new waves in 28 frequency slots. As can be seen from the figure, most of the mixing products are directly superimposed with the original WDM channels. The two channels in the middle of the WDM band (5 and 6), for instance, are superimposed with 29 new waves.

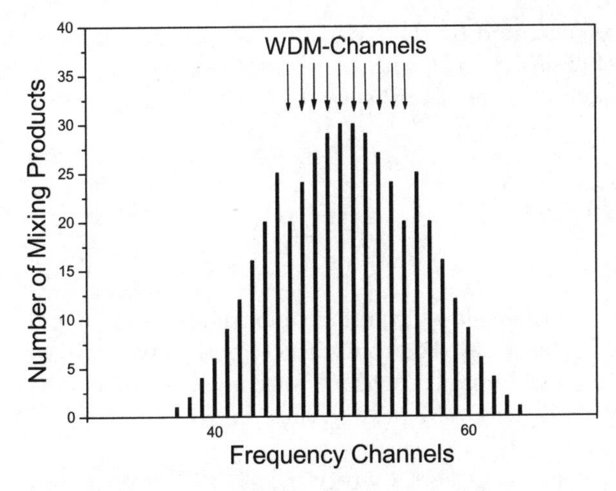

Fig. 7.10. Generation of new mixing products due to the FWM process in a system consisting of 10 WDM channels equally spaced in frequency

Figure 7.11 shows this fact again in detail. The interfering frequencies out of the WDM band can be suppressed by optical filters relatively easy. On the contrary, the interfering mixing products in the WDM band have the same frequencies as the original channels. A passive suppression of these distortions is impossible. Therefore, the mixing products that fall together with the original WDM channels are responsible for a degradation of the system performance.

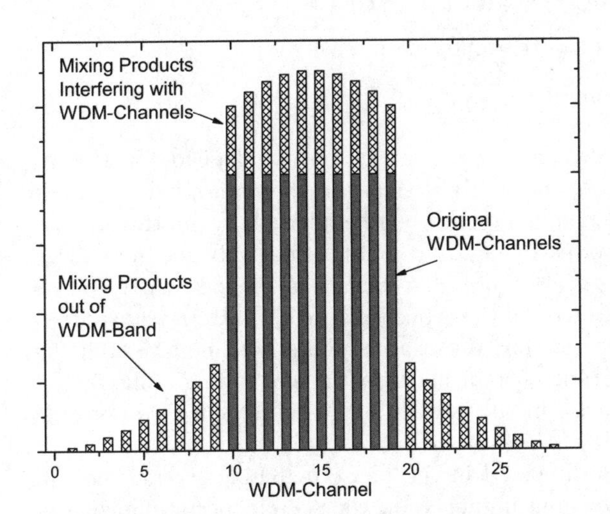

Fig. 7.11. FWM products for a 10 channel WDM system. The out-of-band products can be eliminated by optical filtering

Assume only one wave generated by the FWM process (E_{FWM}) and one wave in the WDM channel (E_{WDM}), both with the same frequency and wave-number. Hence, the two waves can be described as

$$E_{\mathrm{FWM}} = \hat{E}_{\mathrm{FWM}} \mathrm{e}^{\mathrm{j}(kz-\omega t+\varphi_1)} \tag{7.25}$$

and

$$E_{\mathrm{WDM}} = \hat{E}_{\mathrm{WDM}} \mathrm{e}^{\mathrm{j}(kz-\omega t+\varphi_2)}. \tag{7.26}$$

The waves differ only in their slowly varying amplitudes ($\hat{E}_{\mathrm{FWM}}$ and $\hat{E}_{\mathrm{WDM}}$) and their phases (φ_1 and φ_2). For convenience, both waves are assumed to be plane and monochromatic. In the transmission channel, the WDM wave superimposes with the FWM wave, giving rise to

$$E_{\mathrm{S}} = E_{\mathrm{FWM}} + E_{\mathrm{WDM}} = \left(\hat{E}_{\mathrm{FWM}} \mathrm{e}^{\mathrm{j}\varphi_1} + \hat{E}_{\mathrm{WDM}} \mathrm{e}^{\mathrm{j}\varphi_2}\right) \mathrm{e}^{\mathrm{j}(kz-\omega t)}. \tag{7.27}$$

Therefore, the resulting superimposed wave (E_{S}) has the same frequency and wave number as the original waves and only the amplitude differs. The photodiode in the receiver can only detect the intensity of the wave, not its amplitude. The intensity depends, according to (4.88), on the absolute value of the amplitude:

$$\begin{aligned} \left|\hat{E}_{\mathrm{s}}\right|^2 &= \hat{E}_{\mathrm{s}} \hat{E}_{\mathrm{s}}^* \\ &= \left(\hat{E}_{\mathrm{FWM}} \mathrm{e}^{\mathrm{j}\varphi_1} + \hat{E}_{\mathrm{WDM}} \mathrm{e}^{\mathrm{j}\varphi_2}\right) \left(\hat{E}_{\mathrm{FWM}} \mathrm{e}^{-\mathrm{j}\varphi_1} + \hat{E}_{\mathrm{WDM}} \mathrm{e}^{-\mathrm{j}\varphi_2}\right) \\ &= \hat{E}_{\mathrm{FWM}}^2 + \hat{E}_{\mathrm{WDM}}^2 + 2\hat{E}_{\mathrm{FWM}} \hat{E}_{\mathrm{WDM}} \cos\left(\varphi_1 - \varphi_2\right). \end{aligned} \tag{7.28}$$

Therefore, the receiver detects the following intensity

$$I_{\mathrm{s}} = I_{\mathrm{FWM}} + I_{\mathrm{WDM}} + 2\sqrt{I_{\mathrm{FWM}} I_{\mathrm{WDM}}} \cos\left(\varphi_1 - \varphi_2\right). \tag{7.29}$$

The current that the receiver delivers at its output depends on the intensity of the waves. Hence, the received signal in the photodiode is either reinforced or weakened, depending on the relative phase between the original channel and the mixing product. Due to the fact that not only one but a large number of mixing products fall into each WDM channel, the relative phase depends on the superposition of all these phases. The bit pattern transmitted in the channels is random and thus the relative phase is random as well. As a result, the receivers' output current shows a random fading. This fading manifests itself in an increase in the system noise and is hence, responsible for an increase of the BER in the system.

On the other hand, as discussed in the phase matching section, not all mixing products have the same influence on the system performance. According to (7.29), if the intensity of the FWM wave is small compared to the intensity of the wave in the WDM channel, its influence can be neglected. The intensity of the mixing products depends on the phase matching condition.

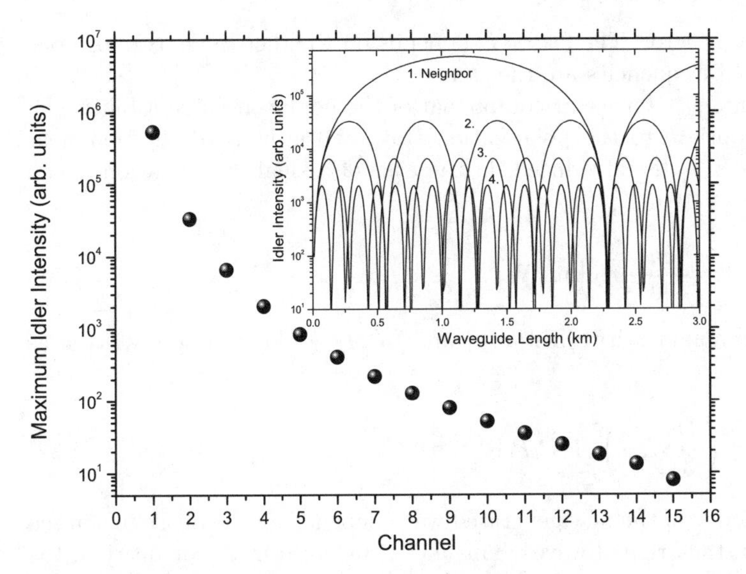

Fig. 7.12. Maximum idler intensity due to the mixing between a WDM channel and its next 15 neighbors for an average wavelength of 1.55μm and a channel spacing of 50 Ghz in pure silica. The inset shows the intensity fluctuation of the mixing products between the first 4 neighbors

The maximum idler intensity for the mixing of a WDM channel with its 15 neighbors is shown in Fig. 7.12. The average wavelength is 1.55μm and the channel spacing is 50 GHz, for the values of dispersion and dispersion slope pure silica was assumed. As can be seen from the Figure, the intensity of the mixing product for the next-but-one neighbor (2) will decrease dramatically when compared to the direct neighbor, at the same time – as shown in the inset – the coherence length is decreased as well. Hence, due to the dispersion, only mixing products between direct neighbors can be generated efficiently in standard single mode fibers. On the other hand, the dispersion in DSF is so small that an interaction between channels farther away can lead to mixing products with a significant intensity as well.

If three waves with the frequencies f_i, f_j and f_k interact with each other and the input signals are not depleted by the generation of the mixing products, the time-averaged optical power as a function of fiber length (L) and phase mismatch (Δk) for the new generated frequency component ($f_{i,j,k}$) can be written (in esu) as [108, 133]

$$P_{ijk}(L) = \frac{1024\pi^6}{n^4\lambda^2c^2}\left(\frac{D\chi_{1111}L_{\text{eff}}}{A_{\text{eff}}}\right)^2 P_iP_jP_k\mathrm{e}^{-\alpha L}\eta, \tag{7.30}$$

Where L_{eff} and A_{eff} are the effective area and length of the fiber; n is the refractive index of the core; λ the wavelength; c is the light velocity in free space; P_i, P_j and P_k are the input powers of the channels; and $D = 1, 3,$ or 6 is a degeneracy factor. The factor D depends on whether three, two, or none of the involved frequencies are the same.

The term χ_{1111} is a scalar component of the nonlinear susceptibility $\chi^{(3)}$ in esu, appropriate to the polarization state of the light. If the nonlinear refractive index n_2 is introduced, in the case of a single polarization it can be expressed as [109]

$$\chi_{1111}\,[\mathrm{esu}] = \frac{cn^2}{480\pi^2} n_2\,[\mathrm{m}^2/\mathrm{W}]\,. \tag{7.31}$$

Using the nonlinearity constant γ and (5.29), (7.30) can be rewritten as [85]

$$P_{ijk}(L) = \left(\frac{D}{3}\gamma L_{\mathrm{eff}}\right)^2 P_i P_j P_k \mathrm{e}^{-\alpha L}\eta. \tag{7.32}$$

The efficiency of the new generated wave depends on the phase mismatch (Δk) between the original waves. This fact is involved in the quantity η, the ratio of the wave power at the fiber output for a given phase mismatch to the output power for matched waves. This ratio can be written as [108]:

$$\eta = \frac{P_{ijk}(L,\Delta k)}{P_{ijk}(L,\Delta k = 0)} = \frac{\alpha^2}{\alpha^2 + \Delta k^2}\left(1 + \frac{4\mathrm{e}^{-\alpha L}\sin^2(\Delta k L/2)}{(1-\mathrm{e}^{-\alpha L})^2}\right) A \tag{7.33}$$

where α is the attenuation constant in km^{-1} and L is the fiber length. The additional factor A is only involved if multiple amplified spans are considered, with N_{A} as the number of amplified spans. It is represented by [85, 110, 116]

$$A = \left(\frac{\sin(N_{\mathrm{A}}\Delta k L/2)}{\sin(\Delta k L/2)}\right)^2 \tag{7.34}$$

with $\Delta k = k_{ijk} + k_k - k_i - k_j$ as the phase mismatch between the involved waves. If the propagation constant is expanded in a Taylor series about ω_k and only terms up to the third order are retained the phase mismatch is [108]

$$\Delta k = \frac{2\pi\lambda_k^2}{c}\Delta f_{ik}\Delta f_{jk}\left[D_{\mathrm{c}} + \frac{\lambda_k^2}{2c}(\Delta f_{ik} + \Delta f_{jk})\frac{\mathrm{d}D_{\mathrm{c}}(\lambda_k)}{\mathrm{d}\lambda}\right] \tag{7.35}$$

where D_{c} is the chromatic dispersion and the frequency difference is related to the absolute value, i.e., $\Delta f_{mn} = |f_m - f_n|$. For a two tone product with a channel spacing of Δf the phase mismatch is then

$$\Delta k = \frac{2\pi\lambda^2}{c}\Delta f^2\left[D_{\mathrm{c}} + \frac{\lambda^2}{c}\Delta f\frac{\mathrm{d}D_{\mathrm{c}}}{\mathrm{d}\lambda}\right]. \tag{7.36}$$

The efficiency against the channel spacing for different fiber length and fiber dispersion is shown in Fig. 7.13. As can be seen, the efficiency has the

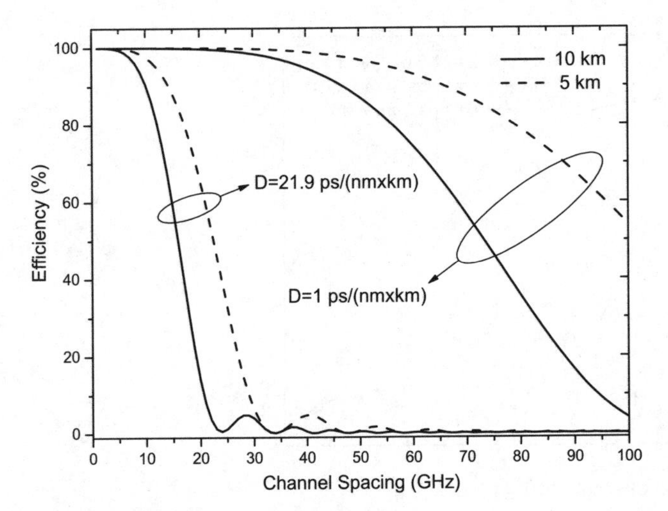

Fig. 7.13. Efficiency of the mixing process at the fiber output versus channel spacing for two different fiber lengths and dispersions. The attenuation constant is $\alpha = 0.2$ dB/km and the considered wavelength is $\lambda = 1550$ nm. The dispersion slope is assumed to be $\mathrm{d}D\,(1550\text{ nm})\,/\mathrm{d}\lambda = 0.065$ ps/(nm$^2\times$km)

expected $\sin x/x$ shape (compare to Fig. 4.18). For a relatively high dispersion of 21.9 ps/(nm×km), which corresponds to the dispersion of pure silica at a wavelength of 1550 nm, the phase mismatch increases very fast with channel spacing and hence the efficiency drops. For a fiber length of 10 km, the efficiency decays to the half for a channel spacing of 16 GHz. In a fiber with a length of 5 km, this happens for a channel spacing of 23 GHz. Whereas for a low dispersion of only 1 ps/(nm×km), the efficiency is also very high for a rather large frequency difference. In the example above for the 10 km fiber, the efficiency decreases by half for a channel spacing of 73 GHz.

At the maximum of the generated power, the phase mismatch is zero and hence, the efficiency $\eta \approx 1$. In this case, if all waves have the same input power, the ratio of the generated power P_{ijk} due to the FWM process to the power P_out of one channel at the fiber output is

$$\frac{P_{ijk}}{P_\mathrm{out}} = \left(\frac{D}{3}\gamma\right)^2 L_\mathrm{eff}^2 P_\mathrm{in}^2. \tag{7.37}$$

As expected from (7.22), the ratio scales with the square of the effective length and the power of the input waves. With a nonlinearity coefficient of $\gamma = 1.3$ W^{-1} km^{-1} and an effective length of $L_\mathrm{eff} \approx 22$ km, one yields $P_{ijk}/P_\mathrm{out} \approx 3.3 \times 10^{-3}\mathrm{mW}^{-2} \times P_\mathrm{in}^2$ if all the waves have different frequencies ($D = 6$). According to this result, the generated wave power at the fiber output comes in the range of the power of the WDM channels for an input power of ≈ 18 mW/channel. But in this case the assumption that the original

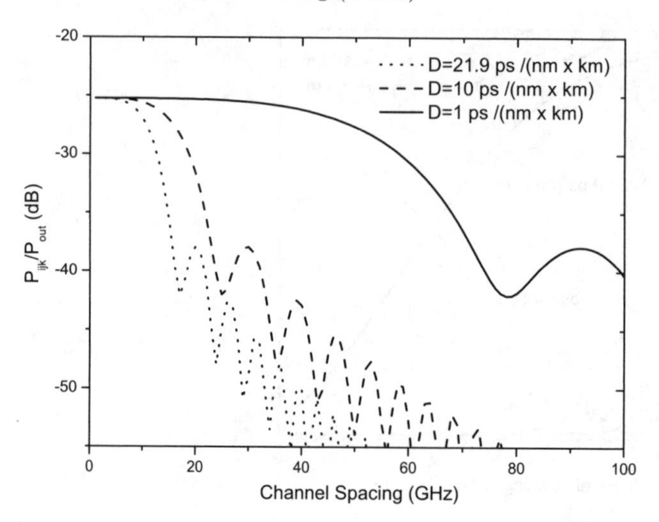

Fig. 7.14. Ratio between the power of the new generated channels and the original channels at the output of a 20km fiber ($\alpha = 0.2$ dB/km; $\lambda = 1550$ nm; $D\,(1550\text{ nm}) = 21.9, 10, 1$ ps/(nm×km); $\mathrm{d}D\,(1550\text{ nm})\,/\mathrm{d}\lambda = 0.065$ ps/(nm^2×km); $\gamma = 1.3\ \mathrm{W}^{-1}\ \mathrm{km}^{-1}$; $L_{\mathrm{eff}} = 22$ km, $P_{\mathrm{in}} = 1$ mW)

channels are not depleted by the FWM process no longer holds and hence, (7.37) is not valid.

Figure 7.14 shows the ratio of the output power of the mixing product to the power of the original channels for three different dispersion values against the channel spacing. The shape shows again a $\sin x/x$ form for the efficiency. In a standard single mode fiber the ratio decreases very fast with channel spacing. Whereas, for a small dispersion the ratio decreases very slowly.

7.5 FWM Suppression

In WDM systems, FWM is a serious problem where the introduction of different carrier wavelengths is limited by the mutual interaction between the channels, especially in dispersion-shifted fibers. Therefore, a large number of proposals have been made to suppress the FWM. For example the suppression can be carried out with a specialized modulation of the channels [111, 112] or with additional devices introduced into the transmission system [113]. Both approaches require making strong changes to the existing systems.

The easiest way to decrease the efficiency of the generation process is the reduction of the channel power in the fiber. As can be seen from (7.37), the ratio between the new generated wave and the original waves depends quadratically on the power of the input channels. But, on the other hand, if the input power is decreased, the signal to noise ratio will be decreased as well, leading to a higher BER of the system. In order to achieve a sufficient

BER, the signal to noise ratio must be preserved and hence, each channel needs adequate power. For instance, a 10Gbit/s system with an amplifier spacing of 100 km operating over 1000 km requires 1 mW/channel [114] and several dB of additional power are typically required for margins [109].

Another possibility of FWM suppression is offered by the phase matching condition, as shown in Fig. 7.7. If the phases between the pump waves on one hand and the signal and idler wave on the other do not match, the efficiency of the generation process is very low and the intensity of the new generated product fluctuates with the fiber length. In WDM systems, the phase matching between the channels depends on the dispersion of the fiber and the frequency spacing of the channels. As can be seen from Figs. 7.13 and 7.14, if the fiber dispersion is rather high, the power of the generated product decreases very fast with the separation of the channels. In the example above, if the minimum channel spacing of 50 GHz of the ITU-T G692 grid is assumed, the ratio between the power of the new wave and the original waves at the fiber output is only -56 dB if the fiber has a dispersion of 21.9 ps/(nm $\times$ km), corresponding to the dispersion of fused silica at 1550 nm. Therefore, for an input power of 1 mW, the power in the original signal channels is about 6 orders of magnitude higher than the power of the FWM product. On the other hand, if the fiber has a low dispersion of only 1 ps/(nm $\times$ km), a channel spacing of 50 GHz results in a ratio of -27.6 dB. Hence, the power in the signal channels is only about 500 times higher in this case.

On the other hand, in order to increase the offered bandwidth of optical transmission systems, the bit rate of every WDM channel has to be increased. For high bit rates a small dispersion of the fiber span is essential (see Sect. 5.3). For a 1000km transmission system with a bit rate of 10 Gbit/s, the net chromatic dispersion must be below 1000 ps/nm. Therefore, the average dispersion of the transmission fiber must be below 1 ps/(nm $\times$ km) [109].

According to the discussion above, the chromatic dispersion of the fiber plays an ambivalent role in high-bit-rate WDM systems. On one hand, a low dispersion offers the possibility for very high bit rates, but on the other hand, it is responsible for the degradation of the system performance due to FWM. An efficient FWM suppression requires a high dispersion, but this high dispersion limits the bit rate of the system. The so called "dispersion management" already mentioned in Sect. 5.4 offers a way out of this dilemma. A dispersion-managed system consists of Fibers with a high positive and a high negative dispersion. Both average each other out so that the total dispersion of the transmission system is low. At the same time each single span has a very high local dispersion, leading to a bad phase matching between the channels, and therefore, a suppression of the FWM.

Another possible way of FWM suppression in dispersion-shifted fibers is based on a method already proposed in the 1950s for the suppression of 3rd-order intermodulation interference in radio systems [115, 135]. If the carrier frequencies of the WDM channels are chosen so that they are unequally spaced, then no four-wave-mixing product is superimposed on any of the transmitted channels. The frequencies of all mixing products are different from the carrier frequencies of the original channels in this case. Therefore, crosstalk due to FWM is suppressed. On the other hand, the generation of mixing products is not suppressed. Hence, the data channels still experience a power penalty due to the FWM. Therefore, the maximum launched power is limited by the channel depletion caused by the FWM generation [135].

An unequal channel spacing in the frequency domain is given if the channels are equally spaced in the wavelength domain, for instance, as can be seen from Fig. 7.1, but this is not sufficient for the prevention of mixing products in the original channels.[2] In order to prevent the superposition between the new generated products and the original channels, the frequency separation of any two channels of a WDM system has to be different from that of any other pair of channels [135]. In other words, the difference between the frequencies of any two channels has to be unique in the system. If N is the number of channels that should be transmitted in the system and n is the number of available frequency slots at which the channels can be transmitted in the optical window of the fiber, then (7.1) can be used to yield the following [135]:

$$n_{ijk} = n_i + n_j - n_k, \tag{7.38}$$

with $k \neq i, j$ and

$$\forall i, j, k \in 1, \ldots, N(k \neq i, j) \,. \tag{7.39}$$

Then the condition above can be expressed as $n_{ijk} \notin (n_1, n_{2,\cdots}, n_N)$. For $n = 10$ out of $N = 80$, for instance, the frequency slots 0, 2, 6, 17, 22, 30, 31, 67, 70, and 77 fulfil the above conditions. The mixing products caused by these channels are shown in Fig. 7.15. A constraint is that the required bandwidth for the WDM system should be minimal. For any value of N and n several optimum solutions can be found with a computer search.

As can be seen from Fig. 7.15, in fact no mixing product superimposes with one of the original WDM channels (compare Fig. 7.15 with the result for an equal channel spacing of 10 WDM channels in Fig. 7.10). But, as already mentioned, this method eliminates only the superposition of the generated channels with the original ones. In the case of dispersion-shifted fibers, a rather high part of the power is still shifted from the WDM channels to the newly generated products. Another disadvantage of this method is that the required bandwidth for the system will be increased.

Another possibility for the FWM suppression was proposed by Xiang and Young [138]. This proposal is based on the fact that the FWM spectrum is

[2] Contrary to the situation in Fig. 7.1 each channel has a bandwidth that depends on the data rate.

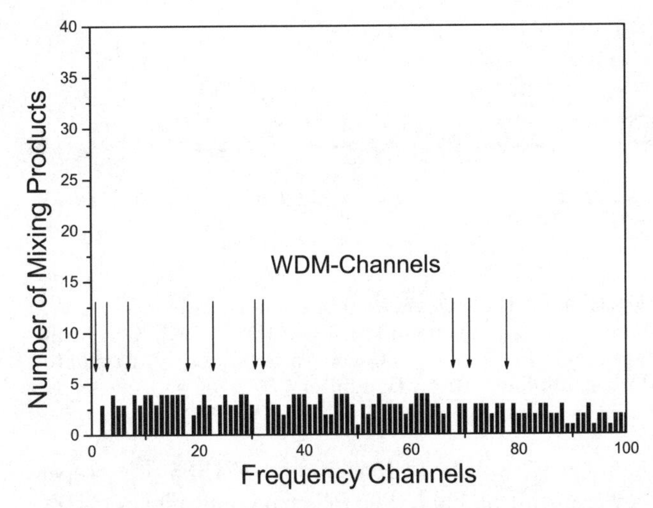

Fig. 7.15. Number of mixing products for an unequal channel spacing. The channel spacing between the WDM channels was chosen so that the condition described in the text is fulfilled. Only the mixing products falling in the range of the 80 time slots are shown in the figure

symmetrical around the point of dispersion zero-crossing in low dispersion fibers. For this technique, again each channel is divided into two subchannels, each of which is symmetrical around the point of zero-crossing of the dispersion. One of the subchannels transmits only logical ones ("1"), whereas the other will exclusively transport logical zeros ("0"). The signals of all users are combined and transmitted over the dispersion-shifted fiber. During their propagation, they experience attenuation and spectrum deformation due to FWM. At the receiver, narrow band filters select the desired user's two wavelengths related to the original WDM channel. The filtered result is detected with a "balanced" detector (the received signal is positive if a "1" is transmitted and is negative otherwise). This form of receiver can eliminate all noise that is experienced by both channels in the same way. Hence, it is possible to reduce the noise influence of FWM in the channels to first order.

7.6 Experimental Results

The experiments to investigate the noise influence of FWM on communication systems with dispersion-shifted fibers were carried out by M. Wolf at the Siemens AG in Munich, Germany [141]. The basic set-up for the experiments is shown in Fig. 7.16.

The 39 semiconductor lasers (λ_1 to λ_{39}) are working in continuos wave (CW) mode and generate constant carrier wavelengths between 1528.77 and 1560.61 nm. The frequency difference between two adjacent lasers is 100 GHz

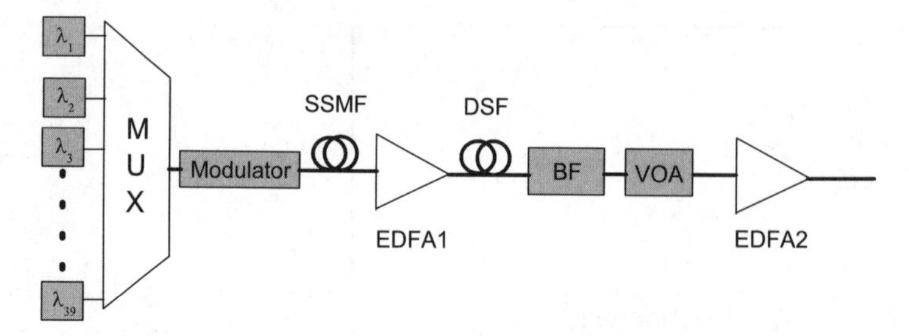

Fig. 7.16. Experimental set-up for the investigation of the noise influence of FWM on communication systems with dispersion-shifted fibers (DSF) (MUX = Multiplexer, SSMF = standard single mode fiber, VOA = variable optical attenuator, EDFA = erbium-doped fiber amplifier, BF = Bandfilter)

(ITU-T Grid). The channels 17 and 21 at 1544.53 and 1541.35 nm, respectively, are dark, i.e., they transmit no light. The following multiplexer (MUX) multiplexes all channels to one external Mach-Zehnder modulator. It modulates all 39 channels at the same time with a 10 Gbit/s NRZ signal. The dispersion in the 5 km long standard singlemode fiber (SSMF) is used to shift the bit sequences in the different wavelength channels to decorrelate them.

The resulting dense WDM (DWDM) signal is increased by an amplifier (EDFA1) and introduced into a 13 km long dispersion-shifted fiber (DSF). The zero point of the dispersion lies in this fiber at a wavelength of 1550 nm and, hence, in the used optical window. Since the signal power is kept rather low, the influence of other nonlinear effects can be neglected.

The spectrum of the signals is measured with an optical spectrum analyzer, not shown in Fig. 7.16. With the bandfilter at the fiber output, distinct channels can be separated to investigate their noise behavior. The variable attenuator behind the bandfilter is adjusted so that the input power at the preamplifier EDFA2 is the same for each channel. For the determination of the noise in each channel, the amplitudes of the signals are scanned synchronously. Due to the fact that the noise due to FWM is assumed to be a stochastic process, two probability distributions will appear: one for the sampling values around the logical zero, and the other around the logical one (see Fig. 7.20). The parameters of the probability distributions are used to determine the noise in the channel.

According to (7.2), 39 channels can cause 28,899 mixing products. The channel allocation of the experimental set-up and the following distribution of the mixing products is shown in Fig. 7.17.

Since the distinct DWDM channels have equidistant frequency spacing, all generated mixing products show the same frequency distance. Most of the mixing products superimpose with the original channels and cause noise there. Above and below the WDM channels, the number of possible combinations decreases symmetrically. Figure 7.17 shows that in the two dark

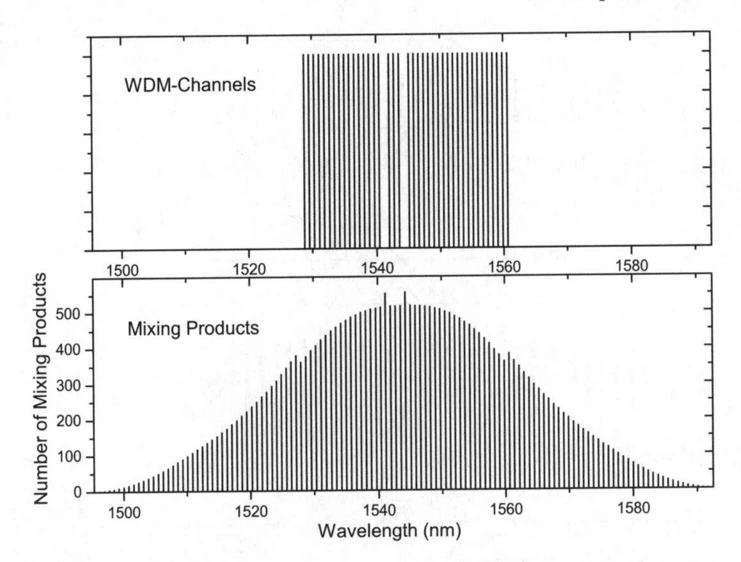

Fig. 7.17. Channel allocation and corresponding distribution of the mixing products for the experimental set-up

channels 17 and 21 at 1541.35 and 1544.53 nm, the greatest number of mixing products can be found (552 and 555[3]).

The power of the new waves generated depends on the phases of the interacting waves. The experimental results for a total input power of 10 mW (top) and 20 mW (bottom) is shown in Fig. 7.18.

As expected, the new frequencies generated show the same frequency spacing as the original WDM comb. The power of the mixing products increases with the input power of the channels because the efficiency of the FWM process is more effective for higher input powers. The mixing products that coincide with the two dark channels are clearly visible in Fig. 7.18. The total power of the mixing products in channel 21 is higher than that in channel 17. The mixing products with frequencies above the original DWDM band are clearly visible as well, with power decreasing as the wavelength increases. Below the DWDM band, no FWM waves are visible although, according to Fig. 7.17, the number of mixing products is distributed symmetrically around the DWDM band.

The power distribution of the FWM depends essentially on the phase matching and, hence, the dispersion of the fiber. Due to the fact that the point of zero-crossing of the dispersion is at 1550 nm in the used fiber, the generation of mixing products near this value is nearly phase-matched and, therefore, particular efficient. For mixing products with a central wavelength

[3]The maximum number of mixing products out of these two channels is 518 in channel 22.

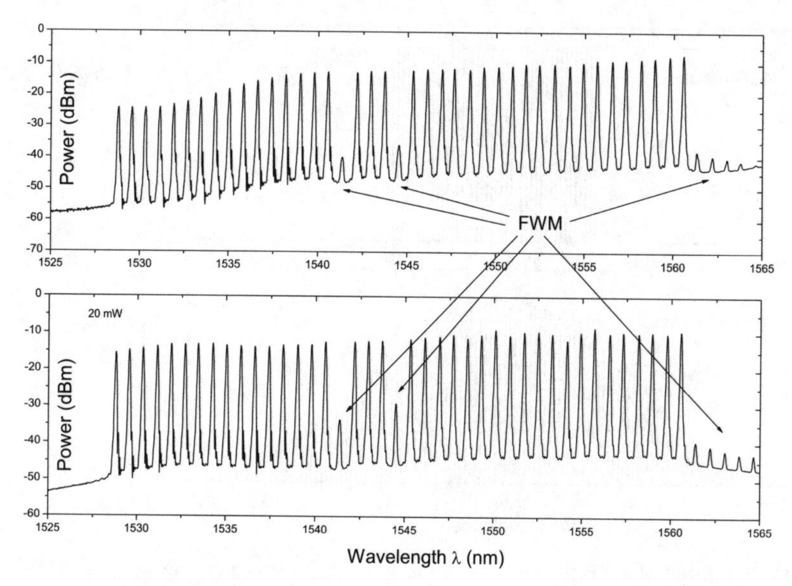

Fig. 7.18. Spectra of the DWDM comb at the output of a 13km dispersion-shifted fiber for a total input power of 10 (*top*) and 20 mW (*bottom*)

far away from the zero-crossing, the efficiency is much lower (channel 17 and FWM below the band).

The FWM products that coincide with the original DWDM channels can not be seen in Fig. 7.18 because they produce noise in these channels only. With the signal sampling method described above, the noise influence to the original channels can be investigated, as shown in Fig. 7.19.

The smallest signal to noise ratio and, therefore, the highest noise can be found, as expected, near the zero-crossing of the dispersion because the phase mismatch is low and the FWM products are generated with a high efficiency in this range. The relative phase between generated FWM and the data stream in the original channel leads to a fluctuation of the intensity that the detector measures and, hence, noise. At both sides of the zero-crossing the signal to noise ratio increases with increasing distance from this point.

The eye diagrams and the corresponding distribution of the amplitude sampling values for a channel far away (13) and near by (24) the zero-crossing are shown in Fig. 7.20. Channel 13 has a carrier wavelength of 1538.9 nm and is, therefore, far away from the zero-crossing of the material dispersion. Hence, the phase matching condition is not fulfilled and the noise generated by FWM is rather small in this channel. The samples of the amplitude are concentrated around the values for a logical "0" and "1" the eye is wide open.

On the other hand, channel 24 has a carrier wavelength of 1548.9 nm and is, therefore, near the point of zero-dispersion in the fiber with a wavelength of 1550 nm. As a result, FWM has a great impact on the noise in the channel.

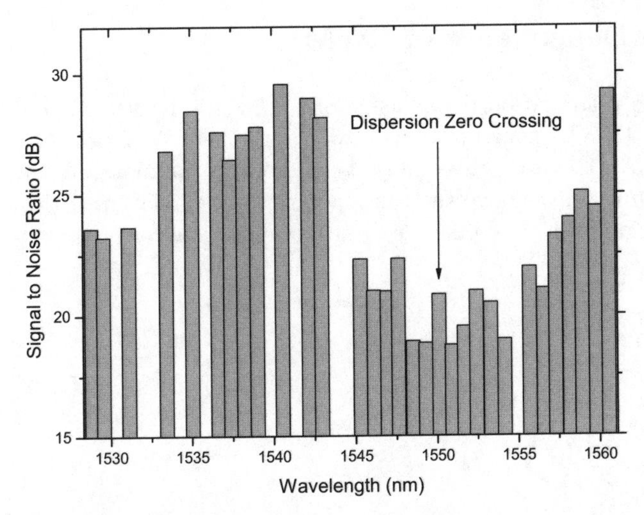

Fig. 7.19. Signal to noise ratio of the distinct channels after the propagation in a dispersion-shifted fiber with a length of 13 km

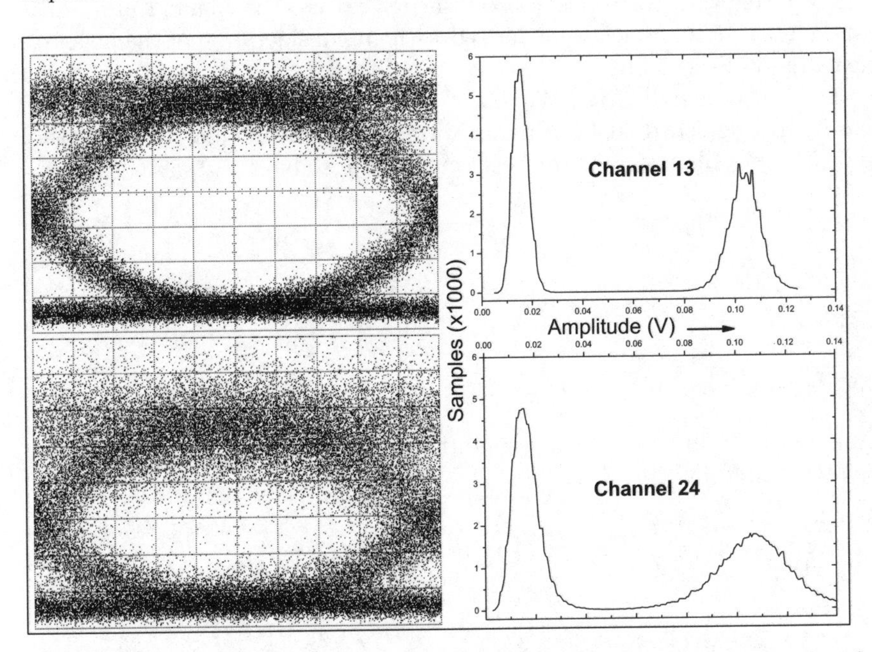

Fig. 7.20. Eye diagram and corresponding distribution of the amplitude values for two channels out of the DWDM comb

The amplitude shows a much broader distribution and the opening range of the eye is decreased. Hence, the number of wrong decisions and, therefore, the BER is increased in this channel.

7.7 Polarization Dependence of FWM

The equations for the polarization dependence of FWM are derived in the Appendix (A.34)–(A.41). To generate a new wave via the FWM process, according to (A.34)–(A.41), three waves must be present in the material. In the easiest case, all three waves have the same polarization direction. If the pump as well as the signal wave are polarized in the x-direction, for instance, we have

$$A_{1x} = 1, \; A_{1y} = 0;$$

$$A_{2x} = 1, \; A_{2y} = 0; \tag{7.40}$$

and

$$A_{3x} = 1, \; A_{3y} = 0.$$

Hence, the differential equation system (DES) is reduced to (A.34)–(A.37) and all terms with the factor 1/3 vanish. As expected, it results to the equations (7.13)–(7.16) which were derived without consideration of the polarization state in Sect. 7.2.

For the perpendicular polarization of the waves, several different cases can be distinguished. If both pump waves are perpendicular to each other and if the signal is parallel to the second pump we have:

$$A_{1x} = 1, \; A_{1y} = 0;$$

$$A_{2x} = 0, \; A_{2y} = 1; \tag{7.41}$$

and

$$A_{3x} = 0, \; A_{3y} = 1.$$

From the DES and (A.37), the generation of an idler wave in x-direction can be represented by A_{4x}:

$$\frac{\partial A_{4x}}{\partial z} = 2\mathrm{j}\gamma P_1 [\left(\frac{1}{2} |A_{4x}|^2 + |A_{1x}|^2 + \frac{1}{3} \sum_{i \neq 1} |A_{iy}|^2 \right) A_{4x} \tag{7.42}$$
$$+ \frac{1}{3} A_{3y}^* A_{1x} A_{2y} \mathrm{e}^{-\mathrm{j}2\Delta k_{\mathrm{W}} z}]$$

From the factor 1/3, it can be deduced that the intensity of the idler ($I \backsim |A_4|^2$) is reduced to 1/9 when compared to the case where all waves have the same polarization. All other possible cases are summarized in Table 7.3. The distinct columns show the responsible differential equation, the polarization of the generated idler wave and the normalized intensity of the idler.

Table 7.3. Polarization and normalized intensity for all possible polarization states of the input waves

Case	Example	Equation	Polarization	I
$A_1 \parallel A_2 \parallel A_3$	$A_{1y}A_{2y}A_{3y}$	(A.41)	y	1
$A_1 \perp A_2 \parallel A_3$	$A_{1x}A_{2y}A_{3y}$	(A.37)	x	1/9
$A_1 \perp A_2 \perp A_3$	$A_{1x}A_{2y}A_{3x}$	(A.41)	y	1/9
$A_1 \| A_2 \perp A_3$	$A_{1x}A_{2x}A_{3y}$	–	–	0

According to Table 7.3, the intensity of the idler is decreased to 1/9 irrespective of the signal polarization if the pump waves are perpendicularly polarized. Only the polarization direction of the generated wave will be changed. If the signal is perpendicular to both pump waves, according to (A.34)–(A.41), no idler is generated in the fiber. This behavior is a direct result of the phase matching condition. The FWM term for this case would be

$$\frac{1}{3}A_{3y}^{*}A_{1x}A_{2x}e^{j(k_{1x}+k_{2x}-k_{3y}-k_{4y})z}. \tag{7.43}$$

In this case, the phase matching depends on the refractive index difference in x- and y- direction. With $\Delta n/n \approx 10^{-7} - 10^{-5}$, the coherence length for this mixing product can be between 4 and 40 cm. Therefore, the term was neglected in the derivation in the Appendix.

As mentioned in Sect. 2.7, due to the birefringence in the fiber, the polarization of the light changes during the propagation of the wave in the fiber. The polarization state of the input waves remains the same only for a few meters. This fact is neglected in the above equations. The influence of an arbitrary alteration of the polarization to the FWM efficiency can only be calculated with a numerical simulation.

7.7.1 Numerical Simulation

The DES in the Appendix (A.34)–(A.41) is a so called "stiff problem" and can be integrated with numerical methods tailored for these tasks. For the integration, the birefringence of the fiber must be taken into account. The birefringence in a standard fiber is completely random (see Sect. 2.7), but it is assumed that it is constant over short segments. For the simulation, this means that the fiber is divided into segments with birefringent axes. The orientation of the birefringent axes is determined by a random generator for every segment. The minimal length of the segments is determined by the constraint that for a further decrease of the segments length, the result of the integration over the entire fiber must remain the same. The fiber with its random axes is shown in Fig. 7.21.

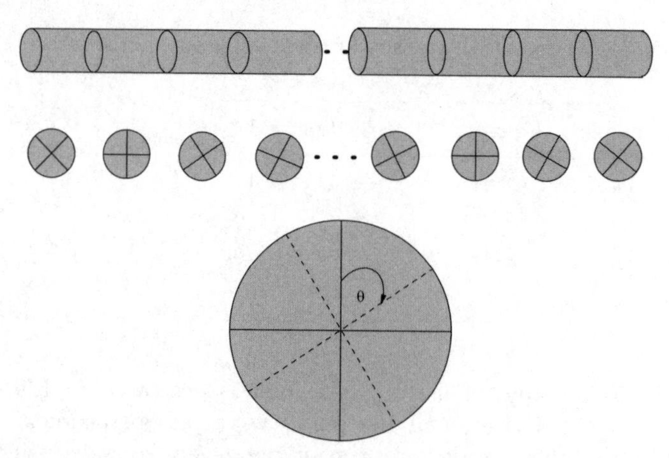

Fig. 7.21. Partitioning of a fiber into axes with a random variation of birefringence and rotation of the axes between two adjacent segments

The equations of the DES that describe the propagation of the amplitudes of the four waves with its x- and y-components are newly separated according to the position of the birefringent axes at the beginning of each segment. The separation is determined by equation [143]:

$$\begin{vmatrix} A'_{nx} \\ A'_{ny} \end{vmatrix} = \begin{vmatrix} \cos\theta & \sin\theta \mathrm{e}^{\mathrm{j}\varphi} \\ -\sin\theta \mathrm{e}^{-\mathrm{j}\varphi} & \cos\theta \end{vmatrix} \begin{vmatrix} A_{nx} \\ A_{ny} \end{vmatrix} \tag{7.44}$$

Equation (7.44) describes a rotation of the birefringent axes of segment $n+1$ with respect to segment n about the arbitrary angle θ. The angle θ varies from 0 to π. The rotation of the coordinate system is also shown in Fig. 7.21.

The angle φ represents an additional phase term. Due to the different refractive indices in the two principle axes, the waves propagate with slightly different velocities. This velocity difference causes a phase shift between the waves and is expressed by the arbitrary angle (φ). With the model, all possible polarizations can be described.

During the propagation of the waves through the nonlinear waveguide, the power of the pump waves drops due to the nonlinearity, whereas the power of the signal and idler waves increases at the same time. Since the attenuation is neglected in the DES, the total power over all waves in the waveguide must remain the same. This fact can be exploited to control the accuracy of the numerical simulation.

7.7.2 Results

If birefringence is neglected, the integration of the DES under consideration of the polarization state must deliver the same results as the deliberation

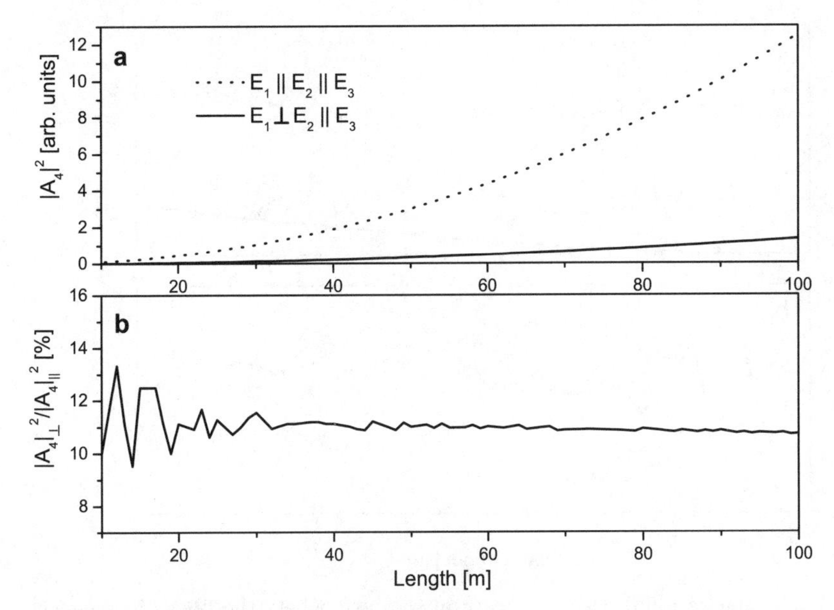

Fig. 7.22. Generation of the idler wave without consideration of birefringence in the fiber. **a** Idler wave for parallel and perpendicular polarization of the input waves. **b** Ratio between perpendicular and parallel polarization. ($\gamma = 0.0176\ \mathrm{m}^{-1}\mathrm{W}^{-1}$,$\Delta n = 1 \times 10^{-7}$, $\Delta f = 55$ GHz, $P_1 = 0.1$ W, $E_1 = E_2 = 1, E_3 = 0.1$)

shown in Table 7.3. In this case the DES, (A.34) to (A.41), is integrated over the fiber length z. The result is shown in Fig. 7.22.

In Fig. 7.22a the idler wave for a perpendicular and a parallel polarization of the pump waves is shown. The power of the idler is increased for both polarizations during the propagation because the distance is much smaller than the coherence length in the fiber. The optical power is translated from the two pump waves to the signal and idler waves. If both pump waves have a parallel polarization the generation of the idler has a higher efficiency as if both waves have a perpendicular polarization. The ratio between these two powers of the generated idler is shown in Fig. 7.22b.

As expected from the DES (Table 7.3), for the case of an orthogonal polarization of the pump waves the power is decreased to 1/9, or 11%, in comparison to the parallel polarization.

The relations under consideration of the birefringence are shown in Fig. 7.23. As mentioned in the former section, the birefringent axes are turned around an arbitrary angle in each segment of the fiber. For Fig. 7.23, the length of the segments was set at 5 m.

Again the power is coupled from the pump to the signal and idler waves during the propagation in the fiber. Contrary to the formerly considered case, for an orthogonal polarization of the pump waves the idler is increased much stronger. The ratio between orthogonal and parallel polarization is 1/4 or

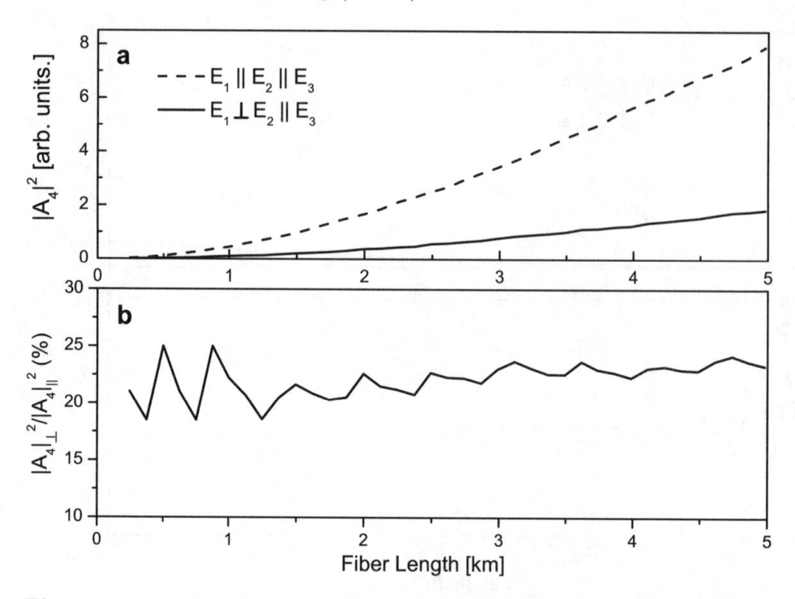

Fig. 7.23. Generation of the idler wave in the fiber where the fiber birefringence is included. **a** Idler for parallel and perpendicular polarization of the input waves. **b** Ratio of the powers for perpendicular and parallel polarization. ($\gamma = 0.00235$ $\mathrm{m^{-1}W^{-1}}$, $\Delta n = 1 \times 10^{-5}$, $\Delta f = 55$ GHz, $P_1 = 0.004$ W, $E_1 = 0.8$ $E_2 = 1, E_3 = 0.7, \Delta\mathrm{Disp} = 5$ m)

25%. Hence, under real conditions – if the birefringence of the fiber is taken into account – the FWM is much less polarization dependent as could be expected from the DES alone.

If the fiber length or the channel separation is increased, the phase mismatch leads to a saturation and to a decrease of the power of the idler wave. For longer distances the idler power is periodic along the propagation distance z. Figure 7.24 shows the idler wave for a fiber length of 26 km for an orthogonal and a parallel polarization where the birefringence in the fiber is taken into consideration. In the upper diagram, after a propagation distance of around 20 km, the idler wave generation comes to a saturation because it is the range of the coherence length in the fiber. For a further increase of the propagation distance, the idler wave would decrease and then increase again and so on. As can be seen from the figure, this fact has no influence on the ratio between the powers of the perpendicular and parallel polarized waves. The ratio still remains at 25% along the hole distance in the fiber.

In the earlier figures the signal wave is always polarized parallel to the second pump wave. Figure 7.25 shows the idler wave against the input angle of the signal wave. The two pump waves are perpendicularly polarized to each other. As can be seen from the figure, the generation of the idler wave is independent of the polarization of the signal wave if both pump waves are orthogonally polarized. This is especially important for applications of

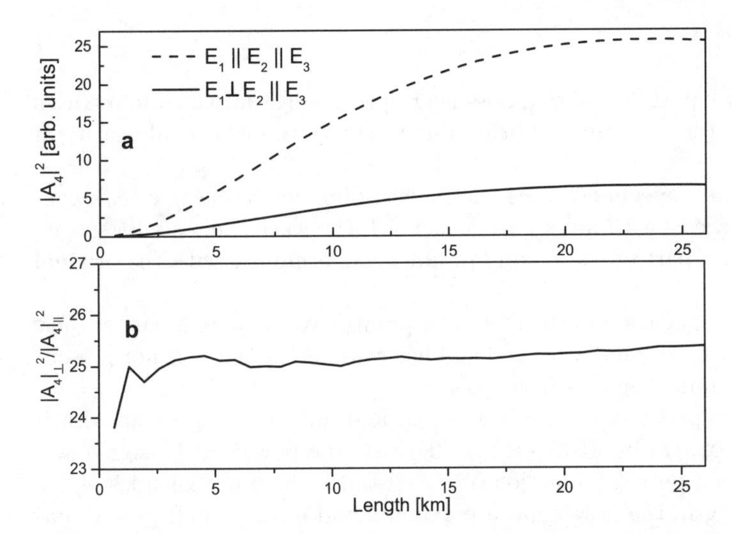

Fig. 7.24. Generation of the idler wave in a 26km long fiber where the fiber birefringence for different input polarizations of the waves is included. **a** Idler wave for orthogonal and perpendicular polarization of the input waves. **b** Ratio between both polarizations. ($\gamma = 0.00235\ \mathrm{m}^{-1}\mathrm{W}^{-1}$,$\Delta n = 1 \times 10^{-7}$, $\Delta f = 3.6$ GHz, $P_1 = 0.004$ W, $E_1 = 1.0$ $E_2 = 1.0$, $E_3 = 1.0$, ΔDisp $= 5$ m)

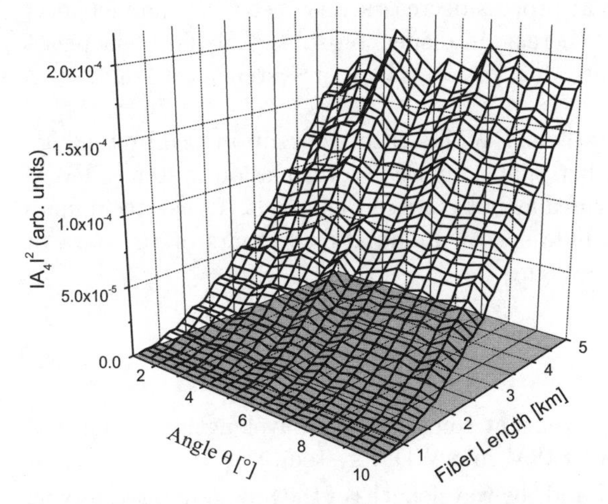

Fig. 7.25. Idler wave against the input angle of the signal wave. The two pump waves are polarized orthogonal to each other ($\gamma = 0.00235\ \mathrm{m}^{-1}\mathrm{W}^{-1}$, $\Delta n = 1 \times 10^{-5}$, $\Delta f = 55.4$ GHz, $P_1 = 0.004$ W, $E_1 = 0.8$ $E_2 = 1.0$, $E_3 = 0.7$, ΔDisp $= 5$m)

the FWM, such as wavelength converters or parametric amplifiers. Due to the birefringence in optical fibers, the input polarization of the signals introduced in such devices is completely random. These devices must, therefore, be polarization-independent.

Summary

- The effect of FWM describes a nonlinear optical effect at which four waves or photons interact with each other due to the third order nonlinearity of the material.
- In a nonlinear waveguide, three waves with different frequencies can generate 9 new waves with mixing frequencies. If the waves are equally spaced in frequency, a part of the mixing products superimpose with the original waves.
- During the propagation in the fiber, two pump waves deliver a part of their optical power to a signal wave and an idler wave. If the signal is not present at the fiber input, it grows from noise.
- In the case of phase matching, the signal and idler waves grow quadratically with propagation distance z. Otherwise, the power of the signal and idler waves is a periodic function of the distance. At odd multiples of the coherence length, the power has a maximum and at even multiples a minimum.
- The signal in the photodiode of the receiver is either reinforced or weakened, depending on the relative phase between the original channel and the superimposed mixing product .
- FWM leads to a reduction of the SNR and, hence, an increase of the BER.
- The efficiency of the FWM process decreases very fast with channel spacing in dispersive fibers. Whereas, in dispersion-shifted fibers, it decreases slowly. Therefore, the introduction of WDM in systems with DSF is a severe problem.
- In dispersion-managed systems, FWM is effectively suppressed by the high local dispersion of the fiber spans. In DSF transmission systems, FWM suppression is possible via unequally spaced channels. Other techniques of FWM suppression include special modulation formats of the data or additional devices in the system.

Exercises

7.1. Write a computer program that computes the wavelength of the generated mixing products due to FWM in a WDM system.

7.2. Determine the number and the wavelengths of the new generated mixing products in a WDM system consisting of 4 channels in the C-band with a frequency spacing of 100 GHz. ($\lambda_1 = 1531.9$ nm, $\lambda_2 = 1532.68$ nm, $\lambda_3 = 1533.47$ nm, and $\lambda_4 = 1534.25$ nm)

7.3. A 5km HNL-DSF ($\gamma = 18$ W^{-1} km^{-1}, $D = 23$ ps/(km×nm), $\alpha = 0.5$ dB/km) is used as a wavelength converter for the C-band. Estimate the ratio between the converted wave and the signal wave at the fiber output if the signal and pump waves have the same input power of 1 mW, the phase mismatch is negligibly small, and all waves have different frequencies.

7.4. Write a computer program that searches N channels out of n possible frequency slots while fulfilling the conditions (7.38) and (7.39).

8. Intrachannel Nonlinear Effects

The backbone of optical transmission systems should provide as much bandwidth as possible for the lowest possible costs in order to make new wide band applications affordable and, therefore, attractive to each subscriber. One possible to increase the transmittable bandwidth of each fiber in the system is the combination of WDM with high-bit-rate time domain multiplexing (TDM) systems. A bit rate of 10 Gbit/s/channel is widely used today whereas the upgrade to 40 Gbit/s/channel and beyond is already planned. With 100 WDM channels, this leads to a transmittable bit rate of 4 Tbit/s per fiber.

The performance of high-bit-rate TDM systems suffers from the nonlinear interactions between the bits in the channels (intrachannel). To reduce this intrachannel interaction it is possible to reduce the pulse widths and to increase the spacing between the pulses so that they only barely overlap. Since the dispersion-compensation distance $L_{\mathrm{map}}\tilde{}T^2/|D|$ decreases quadratically with the bit rate T^{-1} high-bit-rate transmission systems require very short dispersion-compensation distances [90]. In practical systems L_{map} should not be smaller than the distance between the amplifiers. On the other hand, the nonlinear interaction between the pulses increases if the spacing between them is decreased but, it reaches a maximum if pulse width and spacing are equal. If the pulses are closer together, i. e., if they overlap significantly, the interaction decreases again. So, as a key technology for high-bit-rate optical systems, the transmission of return-to-zero pulses in dispersion-managed fiber spans is widely accepted today. In this scheme the very short pulses are propagated in standard single mode fibers with a dispersion of $D = 17$ ps/(nm $\times$ km) or in nonzero dispersion-shifted fibers with $D = 2 - 5$ ps/(nm $\times$ km) [87, 107]. Due to their short temporal width, the pulses have a large spectrum, the dispersion map of the transmission system is much greater than the dispersion length $L_{\mathrm{map}} \gg L_{\mathrm{D}}$. Therefore, the pulses are rapidly dispersed in the fiber, spreading over a very large number of bits. In the dispersion-compensation element of the transmission system, the pulses are compressed back to their original width. In the next span they will spread again and so on.

If we assume a duty cycle of 33% for the 40 Gbit/s RZ system and, therefore, a FWHM pulse width at the fiber input of 1/(3$\times$ bit rate), (5.47) will yield a dispersion length of $L_D \approx 1.15$ km for the standard single mode fiber. The nonlinear length for a 1mW pulse is $L_{NL} \approx 770$ km if we assume a nonlinear coefficient of $\gamma = 1.3$ W^{-1} km^{-1}. Hence, it is clear that the nonlinear length far exceeds the dispersion length. These systems are thus very tolerant against interchannel fiber nonlinearities like FWM and XPM, enabling error free transmission over long distances [188]. The propagation of the pulses can be considered quasilinear. Quasilinear transmission is different from the dispersion-managed soliton (DMS) systems mentioned in the next Chapter. Although they also exploit the technique of dispersion management, in DMS systems, the nonlinearity is essential to preserving the pulse shape. Whereas in quasilinear systems, the nonlinearity can be treated as a very small perturbation.

As an explanation of this quasilinear regime, it can be seen that due to the much longer nonlinear length of the fiber in comparison to the dispersion length, the intensity pattern of the pulses changes rapidly during propagation. Hence, the effect of fiber nonlinearity will be averaged out [89]. At the same time, all dispersion-dependent effects like FWM and XPM are very weak due to the high local dispersion.

Although the interchannel nonlinearities like FWM and XPM are effectively suppressed and the intrachannel effects are reduced in quasilinear transmission, nonlinear interactions between the bits in one channel occur nevertheless. These nonlinear effects have their origin in the overlapping of the pulses due to the high dispersion. Intrachannel effects can lead to a significant system penalty and are a limiting factor of high-bit-rate dispersion-managed systems [87].

In principle, three different intrachannel effects are possible. The first intrachannel effect is of course the SPM, already treated in Sect. 6.1. The SPM depends on the intensity of the pulses and can be effectively suppressed by the dispersion map of the system. The second one is the intrachannel cross-phase modulation (IXPM), leading to a timing jitter of the pulses, and the third is the intrachannel four-wave-mixing (IFWM), resulting in an amplitude jitter and the generation of ghost pulses. Due to the fiber dispersion, many frequency components of the short pulse can interact nonlinearly with each other, leading to a time discrete FWM [94]. With the exception of SPM these effects are a direct result of the strong dispersion management of the system because all of them can only take place as a result of the temporal overlap at least of parts of the pulses during their transmission in the nonlinear waveguide.

While SPM and IXPM can be effectively suppressed by an intelligent system design [88]–[92], the IFWM is a limiting factor of long-haul high-bit-rate transmission [93].

In this chapter, we have a rather simple mathematical description of intrachannel nonlinear effects based on the NSE. This will be followed by a treaty of the impact of intrachannel effects on communication systems. The last section will introduce possible methods for the suppression of these effects.

8.1 Mathematical Description

A nonlinearity that acts on the bits of a bit pattern in one channel is responsible for the interaction between the distinct pulses. The physical relationships here are in principle the same as those for the other nonlinear effects. Hence, it is possible to describe the interaction with the NSE (5.37) as follows

$$\mathrm{j}\frac{\partial u}{\partial z} - \frac{k_2}{2}\frac{\partial^2 u}{\partial T^2} + \mathrm{j}\frac{\alpha}{2}u = -\gamma \left|u\right|^2 u, \tag{8.1}$$

where γ is again the nonlinearity coefficient responsible for the strength of the nonlinear effects, k_2 determines the group velocity dispersion (GVD) and α is the attenuation constant in the fiber. Here, a system with strong dispersion management is under consideration. Therefore, due to the high local dispersion and the relation between the dispersion and nonlinear lengths, we can assume the strength of the nonlinearity is relatively weak. Hence, it is possible to describe the nonlinearity as a weak perturbation of the otherwise linearly propagating pulses. With this assumption the amplitude (u) of the electric field propagating in the quasilinear system can be written as [89, 96]:

$$u\left(z,t\right) = u_{\mathrm{l}}(z,t) + u_{\mathrm{p}}(z,t), \tag{8.2}$$

where u_{l} is the solution of the NSE for the linear case ($\gamma = 0$), whereas u_{p} is the nonlinear perturbation of this linear solution induced by the weak nonlinearity. The linear solution of the NSE was shown in Sec. 5.3; for two Gaussian pulses it leads to the sum of two dispersively spreading pulses with a Gaussian shape. Due to the fact that the nonlinearity is zero the two pulses show no interaction with each other. Therefore, only the nonlinear perturbation is important. If (8.2) is introduced into (8.1) for the perturbation we have

$$\mathrm{j}\frac{\partial u_{\mathrm{p}}}{\partial z} - \frac{k_2}{2}\frac{\partial^2 u_{\mathrm{p}}}{\partial T^2} + \mathrm{j}\frac{\alpha}{2}u_{\mathrm{p}} = -\gamma \left|u_{\mathrm{l}} + u_{\mathrm{p}}\right|^2 \left(u_{\mathrm{l}} + u_{\mathrm{p}}\right). \tag{8.3}$$

According to the prerequisite that the nonlinear part is very small compared to the linear one, it can be neglected on the right hand side. Therefore, the NSE becomes

$$\mathrm{j}\frac{\partial u_{\mathrm{p}}}{\partial z} - \frac{k_2}{2}\frac{\partial^2 u_{\mathrm{p}}}{\partial T^2} + \mathrm{j}\frac{\alpha}{2}u_{\mathrm{p}} = -\gamma \left|u_{\mathrm{l}}\right|^2 u_{\mathrm{l}}. \tag{8.4}$$

For convenience, we assume here that only two pulses propagate in the fiber, such as the bit pattern "11". Hence, the signal at the fiber input is the superposition of two pulses with a time shift to each other. Since both pulses are in the same wavelength channel, they have the same wave-vector and frequency. In order to simplify the derivation drastically, the pulse shape should not be of interest here, although the nonlinearity has in fact an influence on the shape. Solutions for two Gaussian pulses can, for instance, be found in [96] whereas solutions for delta functions as well as solutions for more than two Gaussians are derived in [95]. With the above strong simplification, the signal at the fiber input is

$$u_l = u_1 e^{j\omega(t+\tau)} + u_2 e^{j\omega(t-\tau)}. \tag{8.5}$$

The spacing between the pulses is 2τ and $t = 0$ is set in the center between them. If (8.5) is introduced into (8.4) the right side can be expanded as follows

$$\begin{aligned} -\gamma |u_l|^2 u_l = -\gamma u_l u_l^* u_l = -\gamma\Big(& \left(|u_1|^2 + 2|u_2|^2\right) u_1 e^{j\omega(t+\tau)} \\ & + \left(|u_2|^2 + 2|u_1|^2\right) u_2 e^{j\omega(t-\tau)} \\ & + u_1^2 u_2^* e^{j\omega(t+3\tau)} + u_2^2 u_1^* e^{j\omega(t-3\tau)}\Big). \end{aligned} \tag{8.6}$$

The terms in the first bracket of the right side describe the influence of the nonlinearity on the first pulse at the position $t + \tau$ (the left). From a comparison, for instance, with (6.22), it can be deduced that the first term in the brackets describes the SPM, already treated in Sect. 6.1. The second term is responsible for a mutual interaction between both pulses that interacts on the first pulse. This is the intrachannel XPM (IXPM). The terms in the second bracket describe the same behavior for the pulse at the position $t - \tau$. The last two terms in (8.6) show that two new pulses are generated by the nonlinear process: one at the position $t + 3\tau$ and the other at $t - 3\tau$. These are the ghost pulses. Due to its similarity to the interchannel interaction of waves with different frequencies, this effect is called intrachannel FWM (IFWM).

The fiber nonlinearity has its origin in the distortion of the electron orbit due to the applied field. This light matter interaction is very fast (see Sect. 4.3). Hence, the nonlinearity of the material system, described by the nonlinearity constant (γ) is instantaneous. Therefore, for the occurrence of both the IXPM and IFWM, the product of the amplitudes of the distinct pulses must be greater than zero, e.g., $2|u_2|^2 u_1 \neq 0$, $u_1^2 u_2^* \neq 0$. This means that these effects can only take place if at least parts of the pulses overlap temporally during their propagation in the system. If the dispersion is low and the pulses are not dispersed enough intrachannel effects cannot occur.

8.2 SPM and IXPM

The SPM leads to a phase alteration of the pulse due to its own intensity (see Sect. 6.1). The phase alteration will be translated into a frequency change

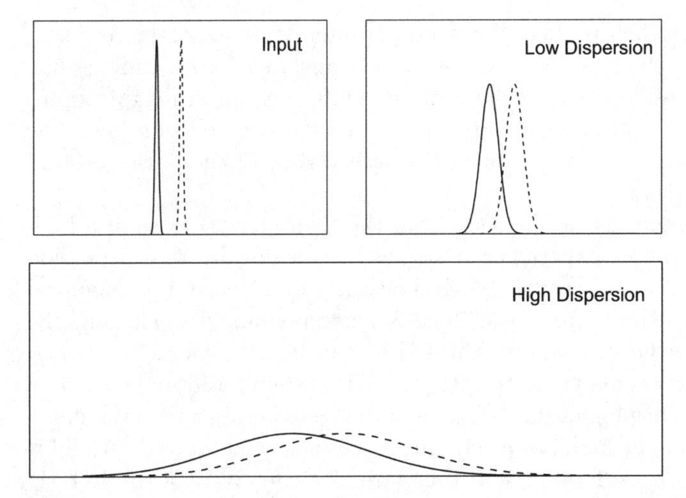

Fig. 8.1. Pulses in two distinct time slots in the case of a high and low dispersion in the transmission system

symmetrically around the pulse center. The slope of the pulse is responsible for this self induced interaction. Hence, especially short pulses with a very fast intensity variation over their time-dependent profile, as used in high-bit-rate systems, will experience a strong frequency broadening. Together with the dispersion of the fiber, the frequency broadening can cause an additional time spread of the pulse. But, as worked out in Sect. 6.1, if this effect is included in the system design it can be cancelled by dispersion management.

The intrachannel XPM for the left pulse at the position $t + \tau$ is described by the term $2j\gamma \left|u_2\right|^2 u_1$. From a comparison with (6.15) it can be seen that the IXPM is the intrachannel analogue to the interchannel XPM described in Sect. 6.2. The nonlinear interaction between the pulses leads to an alteration of the phase of both. The phase alteration will be translated into a frequency change or a chirp, respectively. Contrary to the case of SPM, the frequency alteration is not symmetric. According to (6.32), the frequency alteration that one pulse experiences depends on the temporal derivation of the shape of the other. Figure 8.1 shows an example of the interaction between two pulses.

At the fiber input the two pulses are in their adjacent time slots. Due to the dispersion of the dispersion-managed transmission system, they are temporally broadened during the propagation, then compressed again and so on. A mutual interaction due to IXPM can only happen if at least parts of the pulses are temporally overlapped during their transmission in the system.

According to Fig. 8.1, if the dispersion is rather low, the tailing edge of the first pulse coincides with the leading edge of the second. Therefore, the leading edge of the second pulse is blue-shifted by the first, whereas the tailing edge of the first is red-shifted by the second. Due to the fact that both pulses coincide with their parts of the strongest intensity change, the resulting chirp is large. The chirp of the pulses can be translated into a timing shift of the pulses at the fiber output. The timing shift is equal in magnitude to the product of the frequency shift and the accumulated dispersion [88]. The resulting timing jitter is a variation in the arrival time of the pulses around their mean arrival time.

On the other hand, as can be seen from the figure for the case of a high dispersion, if the pulses experience a strong broadening by the dispersion of the waveguide, their intensity change is rather low. Hence, the temporal derivation is small, the frequency change is correspondingly weak, and the IXPM depends on the dispersion. The IXPM can be effectively suppressed by optimum dispersion management [91, 92]. The system performance can be optimized by a prechirping of the pulses at the system input so that the path-averaged width over the high power nonlinear sections is minimized [91]. This reduces the range of overlapping and therefore, the effectivity of the IXPM.

Another possibility for the reduction of the overlapping range is the relation between timing shift and pulse width of the two pulses. If the ratio is large, i.e., if the time shift between them is much larger than the pulse width, they barely overlap and the IXPM is low. If the ratio is about 1, the IXPM has its maximum, and if the pulse width is larger than the time shift, the effectivity of IXPM will decrease again [90]. This can be understood by the help of Fig. 8.1. A large pulse width corresponds to the figure for a high dispersion. If the pulses are broad and the temporal separation between them low, only the parts of the pulses with a small slope can overlap. Hence, due to the fact that the IXPM depends on the temporal derivation of the power profile, its effectivity is low.

A possibility for a complete cancellation of the timing jitter induced by IXPM was studied by Mecozzi et al. [88] and is based on the incorporation of symmetric links, as illustrated in Fig. 8.2. The dispersion compensation is equally splitted between the transmitter at point A and the receiver at E. At point B the two pulses are spread by the dispersion and overlap significantly. They experience a phase shift due to IXPM, which leads to a frequency shift, and together with the dispersion, a timing jitter. the jitter depends on the sign of the dispersion.

Point D is placed symmetrically with respect to B. Due to the fact that the dispersion has an opposite sign, the pulse overlap at D is identical to that in B. The accumulated dispersion between D and the receiver at point E is opposite to that between A and B. If the pulses at D have the same power as pulses at B and if the dispersion is perfectly symmetrical around C, the timing jitter can be completely cancelled. Hence, this method also requires a symmetric power profile, which is available with Raman amplifiers (see Sect. 13.1). But without the incorporation of Raman amplification, a considerable improvement can be obtained with the symmetric compensation method [88].

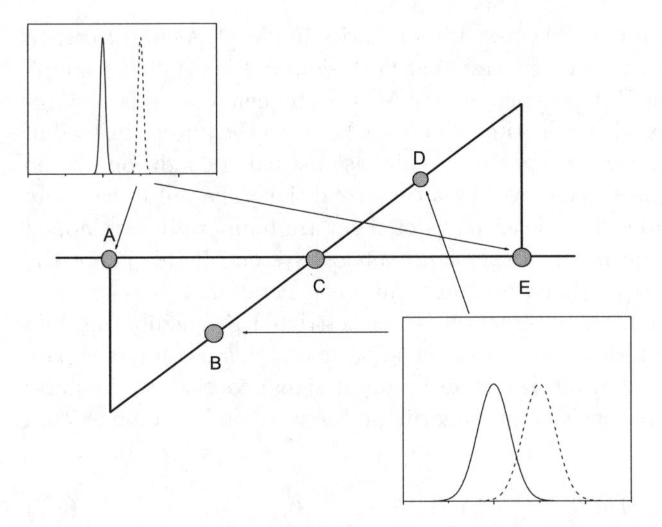

Fig. 8.2. Symmetric cancellation of timing jitter [88]

8.3 IFWM and Ghost Pulses

The last two terms in (8.6) are responsible for the generation of additional pulses, the so called "ghost" pulses and the equation gives an explanation of their position. They are symmetrically arranged around the two original pulses, whereas the time shift corresponds to the time shift between the pulses 1 and 2 ($t + 3\tau$ and $t - 3\tau$). Figure 8.3 shows the temporal positions of the pulses.

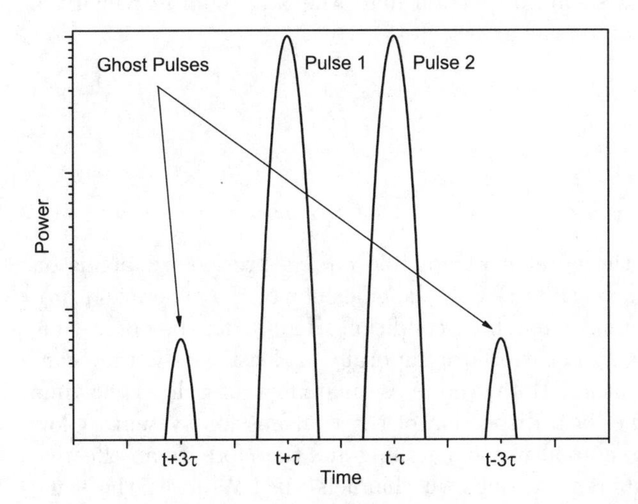

Fig. 8.3. Ghost pulse generation in a short pulse dispersion-managed transmission system. Note the logarithmic scale of the power axes

A comparison with Fig. 7.4 shows the similarity to FWM. As in the case of the interchannel mixing, two new traces at both sides of the original "pump" pulses are generated. But contrary to FWM, the frequency axes is now replaced by a time axes where the time difference between the pumps instead of the frequency determines the spacing. The phase matching condition for the frequency mixing is now taken over by a time condition. The pulses can only interact with each other if at least parts of them are temporally overlapped during the propagation in the dispersion-managed system. If this is not the case, then $u_1^2 u_2^* \approx u_2^2 u_1^* \approx 0$ and no ghost pulse is generated.

On the other hand, the interaction is not restricted to neighboring bits only. If the map strength of the system is large, many pulses can participate in the ghost pulse generation. If the initial input signal consists of the parts u_i, u_j and u_k, the superposition propagating in the waveguide can be written as

$$u_{\mathrm{l}} = u_i \mathrm{e}^{\mathrm{j}\omega(t-\tau_i)} + u_j \mathrm{e}^{\mathrm{j}\omega(t-\tau_j)} + u_k \mathrm{e}^{\mathrm{j}\omega(t-\tau_k)}. \tag{8.7}$$

The introduction of (8.7) into (8.4) leads to a large number of terms. For each initial pulse u_i, u_j and u_k there are SPM terms of the form $|u_n|^2 u_n \mathrm{e}^{\mathrm{j}\omega(t-\tau_n)}$, $(n = i,\ j,\ k)$ and IXPM terms with $2|u_n|^2 u_m \mathrm{e}^{\mathrm{j}\omega(t-\tau_m)}$ $(n, m = i,\ j,\ k;\ n \neq m)$. All other terms are responsible for the IFWM, and can be written as $u_n^2 u_m^* \mathrm{e}^{\mathrm{j}\omega(t-(2\tau_n-\tau_m))}$ and $2u_n u_m u_o^* \mathrm{e}^{\mathrm{j}\omega(t-(\tau_n+\tau_m-\tau_o))}$ with $n, m, o = i,\ j,\ k;\ o \neq n, m$. The temporal distribution of the distinct ghosts for this three-pulse process is shown in Fig. 8.4.

A comparison of the result in Fig. 8.4 with the spectral distribution of a three-tone FWM process in Fig. 7.2 shows in principle the same behavior, except that in the case of IFWM the frequency is replaced by time. If N pulses interact due to a strong dispersion map, the time slots in which the distinct ghost pulses can appear are

$$\tau_{n,m,o} = \tau_n + \tau_m - \tau_o, \tag{8.8}$$

where $o \neq n, m$ and

$$\forall n, m, o \in 1 \ldots N.$$

According to this, the number of possible mixing products depends on (7.2) as well. But, in the case of FWM, the efficiency of the generation process is a matter of the phase matching condition. Here it depends on a time condition. If the pulses do not overlap temporally, at least partly, they cannot interact with each other. Both, the phase matching as well as the time condition depend on the local dispersion of the transmission system. A low local dispersion leads to a good phase matching and therefore, to an effective interaction between different wavelength channels via FWM. At the same time the pulses in one channel will not overlap during transmission and an intrachannel interaction cannot take place. On the other hand, a high local

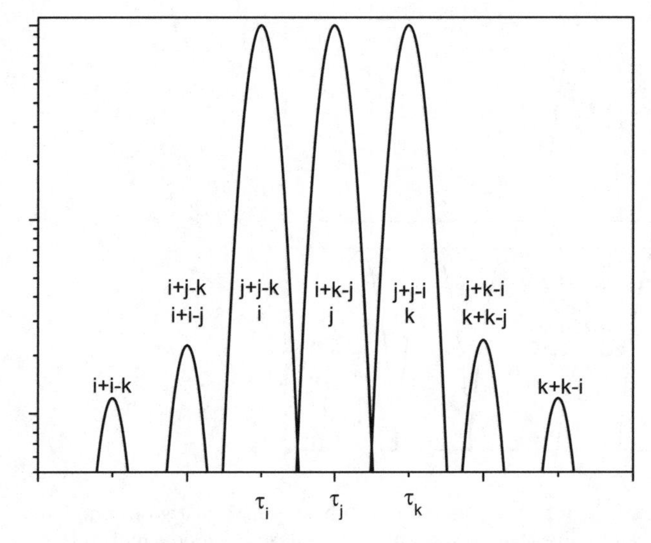

Fig. 8.4. Spatial distribution of the original pulses and the additional ghost pulses generated

dispersion suppresses FWM but it leads to an effective IFWM for high-bit-rate systems.

The IFWM has two important effects on the system performance, both reducing the signal to noise ratio and, therefore, the BER of the system. As can be seen from Fig. 8.4, in the case of three equally spaced pulses, one of the newly generated mixing products always falls in the same time slot of an original pulse. Hence, the power is transferred between the pulses. For the channel that delivers the power, this means that its amplitude is decreased, whereas the amplitude of the one bit in the receiving channel is increased or decreased, depending on the relative phase between the original pulse and the newly generated one. Since many pulses interact with each other in the case of a large dispersion, this effect is arbitrary, leading to stochastic variations of the amplitudes of the "ones", i.e., an amplitude jitter. The amplitude jitter is additional noise that degrades the SNR and therefore, the system performance.

In the other time slots originally carrying a "0" the power transfer leads to the generation of new mark "ones", the ghost pulses. Figure 8.5 shows the possible variations that can produce a ghost for two different bit patterns, namely a "1101" and a "1110111" sequence. As can be seen from the figure, the number of possible variations depends on the length of the bit sequence carrying a mark "1". For the ghost pulse generation in the "1101" pattern, only two combinations ($-1-1-2$; $1-2-1$) are responsible, whereas in the "1110111" sequence 10 variations are possible. In a simulation by Liu et al. [93], the ghost pulse power in a "1110111" pattern grew, for instance, more than 10 times faster than that in a "1110000" pattern. But the growth rate

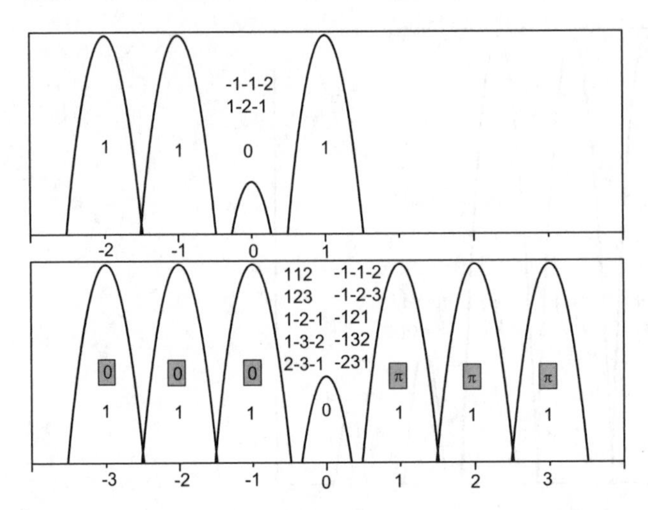

Fig. 8.5. Possible variations for the generation of ghost pulses in two distinct bit patterns. The gray blocks in the lower trace show the phase inversion method

depends on the number of bits that overlap temporally and on the accumulated dispersion, so a longer bit pattern does not necessarily mean stronger ghost pulses. At the same time, as in the case of FWM, the superposition of the combinations depends on the relative phases between them.

In a two-pulse simulation, Johannison et al. [96] showed that the position of the ghost pulses is not stationary at one time slot during propagation in the dispersion-managed system. The ghost pulse center zigzags in time and the shape and position are the same after every dispersion-managed cell. This is caused by an additional frequency shift induced by the pulses that generate the ghosts. Hence, the time location at which the ghosts appear depends on the system parameters and does not coincide with the center of the adjacent bit slots unless the dispersion-managed cell is long enough. The ghost pulse at the zero bit time slot grows resonantly as a result of periodic forcing and temporal phase matching with the signal pulses [95]. In their simulation Johannison et al. found a quadratic increase of the ghost pulse energy with the number of dispersion-managed cells.

8.4 IFWM Suppression

The IXPM can be suppressed by optimum dispersion management, whereas the IFWM remains a severe problem for high-bit-rate dispersion-managed systems. The IFWM is responsible for two effects that decrease the signal to noise ratio and, therefore, degrade the system performance: the amplitude jitter and the ghost pulse generation. Hence, a number of proposals have been made to suppress these nonlinear effects.

If the pulses in an optical time division multiplex system are equally spaced, the nonlinear penalty is maximum since the interaction is resonant [95]. The frequency generation in the case of FWM is similar to the ghost pulse generation due to IFWM. The ghosts can be effectively suppressed if an unequal time spacing is introduced between the bits of a bit pattern [97]. Practically this can be done by adjusting the time delay between the pulses. The number of interacting bits depends on the dispersion of the system. Only bits that temporally overlap participate in the ghost pulse generation. If the time shift of any two bits participating on the ghost pulse generation is unique in the bit pattern (see (7.38) and (7.39)) no ghosts are generated in time slots of the bit sequence. This method slightly reduces the spectral efficiency and cannot completely suppress the amplitude jitter due to IFWM because ghost pulses are still generated, but now they fall into free time slots. Another possibility of ghost pulse suppression is based on the incorporation of duo-binary and modified duo-binary modulation formats [98, 99].

Zweck et al. [136] showed that the power of the ghosts can be decreased if the pulses are split into two subchannels spaced by a frequency shift ($\Delta\Omega$). The subchannel multiplexing of the signals can be produced by using a hybrid optical time and WDM scheme. At the receiver the two subchannels are treated as one single high-bit-rate channel. If two of the pump pulses are in one channel and the third is in the other subchannel the energy of the nonlinear interaction between them decreases exponentially with the channel spacing because the amplitudes of the nonlinear term in (8.4) move apart during propagation [137].

This subchannel multiplexing means that each WDM channel is replaced by a pair of subchannels related to the original channel. The central frequency between two consecutive pulses is changed by $\pm\Delta\Omega/2$. The result is that the central frequency of the pulses in the even time slots is shifted up by $+\Delta\Omega/2$, whereas the central frequency of the pulses in the odd time slots is shifted down by $-\Delta\Omega/2$. For a 40Gbit/s system, two channels with a bit rate of 20Gbit/s each are produced, but the temporal width of the bits still correlates to that of a 40Gbit/s system. At the receiver they are treated as a single 40Gbit/s channel.

For this technique, the efficiency of the IFWM generation depends on the frequency spacing of the subchannels ($\Delta\Omega$). For particular values of $\Delta\Omega$, resonances in the ghost pulse generation power can be seen. The origin of these oscillations lies in the phase change between the mixing channels correlated to the frequency shift ($\Delta\Omega$). The perturbations responsible for the ghost pulse move in and out of phase with $\Delta\Omega$ [136]. If these resonance frequencies are avoided, Zweck et al. found that it is possible to increase the propagation distance by the subchannel multiplexing method. In a subsequent publication, the authors took noise into account and showed that the strength of the resonances decreases due to the noise and that the method performs best with strong dispersion management and Raman amplification [137].

Liu et al. [93] studied a method for the suppression of the worst ghost pulse. The worst case happens if a single "0" occurs between long marker blocks. Their method is based on the phase condition between the different combinations participating on the ghost generation. For the bit sequence at the bottom of Fig. 8.5 "1110111", ten different combinations of the pulses in the marks are able to generate the ghost pulse. As Liu et al. showed, the phase factor of the IFWM process, $0 = \tau_i + \tau_j - \tau_k$ is the same as that of the process $0 = (-\tau_i) + (-\tau_j) - (-\tau_k)$ if all of the marks have the same phase. The phase factors have opposite signs, however, if the marker blocks on the left and right side of the zero have different phases (see Fig. 8.5). At the same time, the growth rate of both processes is the same.

The ten combinations are denoted and sorted according to the above condition in Fig. 8.5. Hence, if the phases of adjacent marker blocks are inverted (as shown with the grey blocks in Fig. 8.5), the combinations have opposite phases and their superposition in the time slot between them is zero. Therefore, this method cannot work for a "1110000" sequence because the required partners are absent. The phase variations in the marker blocks can be realized with duo-binary and modified duo-binary modulation formats [93].

Summary

- Intrachannel nonlinear effects have their origin in the nonlinear interactions inside one high-bit-rate transmission channel.
- These effects can only happen if two or more bits in the channel overlap temporally due to the strong dispersion.
- Three intrachannel effects are possible: SPM, IXPM, and IFWM.
- The IXPM is the intrachannel analogue to the XPM and leads to a timing jitter of the pulses.
- The IXPM can be suppressed by an intelligent system design – prechirping of the pulses and symmetric compensation.
- The IFWM is comparable to the FWM in WDM systems, but frequency is replaced by time.
- The IFWM results in an amplitude jitter and the generation of ghost pulses.
- IFWM is a severe problem in high-bit-rate transmission systems.
- The generation of ghost pulses can be suppressed, at least partly, by an unequal spacing between the bits of a sequence, special modulation formats, the phase inversion between adjacent marker blocks, or the incorporation of subchannels.

Exercises

8.1. Suppose the bit sequence “1111” is generated by a bit pattern generator with 25 ps pulses (FWHM) and a duty cycle of 25%. Determine the number of possible mixing products due to IFWM and their temporal distribution if the phase shift to the first pulse of the sequence is $t_0 - 600$ ps.

8.2. Determine the solution for the right side of (8.4) if (8.7) is introduced.

8.3. Determine the partners with the same phase factor that contribute to the generation of a ghost pulse in the bit sequence “11011” if all marks have the same phase.

9. Solitons

In the year 1834, the Scottish shipbuilding engineer and scientist John Scott Russell observed a natural phenomenon that others had surely watched before but deemed unimportant. He had watched a barge in the Edinburgh–Glasgow channel that was first pulled by horses and then fixed at the river bank. From the bow a water wave was initiated, reaching over the entire channel width. Russell followed the wave at the river bank with his horse and noticed that the wave propagated over several kilometers with a constant velocity and without significant change of its shape [203].

Russel was impressed to such an extent that he began experimental investigations on the propagation of water waves. In 1844, he reported his results at the British Association for the Advancement of Science [204]. As he already noticed correctly, this kind of waves can walk through each other and they keep their shape after a collision. Hence, these waves behave like a kind of particles [205]. Normally, water waves, as well as waves in an optical fiber, experience dispersion, i.e., their shape broadens during propagation. Due to their exceptional behavior, Russel called them solitary waves which was later changed to the term "soliton".

Another example of solitary water waves are the "tsunamis" which can accrue as the result of an earthquake in the deep sea. Most of the tsunamis are generated in the Pacific ocean. At open seas tsunamis have wavelengths of 10 to 100 km and only a small amplitude, therefore, they are rather harmless to ships. But tsunamis propagate very fast and can cross the Pacific in a very short time. If these waves come near the coast, their velocities decrease due to the shallow water. The following water masses accumulate and the wave grows to several meters high in the coastal regions. At the Japanese Pacific coast, tsunamis with heights of up to 40 m have been recorded.

A vital quality of solitons is that they are able to keep their shape over extremely long distances. Therefore, in a distance of several thousand kilometers to the epicenter of the earthquake, tsunamis can do devastating harm. On November 1, 1755, the west coast of Spain, Portugal, and Morocco was hit by such a tidal wave, where 60,000 men died in Lisbon alone. On July 17, 1998 after a seaquake that lasted only a few minutes, a 15 m high Tsunami built up along the west coast of Papua New Guinea. About 1200 men were killed instantly and 6000 were missing, probably washed into the ocean by the flinched water masses [207].

The equation of motion for water waves was developed in 1895 by Korteweg-de Vries, 60 years after Russell. But it took another 75 years before it was solved analytically with the help of the inverse scattering theory in 1967 [208].

Waves with the same features as the tsunamis in water, but with much less dramatic results, can be generated in optical fibers as well. What is much more important here is that they can be exploited for optical telecommunications. In the year 1971, the inverse scattering theory was applied to the NSE by Zakharov and Shabat [211]. In 1973, Hasegawa and Tappert [206] discussed the possibility of the propagation of solitons in optical fibers for the first time and demonstrated their stability using numerical simulations. It was not until Mollenauer [209] gave the experimental proof in 1980 that solitons came to a broad public interest when scientists realized that solitons offer a possibility for the transmission of information with a high capacity over large distances. After that, solitons have received a widespread attention. Today, a large number of publications dealing exclusively with the subject of solitons in optical telecommunications can be found. Therefore, this chapter will only give an overview of the physical background and the basic concepts.

The chapter is organized as follows: First, Sect. 9.1 gives the mathematical description of solitons, starting from the NSE. The next section shows the exceptional behavior of this kind of waves via the example of higher order solitons. Today, the so called dispersion-managed solitons are of special interest. Although they cannot be seen as real solitons in the original sense, these pulses can periodically recover their shape. And, as in the case of solitons, they require a balance between the dispersion and nonlinear effects in the transmission system. A short description of dispersion-managed solitons is given in Sect. 9.6. The remaining sections are devoted to the limits of the incorporation of solitons into optical transmission systems as well as techniques to circumvent these limits.

9.1 Mathematical Description

Solitons have a particular ratio between temporal width and power and are, mathematically, a stable solution of the wave equation. The propagation of waves in a cylindrical nonlinear waveguide and, therefore, optical solitons, can be described by the NSE (see Chap. 5):

$$\mathrm{j}\frac{\partial B}{\partial z} = -\gamma \left|B\right|^2 B + k_2 \frac{1}{2}\frac{\partial^2 B}{\partial T^2}. \tag{9.1}$$

Assuming the attenuation is negligible then only the linear and the nonlinear terms remain. If the soliton is indeed a stable solution of the NSE the effects caused by the linear term must be compensated by the nonlinear term, either at every position z in the waveguide or periodically at defined positions. Otherwise, the pulse spreads in time due to the linear dispersion or it spreads in frequency due to the nonlinear effect of SPM. A mutual compensation takes place if the input pulse complies with certain conditions related to its shape and power.

In the literature, the standard form of the NSE is used in most cases. This can be obtained from (9.1) with some scaling. In the standard form the amplitude of the pulse is related to its input power (P_0) (7.12), whereas the propagation distance is related to the dispersion length (L_D) (5.40) and the time to the temporal width of the pulse (τ_0). Therefore, we have the following for the "soliton units" [51, 210]:

$$A = \frac{B}{\sqrt{P_0}}, \tag{9.2}$$

$$\zeta = \frac{z}{L_\mathrm{D}}, \tag{9.3}$$

and

$$\tau = \frac{T}{\tau_0}. \tag{9.4}$$

If the Equations (9.2) to (9.4) are introduced into (9.1), the NSE is represented by

$$\frac{\mathrm{j}}{L_\mathrm{D}} \frac{\partial A}{\partial \zeta} = -\gamma P_0 \left|A\right|^2 A + \frac{k_2}{2\tau_0^2} \frac{\partial^2 A}{\partial \tau^2}. \tag{9.5}$$

The dispersion length, L_D (5.40), determines the fiber length after which the dispersive effects, for a pulse with the temporal width, τ_0, are important, leading to a broadening in the temporal domain. Whereas the nonlinear length, L_NL (5.41), determines the length after which the nonlinear effects are crucial, leading to a pulse broadening in the frequency domain (see Sect. 5.2). Hence, the ratio between the dispersive and nonlinear lengths,

$$\frac{L_\mathrm{D}}{L_\mathrm{NL}} = \frac{\gamma P_0 \tau_0^2}{|k_2|} = N^2, \tag{9.6}$$

determines if the dispersive or nonlinear effects are detrimental to a fiber with a nonlinearity constant γ for a pulse with the width τ_0 and an input power of P_0 (see Fig. 5.1). If the ratio $L_\mathrm{D}/L_\mathrm{NL}$ is much smaller than 1, the dispersion has a stronger impact on the pulse propagation in the waveguide. Therefore, the temporal pulse width will be broadened. For 25 ps pulses and a carrier wavelength of 1.55 µm, this is the case for input powers much smaller than 25 mW ($\gamma = 1.3\ \mathrm{W}^{-1}\mathrm{km}^{-1}, k_2 = -20.4\ \mathrm{ps}^2\mathrm{km}^{-1}$). On the other hand, if the input power is much higher, the nonlinear effects are more important and it follows a broadening of the pulse spectrum. But if the ratio is $L_{\mathrm{D}}/L_{\mathrm{NL}} = 1$ and hence, both lengths are equal, dispersive and nonlinear effects can compensate each other and a soliton is formed which is neither dispersed in the temporal nor in the time domain.

Using equations (5.41) and (9.6), (9.5) can be expressed as

$$\mathrm{j}\frac{\partial A}{\partial \zeta} = -N^2 |A|^2 A + \frac{1}{2}\mathrm{sgn}(k_2)\frac{\partial^2 A}{\partial \tau^2}, \tag{9.7}$$

where $\mathrm{sgn}(k_2)$ determines the sign of the GVD parameter k_2. In the case of a normal dispersion it is positive, otherwise it is negative. As will be discussed later in this chapter, the standard form of a soliton can only be formed in the anomalous dispersion region of the optical fiber. With

$$u = NA = \sqrt{\frac{\gamma \tau_0^2 P_0}{|k_2|}\frac{B}{\sqrt{P_0}}} = \sqrt{\frac{\gamma \tau_0^2}{|k_2|}}B, \tag{9.8}$$

the case of an anomalous dispersion in (9.7) is represented by

$$\mathrm{j}\frac{\partial u}{\partial \zeta} + |u|^2 u + \frac{1}{2}\frac{\partial^2 u}{\partial \tau^2} = 0 \tag{9.9}$$

This equation corresponds to the time-dependent, dimensionless, nonlinear Schrödinger equation of quantum mechanics. Until now the attenuation in the fiber has been neglected. If it is included, the standard form of the NSE [51, 210] becomes

$$\mathrm{j}\frac{\partial u}{\partial \zeta} + |u|^2 u + \frac{1}{2}\frac{\partial^2 u}{\partial \tau^2} - \mathrm{j}(\alpha/2)u = 0. \tag{9.10}$$

In order to solve the NSE, Zakharov and Shabat applied the inverse scattering theory [211] for the first time in 1971. This theory is based on a spectral transformation. For the solution of the linear problem, the spectral range is defined as one that is independent of the initial conditions. In the nonlinear case, like for the scattering of a particle at a potential in quantum mechanics, a scattering problem is defined by the initial conditions [205].

The results, rather than the details, of the inverse scattering theory are of interest here. In principle, an unlimited variety of possible solutions can be found for the NSE, but only a few are of technical importance. The most important is the class of solutions that shows a hyperbolic secant shape:

$$u(0,\tau) = N\mathrm{sech}(\frac{T}{\tau_0}) = N\mathrm{sech}(\tau) = \frac{2N}{\mathrm{e}^{\tau} + \mathrm{e}^{-\tau}}. \tag{9.11}$$

The intensity of the hyperbolic secant pulse at the fiber input is shown in Fig. 9.1. The pulse width τ_0 and the full-width at half-maximum (FWHM) τ_{FWHM} are in the intensity distribution related via

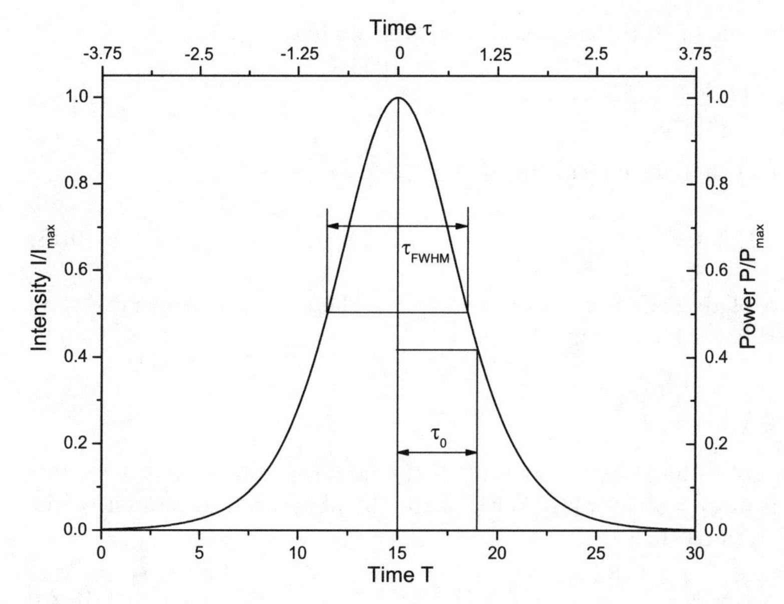

Fig. 9.1. Pulse shape of a hyperbolic secant pulse

$$\tau_{\text{FWHM}} = 2\text{arcosh}(\sqrt{2})\tau_0 = 2\ln\left(\sqrt{2}+1\right)\tau_0 \approx 1.763\tau_0 \tag{9.12}$$

The first order solution ($N = 1$) for the canonical form of the pulse in the fiber is

$$u(\zeta,\tau) = \text{sech}(\tau)\mathrm{e}^{\mathrm{j}\zeta/2}. \tag{9.13}$$

This equation describes a first order or a fundamental soliton. If (9.13) is introduced into the NSE (9.9), neither the shape nor the phase of the pulse should be changed. That the phase remains constant during propagation is shown in the following example [210]. The nonlinear term of the NSE, responsible for a phase alteration is

$$\mathrm{j}\frac{\partial u}{\partial \zeta} = -\left|u\right|^2 u. \tag{9.14}$$

According to Sect. 5.5 with $u = u(\zeta,\tau)$, the nonlinear term generates a phase shift of

$$\mathrm{d}\phi_{\text{NL}} = \left|u(0,\tau)\right|^2 \mathrm{d}\zeta. \tag{9.15}$$

The linear term of the NSE is

$$\mathrm{j}\frac{\partial u}{\partial \zeta} = -\frac{1}{2}\frac{\partial^2 u}{\partial \tau^2}. \tag{9.16}$$

If the right side of (9.16) is expanded with u, we have

$$\frac{\partial u}{\partial \zeta} = \left(\frac{\mathrm{j}}{2u} \frac{\partial^2 u}{\partial \tau^2} \right) u. \tag{9.17}$$

For a real time-dependent function $f(\zeta, \tau)$, the equation

$$\frac{\partial u}{\partial \zeta} = \mathrm{j} f(\zeta, \tau) u \tag{9.18}$$

produces a phase shift of $\mathrm{d}\varphi(\tau) = f(0, \tau)\,\mathrm{d}\zeta$. Therefore, the dispersive term of the NSE has a phase shift of

$$\mathrm{d}\varphi_\mathrm{D} = \left(\frac{1}{2u} \frac{\partial^2 u}{\partial \tau^2} \right) \mathrm{d}\zeta \tag{9.19}$$

Using the pulse shape $u\,(0, \tau) = \mathrm{sech}(\tau)$, the nonlinear phase shift according to (9.15) is $\mathrm{d}\phi_\mathrm{NL} = \mathrm{sech}^2(\tau)\mathrm{d}\zeta$. With (9.19) the phase shift generated by the dispersive term is then

$$\begin{aligned} \mathrm{d}\varphi_\mathrm{D} &= \left(\frac{1}{2\mathrm{sech}(\tau)} \frac{\partial^2 \mathrm{sech}(\tau)}{\partial \tau^2} \right) \mathrm{d}\zeta \\ &= \left[\frac{1}{2} - \mathrm{sech}^2(\tau) \right] \mathrm{d}\zeta. \end{aligned} \tag{9.20}$$

After summation and integration, we obtain the value $\zeta/2$ as a net phase shift of the pulse during its propagation through the waveguide. A comparison with (9.13) shows that neither the dispersive nor the nonlinear term leads to a lasting phase shift of the pulse. Both cancel each other, as shown in Fig. 9.2 where the soliton has a balance between the nonlinear and the linear phase shift and the phase remains constant. If the phase experiences no change along the fiber, the pulse shape remains constant as well. Therefore, the soliton retains its form in the time and the frequency domain during propagation. Figure 9.3 shows a fundamental soliton during its propagation through the waveguide. As can be seen, the shape of the soliton is constant along the propagation distance. This behavior is completely different to the linear case of pulse propagation shown in Fig. 5.8.

If the attenuation is neglected, solitons can propagate over theoretically infinite distances. This fact makes solitons very interesting for the field of optical telecommunications, especially for long haul high-bit-rate transmission systems as can be found in undersea links, for example.

9.2 Higher Order Solitons

The soliton in Fig. 9.3 is called a fundamental soliton. This implies that higher order solitons can be expected in optical waveguides as well. According to (9.6) the power of a soliton in an optical fiber is

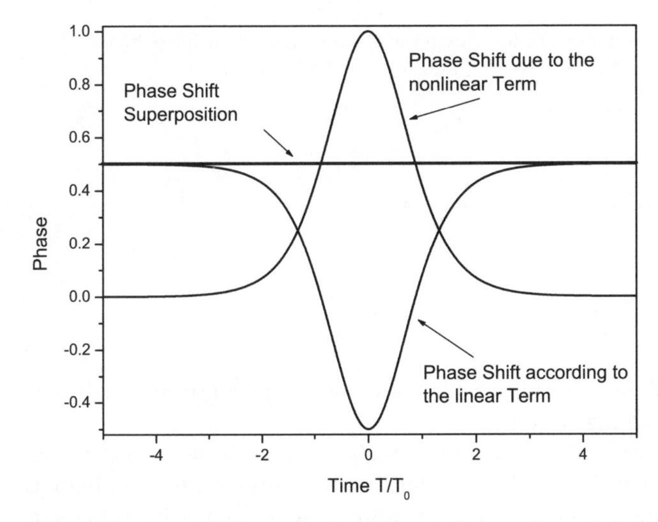

Fig. 9.2. Linear and nonlinear phase shift for a hyperbolic secant in the case of an anomalous waveguide dispersion [210]

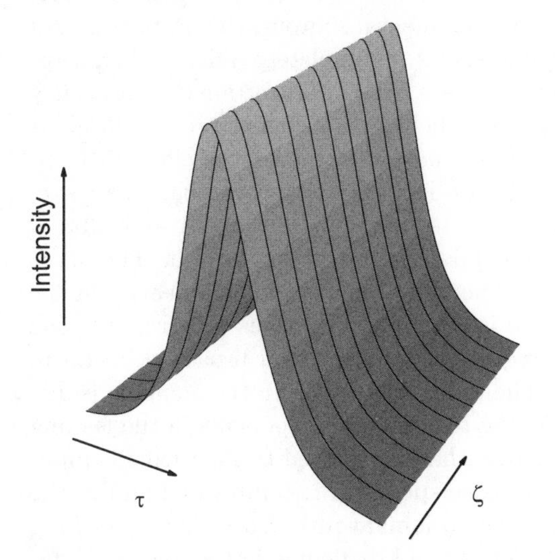

Fig. 9.3. Soliton in a nonlinear waveguide. The shape remains constant

$$P = \frac{N^2 |k_2|}{\gamma \tau_0^2}. \tag{9.21}$$

Let us consider pulses with a temporal width of $\tau_0 = 25$ ps. With $\gamma = 1.3\ \mathrm{W^{-1}km^{-1}}$ and $k_2 = -20.4\ \mathrm{ps^2km^{-1}}$ we obtain an input power of 25 mW for fundamental solitons in the C-band ($\lambda = 1550$ nm). Near the zero-crossing of the GVD (for standard single mode fibers at $\lambda \approx 1.3$ µm, or in the C-band

Table 9.1. Required input powers for the generation of higher order solitons ($\lambda =$ 1550 nm $\tau_0 = 25$ ps, $\gamma = 1.3\ \mathrm{W}^{-1}\mathrm{km}^{-1}$, and $k_2 = -20.4\ \mathrm{ps}^2\mathrm{km}^{-1}$)

N	P(mW)
1	25
2	100
3	225
...	...

for DSF), k_2 is very small and hence, in this range the generation of solitons is possible with a much lower power.[1]

The required power for the generation of higher order solitons is, according to (9.21), N^2 times that for the fundamental one. Therefore, for the higher order solitons with the values above, the input powers shown in Table 9.1 can be calculated.

Figures 9.4 and 9.5 show a second order soliton ($N = 2$) in the time and frequency domain. Contrary to fundamental solitons, the propagation of higher order solitons in a waveguide shows a completely different behavior. Their shape will change during propagation but after a certain distance, they will return to their original shape. The behavior of higher order solitons is useful for the description of the physical nature of the soliton effect [51].

As can be seen from Fig. 9.4, during its propagation through the waveguide, the second order soliton contracts in the time domain and after a certain distance it reaches maximal power and minimal width. The short powerful pulse shows a large slope. Therefore, due to the nonlinearity in the fiber, the leading edge of the pulse is red-shifted whereas the trailing edge is blue-shifted from the central frequency, resulting in a broad spectrum in the frequency domain (Fig. 9.5). The pulse propagates in the anomalous dispersion regime of the fiber. Hence, the red-shifted components in the leading edge of the pulse move slower whereas the blue-shifted trailing wing components are faster. Consequently, the pulse spectrum is compressed and in the time domain, the pulse reaches again its original width. After that everything starts again: first, the nonlinearity leads to a broadening in the frequency domain, then the linear dispersion compresses the spectral components and so on.

The cycle duration of the pulse change for all higher order solitons is [211] $\zeta_0 = \pi/2$. With $\zeta = z/L_\mathrm{D}$, the propagation distance for one cycle is:

$$z_0 = \frac{\pi}{2} L_\mathrm{D} = \frac{\pi}{2} \frac{\tau_0^2}{|k_2|} \approx 0.5054 \frac{\tau_\mathrm{FWHM}^2}{|k_2|}. \tag{9.22}$$

[1] If k_2 is small, higher order dispersion terms must be included for the calculation of the input power.

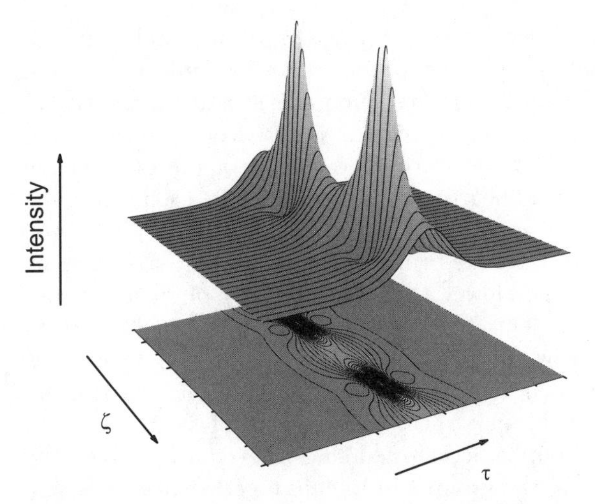

Fig. 9.4. Propagation of a second order soliton in the time domain

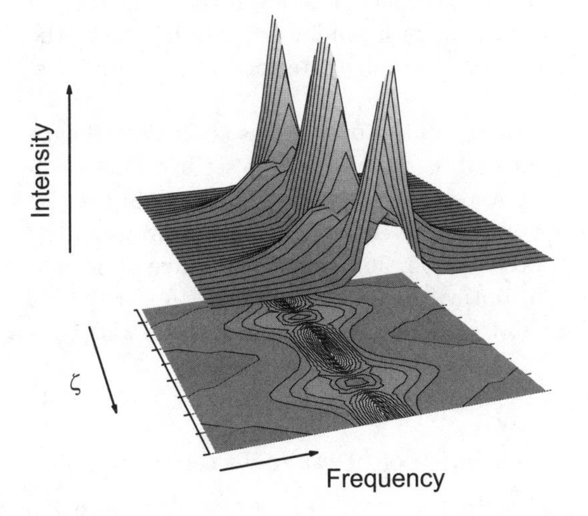

Fig. 9.5. Second order soliton in the frequency domain

A pulse with a temporal width of $\tau_0 = 25$ ps, an input power of 100 mW (according to Table 9.1, $N = 2$), and a carrier wavelength of 1.5 μm will, according to (9.22), receive its original shape back after a propagation distance of 48 km in a standard single mode fiber ($\gamma = 1.3$ $\mathrm{W}^{-1}\mathrm{km}^{-1}$ and $k_2 = -20.4$ $\mathrm{ps}^2\mathrm{km}^{-1}$). If (2.74) is introduced the relation between cycle duration and FWHM width of the pulse is

$$z_0 \approx 0.953 \frac{\tau_{\mathrm{FWHM}}^2}{D\lambda^2}, \tag{9.23}$$

where τ_{FWHM} has to be introduced in ps, D in ps/(nm×km), and λ in µm. Contrary to the behavior of higher order solitons, for a fundamental soliton, neither the nonlinear nor the linear part preponderates at a distinct position. During the entire propagation in the waveguide both effects have an equal strength and compensate each other so that no change of the pulse shape occurs. The interaction between linearity and nonlinearity in a soliton can be described very clearly with a comparison used by many authors.

Suppose there is a group of runners running on an asphalt street. Some are a little bit faster and some slower than the majority of them. Hence, the group will be torn apart. This behavior corresponds to the dispersion in the waveguide. Suppose now that the runners are running on a soft mat. The group will deform the mat due to its mass. As a result, the runners run in a valley indented on the mat – the faster group of them must run uphill, whereas the slower group is accelerated since they can run down the hill. Under certain conditions, the group will remain together like a soliton. The soft mat corresponds to the nonlinearity of the fiber and the strength of the deformation depends on the mass of the runners and the properties of the mat. In principle, this is the same as in a nonlinear material where the strength of the nonlinearity depends on the wave intensity and the material as well.

Until now, we have assumed that the fiber dispersion is anomalous, hence in a standard single mode fiber no soliton like the ones described above can be generated if the wavelength is lower than 1.3 µm. Photonic fibers can provide anomalous dispersion for visible wavelengths so that solitons in the visible part of the spectrum are possible [35, 36]. But in the range of normal dispersion solitons are possible in optical waveguides as well. According to the NSE, we have the following after a change of the sign of the GVD:

$$j\frac{\partial u}{\partial \zeta} - \frac{1}{2}\frac{\partial^2 u}{\partial \tau^2} + |u|^2 u = 0. \tag{9.24}$$

The first order soliton that is a solution of (9.24) fulfills the relation [51]

$$u(\zeta,\tau) = \tanh(\tau)\mathrm{e}^{\mathrm{j}\zeta}. \tag{9.25}$$

Such a soliton is shown in Fig. 9.6. Compared to a soliton for anomalous GVD, it has an inverse intensity distribution. Therefore, this group of solitons is called dark. The intensity distribution can be generated with a relatively broad pulse and an intensity drop in the middle. It is sufficient if the pulse in the back is ≈ 10 times broader than the intensity drop [212]. If the intensity in the middle comes to zero, the soliton is called black; otherwise it is called gray.

A difference between dark and bright solitons is that higher order dark solitons show no periodicity. During their propagation in the waveguide, gray solitons are separated from a higher order dark soliton. Due to the GVD, these gray solitons move apart from the black soliton in the middle which becomes temporally narrower [51, 213, 214].

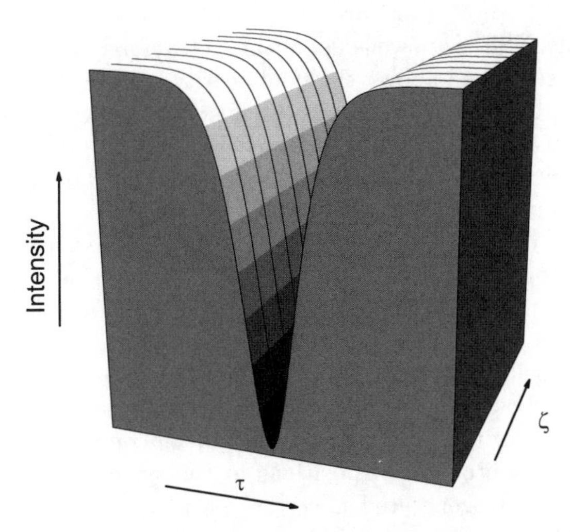

Fig. 9.6. Dark soliton in a waveguide with normal dispersion

9.3 Limits of Solitons

One of the biggest problems in optical telecommunications, especially for long haul high-bit-rate systems like undersea cable, is the fiber dispersion. The pulse spread in time leads to a limitation of the length of the transmission system. If the bit rate is increased, the distance between repeaters must be decreased, which is a severe problem from an economics point of view. One solution is the incorporation of dispersion-shifted fibers with its zero-crossing of the GVD in the C-band. But unfortunately, a low dispersion results in very good phase matching for the FWM and a very low walk-off parameter which increases the efficiency of the XPM. Hence, the changeover of these systems to WDM is rather costly.

Solitons exploit the dispersion and the nonlinearity of the fiber in a natural way so that, theoretically, a transmission over unlimited distances should be possible. But solitons also have their limitations which are discussed in this section.

Solitons are a stable solution of the wave equation. Hence, they are very robust. If the power of the input pulse is, in distinct limits, bigger or smaller than the required N (9.21), a soliton is generated in the fiber. A fundamental soliton can be generated as long as the condition $0.5 < N < 1.5$ is fulfilled [51]. If the power is higher, the excessive power is separated from the pulse by the fiber dispersion and the pulse fits its shape as long as it complies to the condition for a fundamental soliton. Hence, neither the shape nor the input power is a critical parameter for the soliton generation. For communication systems, the separated dispersive parts could be a problem because they can lead to a system penalty [52].

The nonlinear term of the NSE is power-dependent. Therefore, if the power drops below the value mentioned above, the nonlinearity is ineffective and the linear term preponderates. The pulse spreads very fast due to the fiber dispersion. Hence, the pulse must be amplified in time to hold the pulse above the threshold at all times. In principle, two possibilities exist: If the Raman amplification (see Sect. 13.1) is incorporated, it is possible to compensate the attenuation loss over the whole distance in the fiber by a second pump wave that delivers power to the soliton persistently.

The second possibility is the periodical implementation of EDFAs into the transmission system. A problem with the implementation is that the amplifiers have a different group velocity dispersion k_2 than the fiber. Therefore, amplified transmission systems show a strong variation of the signal intensity and the GVD. It follows that linearity and dispersion cannot be completely compensated for a fundamental soliton at all positions in the system. But solitons are so robust that they are generated nevertheless, as long as the path average values for intensity and dispersion in the spans coincide and the condition [210]:

$$L_{\mathrm{D}} \gg L_{\mathrm{A}} \tag{9.26}$$

is fulfilled, with L_{A} as the amplifier distance and L_{D} as the dispersion length. For every span, the dispersive and nonlinear phase shifts compensate each other so that neither the spectral nor the temporal shapes are significantly affected. The periodical arrangement of the amplifiers can be seen as an additional phase matching term.

On the other side EDFAs, as well as all optical amplifiers, add noise to the system. In EDFAs especially, the spontaneous emission leads to a noise floor which is further increased by the amplifier. Therefore, the noise increases with the gain of the EDFA and is called "amplified spontaneous emission"(ASE). This additional noise leads to a limitation of the system performance that is called "Gordon-Haus jitter".

9.3.1 The Gordon-Haus Effect

The task of an amplifier in a long haul transmission system is the compensation of the attenuation in the fiber spans. According to (2.44) the attenuation between two amplifiers with the distance L_{A} is $e^{-\alpha L_{\mathrm{A}}}$. In order to compensate this loss, the amplifier must have a gain of

$$G = \mathrm{e}^{\alpha L_{\mathrm{A}}}. \tag{9.27}$$

The noise penalty due to the amplification can be written as

$$F(G) = \frac{1}{G}\left[\frac{(G-1)}{\ln G}\right]^2. \tag{9.28}$$

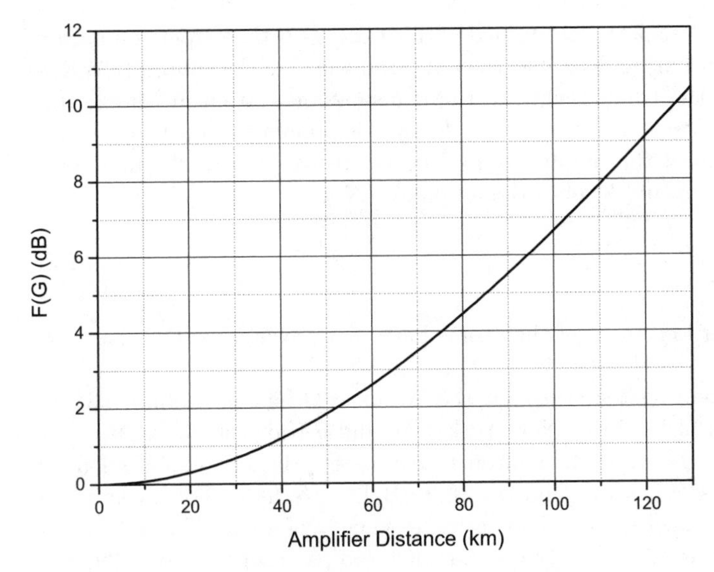

Fig. 9.7. Noise penalty $F(G)$ in dB versus amplifier spacing for $\alpha = 0.2$ dB/km [210]

Therefore, according to (9.27) and (9.28), the noise penalty increases with the distance between the amplifiers and their gain. The noise penalty versus the length of the spans is shown in Fig. 9.7.

The ASE superimposes on the soliton and leads to a stochastic change of the intensity and the phase of the soliton. The intensity fluctuations cause variations in the width of the soliton which are so small that they do not have an impact on the system performance. On the other hand, phase alterations are of particular importance to soliton transmission systems. The fluctuations in the phase will change the carrier frequency of the soliton in a random manner and this frequency alteration is translated, via the GVD, into a change of the arrival time of the soliton. Therefore, the ASE leads to a timing jitter of the soliton.

In principle, this behavior is a direct result of the extraordinary stability of solitons. If a stochastic, but small, signal (the noise) is superimposed on the soliton, it is still a solution of the NSE. But it is now a new soliton with a slightly different amplitude, phase, and carrier frequency. This phenomenon was described for the first time in 1986 and is named after the authors of the publication "Gordon-Haus effect" (GHE) [227]. The alteration of the arrival time in ps is [210]

$$\sigma_{\mathrm{gh}}^2 = 3600 n_{\mathrm{sp}} F(G) \frac{\alpha}{A_{\mathrm{eff}}} \frac{D}{\tau} Z^3. \qquad (9.29)$$

Equation (9.29) requires defined units for the different variables. The effective area A_{eff} is defined in $\mu\mathrm{m}^2$, α is the attenuation constant in km^{-1},

$F(G)$ is the noise penalty determined with (9.28), D is the dispersion parameter in ps/(nm × km), τ describes the temporal width of the soliton (FWHM) in ps, and Z is the whole length of the transmission system in Mm = 1000 km. The excess spontaneous emission factor, or population inversion factor, ($n_{\rm sp}$) is the ratio of the number of excited electrons (N_2) to the number of electrons in the ground level of the amplifier (N_1):

$$n_{\rm sp} = \frac{N_2}{N_2 - N_1}. \tag{9.30}$$

For an amplification, most of the atoms have to be excited so that $N_2 > N_1$ and $n_{\rm sp} \geqq 1$.

A transmission system through the Atlantic Ocean is around 5.3 Mm (from the north of Spain to New York), the pulse width is 25ps, $A_{\rm eff} = 50$ µm^2, and $n_{\rm sp} = 1.5$. For the soliton transmission, dispersion-shifted fibers with $D = 0.5$ ps/(nm×km) and $\alpha = 0.2$ dB/km are used. For an amplifier spacing[2] of 100 km, Fig. 9.7 yields a noise penalty of 6.648 dB (≈ 4.6), and therefore, a variation of the arrival time of ≈ 8.3 ps according to (9.29).

The bit error rate (BER) according to the GHE is the probability that a pulse arrives at the receiver out of its time frame. Hence, for a particular BER, the time frame or the bit period must be so wide that the GHE has no influence. For a BER of 10^{-9}, the bit period has to be $12\sigma_{\rm gh}$ [210]. Therefore, the maximum bit rate is 10 Gbit/s. Figure 9.8 shows the Gordon-Haus jitter and the corresponding data rate for a BER of 10^{-9} and an amplifier spacing of 50 km versus the length of the transmission system.

9.3.2 The Acoustic Effect

Another effect responsible for a timing jitter of the data stream is the so called acoustic effect. The origin of this effect lies in an additional interaction of the solitons with the material. If the soliton propagates through the fiber its electric field interacts with the material, the electronic polarization leads to a mechanic deformation or strain in the material. This effect is called electrostriction (see Sect. 11.2) and describes the converse behavior to the pieco-electric effect. The electrostriction generates an acoustic wave that changes the refractive index for the following pulses. Due to the refractive index alteration that the following pulses will experience, their GVD is altered.

The standard deviation of the acoustic effect in ps in a fiber with $A_{\rm eff} =$ 50 µm^2 [210, 228, 229] is

$$\sigma_{\rm a} \approx 8.6 \,\frac{D^2}{\tau} \sqrt{R - 0.99}\, \frac{Z^2}{2} \tag{9.31}$$

[2] The amplifier spacing in soliton systems is much smaller in most cases.

where D is in ps/(nm×km), τ is in ps, Z is in Mm, and R is the bit rate in Gbit/s. With the values of the above example, the temporal variation is $\sigma_a \approx 3.6$ ps.

Compared to the GHE, the acoustic jitter is smaller. But contrary to the Gordon-Haus jitter, it depends on the bit rate of the system, so it can become important in high-bit-rate systems.

9.4 Increasing the Bit Rate in Soliton Systems

On one side, optical amplifiers are required in soliton transmission systems for the compensation of attenuation losses. On the other, they lead to an additional noise and, therefore, a reduction of the transmission capacity. As can be seen from Fig. 9.8, due to the GHE, the maximum usable bit rate decreases relatively fast with the length of the transmission system. Hence, a suppression of the GHE results in a direct increase of the maximum bit rate.

The origin of the GHE is an additive noise which changes the carrier frequency of the soliton arbitrary and results, via the GVD, in a variation of its arrival time. One approach for a reduction of this effect is the use of frequency filters that define the carrier frequency of the soliton again after each amplifier and, therefore, suppress their statistical variation. Due to the solitons stability the filters will not destroy them. The filters adjust the amplifier bandwidth such that the maximum of the transmission curve falls together

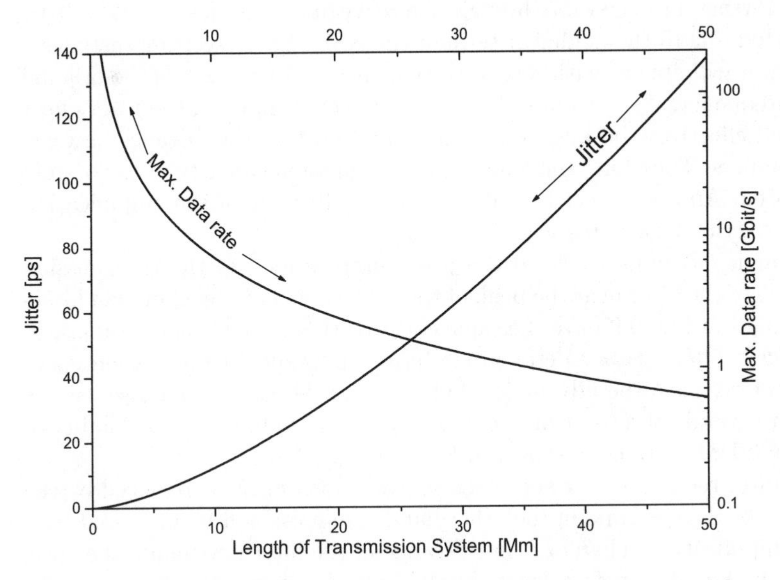

Fig. 9.8. Gordon-Haus jitter and maximum bit rate for an amplifier spacing of 50 km against the length of the transmission system ($A_{\text{eff}} = 50$ μm^2, $n_{\text{sp}} = 1.5$, $D = 0.5$ ps/(nm × km), $\alpha = 0.2$ dB/km, $\tau_0 = 25$ ps, BER $= 10^{-9}$)

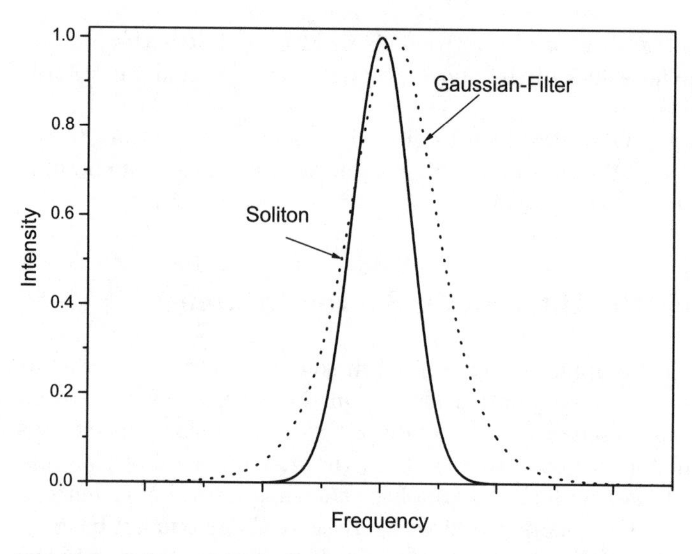

Fig. 9.9. Soliton shifted in its central frequency due to noise influences and transmission curve of a Gaussian filter

exactly with the original peak of the soliton. Figure 9.9 shows a soliton which is shifted in frequency by additional noise and the transmission curve of a Gaussian filter.

The shifted frequency parts will be cut off by the Gaussian filter. During the further propagation through the waveguide, the lost parts will be regenerated due to the nonlinear term of the NSE. After a certain distance a new soliton will appear with a central frequency shifted towards the original central frequency. If such a filter is used after every amplifier, the jitter effect is reduced effectively because it can only build up over the distance between two amplifiers. Therefore, the maximum transmission capacity of the system is increased. Another effect is that due to the filter out of the amplification bandwidth, the noise is reduced.

Of particular importance to the jitter suppression are the transmission properties of the filter near the peak of the soliton. Therefore, Gaussian filters like the ones in Fig. 9.9 have the same effect as the much cheaper etalons, or Fabry-Perot filters. Fabry-Perot filters have a periodical transmission curve and therefore, have the advantage that they can be used simultaneously as filters for several solitons at different wavelengths, enabling the possibility of a soliton WDM system, as shown in Fig. 9.10.

The filter reduces the power of the soliton which must be replaced by the amplifiers of the system. On the other hand, the noise will be increased by a higher amplification. Therefore, in the range of the filter maximum, the noise is rather high and increases exponentially with distance.

Noise reduction is possible if the central frequency of adjacent filters along the transmission system is shifted slightly, as shown in Fig. 9.11. The shifted

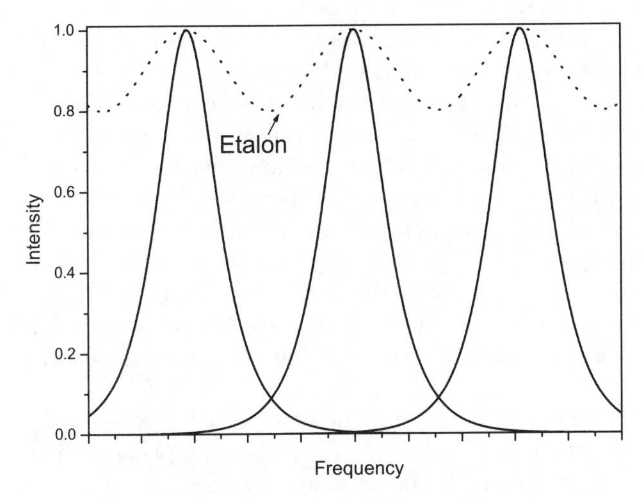

Fig. 9.10. Transmission curve of an etalon and solitons of a soliton-WDM system

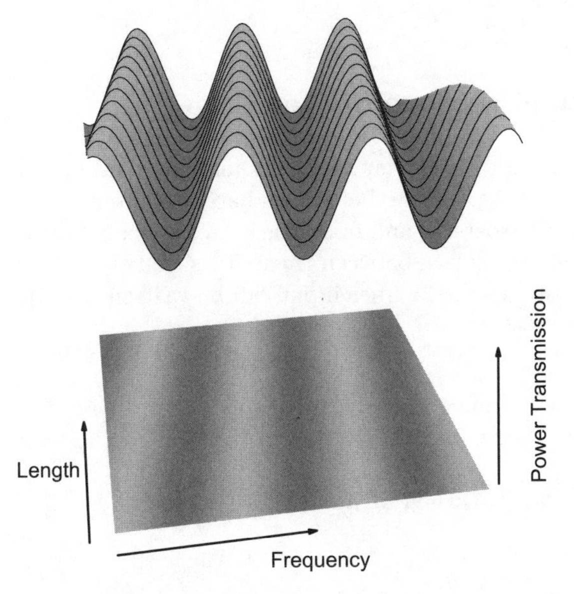

Fig. 9.11. Power transmission curve of frequency filters along the way of a transmission system

filters are called "sliding-frequency guiding filters" [230]. As long as the frequency change from one filter to the next is relatively small, the solitons can follow. After a filter, the solitons are regenerated along the next fiber span. This regeneration is impossible for the noise. Therefore, the transmission system is transparent for the solitons whereas the noise is effectively suppressed.

Not only is the jitter generated by the GHE suppressed by this method, but all noise generated by other sources is suppressed as well.

Another possibility for the suppression of the jitter is the regeneration of the solitons in the time domain. After every amplification, an intensity modulator is introduced into the system. The different modulators are temporally steered so that they only open in the middle of the bit period. If the intensity maximum of a time-shifted soliton passes the modulator a little earlier or later, it is cut off by the modulator. Therefore, the soliton is guided back to the middle of the bit period. In principle, this is the time analogue to the filter method described above. If both methods are used in a transmission system, an error free transmission over unlimited long distances is possible theoretically [231].

The advantage of the time method is that the jitter is corrected directly. On the other hand, a practical implementation is rather complicated. The method is not directly applicable in WDM systems because the control of the modulators requires complex hardware and the modulators generate an additional frequency chirp.

9.5 Mutual Interaction Between Solitons

The temporal jitter is not the only effect that limits the maximum usable bit rate of soliton systems. If the time spacing between adjacent solitons is too small, the same interaction of dispersion and nonlinearity responsible for the generation of a single soliton takes place between them. This leads to a force between the solitons. Two solitons at the fiber input can be written as [51]:

$$u(0,\tau) = \mathrm{sech}(\tau - q_0) + r\mathrm{sech}\left[r\left(\tau + q_0\right)\right]\mathrm{e}^{\mathrm{j}\theta} \tag{9.32}$$

where r is a parameter that determines a relative amplitude and θ is a relative phase between the solitons. Figure 9.12 shows the two solitons at the fiber input.

According to Fig. 9.12, and (9.4) the bit rate of this system is

$$B = \frac{1}{2q_0\tau_0}. \tag{9.33}$$

Therefore, the bit rate depends directly on the minimum distance between the solitons. For $q_0 = 3.5$, as in Fig. 9.12, and a pulse width of 25 ps the bit rate is 5.7 Gbit/s.

Because of the mutual interaction between the pulses, they move together and then away from each other. The interaction between the solitons depends not only on their distance but on their relative phase and amplitude as well. If both pulses have nearly the same amplitude, their motion can be described by an interaction force which exponentially decreases with the distance between them and varies sinusoidally with their relative phase [232]. For the special

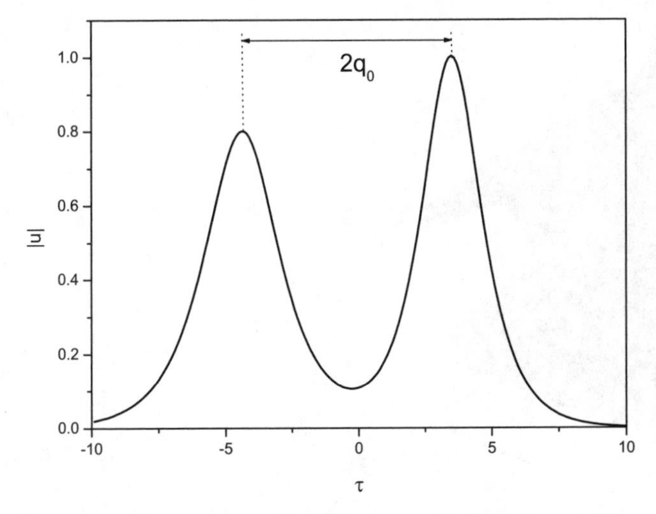

Fig. 9.12. Two solitons at the input of a fiber ($r = 0.8$, $q_0 = 3.5$, $\theta = 0$)

case that $r = 1$ and $\theta = 0$, the distance between the solitons is equal to q at every point ζ in the waveguide [232][3]:

$$q = \frac{\ln\left[\frac{1}{2}\left(1+\cos\left(4\zeta \mathrm{e}^{-q_0}\right)\right)\mathrm{e}^{2q_0}\right]}{2}. \tag{9.34}$$

Figure 9.13 shows two solitons propagating in a waveguide. They have a bond and show a periodic motion during their propagation. The period of this motion is $q(\zeta) = q_0$ and hence

$$\zeta_\mathrm{p} = \frac{\pi}{2}\mathrm{e}^{q_0}. \tag{9.35}$$

This periodicity defines the distance in the waveguide after which the original spacing between the solitons returns or, of course, the distance between two collisions of the solitons. A possibility for the suppression of the interactions is to choose a value for ζ_p much longer than the fiber length. This can be done by a wider distance between the solitons at the fiber input but the result is a reduction of the bit rate. According to (5.40, 9.3, 9.33), and (9.35), the real fiber length (z_p) between two collisions of the solitons is

$$z_\mathrm{p} = \frac{\pi}{2}\frac{\tau_0^2}{|k_2|}\mathrm{e}^{\frac{1}{2\tau_0 B}} \approx 0.953\frac{\tau_\mathrm{FWHM}^2}{D\lambda}e^{\frac{0.88}{B\tau_\mathrm{FWHM}}}. \tag{9.36}$$

where B is in Tbit/s, τ_FWHM is in ps, D is in ps/(nm × km), and λ in µm. For a GVD of $k_2 = -20.4$ ps^2/km, a pulse width of 25 ps, and a bit rate of

[3] According to (9.34), q can have negative values. In that case, the approximation no longer holds.

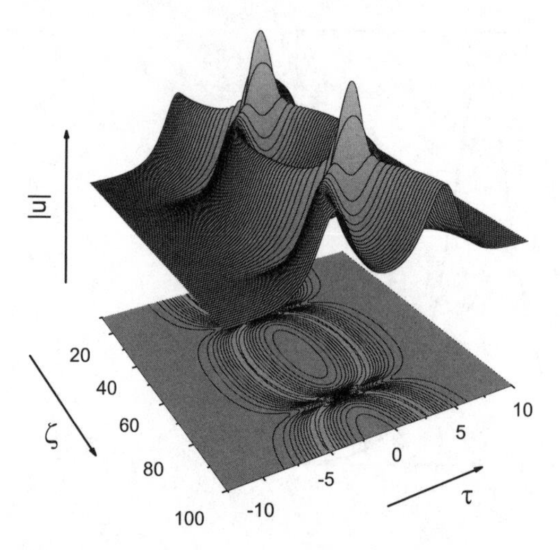

Fig. 9.13. Mutual interaction between two solitons in a waveguide ($r = 0.8$ $q_0 = 3.5$ $\theta = 0$) [51]

5.7 Gbit/s the cycle duration for the collision is 1607 km. If the separation between the solitons at the fiber input is $q_0 = 10$ and the pulse width is $\tau_0 = 4$ ps, the cycle duration is more than 27 000 km. For this example the bit rate is 12.5 Gbit/s.

Another possibility is offered by the relative phase of the Solitons. Under certain conditions, the attractive force will be changed into a repulsive force so that solitons move away during their propagation in the fiber [51].

If two solitons have different frequencies, but are launched at the same time into the fiber, the blue pulse moves through the red one during propagation due to the anomalous dispersion. The pulses will collide and interact with each other due to the nonlinearity of the fiber. Like in the non-soliton case, an additional phase shift due to XPM and the generation of new frequency components due to FWM will occur. But, since the pulses will not propagate in the zero dispersion region of the fiber, they may propagate completely through each other, suppressing the phase alteration induced by XPM (see Sect. 6.2.2). At the same time, the dispersion reduces the phase matching for FWM. If the system design accounts for these phenomena with solitons, the nonlinear effects are only transient. Hence, the pulses will remain unchanged. This potential transparency to each other makes solitons well adapted for WDM [210].

9.6 Dispersion-Managed Solitons

Another kind of periodical solitons are the so called dispersion-managed solitons (DMS). In 1995, DMS were originally proposed to reduce the Gordon-

Haus jitter in long haul high-bit-rate transmission systems [215]. As the name implies, this kind of solitons is only possible together with the incorporation of dispersion management in the transmission system.

The basic idea behind the dispersion management technique is treated in Sect. 5.4. Negative dispersion fibers are periodically introduced into a soliton transmission system consisting of positive dispersion fibers. Due to the dispersion compensation, the overall system dispersion is close to zero and hence, the timing jitter from the Gordon-Haus effect will be significantly reduced[216],[217]. Furthermore, the DMS scheme is robust up to a certain degree against power variations in the amplifiers and the increase of the loss in the spans [218]. Due to the high local dispersion FWM from collisions between solitons in a dense wavelength division multiplexing (DWDM) DMS system is efficiently suppressed as well [219]. The pulse waveform is, of course, distorted in the dispersion compensation segment but it will reshape into a soliton-like short pulse. This variation in pulse shape is periodically repeated during the transmission through the system without generating dispersive wave radiation. Contrary to other DM systems using RZ pulses, the pulse shape is periodically stationary in DMS transmission systems [100].

For the propagation of DMS, the whole system consisting of dispersive fiber spans, dispersion-compensating modules, and amplifiers has to be designed in a way such that the pulse in each individual time slot recovers its width, chirp, and power after each amplification period [58]. Therefore, similar to higher order solitons, DMS show a periodicity during their propagation as well and the period is equal to the amplifier spacing (L_{A}). As in the case of classical solitons, and contrary to the high dispersion systems of Sect. 8, a balance between the dispersive and nonlinear effects is essential for DMS. Hence, dispersion-managed solitons require specific values for their input temporal width and the input power. These values depend on the given dispersion map. The design rules for DMS are not of interest here and can be found in [223].

The dominant nonlinear penalty in a DWDM-DMS system is the cross-phase modulation between solitons in different channels caused by collisions due to the high local dispersion in the spans [219], [224]. The XPM leads to a shift of the carrier frequencies of the solitons and hence, to a timing jitter. If the collisions between the solitons are complete the residual frequency shift is zero (see Sect. 6.2.2). In contrast to a classical soliton transmission, in a DMS system the alternating variation of the high local dispersion causes the colliding solitons to move rapidly back and forth with respect to each other. Due to this rapid alteration of the distance between the solitons the collisions are not complete. In the DMS case the collision length is greater than the distance between collisions, therefore the soliton in one channel can simultaneously collide with several solitons in other channels and initial partial collisions are inevitable [219].

The XPM impact on DWDM-DMS systems can be reduced if adjacent channels are orthogonally polarized (see Sect. 6.2.1) but the improvement is not enough to make the jitter acceptable. The situation can be improved dramatically when guiding filters are introduced into the system design and if a fraction of the fiber based dispersion compensation is replaced by periodic-group-delay dispersion-compensating modules (PGD-DCM) [220], [221]. The effect of the PGD-DCM is a large reduction in the net path over which a colliding pair of pulses interact [222].

DMS were for instance successfully demonstrated in transoceanic systems with massive wavelength division multiplexing [225]. In a test system with PGD-DC modules in an all Raman amplified recirculating loop Mollenauer et al. achieved working distances of ≈ 9000 and $\approx 20,000$ km for uncorrected bit error rates of $< 10^{-8}$ and $< 10^{-3}$, respectively [226]. These distances are very close to those achievable in single channel transmission in the same system.

Summary

- Solitons can be found in many different fields of nature and technique.
- Solitons are waves with a particular shape and power that remain their shape, at least periodically, during propagation and behave like a kind of particles.
- Optical solitons are a stable solution of the NSE.
- The exceptional behavior of solitons is based on the balance between the linear effect of dispersion and the nonlinear effect of SPM.
- Except the dark, solitons can only be found in the anomalous dispersion regime of the fiber.
- Higher order solitons show a periodicity, i.e., their shape will be changed during propagation but they will come back to their original shape after a certain distance.
- Similar to higher order solitons, DMS show a periodicity during their propagation. The period is equal to the amplifier spacing L_{A}. DMS depend on the balance between linear and nonlinear effects over the whole dispersion-managed system.
- The noise of optical amplifiers leads to a timing jitter of the solitons which degrades the system performance.
- The noise-induced timing jitter can be effectively suppressed with "sliding-frequency guiding filters", i.e., the central frequency of adjacent filters along the transmission system is slightly shifted.
- The electrostriction of the material leads to an additional timing jitter that depends on the bit rate.
- If the time spacing between adjacent solitons is to small a force between the solitons leads to a mutual interaction.

Exercises

9.1. Determine the required input power for a first order soliton in the C-band in a standard single mode fiber using $D = 17$ ps/(nm×km), $\gamma = 1.3\ \text{W}^{-1}\text{km}^{-1}$ and a pulse width of $\tau_0 = 25$ ps.

9.2. Estimate the required input power and the cycle duration for a second order soliton with a pulse width of $\tau_0 = 25$ ps in the C-band if a nonzero dispersion-shifted fiber with $D = 3$ ps/(nm × km) and $\gamma = 1.3\ \text{W}^{-1}\text{km}^{-1}$ is used.

9.3. Calculate the maximum length for a soliton transmission system with a BER of 10^{-9} and a bit rate of 2.5 Gbit/s/channel. The solitons have a width of $\tau_0 = 25$ ps. Fiber spans with a dispersion of $D = 0.5$ ps/(nm×km), $A_{\text{eff}} = 50\ \mu\text{m}^2$ and an attenuation of 0.2 dB/km are used. The amplifier spacing is 50 km and $n_{\text{sp}} = 1.5$.

9.4. Estimate the timing jitter due to the acoustic effect for the values of the exercise above (Exc. 9.3).

9.5. Calculate the fiber length for the cycle duration of collisions in a 10 Gbit/s soliton system. The temporal width of the solitons is $\tau_0 = 10$ ps and the incorporated nonzero-shifted fiber has a dispersion of $D = 2$ ps/(nm×km).

10. Raman Scattering

Raman scattering, especially if it is stimulated (SRS), is a very important nonlinear effect because it leads to a decrease of the SNR in WDM systems and it can be exploited for distributed amplification in long haul transmission systems. The spontaneous Raman scattering was discovered, long before the invention of the laser, in the year 1924 by the Indian physicist Sir Chandrasekhara Raman (knighted 1929). During a ship travel, after a visit to a congress in England, Raman admired the wonderful blue colour of the Mediterranean sea. He found the origin of this effect in the scattering of the sun light at the molecules of the water. This leds him to the investigation of light scattering in fluids, bulks, and gases and during its investigations he found that beside the monochromatic wavelength, another frequency-shifted wavelength can be found in the scattered field as well. For the discovery of the effect, which was later named after him, and related cognitions, Raman won the Nobel Prize in physics in 1930.

In the case of a spontaneous Raman scattering, a small part of the incident light with a distinct frequency is transformed into a new wave with lower or higher frequency. During this transformation, the photons interact with the phonons of the medium and the frequency shift is determined by the vibrational modes of the material. Adolf Gustav Smekal had prognosticated such an effect in 1923 and described it with the emission or absorption of motion energy by the photons, Raman was not familiar with Smekal's work.

The transformation efficiency of spontaneous Raman scattering is very low (10^{-6} cm^{-1}). This means that during the propagation of an optical wave in a medium with a thickness of 1 cm, only 1 part in a million is Raman-scattered and translated into a wave with a new frequency. For effective Raman scattering, a distinct threshold must be exceeded to change the spontaneous scattering into a stimulated scattering process. To exceed this threshold, very intensive light fields which are only available via lasers are required. In 1962, two years after the first demonstration of a laser [233], Woodbury and Ng [234] were able to observe stimulated Raman scattering. For their experiments, they used a ruby laser with a so called Q-switch[1] made

[1] A Q-switch changes the optical properties of a laser resonator in a very short time. In this short time window, the allocation inversion of the laser active medium is decayed, leading to a very intensive short pulse.

of a cell filled with nitrobenzene. In the radiated laser field beside the typical ruby laser line at 694.3 nm, an additional line with a frequency of 767 nm was detected. This could not be explained by the fluorescence of ruby but it showed that the difference frequency between both coincides to a Raman vibration mode of the nitrobenzene in the Q-switch. Hence, stimulated Raman scattering was observed for the first time [235].

Contrary to spontaneous Raman scattering, stimulated Raman scattering can transform a large part of the power of the incident field into a new frequency-shifted wave. If the scattering is stimulated, the intensity grows exponentially with the propagation distance in the nonlinear material. Due to a fundamental point of view as well as its potentially high technical significance, SRS was investigated by a number of scientists in the following years. An overview of the results of the early years can be found in [236] and a more direct review is published in [237]. Due to the fact that the frequency shift in the Raman scattering process originates from an interaction with the medium, the scattered wave contains informations about the medium. Therefore, SRS is of great importance in the field of spectroscopy.

In optical transmission systems, Raman scattering causes a kind of cross talk between channels in a WDM system. The channels with a higher carrier frequency deliver a part of their power to the channels with a lower carrier frequency. In terms of wavelength, the channel with the longer wavelength is amplified at the expense of the channel with the shorter carrier wavelength. Raman scattering can reduce the system performance drastically. On the other hand, the same effect can be exploited for the optical amplification of the signals in fibers. Hence, optical fibers can become a Raman amplifier or a Raman laser, which are treated in detail in Sect. 13.1 and Sect. 13.2.

In the next section the physical origin of light scattering as a whole and especially of the Raman scattering is explained. After that the difference between spontaneous and stimulated emission will be shown. In Sect. 10.3 the behavior of Raman scattering and the threshold for stimulated scattering in optical fibers is treated. Whereas, Sect. 10.4 deals with the impact of the SRS on optical communication systems.

10.1 The Scattering of Light

Suppose we have a light beam propagating through a transparent crystal. At the output of the crystal, the measured power of the light is smaller than at its input. This has mainly two reasons: the attenuation due to the imaginary part of the refractive index (see Sect. 2.4) and the deviation of a part of the light beam into all possible directions. The latter phenomenon is called scattering and can have a number of origins. But it is basically caused by an inhomogeneous distribution of distinct optical properties in the bulk of the material, such as the refractive index or the polarizability of the medium.

The air in the atmosphere contains, among other things, particles of dust, aerosols, and drops of water. The dielectric constant and, therefore, the refractive index of these particles differ from that of the surrounding air. Hence, a light beam propagating through the atmosphere will be reflected at the respective boundaries between air and particle. The distribution as well as the position of the boundaries are completely arbitrary. Therefore, the light beam is reflected into any directions or, in other words, scattered in the whole medium. This kind of scattering requires particle dimensions in the range of the wavelength and is called Mie-scattering. The Mie-scattering is independent of the wavelength, i.e., all frequencies in a wavelength band behave in the same manner. This is why the water drops of fog or the fat drops in milk appear to be white.

On the other hand, completely pure media which are free of impurities can show a scattering of light as well. For instance, a gas or a liquid consisting of identical molecules in a thermal equilibrium has an isotropic refractive index over the whole volume of the material. But if only a small clipping of this volume, with dimensions smaller than the wavelength but bigger than the molecules of the material, is considered, the molecule density in the sample is arbitrary. This random density distribution is caused by the thermal motion of the molecules and leads to an arbitrary fluctuation of the refractive index on a scale much smaller than the wavelength. This deviation of the light is called Rayleigh-scattering. Due to the small dimensions the rules of ray optics are no longer valid and the wave character must be included into the consideration.

The scattering of the sun light at the trace gas molecules in the earth's atmosphere is an example of Rayleigh scattering. The intensity of the scattered radiation (I_s) caused by Rayleigh scattering is

$$I_S \sim \frac{1}{\lambda^4}. \tag{10.1}$$

Therefore, the intensity of the scattered radiation depends on the frequency with the fourth power. If we look at the white light of the sun (10.1) shows that in the scattered light the blue part of the spectrum is much more intense than the red. This is one of the reasons for the blue sky and the red evening sun.

Optical fibers are neither a fluid nor a gas and the impurity concentration is very low but, nevertheless, Rayleigh scattering is one of the most important processes in the fiber determining the usable optical windows. The Rayleigh scattering in optical fibers is a result of the inhomogeneous distribution of the molecules and dopants during the fabrication process of the fiber. During the drawing procedure of the preform into fibers, the glass material is melted. Hence, it is a fluid with random density distributions of the molecules. These distributions are frozen when the fiber is cooled down and are unavoidable.

Raman and Brillouin scattering have their origin in an inhomogeneous distribution of optical properties of the medium as well. However, contrary to Rayleigh or Mie scattering, the distribution is not completely arbitrary but are periodical in both cases. For the Raman scattering, the inhomogeneous distribution results from innermolecular oscillations or twists whereas acoustic oscillations are responsible for Brillouin scattering. Hence, the photons interact with the medium, they deliver motion energy to or absorb it from the medium. Since the photon energy is determined by $E = hf$, an energy change is always connected with a frequency change. Therefore, for these two particular cases of scattering, both the direction of the incident wave and its frequency are changed.

10.2 Origin of Raman Scattering

In principle, spontaneous Raman scattering can be observed if a medium is irradiated with monochromatic light and the scattered part is investigated by a spectrometer. In the scattered light not only the original wavelength can be found, but waves with frequencies lower and higher than the frequency of the irradiated field are present as well. The wave originating from the scattering process is called the pump wave, waves with a lower frequency are Stokes waves, whereas waves with a higher frequency are named anti-Stokes waves. The intensity of the Stokes wave is in most cases many orders of magnitude higher than the intensity of the anti-Stokes wave. The distribution of the different frequencies is shown in Fig. 10.1

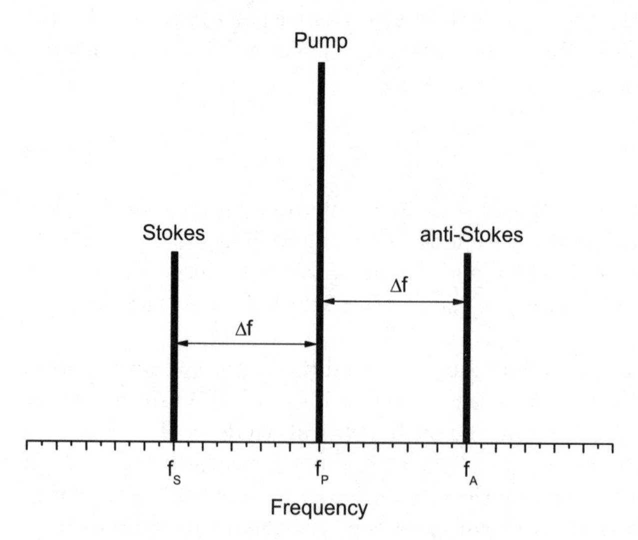

Fig. 10.1. Distribution of the distinct frequencies for the Raman process. Note the similarity to the frequencies generated by FWM in Fig. 7.4

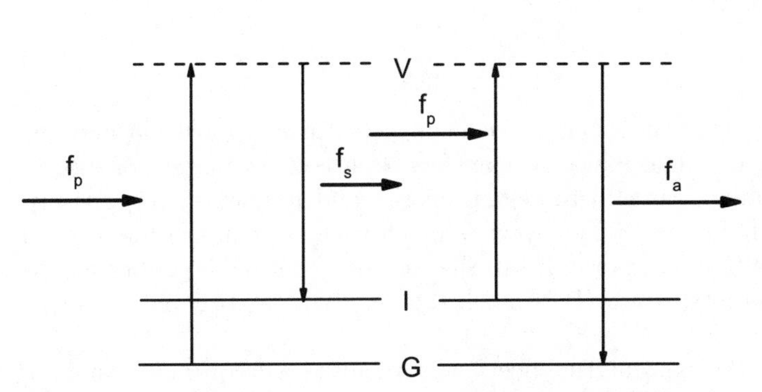

Fig. 10.2. Energy diagram for the description of the Raman Scattering. **(a)** Generation of the Stokes. **(b)** generation of the anti-Stokes wave

The origin of the frequency change lies in an energy exchange between the photons and the medium. The quantum-mechanical energy diagram for the Raman scattering is shown in Fig. 10.2.

Figure 10.2a describes the generation of the Stokes wave. In the first step, an incident pump photon with the energy $E_{\rm p} = hf_{\rm p}$ is destroyed and the molecule will be excited from the ground (G) into a higher virtual energy state (V). The energy difference between the ground and the excited level is equal to the photon energy of the pump photon. In a second step the molecule falls down into an inter level (I) which is generated by periodical oscillations or rotations of the molecule. This decay is accompanied by the emission of a Stokes photon. Due to the fact that V is a virtual state, the destruction of the pump and the generation of the Stokes photon happens simultaneously. The inter state is energetically higher than the ground level. Therefore, the energy of the Stokes photon ($E_{\rm s} = hf_{\rm s}$) is smaller than the energy of the pump. The energy difference between pump and Stokes photon, $\Delta E = h\Delta f$, is equal to the difference between the energy levels I an G. As a result, the frequency of the Stokes wave is about Δf smaller than the frequency of the pump wave:

$$f_{\rm s} = f_{\rm p} - \Delta f. \tag{10.2}$$

The remaining energy, ΔE, is vibrational energy which is delivered to the molecule. In a complete contrast to the other nonlinear phenomena discussed earlier, where the molecule returns to its ground level after the interaction, an energy transfer between photon and molecule takes place here.

For the generation of the anti-Stokes photons, the energy of the pump is equal to the energy difference between the virtual and inter state, V and I in Fig. 10.2b. The pump photon with energy $E_{\rm p} = hf_{\rm p}$ is annihilated, whereas at the same time an anti-Stokes photon with energy $E_{\rm a} = hf_{\rm a}$ is created. The energy of the anti-Stokes photon is about $\Delta E = h\Delta f$ higher and, therefore, its frequency is

$$f_{\mathrm{a}} = f_{\mathrm{p}} + \Delta f. \tag{10.3}$$

In this process, again an energy exchange between the molecule and the photon takes place. The molecule transfers a part of its motion energy to the scattered photon and increases its energy and frequency, respectively. The intensity of the anti-Stokes wave is much smaller than the intensity of the Stokes wave because, normally, in the case of a thermal equilibrium, the allocation of the inter state (I) is much smaller than the occupation of the ground level (G).

Contrary to the parametric effects which caused a frequency change of the wave, such as the FWM at which the energy levels had a virtual nature, real energy levels of the molecule are involved (see Fig. 10.2). The Raman scattering is a result of genuine oscillation and rotation transitions of the molecule and no phase matching condition has to be fulfilled, as was the case for parametric processes. The Raman scattering is intrinsically phase-matched over the energy levels of the molecule.

Suppose three waves with the frequencies f_{s}, f_{p}, and f_{a} are launched into the fiber input with equal optical power, where the relations $f_{\mathrm{s}} = f_{\mathrm{p}} - \Delta f$ (Stokes wave) and $f_{\mathrm{a}} = f_{\mathrm{p}} + \Delta f$ (anti-Stokes wave) are valid, then the Stokes wave (f_{s}) will be amplified by the Raman process. Optical power is transferred from f_{p} to f_{s} and the intensity of f_{s} is increased at the expense of f_{p}. On the other hand, optical power is transferred from the anti-Stokes wave f_{a} to the pump f_{p} at the same time. The intensity of the pump is increased at the expense of the optical power in the anti-Stokes wave. There are two reasons: first, the effectivity of the Stokes process is much higher than the effectivity of the anti-Stokes process. Therefore, effective power transfer is almost always restricted from higher to lower frequencies. Second, the wave f_{a} is now the pump for the wave f_{p} because the relation $f_{\mathrm{p}} = f_{\mathrm{a}} - \Delta f$ is valid. This is a Stokes, not an anti-Stokes process. Therefore, during the propagation in the fiber, the Stokes wave f_{s} is amplified while the anti-Stokes wave f_{a} is weakened by the Raman scattering.

10.2.1 Raman Scattering in the Harmonic Oscillator Model

The Raman scattering can be described on a classical level with the help of the harmonic oscillator model [73, 238]. The basics of the harmonic oscillator are treated in Sect. 2.3. The distinct oscillators are created by the vibratory molecules of the medium which can be simplified as a system of strings and masses. Here we assume that the medium consists of one kind of oscillators only. The oscillator has the mass m, the attenuation constant γ, the string constant k, and hence, the resonance frequency $\omega_0 = \sqrt{k/m}$. The equation of motion for this kind of oscillator is then

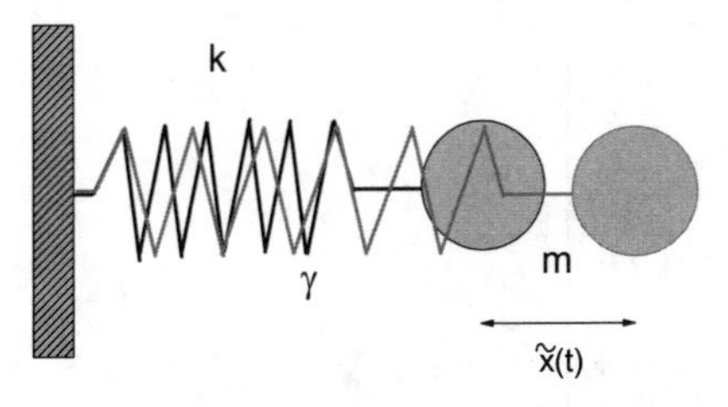

Fig. 10.3. Harmonic oscillator vibrating around its tranquilibrium state for the description of the Raman scattering

$$\frac{\mathrm{d}^2\tilde{x}}{\mathrm{d}t^2} + \frac{\gamma}{m}\frac{\mathrm{d}\tilde{x}}{\mathrm{d}t} + \omega_o^2\tilde{x} = \frac{\tilde{F}(t)}{m}. \tag{10.4}$$

The position coordinate is denoted by $\tilde{x}$, which describes the excursion of the mass from its tranquility state and $\tilde{F}(t)$ indicates the time dependent force acting on the oscillator.

The Raman scattering has its origin in the interaction of the optical field with the vibrational modes of the molecule. Contrary to the oscillator model in Sect. 2.3, the oscillator in Fig. 10.3 already oscillates with its resonance frequency ω_0 independent of an external electric field. Therefore, the position coordinate $\tilde{x}$ has a temporal periodicity. Thus the dipole moment and the macroscopic polarizability of the medium are not constant functions of time, but depend on the periodic coordinate $\tilde{x}$. In Sect. 2.3, the polarizability leads to the refractive index of the medium. Hence, if the dipole already oscillates, the refractive index of the medium is temporally modulated.

If now an electromagnetic wave is launched into this medium, the temporal modulation of the refractive index changes the frequency of the wave. Sidebands are generated, like in an amplitude modulation shifted about the resonance frequency of the oscillators to the original wave $\omega \pm \omega_0$.

In the following, only the generation of the Stokes wave, $\omega_\mathrm{s} = \omega - \omega_0$ will be considered. If the generated Stokes and the original wave propagate further through the medium, they will interfere with each other. The total intensity of the optical field will be modulated corresponding to the fading frequency:

$$I_\mathrm{ges} = I_0 + I\cos(\omega - \omega_s). \tag{10.5}$$

Due to the modulated intensity, the distinct oscillators experience a temporal periodic force which stimulate them to a periodic excursion. The frequency of the excursion correlates to the fading frequency but, according to the discussion above, the fading frequency is the resonance frequency of the oscillators (ω_0). This means that the oscillators are, during the propagation of the fields, driven by the fading between the waves to oscillate with the resonance frequency. Both processes amplify each other, as shown in Fig. 10.4.

If the monochromatic laser beam hits a dipole oscillating with the frequency ω_0 at the input layers of the medium, a new wave will be generated

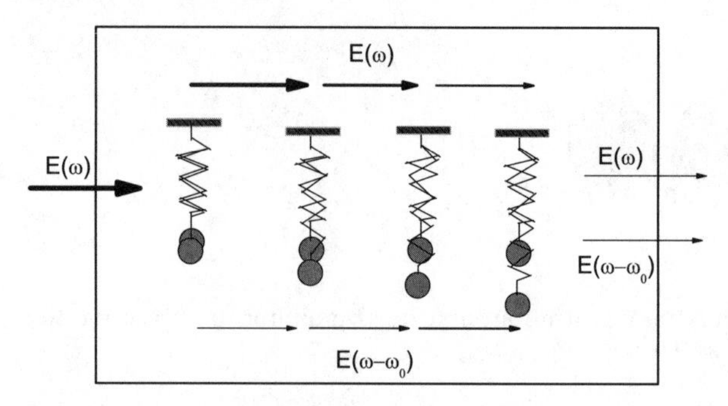

Fig. 10.4. Raman scattering for the generation of the Stokes wave in a medium consisting of identical oscillators

which has a frequency lower by ω_0. The fading frequency between both drives the dipoles in succeeding layers to a stronger oscillation at this frequency. This leads to a stronger Stokes wave, which drives the dipoles to a stronger oscillation again and so on. After exceeding of a certain threshold, a large portion of the optical power of the pump is translated to the Stokes wave.

Due to the Raman interaction, the macroscopic polarizability and, therefore, the susceptibility of the medium is changed. In the case of a resonance with the oscillators, the Stokes wave is reinforced, which means the susceptibility for the Stokes must have a negative imaginary part.[2] If this condition is introduced into the Maxwell equations, the coupled differential equation system which describes the position depending amplitudes of the pump (B_p) and Stokes wave (B_s) can be derived [51]:

$$\begin{aligned}\frac{\partial B_\mathrm{p}}{\partial z} &= \mathrm{j}\gamma_\mathrm{p}\left(|B_\mathrm{p}|^2 + 2|B_\mathrm{s}|^2\right)B_\mathrm{p} - \frac{g_\mathrm{p}}{2}|B_\mathrm{s}|^2 B_\mathrm{p} \\ &\quad - \frac{1}{v_\mathrm{gp}}\frac{\partial B_\mathrm{p}}{\partial t} - \frac{\mathrm{j}}{2}k_{2\mathrm{p}}\frac{\partial^2 B_\mathrm{p}}{\partial t^2} - \frac{\alpha_\mathrm{p}}{2}B_\mathrm{p}\end{aligned} \tag{10.6}$$

and

$$\begin{aligned}\frac{\partial B_\mathrm{s}}{\partial z} &= \mathrm{j}\gamma_\mathrm{s}\left(|B_\mathrm{s}|^2 + 2|B_\mathrm{p}|^2\right)B_\mathrm{s} + \frac{g_\mathrm{s}}{2}|B_\mathrm{p}|^2 B_\mathrm{s} \\ &\quad - \frac{1}{v_\mathrm{gs}}\frac{\partial B_\mathrm{s}}{\partial t} - \frac{\mathrm{j}}{2}k_{2\mathrm{s}}\frac{\partial^2 B_\mathrm{s}}{\partial t^2} - \frac{\alpha_\mathrm{s}}{2}B_\mathrm{s},\end{aligned} \tag{10.7}$$

where $\gamma_{\mathrm{p,s}}$ is the nonlinearity coefficient for the two waves, $v_{\mathrm{gp,s}}$ denotes the group velocity, $\alpha_{\mathrm{p,s}}$ is the attenuation at the particular wavelength, $k_{2\mathrm{p,s}}$ is the group velocity dispersion, and $g_{\mathrm{p,s}}$ is the gain due to the Raman scattering.

From a comparison between the differential equation system (10.6), (10.7), and (5.37), the influence of the distinct terms on the amplitudes of the pump

[2]The real part of the susceptibility leads to the refractive index, whereas the positive imaginary part results in attenuation losses in the medium.

and Stokes wave can be determined. The term $\frac{1}{v_{\mathrm{gp,s}}} \frac{\partial B_{\mathrm{p,s}}}{\partial t}$ describes the influence of the group velocity to the amplitude. For the derivation of the NSE, it is cancelled by the introduction of a moving frame. With $\frac{j}{2} k_{2\mathrm{s,p}} \frac{\partial^2 B_{\mathrm{s,p}}}{\partial t^2}$, the dispersion of the medium is included and $\frac{\alpha_{\mathrm{s},p}}{2} B_{\mathrm{s,p}}$ denotes the decay of the amplitude due to the attenuation in the waveguide. The terms in the brackets lead to the nonlinear effects of self- and cross-phase modulation and, therefore, an alteration of the phase of the waves.

Therefore, the only new contribution responsible for the Raman scattering is $\frac{g_{\mathrm{s,p},}}{2} \left|B_{\mathrm{p,s}}\right|^2 B_{\mathrm{s,p}}$. Since these terms are real, they lead to a change of the amplitudes. In (10.6) the Raman term is negative, whereas in (10.7) it is positive. This means that the amplitude of the pump wave decreases during their propagation through the waveguide, while at the same time the amplitude of the Stokes wave increases. According to this, the alteration of the amplitude of pump and Stokes waves are:

$$\frac{\partial B_{\mathrm{p}}}{\partial z} = -\frac{g_{\mathrm{p}}}{2} \left|B_{\mathrm{s}}\right|^2 B_{\mathrm{p}} - \frac{\alpha_{\mathrm{p}}}{2} B_{\mathrm{p}} \tag{10.8}$$

and

$$\frac{\partial B_{\mathrm{s}}}{\partial z} = \frac{g_{\mathrm{s}}}{2} \left|B_{\mathrm{p}}\right|^2 B_{\mathrm{s}} - \frac{\alpha_{\mathrm{s}}}{2} B_{\mathrm{s}}. \tag{10.9}$$

If the attenuation in the fiber (the second term in (10.9)) and the depletion of the pump wave (10.8) is neglected, the amplitude of the Stokes wave is

$$B_{\mathrm{s}}(z) = B_{\mathrm{s}}(0) \mathrm{e}^{g_{\mathrm{s}}/2 \left|B_{\mathrm{p}}\right|^2 z}, \tag{10.10}$$

where $B_{\mathrm{s}}(0)$ is the Stokes amplitude at $z = 0$, and is not necessarily the fiber input. Under the simplified conditions, the Stokes wave shows an exponential growth during its propagation through the waveguide. But, on the other hand, a Stokes wave can only be generated if $B_{\mathrm{s}}(0) \neq 0$. This means that the exponential growth starts from a position where a wave with the correct frequency shift is present. Or, it can start growing at a position where some dipoles already oscillate with the right frequency. Therefore, the initial Stokes wave is always required and can be generated by spontaneous scattering or noise in the fiber. If the pump power is high enough, the position where it is generated is the fiber input, but it can also be launched into the fiber input already.

The growth rate of the Stokes wave depends, according to (10.8) and (10.9), on the gain of the Stokes wave (g_{s}) and the gain of the pump wave (g_{p}). The Raman coefficient for the pump is slightly different from that for the Stokes wave because the pump is frequency-shifted (for the maximum of the gain in optical fibers at about 15 THz). The Raman gain coefficient in fibers scales, to first order, with the inverse pump wavelength. Therefore, the Raman coefficient of the pump is related to that of the probe via:

$$g_{\mathrm{P}} = \frac{\omega_{\mathrm{P}}}{\omega_{\mathrm{S}}} g_{\mathrm{S}}. \tag{10.11}$$

According to (10.10), the pump wave's amplitude depends on the intensity of the pump $|B_{\mathrm{p}}|^2 \sim I_{\mathrm{p}}$. Therefore, if the amplitude increases exponentially, the power and the intensity of the Stokes-wave must show an exponential growth along the z-direction as well. At the same time the pump must lose a part of its intensity (negative sign in (10.8)) corresponding to the intensity that the Stokes wave will receive. The attenuation in the fiber acts on the intensity of the pump as well as on the intensity of the Stokes wave, thus decreasing both. The attenuation constant is different for the two wavelengths (α_{s} and α_{p}), however.

The intensity is connected to the power via the effective area of the waveguide. Hence, if (10.9) and (10.8) should be expressed as a function of the intensities the Raman gain coefficient depends on the effective area of the fiber, as follows:

$$g_{\mathrm{R}} = g_{\mathrm{S}} \times A_{\mathrm{eff}}, \tag{10.12}$$

with g_{R} as the Raman gain. At the same time, the relative polarization between pump and Stokes wave has, of course, an influence on the efficiency of the Raman scattering as well. The differential equation system, describing the intensities of both waves under the influence of Raman scattering during its propagation through the medium is then represented by.

$$\frac{\mathrm{d}I_{\mathrm{s}}}{\mathrm{d}z} = \frac{g_{\mathrm{R}}}{K_{\mathrm{s}}} I_{\mathrm{p}} I_{\mathrm{s}} - \alpha_{\mathrm{s}} I_{\mathrm{s}} \tag{10.13}$$

and

$$\frac{\mathrm{d}I_{\mathrm{p}}}{\mathrm{d}z} = -\frac{\omega_{\mathrm{p}}}{\omega_{\mathrm{s}}} \frac{g_{\mathrm{R}}}{K_{\mathrm{s}}} I_{\mathrm{p}} I_{\mathrm{s}} - \alpha_{\mathrm{p}} I_{\mathrm{p}}, \tag{10.14}$$

where K_{s} is a factor that includes the relative polarization between the pump wave and the Stokes wave. The Raman gain has its maximum if $K_{\mathrm{s}} = 1$, the case where both waves have an identical polarization. Standard single mode fibers show birefringence and hence, the polarization state of the waves is arbitrary. In this case, the polarization factor is $K_{\mathrm{s}} = 2$ [241].

10.3 Raman Scattering in Optical Fibers

As mentioned in the previous sections, the Raman scattering depends on particular material resonances. In crystalline media, these resonances can show a very narrow bandwidth. Therefore, the generated Stokes or anti-Stokes wave reflects material-specific properties. This fact has made the Raman scattering a widely used tool in the field of spectroscopy.

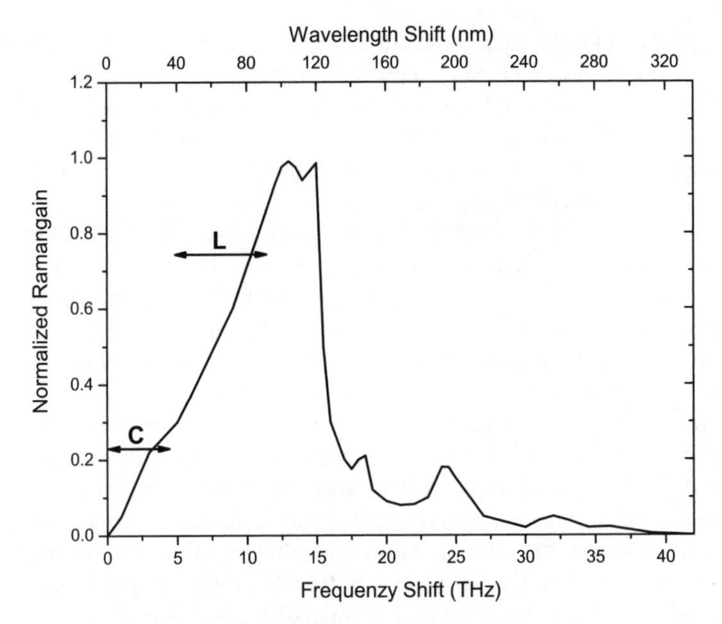

Fig. 10.5. Normalized Raman gain against the frequency shift between the pump and Stokes waves in a fiber consisting of silica glass [241]

On the other hand, the basic material of optical fibers is glass which is not crystalline but amorphous. The resonance frequencies of the molecular vibration modes in glasses are overlapped with each other and form rather broad frequency bands. Therefore, optical fibers show Raman scattering over a relatively wide frequency range. The Raman gain $g_{\mathrm{R}}(f)$ depends mainly on the material composition of the fiber core and the contained dopants [242].

A typical spectrum of the Raman gain in optical fibers consisting of silica glass is shown in Fig. 10.5. The Figure shows the gain versus the frequency shift between pump- and Stokes wave normalized to the maximum. The Raman gain has one main and several side maxima. For a pump wavelength of 526 nm and a parallel polarization between the pump and Stokes waves the maximum of the Raman gain in pure quartz glass is $g_{\mathrm{R\,max}} \approx 1.9\times10^{-13}$ m/W [243]. As discussed in the previous section, the gain is inversely proportional to the wavelength of the pump and depends on the polarization of the waves. The Raman gain for a parallel polarization between the pump and Stokes waves ($g_{\parallel}$) is much greater than that for a perpendicular polarization between both ($g_{\perp}$) [244]. In an optical fiber, the birefringence leads to an arbitrary alteration of the polarization states of the waves during propagation. If the fiber is long enough, so that the polarization of pump- and Stokes wave are randomized completely, the effective gain in the fiber is [245]

$$g_{\mathrm{R}} = 1/2\left(g_{\parallel} + g_{\perp}\right). \qquad (10.15)$$

Therefore, with (10.11), the maximum Raman gain at a pump wavelength of 1.4 μm in pure quartz glass can be calculated to be $g_{\mathrm{R\,max}}(1.4) = 1.9 \times 10^{-13}$ m/W×1/2 (0.526/1.4) $= 3.57 \times 10^{-14}$m/W, whereas the gain for a pump wavelength of 1.55 μm is $g_{\mathrm{R\,max}}(1.55) = 3.2 \times 10^{-14}$m/W. Contrary to quartz, optical fibers consist of silica glass doped with different materials such as GeO_2. Of course, the dopants have an influence on the gain in the fiber as well. Mahgerefteh et al. [245] obtained for the Raman gain in a standard single mode fiber experimentally a value of $g_{\mathrm{R\,max}} = 3.1 \times 10^{-14}$m/W for a pump wavelength of 1.55 μm. In a microstructured or holey fiber, the Raman gain can be much higher. Yusoff et al. [246] measured, for instance, a gain of $g_{\mathrm{R\,max}} = 7.6 \times 10^{-14}$m/W in a highly nonlinear holey fiber with an effective area of 2.85 μm^2.

The gain depends on the imaginary part of the nonlinear susceptibility, as mentioned in Sect. 10.2.1. The inverse proportional scaling of the Raman gain on the wavelength of the pump is only valid to first order. A detailed analysis of this dependence with results for several commercially available fibers can be found in [247]. Basically, the Raman gain coefficient (g_{S}) depends on the effective area of the fiber therefore. It is higher for fibers with a small A_{eff}, such as LEAF and TrueWave, than it is for standard single mode fibers.

As can be seen from Fig. 10.5, if the smaller peaks in the tails are included, the Raman gain has a wide bandwidth ($\approx$ 40 THz) and shows a broad maximum between 9 and 15.5 THz. The maximum value of the gain can be found for pure silica glass at a frequency shift of 13.2 THz, or $\approx$ 105 nm, respectively. For commercially available standard single mode fibers, a similar value is valid ($\Delta\lambda = 112$ nm for a pump wavelength of 1550 nm [245]).

Until now, no analytical function for the determination of the Raman gain has been given in the literature. For the calculation of the threshold between spontaneous and stimulated emission, usually only the main maximum is considered. The frequency dependence of the gain will be approximated by a Lorentz function [239].

The pump wave and the scattered Stokes wave show relatively high frequency shifts. As already mentioned, for Raman scattering no phase matching condition is required but, nevertheless, this high frequency shift in dispersive fibers leads to a walk-off between pump and Stokes wave if the pulses are short. Therefore, the power of the scattered Stokes wave depends on the temporal duration of the pulses [245]. A shorter pump pulse leads to a decrease of the interaction length and results in a smaller output power.

Raman scattering can be seen in the propagation direction of the pump and in the opposite direction. In optical transmission systems, only the forward Raman scattering (in direction of the pump wave) is of importance. The backward Raman scattering can be exploited for a signal amplification in fiber Raman amplifiers.

If waves with an anti-Stokes frequency are launched into the fiber input, they will be weakened by the Raman scattering. Therefore, they experience

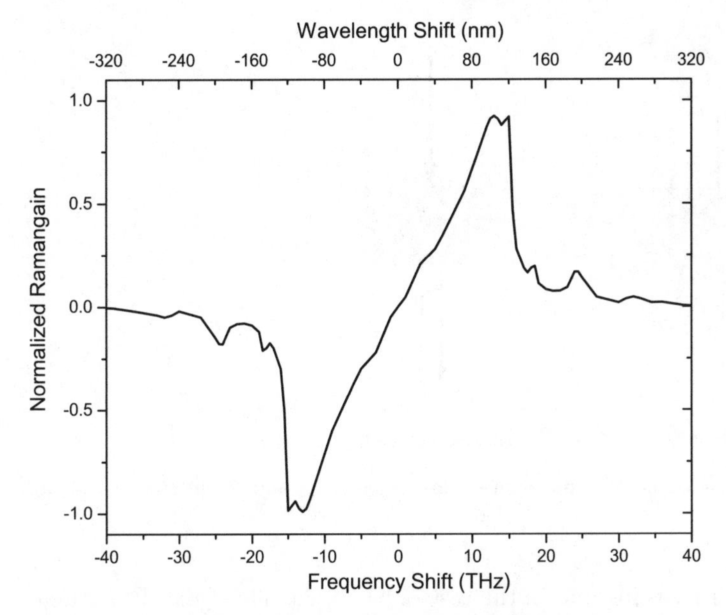

Fig. 10.6. Qualitative behavior of the normalized Raman gain versus a positive and negative frequency shift between the pump wave and the Stokes wave in a standard single mode fiber [247]

a negative Raman gain. Figure 10.6 shows the Raman gain for a positive and a negative frequency shift in relation to the pump frequency.

The Raman gain is not symmetrical around the pump wavelength, as could be assumed in a first approximation. The asymmetry results from the inverse proportionality of the Raman gain against the wavelength. The loss on the anti-Stokes side is larger than the gain on the Stokes side. Hence, a wave with a shorter wavelength loses more power to the pump wave than the pump gives to a wave with a higher wavelength.

10.3.1 Spontaneous and Stimulated Scattering

Spontaneous and stimulated scattering show, in principle, the same relation to each other as spontaneous and stimulated emission. However, in the case of scattering, no allocation inversion is involved. The basic differences between a spontaneous and a stimulated scattering are shown in Fig. 10.7.

If the intensity of the incident field is below a certain threshold, spontaneous scattering appears. As shown on the left side of Fig. 10.7, a relatively weak Stokes wave is radiated into all directions. This is, of course, not the case in optical fibers. Since only the forward and backward direction is possible here, the field radiated into other directions is constrained into the basic directions due to the waveguide characteristic of the fiber.

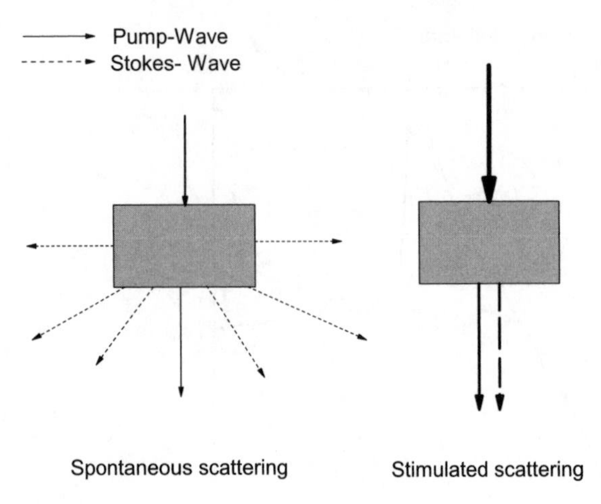

Fig. 10.7. Basic differences between spontaneous (*left*) and stimulated scattering (*right*)

If the intensity is high, as in the case on the right side of Fig. 10.7 stimulated scattering occurs after it exceeds the threshold. The Stokes wave is very intense and shows a similar spatial distribution as the pump wave. Figure 10.7 only shows the generation of the Stokes wave into the forward direction, but a wave into the opposite direction can be generated as well.

The transition between spontaneous and stimulated scattering can be described with the model of the harmonic oscillator. If, at the input layers of the medium, the pump wave hits a dipole oscillating with its resonance frequency, an additional Stokes wave will be generated by the dipole. The dipole emits this Stokes wave with a radiation pattern typical for a dipole. The radiation pattern shows a $\sin\theta$ dependence.[3] If the intensity of the pump is low, the intensity of the radiated Stokes wave is low as well. The fading frequency between both is unable to excite following dipoles to a strong vibration at the resonance frequency. If the pump wave hits another oscillating dipole during its propagation through the material, it emit a Stokes wave as well. The radiation pattern of the second dipole shows again a $\sin\theta$ dependence and has different phase than the first one. Hence, no coherent superposition with the Stokes field radiated by the first dipole can occur. The intensity, as well as the direction of the superposition, depends on the arbitrary phase difference between the two dipoles. In consequence, a weak Stokes wave is radiated into arbitrary directions.

On the other hand, if the intensity of the pump wave is higher than a particular threshold, then the wave scattered at the first dipole is relatively intense. Pump and emitted Stokes waves propagate through the material

[3]In a direction parallel to the dipole the angle θ is zero or π. Perpendicular to the dipole the angle is $\theta = \pi/2$. Hence, the maximum of the field is radiated perpendicular to the dipole, whereas no field is radiated parallel to it.

and the following dipoles are driven by the difference frequency between both waves to oscillate at their resonance frequency. Now the phase of the excited dipoles is determined by the incident wave. The Stokes wave of the following dipole superimposes coherently in the forward direction with the Stokes wave of the first dipole. Therefore, the intensity grows and the next dipoles are driven stronger to oscillate at their resonance frequency. In the stimulated case, the Stokes wave grows exponentially with the propagation through the material and is only scattered into one direction.

10.3.2 Threshold of Raman Scattering in Optical Waveguides

The intensities of the pump and Stokes waves under the influence of Raman scattering in an optical fiber are described by the differential equation system (10.13) and (10.14). If, in a first approximation, it is assumed that the intensity depletion of the pump due to the Raman interaction is small, the first term on the right side of (10.14) can be neglected. In this case the intensity distribution along z depends only on the attenuation of the fiber:

$$I_{\mathrm{p}}(z) = I_{\mathrm{p}}(0)\mathrm{e}^{-\alpha_{\mathrm{p}} z}. \tag{10.16}$$

If (10.16) is introduced into (10.13), the intensity of the Stokes wave is

$$I_{\mathrm{s}}(z) = I_{\mathrm{s}}(0)\exp\left(\frac{g_{\mathrm{R}}}{K_{\mathrm{s}}} I_{\mathrm{p}}(0) L_{\mathrm{eff}} - \alpha_{\mathrm{s}} z\right), \tag{10.17}$$

where L_{eff} determines the effective interaction length

$$L_{\mathrm{eff}} = \frac{1 - \mathrm{e}^{-\alpha_{\mathrm{p}} z}}{\alpha_{\mathrm{p}}}, \tag{10.18}$$

and g_{R} is the Raman gain for co-polarized pump and Stokes waves (for standard single mode fibers $2 \times 3.1 \times 10^{-14}$ m/W). Therefore, the exponential growth of the Stokes wave is not unlimited for unlimited long waveguides. The effective length and, therefore, the intensity of the generated Stokes wave, have limiting values for an increasing distance z, as shown in Fig. 10.8. For nonamplified systems, we have

$$\lim_{z \longrightarrow \infty} L_{\mathrm{eff}} = \frac{1}{\alpha_{\mathrm{p}}}. \tag{10.19}$$

If the threshold for stimulated scattering is defined as the input intensity value of the pump wave for which the Stokes wave shows a growth in the fiber I_{pG}, the amplification due to the Raman process must exceed the attenuation loss of the Stokes wave. From (10.17) we obtain

$$\frac{g_{\mathrm{R}}}{K_{\mathrm{s}}} I_{\mathrm{pG}} L_{\mathrm{eff}} \gg \alpha_{\mathrm{s}} z. \tag{10.20}$$

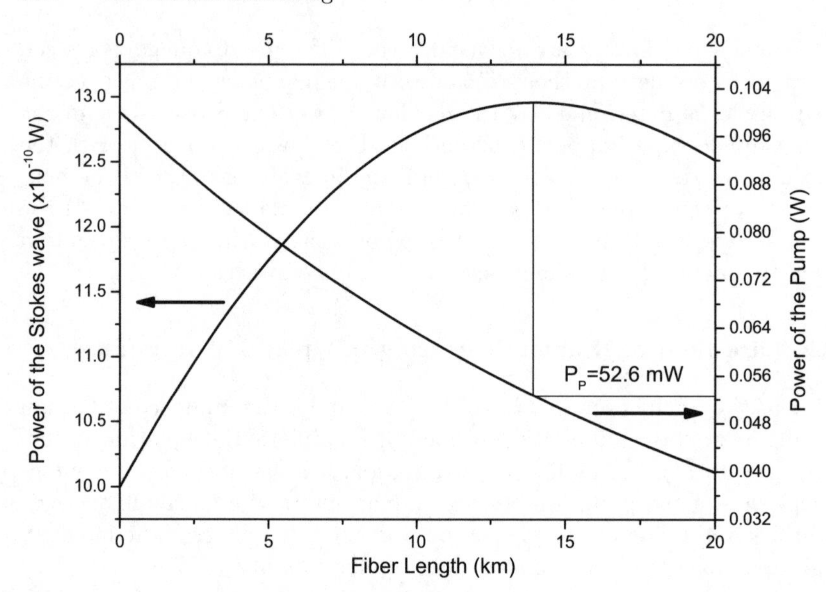

Fig. 10.8. Computed power of a pump wave and a Stokes wave along a fiber with simulation parameters described in the text

Using $P_{0G} = I_{pG} A_{eff}$, the inequality in (10.20) gives

$$\frac{g_R}{K_s} P_{0G} L_{eff}/A_{eff} \gg \alpha_s z. \tag{10.21}$$

If the fiber length is small we can assume that the effective length corresponds approximately to the fiber length, $L_{eff} \approx z$. The threshold of stimulated Raman scattering (SRS) is then

$$P_{0G} \gg \frac{\alpha_S A_{eff} K_s}{g_R}. \tag{10.22}$$

If the attenuation constant in a fiber with an effective area of $A_{eff} = 80\ \mu m^2$ is $\alpha = 0.2$ dB/km and the Raman gain is $g_R = 7 \times 10^{-14}$ m/W, we obtain a critical pump power of $P_{0G} \gg 52.6$ mW for a parallel polarization of the waves ($K_S = 1$). Figure 10.8 shows the computed power of a pump and a Stokes wave in a fiber with the above mentioned parameters.

The pump wave has a frequency of $\omega_p = 1.55\ \mu m$ and the intensity of the pump at the fiber input is $I_p(0) = 1.25 \times 10^9$ W/m^2. With an effective area of $A_{eff} = 80\ \mu m^2$, this corresponds to a pump power of $P_p = 100$ mW. The frequency of the Stokes wave is $\omega_s = 1.65\ \mu m$. According to (10.13), a wave with the right frequency shift must be already present in the fiber in order to generate a Stokes wave. In the example this original wave is generated by spontaneous scattering or by an already present noise floor. For the intensity of the initial wave we assume a value of $I_s(0) = 12.5$ W/m^2, which corresponds to a power of $P_s = 10 \times 10^{-10}$ W.

As can be seen from Fig. 10.8, the intensity of the Stokes wave increases very strongly from 0 to 5 km in this range the inequality (10.21) is valid. If the waves propagate further, the intensity of the pump decreases. The growth rate of the Stokes wave is smaller and comes to a maximum after a propagation distance of 14 km. After this the attenuation (the right side in (10.21)) exceeds the amplification and the Stokes wave will decrease.

Note that, in spite of its strong increase at the first kilometers in the fiber, if the initial Stokes wave is generated from noise, its maximum power is still 8 orders of magnitude smaller than the pump power. For a strong measurable effect, the intensity of the Stokes wave at the fiber output has to be comparable to the intensity of the pump. Therefore, the forward threshold for Raman scattering in optical fibers is defined as the input pump power at which the output powers for pump and Stokes wave are equal. This value can be estimated as [239]

$$P_{\mathrm{th}} = 16\frac{K_{\mathrm{S}} A_{\mathrm{eff}}}{g_{\mathrm{R}} L_{\mathrm{eff}}}. \tag{10.23}$$

Equation (10.23) is an approximation and is only valid under the conditions that the power transfer of the pump to the Stokes wave due to the Raman process is negligible, the effective area for pump- and Stokes wavelength are equal, the Raman gain can be approximated by a Lorentz function, and the initial Raman signal is generated by spontaneous scattering only, i.e., no wave with the Raman frequency shift is launched into the fiber.

The threshold for the stimulated Raman scattering is $P_{\mathrm{th}} \approx 1$ W in polarization-maintaining fibers ($K_{\mathrm{S}} = 1$) with an effective area of $A_{\mathrm{eff}} = 80\ \mu\mathrm{m}^2$, a Raman gain of $g_{\mathrm{R}} \approx 7 \times 10^{-14}$ m/W, and an effective length of $L_{\mathrm{eff}} \approx 22$ km. Whereas, in standard single mode fibers, due to the arbitrary distribution of the polarization states of pump and Stokes wave ($K_S = 2$), it has a threshold of $P_{\mathrm{th}} \approx 2$ W.

Figure 10.9 shows pump waves with an intensity in the range between 1×10^{10} and 2×10^{10} W/m^2. This corresponds to a power range from 800 mW to 1.6 W if an effective area of $A_{\mathrm{eff}} = 80\ \mu\mathrm{m}^2$ is assumed.

As can be seen from Fig. 10.9, the Stokes power exceeds that of the pump at the fiber output for the first time. For a pump intensity of $I_{\mathrm{p}} \approx 1.5 \times 10^{10}$ W/m^2, this corresponds to a pump power of 1.2 W. The higher threshold comes from the fact that the effective length is shorter and that the whole differential equation system (10.13) and (10.14) was used for the simulation. Hence, the pump power depletion is included.

If the intensity of the pump wave is higher than the threshold, the power of the Stokes wave at the end of the fiber is greater than the output power of the pump. Figure 10.10 shows this behavior for a pump intensity of $I_{\mathrm{P}} = 1.9 \times 10^{10}$ W/m^2 and an input power of $P_{\mathrm{P}} = 1.5$ W.

The figure shows clearly the intensity loss of the pump due to the power transfer to the Stokes wave. After a propagation distance of 20 km, the Stokes

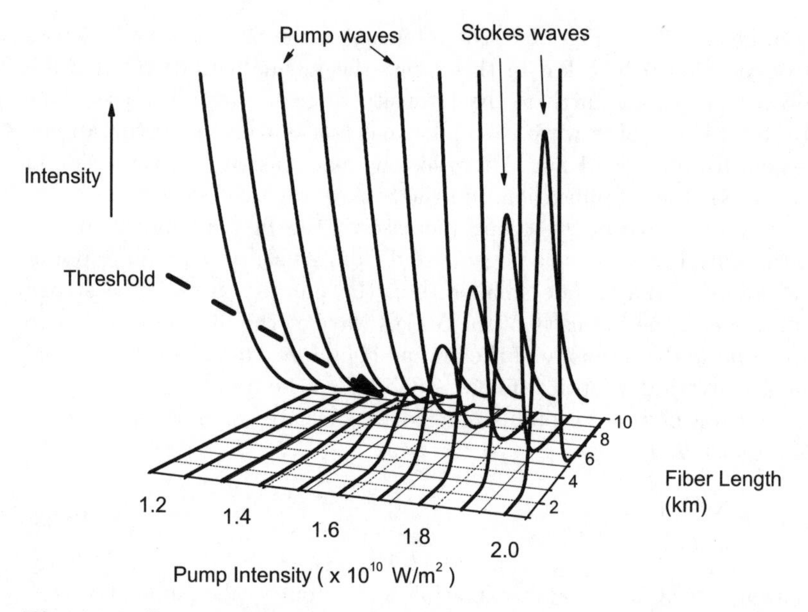

Fig. 10.9. Pump waves with different intensities and generated Stokes waves in a polarization-maintaining fiber ($g_{\mathrm{R}} \approx 7 \times 10^{-14}$ m/W, $A_{\mathrm{eff}} = 80\ \mu\mathrm{m}^2$, $\alpha = 0.2$ dB/km)

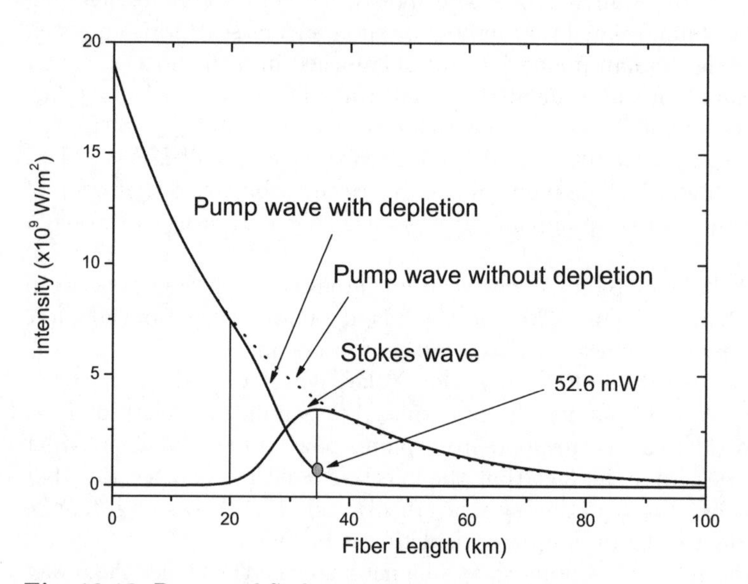

Fig. 10.10. Pump and Stokes waves in an optical fiber whose pump depletion with the parameters of Fig. 10.9 was taken into consideration and an input pump power above the threshold for stimulated scattering

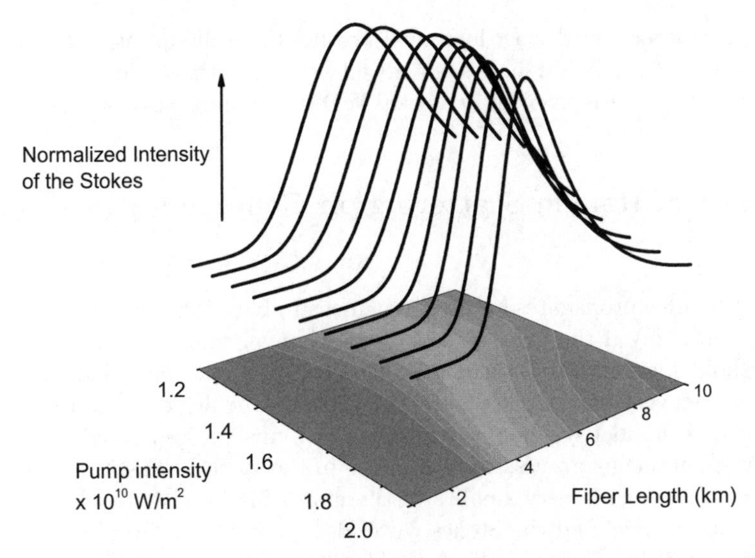

Fig. 10.11. Normalized Stokes waves created by pump waves with different intensities above the threshold of stimulated scattering

wave increases very strongly and the depletion of the pump can be no longer neglected. At a distance of 29 km, the intensity of the Stokes exceeds that of the pump and after a distance of 34.6 km, the pump intensity is no longer strong enough to amplify the Stokes wave further. Over this distance the intensity of the pump wave has decayed to a value of 0.657×10^9 W/m^2. For an effective area of 80 μm^2, this corresponds to a power of 52.6 mW and, hence, the threshold for the growth of the Stokes wave (10.22). If both waves propagate further, the attenuation in the fiber is stronger than the power transfer between pump and Stokes wave. Due to the fact that the power contribution of the Stokes wave is negligibly small, the total power in the system (pump + Stokes) corresponds to the pump power without depletion. Figure 10.11 shows the normalized intensity of Stokes waves created by pump waves with different intensities in a fiber with the same parameters.

As can be seen from Fig. 10.11, the maximum of the Stokes is shifted to smaller fiber lengths with increasing pump power. This effect has its origin in the strong decay of the pump power due to the power transfer to the Stokes wave. The pump wave reaches its maximal value at which the fiber attenuation exceeds the power transfer to the Stokes wave (52.6 mW for this example) for shorter propagation distances.

Powers in the range of the threshold for forward Raman scattering in fibers (1 W and more) are normally not reached in optical communication systems. Hence, the threshold is especially important for applications of the SRS, like the Raman laser and amplifier, treated in detail in Sects. 13.1 and 13.2. The threshold can be decreased orders of magnitude if hollow core photonic crystal fibers filled with hydrogen are used as cells for stimulated Raman scattering [240]. Nevertheless, even well below the threshold, Raman scattering can be a severe problem in optical WDM communication systems.

10.4 Impact of Raman Scattering on Communication Systems

If merely one channel propagates in the fiber a measurable effect of the Raman scattering occurs only if the input power of the channel comes in the range of the threshold for forward Raman scattering. Due to the fact that this threshold requires very high input powers, the SRS can be neglected in these systems. For one-channel systems, the Stokes wave must be generated from an arbitrary spontaneous process or from noise in the system. The intensity of the Stokes wave (I_S) is very small and therefore, the intensity exchange between the pump wave and the Stokes wave, described by the first term of the right side of (10.14), is very ineffective. Therefore, the power depletion of the one-channel system due to Raman scattering is negligibly small.

A completely different behavior can be seen in WDM systems. Here the initial Stokes wave is not generated by a spontaneous process because waves with the right frequency shift are already present at the fiber input. Furthermore, the input power of the Stokes wave is equal to the input power of the pump if a logical "1" is transmitted in both channels. Note, that equal powers of the pump and Stokes waves at the fiber output form the basis of the definition of the threshold for forward scattering. Of course their intensities are equal as well $I_S \approx I_P$. Therefore, neither the first term in (10.13) nor the first term in (10.14) can be neglected in this case. As a result, the channels with a higher wavelength are amplified (10.13) at the expense of the channels with a lower wavelength (10.14).

As can be seen from Fig. 10.6, the Raman scattering has a very broad gain bandwidth and acts on channels that can have a frequency shift of more than 30 THz or 240 nm. This value is smaller than the transparency range of an optical fiber, which can be determined to be 50 THz. In the 35 nm broad C-band, the channels at 1530 nm can lead, for instance, to an amplification of the channels at 1560 nm. The interaction between the C- and L- or S- and L-band is more efficient because in this case the Raman gain is much higher.

The influence of the Raman scattering depends on the input power of the WDM channels and their frequency shift. Contrary to the case of FWM, Raman scattering is not a parametric process. The medium is actively involved in the interaction and the momentum conservation or phase matching condition is always fulfilled. As long as the frequencies are inside the spectrum of the Raman gain, an interaction can take place. If the power of the channels is above the growth threshold of the Stokes wave (10.22), the influence is dramatic, as can be seen from Fig. 10.12.

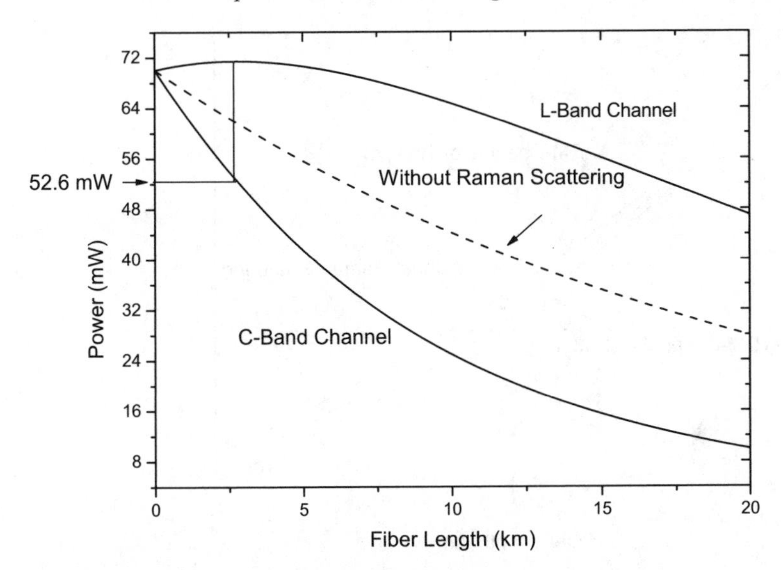

Fig. 10.12. Raman scattering for two channels with an input power of 70 mW out of the C- and L-band ($A_{\text{eff}} = 80\ \mu\text{m}^2$, $g_R = 7 \times 10^{-14}$ m/W, $\alpha = 0.2$ dB/km, $K_s = 1$)

For a fiber with an effective area of $A_{\text{eff}} = 80\ \mu\text{m}^2$, a Raman gain of $g_R = 7 \times 10^{-14}$ m/W, and an attenuation constant of $\alpha = 0.2$ dB/km (10.22) yields a threshold of 52.6 mW. The input power of both WDM channels for Fig. 10.12 is 70 mW. Due to the fact that the power of the pump channel is above the growth threshold, the channel at the Stokes wavelength increases up to a propagation distance of ≈3 km. Here, the power of the Stokes channel exceeds its input power at about 1.5 mW. At the same time, the power loss of the pump channel is very strong as well. For a propagation distance of 3 km, the pump power is about 9 mW below the value it would have without Raman scattering.

If both waves come from the same band, the Raman gain is smaller and hence, the threshold higher. For $g_R = 3 \times 10^{-14}$ m/W we obtain a threshold of ≈ 123 mW. But input powers of a few mW can lead to a non-negligible influence on WDM communication systems as well. The Raman scattering between two waves out of the C-band at 1.53 and 1.56 μm is shown in Fig. 10.13.

Both channels have a power of 10 mW at the fiber input. Because this is well below the growth threshold of 123 mW, the attenuation in the fiber is always much higher than the power transfer between pump and Stokes wave. Nevertheless, an influence of the Raman scattering can be seen. The channel with the higher wavelength is amplified at the expense of the channel with the lower wavelength.

The amplification factor of the Stokes wave, or the channel with the higher wavelength, is the ratio between the intensity of the Stokes wave at the fiber

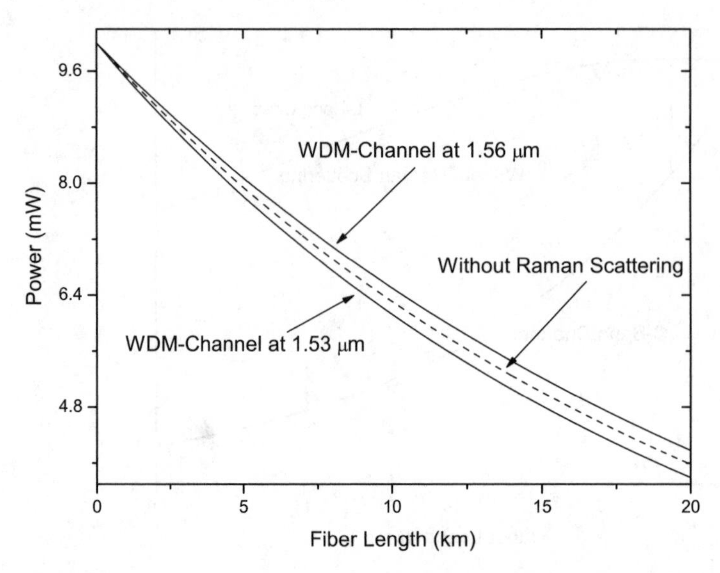

Fig. 10.13. Influence of Raman scattering on two WDM channels. The power of both channels at the fiber input is 10 mW. ($g_R = 3\times10^{-14}$ m/W, $\alpha = 0.2$ dB/km

output $I_s(L)$ and the intensity it would have without the Raman scattering. In the latter case, only the attenuation has an influence, as follows:

$$I_s(L) = I_s(0)e^{-\alpha_s L}. \tag{10.24}$$

The ratio between the two intensities is the gain of the Raman amplification (not to be confused with the Raman gain, which is a material-dependent constant):

$$G_R = \frac{I_s(L)}{I_s(0)e^{-\alpha_s L}}. \tag{10.25}$$

At the same time, the pump wave, or the channel with the lower wavelength, loses power to the Stokes channel. The depletion factor of the pump wave is

$$D_R = \frac{I_p(L)}{I_p(0)e^{-\alpha_p L}}. \tag{10.26}$$

Figure 10.14 shows the amplification and depletion factor for a transmission system consisting of two channels in fibers with different length.

As expected, according to the differential equations (10.13) and (10.14), the amplification factor grows with increasing fiber length and signal power, whereas the depletion factor shows the opposite behavior. For an input power of 20 mW, the channel with the higher carrier wavelength has ≈ 1.2 times as much power as it would have without Raman scattering after a propagation

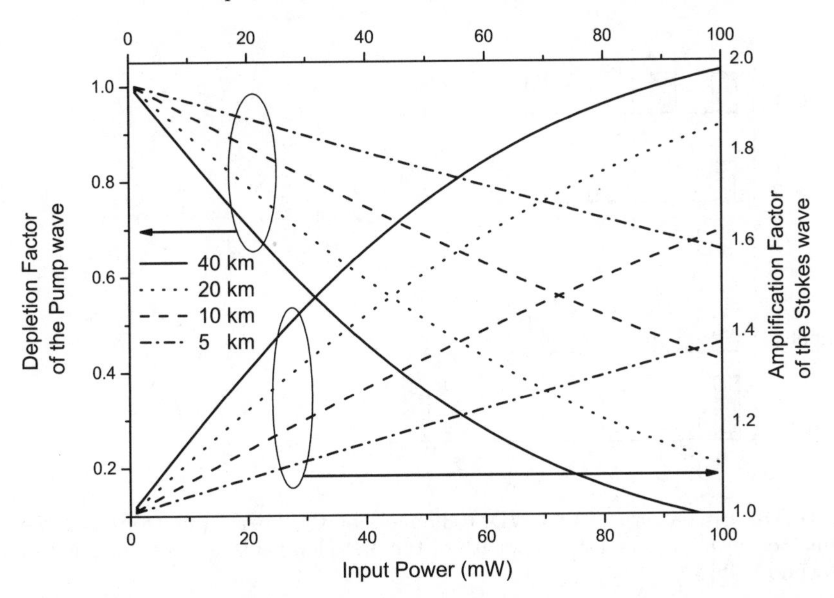

Fig. 10.14. Amplification and depletion factor for a two channel transmission system versus input power ($\alpha_{\mathrm{p}} = \alpha_{\mathrm{s}} = 0.046\ \mathrm{km}^{-1}$, $g_{\mathrm{R}} = 7 \times 10^{-14}$ m/W, $A_{\mathrm{eff}} = 80\ \mu\mathrm{m}^2$)

distance of 20 km. Therefore, the power of the channel with the lower carrier is further decreased by a factor of 0.8. For an input power of 100 mW, the amplification and depletion factors in a 20 km fiber are ≈ 1.8 and ≈ 0.2, respectively. If the channels propagate in longer fibers, already relatively weak input powers show a strong cross talk between the channels. If the power is increased, it comes to a saturation, as can be seen from the 40 km trace. The influence of the Raman scattering on the bits in the distinct channels is shown in Fig. 10.15.

Assume channel 1 has a carrier wavelength of 1550 nm, whereas the carrier wavelength of channel 2 is 1600 nm. At the fiber input, both channels are launched with the same input power. As can be seen from Fig. 10.15, if both channels transmit a logical "1", it is amplified in channel 2 at the expense of the "1" in channel 1. Whereas, if one of the channels contains a logical "0", no interaction between the channels can be seen. Therefore, the Raman scattering shows a logical "AND" functionality.

If the dispersion is taken into consideration, the interaction length is smaller due to the walk-off between the pulses in both channels. Hence, the amplification and depletion between the "1" bits is decreased as well. On the other hand, AND combinations between bits that do not temporally overlap at the fiber input are possible.

For a transmission system with a large number of WDM channels, the behavior is more complicated, but qualitatively the same result can be seen.

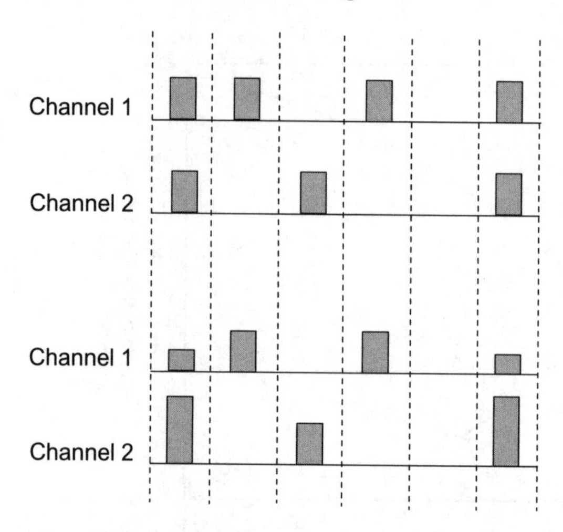

Fig. 10.15. Bit pattern of two WDM channels at the fiber input (*top*) and the output (*bottom*). The carrier wavelength of the first channel is smaller than that of the second [248]

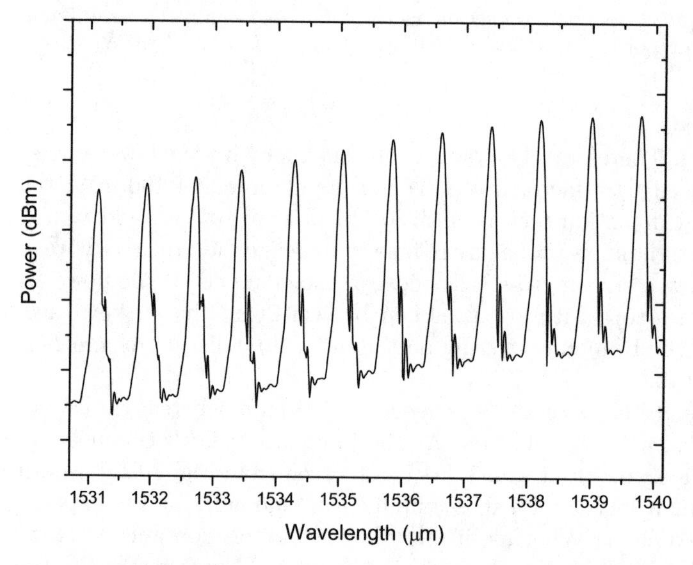

Fig. 10.16. Qualitative power distribution at the fiber output due to Raman scattering for a 12-channel WDM system

The "1" bits in the channels with a higher carrier wavelength are amplified while the marks which are transmitted by a smaller wavelength are weakened. Figure 10.16 shows this behavior qualitatively for a system with 12 WDM-channels. The channels in the upper part of the spectrum are amplified, whereas the channels in the lower part lose their power.

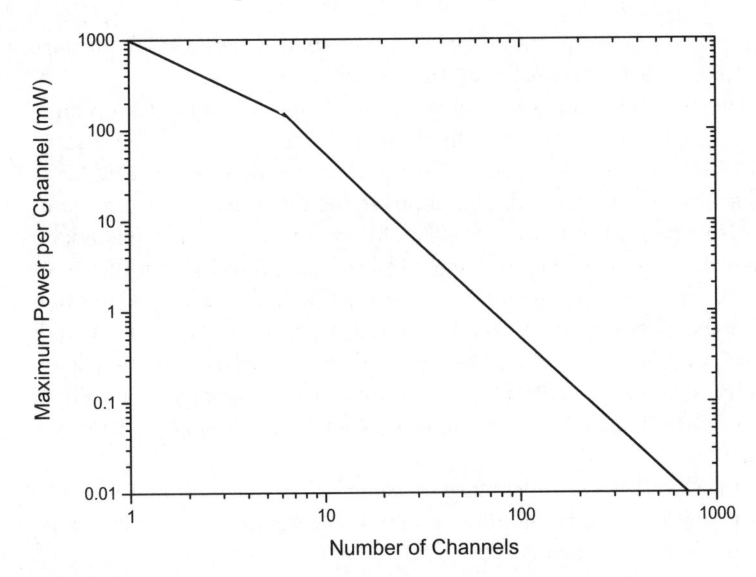

Fig. 10.17. Maximum input power per channel against the number of channels if the decrease by SRS should be smaller than 1 dB ($\lambda = 1550$ nm, $\alpha = 0.2$ dB/km, $A_{\text{eff}} = 50\ \mu\text{m}^2$, $L_{\text{eff}} = 22$ km) [248]

The decrease of the power in the lower wavelength channels is a problem in optical communication systems because it decreases the SNR and leads to a higher BER. This decrease can be estimated if the gain profile of Fig. 10.5 is approximated by a triangle in the range from 0 to 15 THz [249]. In a WDM system with N channels, a channel spacing of Δf and a power per channel of P, no channel is decreased by more than 1 dB if the product of the whole optical power (NP) and the entire required optical bandwidth in the fiber ($(N-1)\Delta f$) is smaller than 500 GHz $\times$W [248]:

$$[NP]\,[(N-1)\Delta f] < 500\ \text{GHz}\ \times \text{W}. \tag{10.27}$$

Figure 10.17 shows the maximum power per WDM-channel versus the number of channels for a frequency spacing of 10 GHz. If only a small number of channels propagates in the fiber, the power of all N channels contribute to the SRS and the maximum power decreases with $1/N$. But, if the number is increased, a $1/\text{N}^2$ dependence develops because the occupied bandwidth increases and the interchannel interactions become significant [248].

For a particular BER, determined by the system, the SNR must be above a distinct lower value. This SNR determines the capacity of the transmission system. If the system contains no optical amplifiers, i.e., in regenerated systems, the SRS leads to a power decrease of the channels with a lower carrier wavelength and, therefore, a reduction of the SNR and an increase of the BER. This results in a smaller transmission capacity of the system. Longer transmission systems require more signal power for the same SNR and are more susceptible to Raman scattering than shorter systems.

In long haul transmission systems, a large number of optical amplifiers increases the signal power after each fiber span. At the same time, the accumulated noise is increased and additional noise is included by the amplifier. The Raman scattering reduces the signal power for the lower frequency channels, but it also acts on the noise power in this channel. Since noise is added periodically over the entire system length by each amplifier, the influence of the SRS on the noise power is smaller than on the signal power. For small degradations, the SRS impact on the fractional depletion of the noise is half that of the signal [250]. Therefore, the SRS reduces the SNR and the capacity in amplified systems as well. The limitations on the capacity of amplified systems imposed by the SRS can be determined following an estimation described in [250].

If, in a WDM system, the decrease of the SNR in the channel with the smallest wavelength has to be smaller than 0.5 dB, the number of the transmittable channels n can be calculated [250]:

$$n\,(n-1) < \frac{8.7 \times 10^{12}\ \mathrm{Hz} \times \mathrm{W} \times \mathrm{km}}{P f L_{\mathrm{eff}}}, \tag{10.28}$$

where f denotes the frequency spacing between the wavelength channels and L_{eff} is the effective interaction length for an amplified system (4.101). The BER of the system is determined by the noise in the system and the input power of the channels. The average minimum input power of each channel is denoted as P in (10.28). The input power is the product between the required SNR for a particular BER (R) and the noise power in the channel; $P = RN$. The noise power in the system is

$$N = 2mh\nu n_{\mathrm{sp}} B_o (G-1), \tag{10.29}$$

where m is the number of amplifiers, $h\nu$ is the photon energy with the Planck constant $h = 6.626 \times 10^{-34}$Js, ν is the frequency of the photon, n_{sp} determines the excess noise factor of the amplifiers (the additional noise introduced by the amplifiers), B_{o} is the optical bandwidth of the input filter in the receiver, and $G = E\mathrm{e}^{\alpha L_{\mathrm{A}}}$ is the gain of each amplifier with E as the additional loss due to the coupling of the amplifiers into the transmission link and L_{A} as the amplifier spacing. Hence, the average input power in each channel can be written as

$$P = 2Rh\nu n_{\mathrm{sp}} B_{\mathrm{o}} L / L_{\mathrm{A}} \left(E\mathrm{e}^{\alpha L_{\mathrm{A}}} - 1 \right). \tag{10.30}$$

Let us consider an example [250]. The bit rate in each channel of a WDM transmission system is 2.5 Gb/s. The channel spacing is 0.5 nm at a carrier wavelength of 1550 nm ($f = 62$ GHz) and the attenuation constant, including all splices, is 0.25 dB/km. For the excess noise factor and the excess input

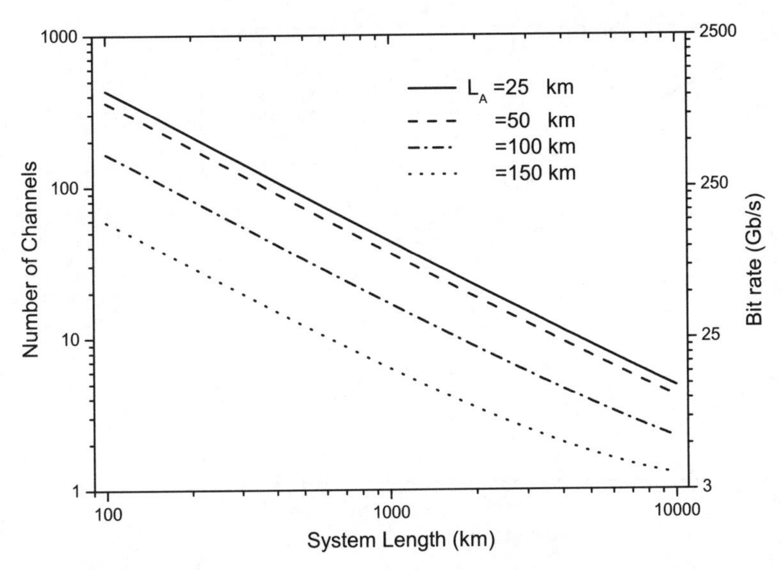

Fig. 10.18. Number of channels and total bit rate as a function of the system length for different amplifier spacing. ($\alpha = 0.25$ dB/km, $f = 62$ GHz, $\lambda = 1.5$ μm, $R = 90$, $E = 2$, $n_{\text{sp}} = 2$, $B_{\text{o}} = 10$ GHz [250])

coupling loss factor of the amplifiers we assume a value of two and 10 dB of transmitter/receiver margin ($R = 90$) is required. If the optical bandwidth of the filter at the receiver input is assumed to have 4 times the data rate we have $B_{\text{o}} = 10$ GHz. Figure 10.18 shows the number of channels against the system length and the corresponding total bit rate for 4 different amplifier spacings.

As can be seen from the figure, if the decrease of the SNR in the channel with the smallest wavelength has to be smaller than 0.5 dB, the number of transmittable channels and therefore the bit rate of the system decreases drastically with the system length.

For systems with arbitrary bit rates, channel spacings, and filter widths proportional to the bit rate, the total capacity of the system (C) can be calculated from (10.28) to (10.30) with $n \gg 1$[250]:

$$C < \sqrt{\frac{1.8 \times 10^{11}\text{Hz} \cdot \text{W} \cdot \text{km}}{Rh\nu n_{\text{sp}}(L/L_{\text{A}})^2 \left(\frac{1-\mathrm{e}^{-\alpha L_{\text{e}}}}{\alpha}\right)(E\mathrm{e}^{\alpha L_{\text{A}}} - 1)}}. \tag{10.31}$$

We assume here that the channel spacing is six times and the optical filter width four times the bit rate. Figure 10.19 shows the total bit rate versus the system length for this example. Therefore, due to the Raman scattering, the total capacity is less than 100 Gbit/s even in densely packed WDM systems for transcontinental, nondegenerated, amplified links [250].

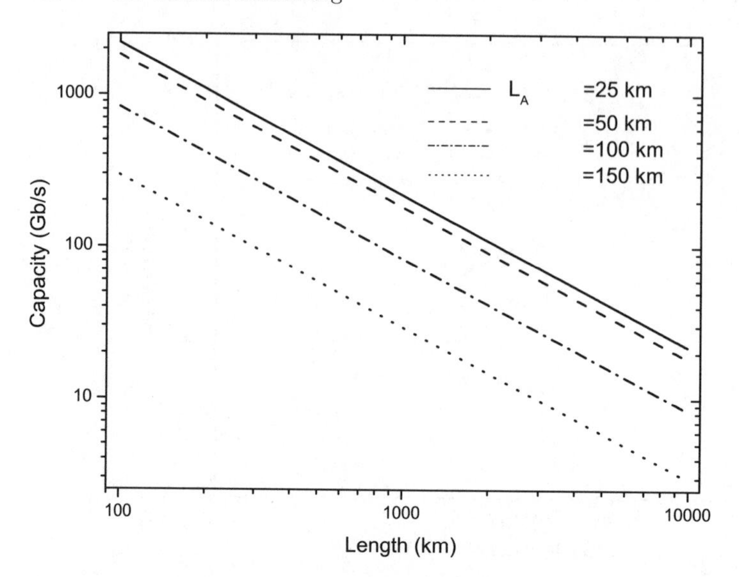

Fig. 10.19. Maximum capacity of a WDM system with a channel spacing of six and an optical filter width of four times the bit rate. ($\alpha = 0.25$ dB/km, $\lambda = 1.5$ µm, $R = 90$, $E = 2$, $n_{sp} = 2$ [250])

The calculations include neither the dispersion of the system, leading to a walk-off between the bits in the WDM channels, nor the arbitrary distribution of the marks in each channel. The Raman degradation is reduced if some of the channels transmit spaces.

The degradation of WDM systems due to Raman scattering can be compensated by a method called mid-span spectral inversion (MSSI, see Sect. 14.5). The MSSI leads to an inversion of the whole WDM spectrum in the middle of the fiber transmission link. Hence, the entire WDM comb is spectrally inverted. The channels with the higher wavelengths, formerly amplified by the SRS, have now a lower carrier wavelength and deliver power to the channels which were formerly depleted. As a result, the same SRS leads now to a balance of the power in the distinct WDM channels in the second half of the fiber span.

Summary

- Scattering is essentially caused by an inhomogeneous distribution of distinct optical properties in the bulk of a material and results in the deviation of a part of the incident light.
- If the inhomogeneous distribution is in the range of the wavelength, it is called Mie scattering. If it is much smaller it is called Rayleigh scattering. Both processes do not change the frequency of the wave.

- Raman scattering is a result of the interaction between the optical wave and the molecules of the material. Therefore, no phase matching condition has to be fulfilled.
- The scattered wave could be shifted down (Stokes) or up (anti-Stokes) in frequency. In optical fibers, the intensity of the downshifted wave is much higher.
- The Raman gain in optical fibers is very broad (≈ 40 THz).
- The maximum of the frequency shift is at ≈ 13 THz (105 nm).
- The Raman gain is inversely proportional to the wavelength of the pump. It depends on the polarization of the waves and has a value of $g_{\mathrm{R\,max}} = 3.1 \times 10^{-14}$m/W [245] in a standard single mode fiber for a pump wavelength of 1.55 μm and a completely randomized polarization of the pump wave and the Stokes wave.
- The threshold for forward Raman scattering in optical fibers is very high (≈ 2 W).
- Even well below the threshold, Raman scattering causes a kind of cross talk between channels in a WDM system if both channels transmit a logical "1". The channels with a higher carrier frequency deliver a part of their power to the channels with a lower carrier frequency. Or, in terms of wavelength, the channel with the higher wavelength is amplified at the expense of the channel with the lower carrier wavelength. This reduces the SNR for the lower wavelength channels and therefore the capacity of the system.
- The influence of the SRS to transmission systems can be compensated by the mid-span spectral inversion.

Exercises

10.1. Mahgarefteh et al. [245] measured the maximum Raman gain coefficient for an arbitrary polarization between pump and probe in standard single mode fibers. At a pump wavelength of 1.55 μm, he obtained a value of $g_{\mathrm{s}} = 0.39\ \mathrm{W^{-1}km^{-1}}$. Calculate the Raman gain (g_{R}) at a wavelength of 1.3 μm for an effective area in the fiber of $A_{\mathrm{eff}} = 80\ \mathrm{\mu m^2}$.

10.2. The measurement of the Raman gain coefficient in a non-polarization-maintaining optical fiber, with an effective area of $A_{\mathrm{eff}} = 90\ \mathrm{\mu m^2}$ and an attenuation constant of $\alpha = 0.25$ dB/km, gives a value of $g_{\mathrm{s}} = 0.5\ \mathrm{W^{-1}km^{-1}}$ at a wavelength of 1460 nm. Determine the threshold for stimulated scattering and the forward Raman scattering if the fiber is independently long.

10.3. Five km of the fiber in Exc. 10.2 are used as a Raman amplifier. Estimate the required input power of the pump for which the output power of Stokes and pump waves are equal.

10.4. Two WDM channels at the beginning and end of the C-band at $\lambda_1 = 1528.77$ nm and $\lambda_2 = 1560.61$ nm are transmitted in a standard single mode fiber with a maximum Raman gain (polarization averaged) of $g_{\mathrm{R}} = 3.12 \times 10^{-14}$ m/W, an attenuation constant of $\alpha = 0.25$ dB/km and an effective area of $A_{\mathrm{eff}} = 80\ \mu\mathrm{m}^2$. Above which power level does one of the channels show a growth during propagation? What is the propagation distance for lasting growth if the input power is 1 W ?

10.5. Determine the maximum power per channel in a WDM system reaching over the entire C-band, from $\lambda_1 = 1528.77$ nm to $\lambda_2 = 1560.61$ nm with a channel spacing of 50 GHz, if no channel should experience a penalty of more than 1 dB.

11. Brillouin Scattering

In the 1920s the french physicist Leon Brillouin investigated the scattering of light at acoustic waves. This effect was later named after him. As in the case of Raman scattering, strong scattered fields require high intensity light sources only available with lasers. For small intensities the scattered part of the field is very weak. Hence, Brillouin scattering was out of practical interest until the sixties.

Brillouin scattering is, like Raman scattering, a result of an interaction between the incident light field and the material. However, contrary to the Raman case, no vibrational modes of the molecules but density fluctuations of the medium are responsible for the scattering. These density fluctuations propagate through the medium with the velocity of sound and can be seen as acoustic waves or phonons. They have their origin in thermo-elastic motions of the molecules in the optical material. This motion is a superposition of many monochromatic, plane, acoustic waves, each of which responsible for a spatial and temporal density modulation in the optical material [129]. The periodic density modulation results in a refractive index modulation and causes a diffraction of the incident waves. If the density modulation moves with a relative velocity to the incident wave, the scattered wave experiences a frequency shift due to the Doppler effect. If the origin of the density modulation is the incident lightwave by itself, the effect is called stimulated and a large part of the power is transferred to the scattered wave. Otherwise, it is called spontaneous.

In the field of quantum mechanics, the photon is annihilated by the Brillouin scattering process while a new photon with a different frequency (Stokes) and a phonon is created. The phonon adds to the acoustic wave in the material. Since acoustic phonons with a corresponding frequency and energy are involved in the process, the frequency shift between the incident and scattered waves is much smaller than in the case of Raman scattering.

Contrary to all other nonlinear effects, already relatively weak intensities are sufficient to initiate a stimulated Brillouin scattering (SBS) in optical fibers. The SBS is the nonlinear effect with the smallest threshold which can be orders of magnitude smaller than the threshold of the SRS. The SBS detracts power from the signal waves and increases the noise in the system. It can therefore be responsible for a decrease of the SNR and increase of the BER. Much more important is the fact that, if the threshold of SBS is exceeded, the signal power cannot be increased further and all excess power is simply scattered back (see Fig. 11.8). Therefore the SBS determines the maximum launchable power for optical communication systems. Furthermore, since the scattered wave moves in the opposite direction to the incident wave, it can cause the destabilization or even destruction of the laser diodes if it couples into their resonators. The last effect can be avoided if, for instance, Faraday rotators[1] are used. Like diodes, they allow the propagation of light into only one direction. The threshold of the SBS can be increased relatively easy if the transmitted signal has a higher bandwidth than the intrinsic Brillouin gain. Hence, with intelligent system design, the effects of SBS can be effectively suppressed.

In this chapter, the physical origin of the Brillouin scattering is presented first and it will be shown how a wave diffracts on an acoustical density modulation. After that is an explanation of the differences between stimulated and spontaneous scattering followed by an introduction to the Brillouin gain. Section 11.5 then discusses the impact of the SBS on optical communication systems and the suppression of the SBS. The last Section shows some interesting applications of the SBS, whereas the SBS as an optical filter, amplifier and laser is presented in a separate chapter.

11.1 Scattering of Optical Waves at Sound Waves

Brillouin scattering is the result of the deviation of an optical wave on a density modulation in the material. Figure 11.1 shows schematically two density modulations moving with a relative velocity and angle to the incident field. The density fluctuation is caused by an acoustic wave with the sound velocity $\boldsymbol{v}_{\mathrm{A}}$. In the upper part of the figure the wave moves up and in the bottom it moves down. The angle and frequency of the incident field and hence, its wave vector $\boldsymbol{k}_{\mathrm{E}}$ are equal in both cases. The scattered wave has the wave vector $\boldsymbol{k}_{\mathrm{S}}$.

The scattering process requires that energy as well as momentum are conserved during the interaction. The condition of momentum conservation is shown in the vector diagram on the right side of Fig. 11.1 for both cases. As can be seen, the condition

$$\boldsymbol{k}_{\mathrm{S}} = \boldsymbol{k}_{\mathrm{E}} \pm \boldsymbol{k}_{\mathrm{A}} \tag{11.1}$$

[1]Faraday rotators exploit the magneto-optic effect. The refractive index of the material depends on the magnetic field. Similar to the electro-optic Kerr effect, the refractive index alteration can be used for a polarization rotation of the wave. The direction of rotation is independent of the propagation direction relative to the electric field. Hence, if the optical wave passes a 45° rotator, two times the resultant rotation is 90°.

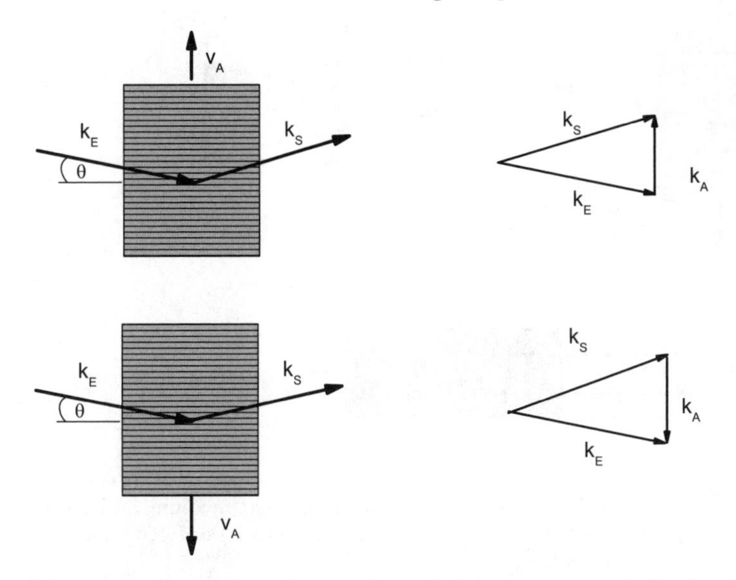

Fig. 11.1. Scattering of an incident lightwave at a periodic density modulation moving through the material with the sound velocity v_A

is valid. According to the conservation of energy the frequency of the scattered wave is

$$f_S = f_E \pm f_A. \tag{11.2}$$

As expected from the Doppler effect, the frequency of the scattered wave decreases if the density modulation moves away (Fig. 11.1, bottom) and it increases if the density modulation comes nearer (Fig. 11.1, top).

Figure 11.2 shows the vector diagram and the corresponding angles for a density modulation moving away from the incident field. For the distinct vectors, it follows from (11.1) that $\boldsymbol{k}_S = \boldsymbol{k}_E - \boldsymbol{k}_A$. The density modulation is caused by the acoustic wave in the medium. Therefore, its frequency is much smaller than the frequency of the incident optical wave. Hence, it is possible to assume that the absolute values of the wave vectors for the incident and the scattered wave are approximately equal $(|\boldsymbol{k}_S| \approx |\boldsymbol{k}_E|)$. In this case the angle θ is the half angle between the vectors of the pump and Stokes waves, as can be seen from Fig. 11.2b. The absolute value of the wave vector of the acoustic wave is then

$$|\boldsymbol{k}_A| = 2\,|\boldsymbol{k}_E| \sin\theta. \tag{11.3}$$

On the other hand, the wave vector is determined by its frequency and the velocity of sound in the material such that $|\boldsymbol{k}_A| = \omega_A/v_A$. The absolute value of the wave vector of the incident optical wave is $|\boldsymbol{k}_E| = 2\pi n/\lambda_E$, with λ_E as the wavelength of the optical wave in free space and n as the refractive index

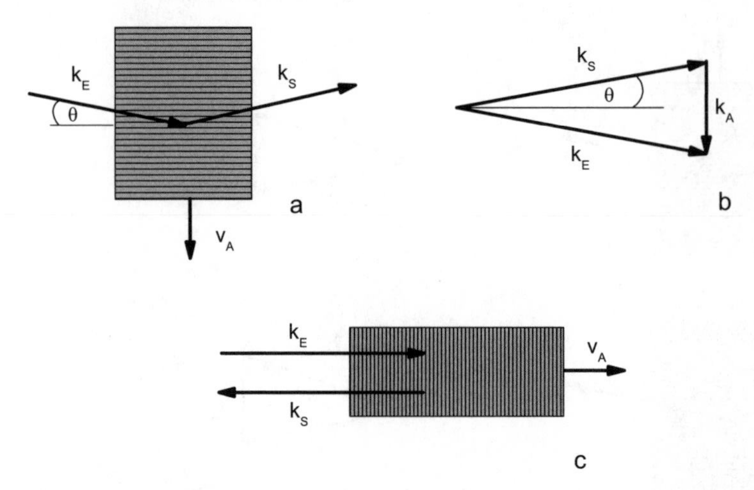

Fig. 11.2. Scattering at a density modulation of an optical medium and corresponding vector diagram if the wave moves under an angle (**a**,**b**) or in the opposite direction to the density modulation (**c**)

of the medium. Accordingly using (11.3) for the frequency of the acoustic wave allows us to calculate the frequency shift of the backscattered wave:

$$f_{\mathrm{A}} = \frac{2v_{\mathrm{A}}n}{\lambda_{\mathrm{E}}}\sin\theta. \tag{11.4}$$

As can be seen from (11.4) the frequency shift depends on the angle between the pump beam and the grating, or the angle between the incident and the scattered waves (2θ). In the forward direction, no frequency shift occurs because $\theta = 0$ and, hence, $f_{\mathrm{A}} = 0$. On the other hand, if the incident and the scattered waves propagate in opposite directions, then $2\theta = \pi$ and the frequency shift is at a maximum. In optical fibers, only the forward and backward directions are possible. Therefore, the Brillouin-scattered wave and the pump wave move in opposite directions[2]. Since $\sin\pi/2 = 1$ the frequency shift of the scattered wave is

$$f_{\mathrm{A}} = \frac{2v_{\mathrm{A}}n}{\lambda_{\mathrm{E}}}. \tag{11.5}$$

Pure silica glass has a sound velocity of $v_{\mathrm{A}} = 5.96$ km/s [279] and a refractive index of $n = 1.44$ at a wavelength of 1.5 µm. The corresponding frequency shift is then $f_{\mathrm{A}} \approx 11.4$ GHz. Since the density modulation moves away, the scattered wave (f_{S}) has a smaller frequency than the incident wave.

$$f_{\mathrm{S}} = f_{\mathrm{E}} - f_{\mathrm{A}} \tag{11.6}$$

[2]If the density modulation moves opposite to the pump, a spontaneous Brillouin scattering is possible but it cannot lead to a stimulated scattering.

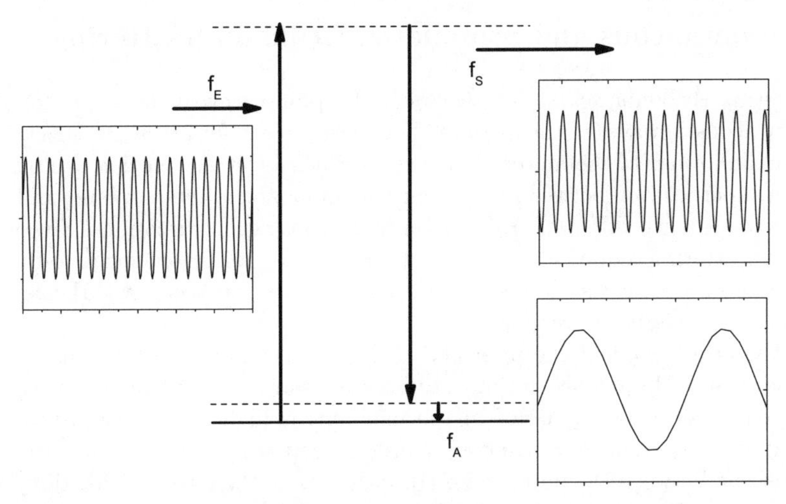

Fig. 11.3. Quantum mechanical model of Brillouin scattering. One photon of the incident wave is annihilated and creates simultaneously a photon of the scattered wave and a phonon

As in the case of Raman scattering, the incident wave is called the pump and the scattered wave is the Stokes wave. Figure 11.3 shows the energy diagram of the described frequency condition (11.6). In a quantum mechanical picture, the incident pump photon is annihilated and generates a Stokes photon with a smaller energy. Due to energy conservation, the rest of the energy is required for the generation of a phonon. Contrary to the case of Raman scattering, no real energy states of the molecule are involved in Brillouin scattering, as can be seen from the figure. Therefore, the conservation of energy and momentum must be fulfilled by the frequencies and wave numbers of the involved waves.

The described connections are valid if only one longitudinal acoustic mode (propagating in the same direction as the optical field) is present in the fiber. In a cylindrical waveguide like an optical fiber, beside the axial longitudinal sound wave, transverse acoustic modes, radial, flexural or torsional are possible as well. These modes can cause acousto-optic effects with their own resonance frequencies [252]. But the longitudinal mode shows the strongest influence on the Brillouin scattering.

Contrary to SRS no anti-Stokes wave with $f_{\mathrm{AS}} = f_{\mathrm{E}} + f_{\mathrm{A}}$ is generated in the case of SBS because this process would require an energy transfer from the sound wave to the photon. But if a wave with the frequency f_{AS} is launched into the fiber output, it transfers power to the pump wave (with frequency f_{E}), because both have a frequency difference of f_{A}. Therefore, the wave with frequency f_{AS} is the pump wave for f_{E}.

11.2 Spontaneous and Stimulated Brillouin Scattering

Spontaneous Brillouin scattering occurs if the pump wave is scattered on arbitrary acoustic waves in the medium. These waves are, for instance, caused by thermal motions of the material molecules. Their arbitrary motion can be seen as a superposition between a large number of monochromatic, plane, and acoustic waves with different frequencies and propagation directions. Since Brillouin scattering requires the conservation of energy and momentum, only a few of the acoustic waves can generate a scattered Stokes wave. Hence, spontaneous scattering is very weak.

On the other hand, if the intensity of the pump wave is so high that – as in the case of Raman scattering – the power transfer to the Stokes wave (generated at an arbitrary point in the medium) is higher than the attenuation it will experience, a stimulated process can occur. In this case the superposition between the pump wave (propagating in the forward direction) and the backscattered Stokes wave will cause a fading with a frequency that corresponds to the acoustic wave $f_A = f_S - f_E$ (Fig. 11.4). Hence, the superposition results in an intensity modulation propagating into the direction of the pump wave. The intensity modulation will be translated into a density modulation of the medium via electrostriction.[3] Therefore, the fading leads to an amplification of the acoustic wave. A stronger acoustic wave causes a stronger Stokes wave which results in a stronger intensity modulation, and so on. If the intensity of the pump and, therefore, the intensity of the generated Stokes wave, exceed a particular threshold, the positive feedback effect between the Stokes and the acoustic wave of the medium can result in an exponential increase of the Stokes wave. Since the pump by itself is responsible for an amplification of the effect, the process is called stimulated.

Stimulated Brillouin scattering is the interaction between the pump wave, the generated Stokes wave and the acoustic wave in the fiber. This interaction can be described by the nonlinear wave equation again. On the other hand, the differential equation system describing the behavior of the pump and Stokes waves can be derived from a relatively easy deliberation. The intensity of the Stokes wave (I_S) depends on the medium (denoted by the Brillouin gain, g_B) the intensity of the pump wave (I_P) and the attenuation the wave will experience in the material. Since the Stokes wave is generated in backward direction it grows exponentially with negative z, indicated by a

[3]Every isolator that is positioned in an electric field reacts with an electronic polarization on the electric force it will experience. This polarization leads to a mechanical deformation or a mechanical strain in the medium, respectively. This means, the medium is denser in the areas of a high field strength and tenuous in the range where the electric field is rather low. The effect is named eletrostriction, it describes the converse behavior to the piezoelectric effect, i.e. a mechanical deformation of a crystal results in an electric field.

The above mechanical deformation as a result of the electric field causes a periodical density variation or an acoustic wave in the medium. With the density modulation of the material, its refractive index is modulated as well.

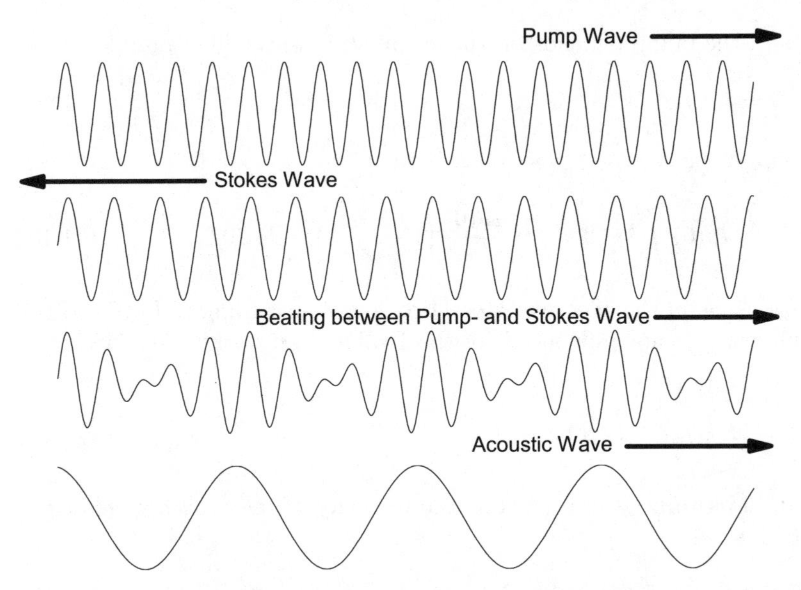

Fig. 11.4. Generation of an acoustic wave due to the superposition between the pump wave, propagating in forward direction, and the backscattered Stokes wave for a fixed time ($t = const$)

negative sign, as in

$$\frac{\mathrm{d}I_\mathrm{S}}{\mathrm{d}z} = -g_\mathrm{B} I_\mathrm{P} I_\mathrm{S} + \alpha I_\mathrm{S}. \tag{11.7}$$

Due to the Brillouin process, the pump wave transfers a part of its power to the Stokes wave ($-g_\mathrm{B}$) at the same time it is attenuated in the material. The frequency shift between the pump and Stokes waves is rather small ($f_\mathrm{A} \approx$ 11.4 GHz). Therefore, the frequency of the Stokes wave is approximately equal to the frequency of the pump wave. This means the fiber attenuation and the Brillouin gain are approximately equal for the pump and Stokes waves, i.e, $\alpha_\mathrm{P} \approx \alpha_\mathrm{S}$, $g_\mathrm{Bs} \approx g_\mathrm{Bp}$. The intensity of the pump wave can then be described by

$$\frac{\mathrm{d}I_\mathrm{P}}{\mathrm{d}z} = -g_\mathrm{B} I_\mathrm{P} I_\mathrm{S} - \alpha I_\mathrm{P}. \tag{11.8}$$

Note the similarity between the equation system describing the Brillouin scattering, (11.7) and (11.8), and that responsible for Raman scattering, (10.13) and (10.14). The equations (11.7) and (11.8) are only valid for approximately monochromatic waves or relatively long pulses. Furthermore, the polarization alteration of the waves during their propagation, due to the birefringence of the fiber, is not taken into consideration either.

If, for small intensities, the depletion of the pump (caused by the Brillouin process) is very small, the first term in (11.8) can be neglected. Hence, the intensity of the pump depends on the attenuation in the fiber only:

$$I_{\mathrm{P}}(z) = I_{\mathrm{P}}(0)\mathrm{e}^{-\alpha z}. \tag{11.9}$$

The pump intensity at the position L is then

$$I_{\mathrm{P}}(L) = I_{\mathrm{P}}(0)\int_0^L \mathrm{e}^{-\alpha z}\mathrm{d}z = \frac{I_{\mathrm{P}}(0)}{\alpha}\left(1 - \mathrm{e}^{-\alpha L}\right) = I_{\mathrm{P}}(0)L_{\mathrm{eff}}, \tag{11.10}$$

where L_{eff} is again the effective interaction length according to (4.97). If the pump intensity is not influenced by the Brillouin scattering, (11.7) can be written as

$$\frac{\mathrm{d}I_{\mathrm{S}}}{\mathrm{d}z} = \left[-g_{\mathrm{B}}I_{\mathrm{P}}(z) + \alpha\right] I_{\mathrm{S}}. \tag{11.11}$$

If (11.10) is introduced into (11.11), the intensity of the Stokes wave at the position L is

$$I_{\mathrm{s}}(L) = I_{\mathrm{s}}(0)\mathrm{e}^{-g_{\mathrm{B}}P_0L_{\mathrm{eff}}/A_{\mathrm{eff}}+\alpha L}. \tag{11.12}$$

The intensity of the Stokes wave at the fiber input $I_{\mathrm{s}}(0)$ is therefore, [239]

$$I_{\mathrm{s}}(0) = I_{\mathrm{s}}(L)\mathrm{e}^{g_{\mathrm{B}}P_0L_{\mathrm{eff}}/A_{\mathrm{eff}}-\alpha L}. \tag{11.13}$$

According to (11.13), an initiating wave with the right frequency shift is required at the position L for the generation of the Stokes wave. This initiating Stokes wave $I_{\mathrm{s}}(L)$ can be a result of spontaneous scattering or it can be generated from the noise floor in the fiber. Note that in (11.13), the influence of the polarization between the pump and Stokes waves on the Brillouin process is not included. As will be discussed in Sect. 11.4, the polarization dependence results in an additional factor of 1/2 in the first term of the exponential function in a standard single mode fiber.

Figure 11.5 shows the power of the Stokes wave at the input of a polarization-maintaining fiber against the distance to the initiating Stokes wave (here the fiber length) for a pump power of 1.3 and 1.5 mW. The power of the initiating Stokes wave is in both cases 1×10^{-10} W. As can be seen from the figure, the power of the backscattered wave growth very strong if the fiber is longer than 20 km and reaches a maximum at around 65 km. For longer fibers the efficiency of the generation decreases again because the attenuation in the fiber is higher than the amplification due to the Brillouin process.

For pump wave powers above the threshold (11.13) is no longer valid because the depletion of the pump wave, due to the power transfer to the Stokes, cannot be neglected.

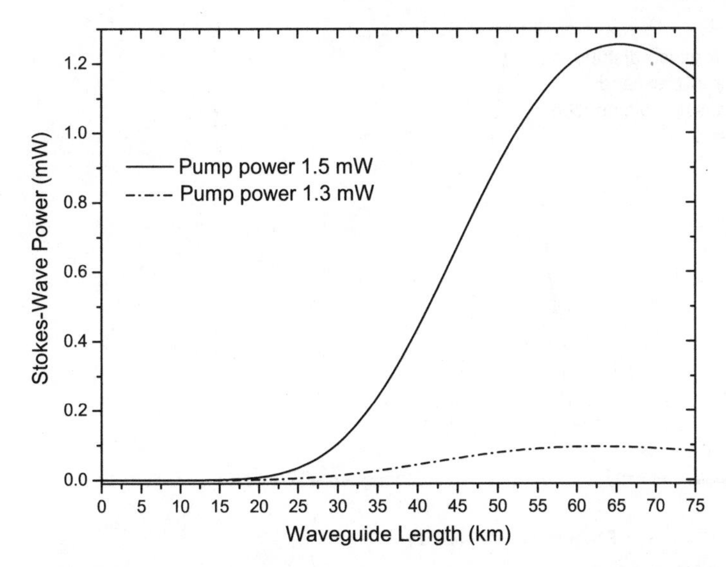

Fig. 11.5. Power of the Stokes wave at the input of a polarization-maintaining fiber against the fiber length for two different pump powers of 1.3 and 1.5 mW. Power of the initiating Stokes wave at the position L ($P_s(L) = 1 \times 10^{-10}$ W , $A_{\text{eff}} = 80\ \mu\text{m}^2$, $\alpha = 0.2$ dB/km, $g_B = 5 \times 10^{-11}$ m/W)

11.3 The Brillouin Gain

As in the case of Raman scattering, the intensity of the Brillouin-scattered wave depends on a material-specific parameter, the Brillouin gain coefficient (g_B). The Brillouin gain coefficient of a fiber is determined by three important parameters: the frequency shift between the pump and Stoke waves (f_A), the peak Brillouin gain ($g_{B\,\max}$) and the linewidth of the distribution (Δf_A).

11.3.1 Spectral Distribution

Contrary to the Raman gain, the Brillouin gain coefficient has a very narrow bandwidth and its maximum determines the frequency shift or the frequency of the acoustic wave. In most cases, the distribution is approximated by a Lorentzian function [254]:

$$g_B(f) = \frac{1}{1 + \left[(f - f_A) / (\Delta f_A/2)\right]^2} g_{B\,\max}. \tag{11.14}$$

On the other hand, (11.14) is only valid for narrow pump spectra (quasi-monochromatic waves). If the pump pulses are temporally short, their spectrum will be correspondingly large and the gain curve merges into a Gaussian distribution [262]:

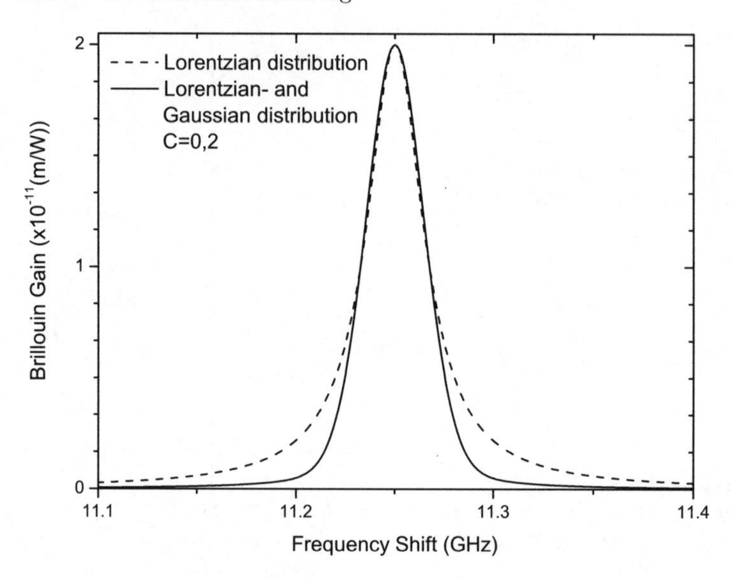

Fig. 11.6. Brillouin gain versus frequency shift for a Lorentzian- and a Lorentzian-Gaussian- distribution ($g_{\mathrm{B\,max}} = 2\times10^{-11}$ m/W, $\Delta f_{\mathrm{A}} = 35$ MHz, $f_{\mathrm{A}} = 11.25$ GHz)

$$g_{\mathrm{B}}(f) = \left\{ C \frac{1}{1 + \left[(f - f_{\mathrm{A}}) / (\Delta f_{\mathrm{A}}/2) \right]^2} + (1 - C) \exp\left[- \ln(2) \frac{(f - f_{\mathrm{A}})^2}{(\Delta f_{\mathrm{A}}/2)^2} \right] \right\} g_{\mathrm{B\,max}} . \tag{11.15}$$

Equation (11.15) is the superposition between a Lorentzian and a Gaussian distribution, whereas both have the same width Δf_{A}, C is a constant determining the contingent of each function on the whole result, $(1 - C)$ and C. Figure 11.6 shows a comparison between the two distributions of the Brillouin gain for a FWHM width of 35 MHz and a frequency shift of 11.25 GHz.

Beside the longitudinal acoustic modes which propagate with a sound velocity corresponding to the frequency shift – determined in (11.5) – along the axes of an optical fiber, additional transversal vibrational modes in the core and cladding exist in a cylindrical waveguide. These modes depend on the density of the material of the core and the cladding, and on the interface between them and have their own resonance frequencies. The transversal modes can also scatter light [265] so the Brillouin gain distribution has several maxima. Table 11.1 shows the resonance frequencies of the Brillouin gain for different fiber types. In a recent measurement in a photonic crystal fiber with a partially germanium-doped core, several subpeaks were found as well. The authors believe that the subpeaks are caused by an interaction between light-wave and guided modes of longitudinal acoustic waves in the graded Ge-doped region, the silica region and the microstructured cladding [266]. The differences in the Brillouin spectra for the fibers are a direct result of

Table 11.1. Brillouin gain resonance frequencies in GHz for different fiber types measured with a DFB laser at 1552 nm (AW=AllWave, TW=TrueWave, DCF=dispersion-compensating fiber, HG=highly Ge-doped), [267]

Fiber type	Main res.	2nd res.	3rd res.	4th res.	5th res.
AW	10.83	10.95	11.05	11.2	–
TW	10.7	10.82	10.9	–	–
DCF	9.77	10.28	10.68	10.78	10.89
HG	8.9	9.65	10.6	11.21	11.25

Table 11.2. Experimental results for the threshold (P_0^{S}), the bandwidth for spontaneous ($\Delta f_{\mathrm{A}}^{\mathrm{spon}}$) and stimulated scattering ($\Delta f_{\mathrm{A}}^{\mathrm{stim}}$) and the frequency shift f_{A} for four different fiber types at a pump wavelength of 1552 nm. The attenuation constant and effective length are also given for each fiber (AW=Allwave, TW=Truewave, DCF=dispersion-compensating fiber and HG=highly germanium doped fiber) [267]

Fiber	P_0^{S} (dBm)	$\Delta f_{\mathrm{A}}^{\mathrm{spon}}$ (MHZ)	$\Delta f_{\mathrm{A}}^{\mathrm{stim}}$ (MHz)	f_{A} (GHZ)	L_{eff} (km)	α (dB/km)
AW	5.2	86	11.4	10.81	15.3	0.186
TW	4.8	88	12.4	10.85	16.1	0.18
DCF	8.2	92	12.6	9.75	4.4	0.64
HG	19	98	14	8.9	0.3	4.5

their special waveguide characteristics (see Fig. 3.5). While TrueWave (TW) and AllWave (AW) fibers have relatively large core diameters (7.2 µm for TW, 8.1 µm for AW), the two other fibers have very narrow cores (1.7 µm for DCF, 2.5 µm for HG). Furthermore, they are different in their refractive index progressions, DCF, HG and AW have uniform refractive index progressions in the core, whereas the TW fiber shows a parabolic distribution with a minimum in the middle.

On the other hand, the peaks of the additional resonances of the core and cladding modes decrease for increasing input powers. For input powers higher than the threshold of stimulated scattering the main resonance peak increases with input power whereas the other resonance peaks decrease and eventually vanish [267].

The bandwidth of the main resonance depends on the power and duration of the pump pulse. If the power exceeds the threshold for stimulated scattering, the Brillouin gain coefficient will become narrower (see Table 11.2). The Brillouin-linewidth pulse-duration relationship can be divided into three cases [262]:

- If the pulse duration is much longer than the lifetime of the phonons (10 ns) the linewidth of the gain coefficient is the natural one which depends on the material. At a pump wavelength of 1550 nm, this natural spectral width is $\approx 35 - 40$ MHz [239, 262] in pure silica glass. The line shape shows a Lorentzian profile.
- If the pulse duration falls in the range of the phonon lifetime, the linewidth is increased up to a maximum of $90 - 140$ MHz at 5 ns. Furthermore, as a result of the convolution of many Lorentzian profiles the line shape becomes more Gaussian.
- For a further reduction of the pulse duration, the linewidth is decreased very sharply and for a pulse duration of 1 ns, again the natural linewidth can be seen.

Due to this behavior, the Brillouin linewidth in optical communication systems depends on the modulation format (see Sect. 12.5). If the spectrum of the modulated pump is much broader then the intrinsic SBS bandwidth, the SBS gain follows the pump spectrum directly.

11.3.2 Frequency Shift

The frequency shift of the first maximum (f_A) depends on the wavelength of the pump, the sound velocity of the longitudinal mode and the refractive index of the material, as can be seen from (11.5). Furthermore, the frequency shift depends very strong on environmental conditions like temperature and strain in the fiber (see Sec. 11.6). Both the sound velocity and the refractive index depend, beside other things, on the dopant concentration in the fiber. A higher dopant concentration reduces the sound velocity in the material but, at the same time, it increases the refractive index. The decrease of the acoustic velocity with dopant concentration is higher then the increase of the refractive index. Therefore, if the concentration is increased, the frequency shift is decreased. This behavior can be seen in Table 11.1, the fibers with the highest GeO_2 concentration (HG and DCF) show the smallest Brillouin shift. On the other hand, the GeO_2 concentration in TW and AW fibers is nearly the same, at 5.4% and 4.6%, respectively. Nevertheless, Yeniay et al. [267] measured the difference in the frequency shift shown in the table. The authors attributed this to the increase of acoustic velocity due to the removal of OH-ions in AW fibers. The waveguide characteristics of the fiber have an influence on the frequency shift of the Brillouin scattering as well [264].

Shiraki et al. [263] measured the dopant concentration dependence of the Brillouin frequency shift for fibers with an fluorine and GeO_2 codoped silica core at 1550 nm. They have found the empirical equation:

$$f_A = 11.045 - (0.277\,[\mathrm{F}] + 0.045\,[\mathrm{GeO_2}]) \qquad (\mathrm{GHz})\,, \tag{11.16}$$

where [F] and $[GeO_2]$ are the concentrations of the respective dopants in wt%. According to (11.16), a change of the F concentration alters the frequency of the acoustic wave at about 277MHz/wt%, whereas an increase of the $Ge0_2$ concentration results in a frequency shift of 45 MHz/wt% only. For an optical fiber with a F concentration of 1.5 wt% and a $Ge0_2$ concentration of 15 wt%, the Brillouin frequency shift is $f_A = 9.954$ GHz. Therefore, with a slight change of the F concentration, a strong shift of the middle frequency of the curve in Fig. 11.6 is possible, making it possible to suppress SBS in optical transmission systems with such fibers [263].

Niklès et al. [328] have measured the GeO_2 dopant concentration dependence of the Brillouin shift in standard single mode fibers at a pump wavelength of 1320 nm. They have found a decrease of -94 MHz/wt% for an increasing GeO_2 concentration in the core.

11.3.3 Gain Maximum

The waveguide characteristics, the material properties, and the linewidth of the pump source have an influence on the third parameter of the Brillouin gain coefficient, the maximum of the Brillouin gain $g_{\mathrm{B\,max}}$. For pulses with a temporal duration much longer than the phonon lifetime, it is [239]

$$g_{\mathrm{B\,max}} = \frac{\pi f_{\mathrm{A}}^2 M_2}{c \Delta f_{\mathrm{A}} 2n}. \tag{11.17}$$

Here n is the refractive index of the material and M_2 the elasto-optic figure of merit [239]:

$$M_2 = \frac{n^6 p^2}{\rho v_{\mathrm{A}}^3}, \tag{11.18}$$

where p describes the elasto-optic constant, ρ is the material density, and v_{A} is the velocity of the acoustic wave in the medium. Due to the fact that most of the values in (11.18) depend on the material properties in the fiber, the dopant concentrations play an important role here. With the Equations (11.5) and (11.18), (11.17) can be rewritten as [51, 271]:

$$g_{\mathrm{B\,max}} = \frac{4\pi n^8 p^2}{c \lambda_{\mathrm{p}}^3 \rho f_{\mathrm{A}} \Delta f_{\mathrm{A}}}, \tag{11.19}$$

where λ_{p} is the pump pulse wavelength. With a spectral width of 35 MHz, a frequency shift of 11.1 GHz, an elasto-optic figure of merit of $M_2 = 1.51 \times 10^{-15}$ s^3/kg [239], and a refractive index of $n = 1.44$ at 1550 nm (11.17) yields a maximum value of the Brillouin gain of $g_{\mathrm{B\,max}} \approx 2 \times 10^{-11}$ m/W for bulk silica glass. If this value is compared to the Raman gain ($g_{\mathrm{R\,max}} \approx 3.1 \times 10^{-14}$ m/W), we can see that the Brillouin gain is around three orders of magnitude higher. In consequence, the threshold of SBS should be much below that of SRS.

As discussed in Sect. 4.6, nonlinear effects that are based on an electronic polarization of the molecules or atoms of the material are very fast. On the other hand, the Brillouin scattering is not instantaneous, i.e., the involved acoustic grating and the phonons each has a life time. The ratio between the temporal duration of the excitation and the life time of the phonons has a strong influence on the efficiency of the scattering. If the temporal duration of the pump pulses is much higher then the life time of generated lattice vibrations, then Δf_{A} and $g_{\mathrm{B\,max}}$ depend mainly on the material parameters. Whereas, if the duration of the pump pulses is in the range of the life time of the phonons, its maximum is decreased [262]. Therefore, from a distinct duration of the pump pulses Δf_{P}, the Brillouin gain depends on the temporal width of the pulses. In this case (11.18) and (11.19) must be expanded [254, 271] :

$$g_{\mathrm{B\,max}} = \frac{\pi f_{\mathrm{A}}^2 M_2}{c\Delta f_{\mathrm{A}} 2n} \left(\frac{\Delta f_{\mathrm{A}}}{\Delta f_{\mathrm{A}} \otimes \Delta f_{\mathrm{P}}} \right), \tag{11.20}$$

where $\otimes$ denotes the convolution operation between the pulses. For Gaussian pulses, we have

$$\Delta f_{\mathrm{A}} \otimes \Delta f_{\mathrm{P}} = \sqrt{\Delta f_{\mathrm{A}}^2 + \Delta f_{\mathrm{P}}^2}, \tag{11.21}$$

whereas for a Lorentzian distribution, the corresponding equation yields:

$$\Delta f_{\mathrm{A}} \otimes \Delta f_{\mathrm{P}} = \Delta f_{\mathrm{A}} + \Delta f_{\mathrm{P}}. \tag{11.22}$$

For a Gaussian as well as for a Lorentzian distribution, (11.20) results in (11.19) as long as $\Delta f_{\mathrm{A}} \gg \Delta f_{\mathrm{P}}$ for a CW excitation for example. The Brillouin gain for three pump pulses with different spectral width while spectral broadening is neglected is shown in Fig. 11.7.

Below the threshold the spectral power distribution of the backscattered Stokes wave, $S_{\mathrm{p}}(f)$ shows a Gaussian shape. If the power of the pump is increased, the spectrum becomes inhomogeneous, it is strongly broadened, and it experiences a collapse in the center (hole burning) [272, 273]. The power of the backscattered wave has strong statistical fluctuations near and above the threshold. The pump waves's power depletion leads to a feedback between the waves propagating in opposite directions and can lead to a decrease of the power fluctuations [274].

Beside the above mentioned influences on the Brillouin gain coefficient, it also depends on environmental conditions like temperature, mechanical strain, or other distortions of the fiber. This fact can be exploited for distributed measurements of temperature and strain [255, 261], which will be treated in Sect. 11.6, but is, of course, a disadvantage for other applications of the Brillouin scattering. On the other hand, the temperature dependence can be exploited for a cheap and effective increase of the SBS threshold [275].

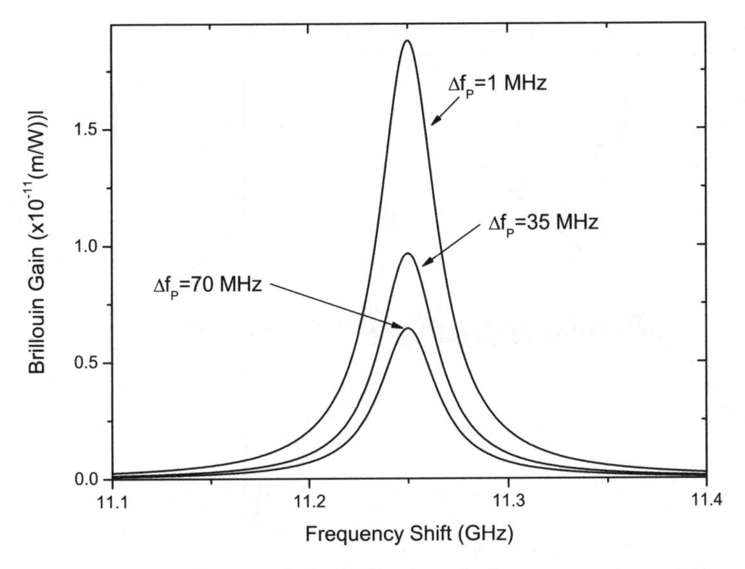

Fig. 11.7. Distribution of the Brillouin gain for pump pulses with a spectral width of 1, 35, and 70 MHz. The spectral broadening of the gain was neglected ($\Delta f_{\mathrm{A}} =$ 35 MHz, $f_{\mathrm{A}} = 11.25$ GHz, $M_2 = 1.51 \times 10^{-15}$ s^3/kg, $n = 1.44$)

11.4 Threshold of Brillouin Scattering

Figure 11.8 shows the optical power at the input and output of a 50 km standard single mode fiber measured by T. Höppner [253]. A narrow linewidth (1 MHz) laser diode oscillating at a wavelength of 1537 nm was used as the pump source. As can be seen from the figure, the input and output power grow almost linearly with the current of the laser diode at low intensities, but if the input power exceeds a distinct value there is no additional increase in the output power. The spontaneous process of the scattering at the sound waves in the fiber becomes stimulated. All excess power launched into the fiber is scattered back by the stimulated Brillouin scattering. The value that determines this change of the physical behavior is the threshold.

As can be seen from (11.13), the generation of the Stokes wave requires a source ($I_{\mathrm{S}}(L)$) which initiates the process somewhere in the fiber. There are two different models, each with a different source that can be used for the description of the SBS and the threshold of the process [267]. The model by Zel'dovich et al. [268] postulates a non-fluctuating, thin source at the end of the medium which generates the backscattered Stokes wave. In the second model by Boyd et al. [269], the source has a spatial distribution and fluctuation. The predictions of both models are in good agreement with each other and can be verified by experiments. In the following the first, much easier, model is assumed.

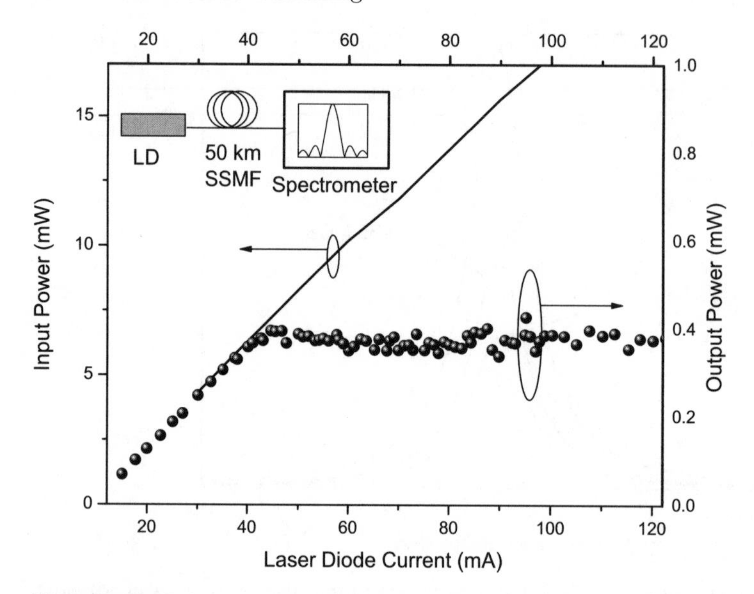

Fig. 11.8. Optical input and output power of a 50 km fiber versus laser diode current. The inset shows the experimental set-up (LD=laser diode at 1537 nm, SSMF=standard single mode fiber). The power was measured with a spectrometer and is the peak of the distribution

Several definitions of the threshold for the pump power in optical fibers are given in the literature [254]. It has been defined as:

- The input pump power that is equal to the power of the backscattered Stokes wave [239, 263].
- The input pump power at which the power of the backscattered Stokes wave is equal to the transmitted power [276, 277].
- The input pump power that is required for a strong increase of the Stokes power or a strong decay of the pump [278, 279].
- The input pump power at which the backscattered power at the fiber input corresponds to 1 % of the pump input power [280, 281].

Assuming the source is localized and the decay of the pump power due to the Brillouin process is negligible, the first definition for the critical power of the Brillouin scattering results to [239]

$$P_0^{\mathrm{S}} = 21 \frac{K_{\mathrm{B}} \times A_{\mathrm{eff}}}{g_{\mathrm{B}} L_{\mathrm{eff}}}. \tag{11.23}$$

Here and in the following, g_{B} is the peak value of the Brillouin gain coefficient (g_{Bmax}), K_{B} is a factor that includes the polarization dependence of pump and Stokes wave with a value between 1 and 2. For polarization-maintaining fibers, if pump and Stokes wave have the same polarization then

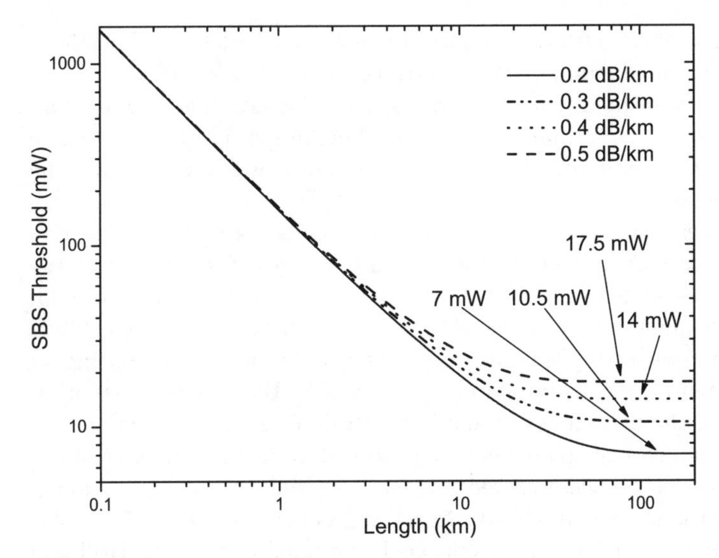

Fig. 11.9. Threshold of stimulated Brillouin scattering versus fiber length in a standard single mode fiber with different attenuation ($K_\mathrm{B} = 2$, $g_\mathrm{B} = 2\times10^{-11}$m/W, $A_\mathrm{eff} = 80$ µm^2)

$K_\mathrm{B} = 1$. If both waves have completely random polarization $K_\mathrm{B} = 1.5$ [251], and conventional single mode fibers show a value of 2 for K_B[282].

Equation (11.23) is the threshold value for SBS commonly used in the literature. Nevertheless, an intercomparison of the European Cooperation for Scientific and Technical research (COST) led to a recommendation to the ITU to adapt the threshold [254, 283]:

$$P_0^\mathrm{S} = 19\frac{K_\mathrm{B} \times A_\mathrm{eff}}{g_\mathrm{B} L_\mathrm{eff}}. \tag{11.24}$$

Figure 11.9 shows the threshold of the Brillouin scattering, according to (11.24), for a standard single mode fiber with different attenuation constants against the fiber length.

The threshold decreases with increasing fiber length and after a length that will be defined by the attenuation in the fiber, it approaches a limiting value asymptotically, according to the behavior of the effective length. With an effective area for standard single mode fibers of $A_\mathrm{eff} = 80$ µm^2, an attenuation constant of 0.2 dB/km, and a corresponding effective interaction length of 21.7 km, we obtain a threshold of $P_0^\mathrm{S} \approx 7$ mW (8.45 dBm) if we assume a Brillouin gain of $g_\mathrm{B} = 2 \times 10^{-11}$ m/W and a polarization factor of $K_\mathrm{B} = 2$. In long conventional fibers, the threshold of SBS for a narrow linewidth source is about 6 dBm [270]. Thresholds for different fiber types can be found in Table 11.2. A threshold of only 4.8 dBm ($\approx$ 3 mW) was measured in a TrueWave fiber with an effective length of 16.1 km. Note that the limiting value of the effective length in this fiber is ≈ 24 km. Therefore, the minimum threshold can be roughly estimated to be ≈ 2 mW.

The Brillouin threshold is about three orders of magnitude smaller than the threshold required for Raman scattering. Furthermore, if reflections are present, a cavity is built and laser oscillations at power levels below the threshold can occur.

As can be seen from (11.24), the threshold decreases with the effective length of the system. Amplified optical transmission systems have very large effective lengths as shown in Sect. 4.9 and, therefore, a very small threshold. An optical transmission system consisting of standard single mode fibers ($\alpha = 0.2$ dB/km) with 100 optical amplifiers and an amplifier spacing of 60 km would then have a threshold of $P_0^{\mathrm{S}} \approx 75\ \mu\mathrm{W}$. But, contrary to other nonlinear effects, the threshold of stimulated Brillouin scattering cannot decrease with the number of spans because practical optical amplifiers contain optical isolators, so the backscattered Stokes wave will not accumulate along the system and the threshold is that of each individual span $P_0^{\mathrm{S}} \approx 7.5$ mW.

On the other hand, as already mentioned, the maximum of the Brillouin gain depends on the temporal duration of the pump pulses. Therefore, the threshold depends on the pulse duration as well. Equation (11.20) then leads to the following for the threshold when we take the pump linewidth into consideration [271]:

$$P_0^{\mathrm{S}} = 21 \frac{K_{\mathrm{B}} \times A_{\mathrm{eff}}}{g_{\mathrm{B}} L_{\mathrm{eff}}} \frac{\Delta f_{\mathrm{A}} \otimes \Delta f_{\mathrm{P}}}{\Delta f_{\mathrm{A}}}. \tag{11.25}$$

Figure 11.10 shows the threshold for a standard single mode fiber as a function of the temporal duration of the pulses or their spectral width. The calculation assumed that the pump pulse has a Lorentzian shape.

The behavior of SBS distinguishes in different types based on the differences in their waveguide characteristics. Table 11.2 shows experimental results, according to Yeniay et al. [267], for the threshold, the bandwidth for spontaneous and stimulated scattering, and the frequency shift for four different fiber types. The measurements were carried out with a DFB-laserdiode at a wavelength of 1552 nm.

As can be seen from the table, the bandwidth of the spontaneous scattering is rather broad in general and is different for the four fiber types. If the pump power is increased over the threshold, a stimulated scattering occurs and the bandwidth will become narrower. The threshold of the SBS shows very strong differences in the different fibers. It reaches from only 3 mW for TW up to ≈ 80 mW for highly germanium-doped fibers.

11.5 SBS's impact on Communication Systems

As shown in the last Section, the threshold of Brillouin scattering is much smaller than that of Raman scattering. Only a few mW – reachable with

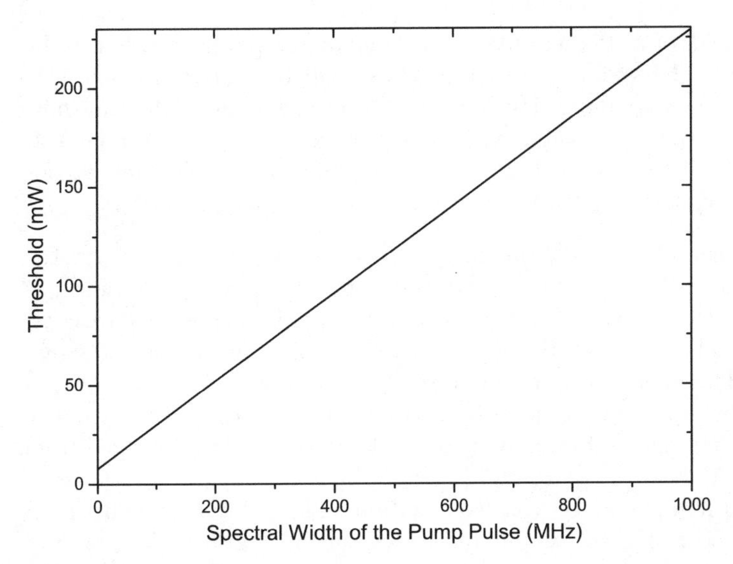

Fig. 11.10. Brillouin scattering threshold versus spectral width of the pump pulses ($K_B = 2$, $\alpha = 0.2$ dB/km, $g_B = 2 \times 10^{-11}$m/W, $A_{eff} = 80$ µm^2, $\Delta f_A = 35$ MHz, $L_{eff} = 21.7$ km)

standard laser diodes – are sufficient to initiate a stimulated scattering. If SBS is generated in a transmission system, three main effects that are detrimental to system performance can occur.

- The pump wave, propagating in forward direction, transfers optical power to the backscattered Stokes wave. Hence, the Stokes wave is amplified at the expense of the pump. The power loss degrades the SNR of the pump and leads to an increase of the BER. If the backscattered Stokes wave couples into the resonator of the laser diode, it can destabilize or even destroy it. This can be suppressed by the incorporation of optical isolators.
- In WDM systems, Brillouin scattering can cause a kind of cross talk far below the threshold for stimulated scattering. As in the case of SRS, if a channel with the right frequency shift is already present in the fiber, the mutual interaction can be very strong even for low intensities. The channel with the Stokes frequency is amplified at the expense of the channel with the pump frequency. But, contrary to SRS, very strict conditions must be fulfilled for such an interaction for the case of SBS. First, the second channel with the Stokes frequency shift must propagate in an opposite direction to the pump wave. Second, the Brillouin gain is very narrow and the shift is rather small ($f_A \approx 11$ GHz, $\Delta f \approx 35$ MHz for pure silica), so the frequency shift must completely comply to the channel spacing. Therefore, this kind of interaction can be suppressed with a slight change of the frequency shift between the channels.

- The threshold of SBS determines the maximum power which can be launched into the system. All excess power will be scattered back. This again limits the maximum signal to noise ratio and it limits the transmission distance that is possible without amplification. In highly nonlinear fibers, SBS will become a limiting factor which can be detrimental for nonlinear applications even for relatively short fiber lengths [275].

On the other hand, the Brillouin gain coefficient – responsible for the effectivity of the SBS – is narrow and dependent on the pump pulse width, as shown in Sect. 11.3. In optical transmission systems, the narrow band carriers, delivered by stabilized laser diodes, are modulated and depend on the signal parameters. This modulation results directly in an increase of its spectrum and the broader spectrum, in turn, has an influence on the effectivity of the Brillouin process. The following derivation of the threshold is adapted from Aoki et al. [271].

First, let's assume we have a fixed unlimited bit pattern "1010101...". According to Sect. 3.4, the modulated waves can be expressed as a sum of equidistant spectral components so the pump wave can be represented by

$$E_{\mathrm{p}}(z,t) = \frac{1}{2}\sum_{n} E_{\mathrm{p}n}(z)\,\mathrm{e}^{\mathrm{j}(\omega_{\mathrm{p}n}t - k_{\mathrm{p}n}z)} + \mathrm{c.c.}. \tag{11.26}$$

The modulated pump generates, via the Brillouin effect in the optical fiber, a backscattered Stokes wave. The Stokes wave again consists of distinct frequency components, as follows:

$$E_{\mathrm{s}}(z,t) = \frac{1}{2}\sum_{n} E_{\mathrm{s}n}(z)\,\mathrm{e}^{\mathrm{j}(\omega_{\mathrm{s}n}t + k_{\mathrm{s}n}z)} + \mathrm{c.c.}. \tag{11.27}$$

The superposition of both waves, $E(z,t) = E_{\mathrm{p}}(z,t) + E_{\mathrm{s}}(z,t)$, must fulfil the nonlinear wave equation in the fiber. If the transversal component as well as the attenuation is neglected, the nonlinear wave equation (5.12) becomes:

$$\frac{\partial^2 E}{\partial z^2} - \frac{n^2}{c^2}\frac{\partial^2 E}{\partial t^2} - \frac{\alpha n}{c}\frac{\partial E}{\partial t} = \frac{4\pi}{c^2}\frac{\partial^2 P_{\mathrm{NL}}}{\partial t^2}, \tag{11.28}$$

where P_{NL} describes the nonlinear polarization of the medium. Since only Brillouin scattering is of interest here all other nonlinear effects are neglected. The electric field causes, via the electrostriction, an acoustic wave or phonons in the medium. Hence, the nonlinear polarization can be represented by:

$$P_{\mathrm{NL}}(z,t) = \frac{\gamma_{\mathrm{e}}}{4\pi\rho_0}\rho(z,t)E(z,t), \tag{11.29}$$

where $\rho(z,t)$ denotes the phonons propagating in the waveguide or the deviation in the material density from its equilibrium value, ρ_0, and γ_{e} describes the electrostrictive coupling. Since the generated phonons are a wave in the waveguide they must also fulfil a wave equation which should be not of interest here.

If all parts are joined together, the result is an equation which describes the Brillouin scattering of a modulated pump wave in an optical fiber. Under consideration of the phase matching requirements for different frequency components and if it is assumed that the fundamental modulation frequency for ASK and PSK ($f_{\mathrm{m}} = 1/2T$) as well as the frequency difference between the frequencies of the FSK $\Delta F = f_1 - f_2$ (Sect.3.4) is much higher than the bandwidth of the Brillouin gain (Δf_{A}), it follows for the intensities of the distinct spectral components of pump and Stokes wave the coupled intensity equation [271]:

$$\frac{\mathrm{d}I_{\mathrm{S}n}}{\mathrm{d}z} = -g_{\mathrm{B}} I_{\mathrm{P}n} I_{\mathrm{S}n} + \alpha I_{\mathrm{S}n} \tag{11.30}$$

and

$$\frac{\mathrm{d}I_{\mathrm{P}n}}{\mathrm{d}z} = -g_{\mathrm{B}} I_{\mathrm{P}n} I_{\mathrm{S}n} - \alpha I_{\mathrm{P}n} \tag{11.31}$$

A comparison between (11.30) and (11.31) and the two equations for the Brillouin scattering of a monochromatic pump wave, (11.7) and (11.8), shows that the different frequency components of the modulated wave will not influence each other. The Stokes wave is independently generated for each frequency component of the pump. This result means that as long as no single frequency component of the pump spectrum exceeds the threshold for SBS, the intensity of the modulated wave is smaller than the threshold. The SBS threshold for modulated light can then be defined as the power level at which the largest frequency component reaches the threshold level for CW light [271]. Therefore, the consideration can be limited to the spectral component with the strongest amplitude. For WDM systems this means there is no interaction between the channels in the sense of SBS. Hence, if each frequency component in each individual channel is below the threshold, stimulated Brillouin scattering will not occur.

The comparison between the spectra of the distinct modulation formats in Fig. 3.8 and the spectrum of a monochromatic wave shows that in the case of ASK, the strongest spectral component of the modulated pump is only half as strong as the spectral component of an unmodulated wave. Therefore, the threshold of an amplitude-modulated wave with a fixed bit pattern is twice that of the CW case. Likewise, the threshold of the frequency modulation is four times that of a CW pump, while the threshold of the phase modulated signal is increased 2.5 times (Fig. 3.8).

If the half bit rate of the modulated signal for ASK, PSK, and the frequency spacing for FSK is much higher than the bandwidth of Brillouin scattering (for pure silica glass $f_{\mathrm{m}}, \Delta f \gg 35$ MHz), (11.24) then yields a threshold for the unmodulated pump of $P_0^{\mathrm{S}} \approx 7$ mW for standard single mode fibers with the values given in Sect. 11.4 ($K_{\mathrm{B}} = 2$, $A_{\mathrm{eff}} = 80\ \mu\mathrm{m}^2$, $\alpha = 0.2$ dB/km,

$L_{\rm eff} = 21.7$ km, $g_{\rm B} = 2 \times 10^{-11}$ m/W). Therefore, the threshold for ASK, PSK, and FSK modulation is $P^{\rm S}_{\rm 0ASK} \approx 14$ mW, $P^{\rm S}_{\rm 0PSK} \approx 17.5$ mW and $P^{\rm S}_{\rm 0FSK} \approx 28$ mW, respectively.

If the modulation of the pump shows no fixed bit pattern, the estimation of the thresholds for the different modulation formats is more complicated because the spectra of the pump and Stokes waves cannot be expressed as a superposition of distinct equidistant frequency components. Nevertheless, the spectrum is concentrated around the carrier frequency and the carrier is the component with the highest amplitude. The time-dependent amplitude of a NRZ-ASK modulation is, for instance [248]

$$E(t) = E_0 \left(1 - [1 - m(t)] \left[1 - (1 - k_{\rm a})^{1/2}\right]\right) \tag{11.32}$$

where $m(t)$ is the binary data stream with two possible states "0" and "1", both of which have equal probability, and $k_{\rm a}$ is the modulation index of the intensity modulation ($0 < k_{\rm a} \leq 1$). According to [284], the Brillouin gain in the case of an ASK modulation with equal probability of the values "0" and "1" and under the assumption that the pump is not depleted by the Brillouin process is then:

$$g_{\rm ASK} = g_{\rm B} \left[\left(1 - \frac{a}{2}\right)^2 + \frac{a^2}{4}\left(1 - \frac{B}{\Delta f_{\rm A}}\left(1 - {\rm e}^{-\Delta f_{\rm A}/B}\right)\right)\right], \tag{11.33}$$

where $a = 1 - \sqrt{1 - k_{\rm a}}$ and the bit rate is $B = 1/T$. According to (11.33), the smallest Brillouin gain for ASK modulation is achieved if the modulation index is $k_{\rm a} = 1$ ($a = 1$). This is the case if the amplitude of the carrier is completely modulated, i.e., it is switched between zero and a fixed value. For this on-off keying of the amplitude, (11.33) yields

$$g_{\rm ASK}\left(k_{\rm a} = 1\right) = g_{\rm B} \left[\frac{1}{2} - \frac{B}{4\Delta f_{\rm A}}\left(1 - {\rm e}^{-\Delta f_{\rm A}/B}\right)\right] \tag{11.34}$$

The equations (11.33) and (11.34) are only valid if the carrier frequency of the optical wave is not influenced by the modulation, this is only given for an external modulation of the carrier.[4]

In the case of PSK the digital baseband signal modulates the phase of the carrier, as described by [248]

$$E(t) = E_0 {\rm e}^{{\rm j}k_{\rm p}m(t)}, \tag{11.35}$$

with $k_{\rm p}$ as the PSK modulation index. Under the same assumptions as for ASK, the Brillouin gain for PSK is [284]:

[4]If the carrier is directly modulated, for instance via its current, it shows an additional frequency modulation (chirp).

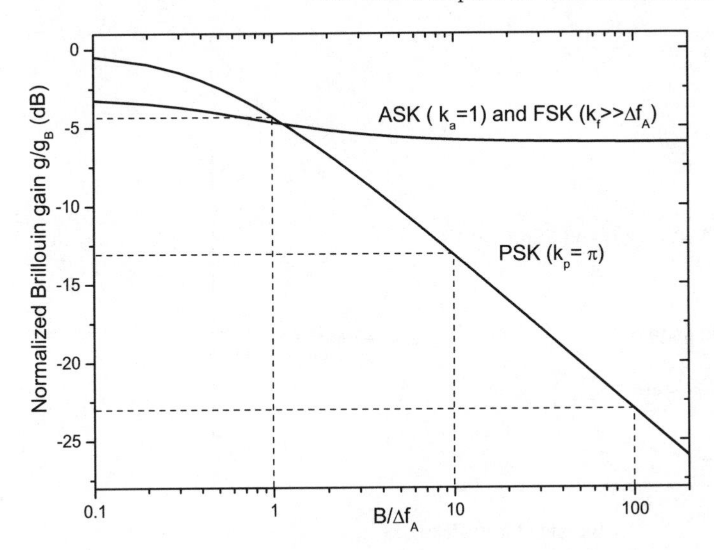

Fig. 11.11. Brillouin gain of the basic binary modulation formats, normalized to the CW gain, against the ratio between bit rate and band width of CW Brillouin scattering

$$g_{\mathrm{PSK}} = g_{\mathrm{B}}\left[\frac{1}{2}\left(1+\cos k_{\mathrm{p}}\right) + \frac{1}{2}\left(1-\cos k_{\mathrm{p}}\right)\left(1-\frac{B}{\Delta f_{\mathrm{A}}}\left[1-\mathrm{e}^{-\Delta f_{\mathrm{A}}/B}\right]\right)\right]. \tag{11.36}$$

According to (11.36), the smallest gain is achieved if the modulation index of the PSK signal is $k_{\mathrm{p}} = \pi\,(2n+1)$. For a phase shifting between 0 and π we have:

$$g_{\mathrm{PSK}}\left(k_{\mathrm{p}}=\pi\right) = g_{\mathrm{B}}\left(1-\frac{B}{\Delta f_{\mathrm{A}}}\left[1-\mathrm{e}^{-\Delta f_{\mathrm{A}}/B}\right]\right). \tag{11.37}$$

If the modulation index of the FSK, $k_{\mathrm{f}} = \omega_1 - \omega_2$, is much higher than the bit rate, the spectrum can be seen as the superposition between two ASK spectra with $k_{\mathrm{a}} = 1$ (see Fig. 3.8). Under this condition, the Brillouin gain of FSK [248, 284] is

$$g_{\mathrm{FSK}}\left(g_{\mathrm{f}} \gg \Delta f_{\mathrm{A}}\right) = g_{\mathrm{B}}\left[\frac{1}{2}-\frac{B}{4\Delta f_{\mathrm{A}}}\left(1-\mathrm{e}^{-\Delta f_{\mathrm{A}}/B}\right)\right]. \tag{11.38}$$

A comparison of the equations (11.34), (11.37), and (11.38) shows that the PSK Brillouin gain has the strongest decay with increasing bit rate. Figure 11.11 shows the gain of the binary modulation formats, normalized to the Brillouin gain for unmodulated light ($g_{\mathrm{ASK}}/g_{\mathrm{B}}$, $g_{\mathrm{PSK}}/g_{\mathrm{B}}$, $g_{\mathrm{FSK}}/g_{\mathrm{B}}$), against the ratio of the bit rate to the bandwidth of the CW Brillouin gain ($B/\Delta f_{\mathrm{A}}$ and $1/(Tf_{\mathrm{A}})$, respectively).

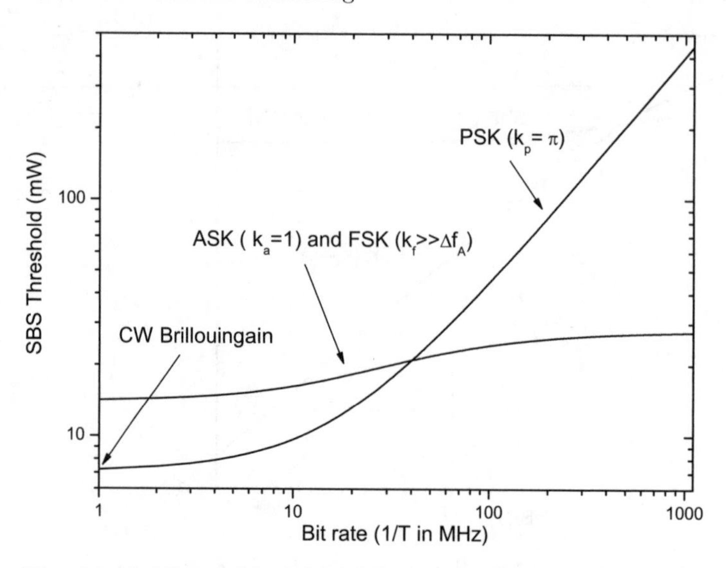

Fig. 11.12. Threshold of the Brillouin gain for binary modulation formats versus bit rate ($L_{\text{eff}} = 21.7$ km, $A_{\text{eff}} = 80$ µm^2, $g_{\text{B}} = 2 \times 10^{-11}$ m/W, $K_{\text{B}} = 2$, $\Delta f_{\text{A}} =$ 35 MHz)

As can be seen from the figure, the Brillouin gain is reduced between ≈ -3 dB for small bit rates and ≈ -6 dB for high bit rates for ASK and FSK modulation formats. A much stronger decay of the Brillouin gain is shown by the phase modulation where each increase of the bit rate by a factor of 10 results in a decrease of the Brillouin gain by 10 dB.

The decrease in the Brillouin gain leads directly to an increase of the Brillouin threshold. The threshold in long fibers ($L_{\text{eff}} = 21.7$ km) for different modulation formats against the bit rate is shown in Fig. 11.12. The threshold of SBS for a PSK modulated signal with a bit rate of 1 Gbit/s is 300 mW, for example.

Another method to increase the SBS threshold is the application of a temperature distribution along the fiber. With several discrete fixed temperatures evenly divided between a minimum and a maximum temperature, the SBS threshold was increased to 8 dB in a short HNLF [275] by winding the fiber onto eight spools with a temperature difference of >350 °C.

11.6 Brillouin Scattering for Distributed Temperature and Strain Sensors

For narrow linewidth pump lasers, SBS is the nonlinear effect with the smallest threshold so that even diodes with rather small output power are sufficient for its generation. At the same time, the effect is controllable with a modulation of the pump. Hence, SBS has many interesting applications. Such as the

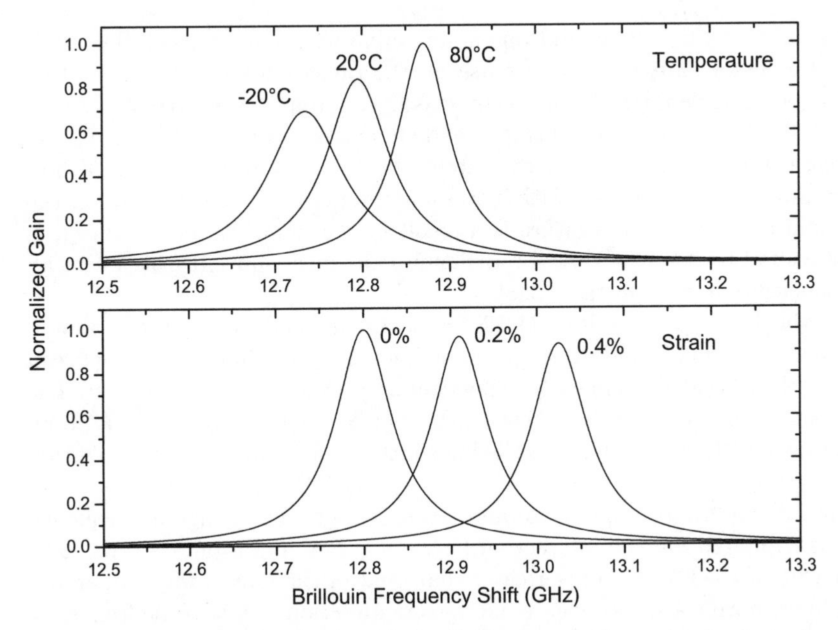

Fig. 11.13. Principal behavior of the Brillouin gain in a standard single mode optical fiber for different environmental conditions. The curves were calculated after measured data published by Niklés et al. [328] for a pump wavelength of 1320 nm and 60 m fiber with a GeO_2 concentration of 8.24 wt.%

combination of several laser beams from different sources in a multi-mode fiber to only one spatially coherent beam with a high power [316, 317]. The exploitation of SBS in optical telecommunications (as optical filter, for the clock recovery, as nonlinear optical amplifier, and for the phase conjugation) is discussed in special chapters (12.4, 12.5, 14). In this Section, we are interested in the exploitation of the SBS for distributed sensors.

Since Brillouin gain is caused by the scattering of a pump wave on acoustical waves, it depends on every alteration of the material density. This density alteration can be caused by environmental influences like temperature or a mechanical distortion of the fiber [255, 261]. Therefore, a measurement of the Brillouin gain in the fiber gives information on the temperature or mechanical stress along the fiber. This opens the possibility for the construction of distributed temperature and strain sensors [325, 326]. Applications of these sensors are, for instance, temperature measurements along streets with the possibility to provide information on the risk of glazed frost at distinct positions. Applications for high temperature sensing up to 850°C are also possible [327]. Another field is the persistent monitoring of the mechanical stress of bridges, dams or pipelines.

In the upper part of Fig. 11.13 the Brillouin gain in an optical fiber for three different temperatures is shown. The Brillouin gain coefficient changing its frequency shift, shape, and maximum value with temperature. The Brillouin frequency shift linearly increases in the range between -30 and 100°C. The slope coefficient $\mathrm{d}f/\mathrm{d}T$ of this increase at a pump wavelength of 1320 nm is 1.36 MHz/°C in a standard single mode fiber. The slope coefficient slightly decreases for a higher GeO_2 core content. At the same time the linewidth of the Brillouin gain decreases with increasing temperature. The increase of the Brillouin gain with temperature is a result of the linewidth narrowing, it is proportional to $\Delta f_{\mathrm{A}}^{-1}$ so that the Brillouin gain spectrum integrated over all frequencies remains constant [328].

In the lower part of Fig. 11.13 the strain dependence of the Brillouin frequency shift in a standard single mode fiber is shown. In the range between $0 < \varepsilon < 1\%$ the strain is linearly dependent on the elongation of the fiber. The frequency shift increases with 594.1 MHz per percent elongation at a pump wavelength of 1320 nm, whereas the linewidth remains unchanged with strain [328].

In a distributed temperature and strain sensor, the change in frequency shift is exploited via the interaction between a CW and a pulsed wave. The frequency of the CW beam is greater than that of the pulsed wave. In an optical fiber, both waves propagate in opposite directions [260]. If the frequency difference between the CW wave and the pulse corresponds to the Brillouin frequency shift at any position in the fiber, the pulse will be amplified while the CW wave is weakened. But, according to Fig. 11.13, the frequency shift depends on the temperature. Therefore, it is possible to measure the environmental temperature of the fiber if the amplification of the pulse is recorded as it changes with the frequency difference between both waves. At the same time, the delay between the transmitted pulse and the received amplification of the Stokes wave contains information about the position of the measured temperature.

The temperature and strain sensitivity of the SBS frequency shift cannot be separated from each other. In order to achieve results for both, it is possible to separate the fiber into two parts with the same length. One of these parts measures the temperature and strain distribution along the distance z, whereas the other part is isolated from the strain influences and measures only the temperature along the same way [260]. Another possibility is offered by the different linewidth dependence of temperature and strain. The linewidth is decreased if the temperature is increased, whereas it remains unchanged for increasing strain. This could decorrelate temperature from strain effects [328].

As mentioned in Sect. 11.4, the gain of the Brillouin scattering depends on the temporal duration of the pump pulses. If the duration is decreased, the maximum of the gain is decreased as well. If the pump pulse duration is in the range of the phonon life time ($\approx$ 10 ns), the band width of the gain is twice that for CW excitation. This fact limits the spatial distribution of such a sensor to $\approx$ 1 m ($\Rightarrow$ 10 ns) [329, 330]. On the other hand, Bao et al.

[262] measured a decreasing of the Brillouin bandwidth for very short pulses smaller than 5 ns, so distributed temperature and strain sensors with spatial solutions in the centimeter or millimeter range could be possible.

Summary

- Brillouin scattering is a result of the interaction between the optical wave and an acoustical wave of the material. A strong pump wave generates a time-dependent refractive index modulation in the material via the effect of electrostriction.
- A part of the pump wave is backscattered at the refractive index modulation; this is the Stokes wave. Due to the relative velocity between the modulation and the pump, the Stokes wave is down-shifted in frequency. The pump wave and the Stokes wave propagate into opposite directions in the fiber.
- Above a distinct threshold, the scattering is stimulated and the Stokes wave shows an exponential growth along the fiber. The threshold is determined by the Brillouin gain coefficient.
- The Brillouin gain coefficient has a very narrow band width ($\approx$ 35 MHz for pure silica). It depends on the temperature, mechanical stress, the endowment of the fiber core, waveguide characteristics, and the temporal width of the pump wave.
- The maximum of the Brillouin gain coefficient has a frequency shift of about 11 GHz (for pure silica) to the pump wave. Due to the nature of the effect it depends, like the gain, on the fiber and the pump wave properties.
- The threshold of stimulated Brillouin scattering determines the maximum launchable power for an optical transmission system.
- For a CW excitation, the threshold for stimulated Brillouin scattering is very low ($\approx$ 7 mW, 8.45 dBm for a standard single mode fiber).
- As other nonlinear effects the threshold depends on the effective length. But, since optical isolators are incorporated in practical transmission systems, the effect cannot accumulate over the spans of an amplified system.
- In a WDM system, the channels will not interact with each other in the sense of SBS. If the power of each individual channel is below the threshold, stimulated Brillouin scattering will not occur.
- A modulation of the pump wave can significantly increase the threshold. This increase depends on the modulation format used.
- In the case of PSK, each increase of the bit rate by a factor of 10 results in an increase of the Brillouin threshold by 10 dB.
- Other possible methods for the threshold increase are based on nonuniform SBS properties along the fiber length. Therefore, the threshold can be increased by changing the core dimension [331], dopant concentration [263], strain [332], or temperature [275] along the fiber.

Exercises

11.1. Silica glass with a GeO_2 concentration of 20 wt% has a sound velocity of ≈ 5300 m/s and a refractive index of ≈ 1.48. Estimate the frequency shift of the scattered wave if this material is used in the core of an optical fiber and the pump has a wavelength of 1550 nm.

11.2. Estimate the Brillouin frequency shift for a fiber with a F concentration of 1.7 wt% and a GeO_2 concentration of 16 wt%.

11.3. Determine the power of a Stokes wave for a 5 mW pump signal generated in a fiber with a polarization averaged maximum Brillouin gain of 1×10^{-11} m/W and a linewidth of 40 MHz. The fiber parameters are $\alpha = 0.25$ dB/km, $L = 50$ km, and $A_{\text{eff}} = 50\ \mu\text{m}^2$. The initial Stokes wave is assumed to be generated from thermal noise only, with an equivalent noise temperature of 300 K.

11.4. Calculate the maximum Brillouin gain for a material with an elasto-optic figure of merit of $M_2 = 1.7 \times 10^{-15}\ \text{s}^3/\text{kg}$ [239] and a refractive index of $n = 1.5$ at 1550 nm if the pump pulse has a Lorentzian shape and a linewidth of 200 MHz. The Brillouin gain coefficient has a spectral width of 11.4 MHz and a frequency shift of 10.8 GHz.

11.5. Determine the maximum Brillouin gain for the fibers listed in Table 11.2 under the assumption that (11.24) is valid for the threshold and the effective area of each fiber is simply determined by the core diameter.

11.6. Estimate the Brillouin threshold in a TrueWave fiber for a 10Gbit/s ASK($k_{\text{a}} = 1$), FSK($g_{\text{f}} \gg \Delta f_{\text{A}}$), and PSK($k_{\text{p}} = \pi$) signal.

Part III

Applications of Nonlinear Effects in Telecommunications

12. Optical Signal Processing

In the early days of optical communications, signal processing was provided electrically. This means that switching and routing of signals was operated in an electrical domain whereas only transmission over large distances was done optically. In order to reduce costs and to increase reliability, signal processing should be carried out in the optical domain in future optical transport networks. This is mainly due to two reasons; with increasing bit rate the complexity of the electrical domain increases significantly and, as will be shown in this Chapter, optical signal processing is much simpler.

With increasing bit rates, electrical signal processing reached its limits because it depends on the mobility of electrons in semiconductor materials. All optical signal processing, on the other hand, exploits photons propagating with the speed of light. Therefore, only optical switches and demultiplexers have the potential to operate in the femtosecond (10^{-15} s) time domain. Furthermore, with increasing bit rate the propagation impairments as a result of noise accumulation, interchannel interactions, fiber dispersion, and nonlinear effects will increase drastically. These impairments result in signal degradations in the amplitude and time domain. Signal degradation in the amplitude domain leads to random fluctuations in the power levels of marks (logical "1") and spaces (logical "0"). In the time domain the time slots of the pulses change randomly (jitter). To overcome the limitations due to these signal degradations in-line optical signal regeneration is essential. The basic functions of such an in-line regenerator are reamplifying, reshaping and retiming (3R), all of these can be done in the optical domain by nonlinear optical devices.

Optical transport networks have a hierarchical organization, as shown in Fig. 3.10. The data or speech signals from a local area network (LAN) are multiplexed in a metropolitan area network or regional network with data streams from other LANs. Different regional networks are then again combined in a backbone network, connecting them with each other. Key elements in such structures are switching elements, multiplexers, demultiplexers, and wavelength converters based on various optics principles. In this chapter, optical signal processing devices based on nonlinear optical effects are introduced and their working principles will be described in detail.

The devices exploit fiber-based or non-fiber-based nonlinear effects, such as those in SOAs. The latter devices have the advantage that they require only low pump powers and they offer the possibility to design small, compact, or even integrated devices. Due to the resonant character of the nonlinearity, they are rather slow, however. In contrast, the exploitation of nonlinear effects in standard single mode fibers requires several kilometers of fiber to achieve realistic operating powers. Furthermore, problems with the dispersion and walk-off in the fibers can limit the performance of such devices. In recent years, the development of fibers with a high nonlinearity, like HNL or highly nonlinear microstructured fibers with nonlinear coefficients up to 30 times higher than in SSMF, offered the possibility to build fiber-based signal processing devices from only a few meters of these fibers. This makes them a serious alternative to resonant nonlinear devices.

The first section shows nonlinear sources that deliver a broad spectrum of wavelengths exploitable as carriers for WDM systems. In Sect. 12.2, devices that open the possibility to change the carrier wavelength in the network are presented. This degree of freedom is a standard feature of classical electric signal processing, but a severe problem in all optical networks. If the assigned wavelength cannot be changed in the links along the route, the capacity of the network is reduced and a blocking of wavelengths over the whole network occurs in the case of a failure.

Section 12.3 deals with the possibility of all-optical switching. Switching is carried out in add-drop multiplexers and in cross-connects all over the network. The possibility to switch channels, or maybe packets, is a basic requirement of a network. Without it, the network is only a link. Electro-optical switching will arrive at its limits because for bit rates beyond 40 Gbit/s, the switches have to be equipped with hundreds or even thousands of ports. The complexity of electrical switches will strongly increase with increasing bit rate and number of channels. Optical switching has, instead, a number of advantages over electrical switching [160, 187]. Switching in the optical domain will save costs, power, and required space, and optical switches offer a granularity that would be impossible with electro-optic switches. They can switch entire fibers, single wavelength, bands of wavelength, or even packets.

The last two sections introduce applications of nonlinear effects in all-optical clock recovery, a basic functionality of repeaters, and in the optical filtering of optical signals.

12.1 Spectrum Slicing and Nonlinear WDM Sources

In order to further increase the bit rate and, therefore, the capacity of optical time division multiplex (OTDM) systems, sources that can deliver very short pulses are essential. With the effect of SPM, it is possible to shorten pulses with a temporal duration of a picosecond (ps $= 10^{-12}$ s) down to ultrashort pulses with only a few femtoseconds (fs $= 10^{-15}$s).

According to the Fourier theorem for bandwidth-limited pulses, the temporal duration (τ) and the energy uncertainty or bandwidth (Δf) are in inverse proportion to each other. Therefore, very short temporal pulses are very broad in the frequency or wavelength domain and pulses with a small spectral width are very broad in the time domain. With this natural contiguity, it is possible to determine the spectral width of a pulse with a variation of its duration. This is the background of the so called "spectrum-slicing" technique.

As mentioned in Sect. 6.1.1, a temporally short pulse will be strongly broadened in an optical fiber due to the effect of SPM. If the fiber is long enough, or the intensity slope of the pulse strong enough, it accrues a so called "supercontinuum" [105]. Generally, supercontinuum generation can involve a rather complicated combination of nonlinear physical processes. If the fiber is operated in the dispersion minimum XPM and FWM are responsible for the spectral broadening as well. Furthermore, if the generated bandwidth exceeds the Raman gain spectrum, the effects of Raman scattering must be included into the consideration. If the fiber is operated in the anomalous dispersion regime soliton generation and fission can play an important role for the initial broadening mechanism [106].

Contrary to the aforementioned Fourier theorem, due to the dispersion of the fiber, the pulse shows an additional frequency modulation – a chirp. Therefore, the pulse is no longer bandwidth-limited. As a result, its duration is not only determined by its spectral width, the dispersion is responsible for an additional broadening of the pulse duration. This additional broadening can be compensated by an element with oppositional dispersion, for instance a dispersion-compensating fiber (DCF). At the end of the line, a filter with variable bandwidth slices a part of the spectrum (spectrum-slicing). If the sliced spectrum is greater than the spectrum of the initial pulse, its temporal duration is smaller. It is wider otherwise.

The fundamental set-up of this technique is shown in Fig. 12.1. The laser delivers pulses with a temporal duration of 4.8 ps, a wavelength of 1565 nm, and a repetition rate of 6.3 GHz. The following EDFA amplifies the pulses to around 1.5 W. This power is sufficient to produce a supercontinuum with a spectral width of around 100 nm in a 3-km fiber. The central wavelength of the spectrally-broadened pulses is 1550 nm. With this set-up Morioka et al. [104] were able to tune the pulse width to somewhere between 370 fs and 11.3 ps.

Not only can nonlinear optics increases the capacity of OTDM systems, different wavelengths or frequencies act as carriers for the signals in a WDM System. In today's WDM networks one DFB laser diode is used for every wavelength channel, and a subsequent modulator changes the intensity, frequency, or phase of the carrier as a function of the data signal. The wavelength of the DFB depends on the environmental temperature and its current, so every DFB needs a wavelength-stabilizing mechanism. This approach is cost

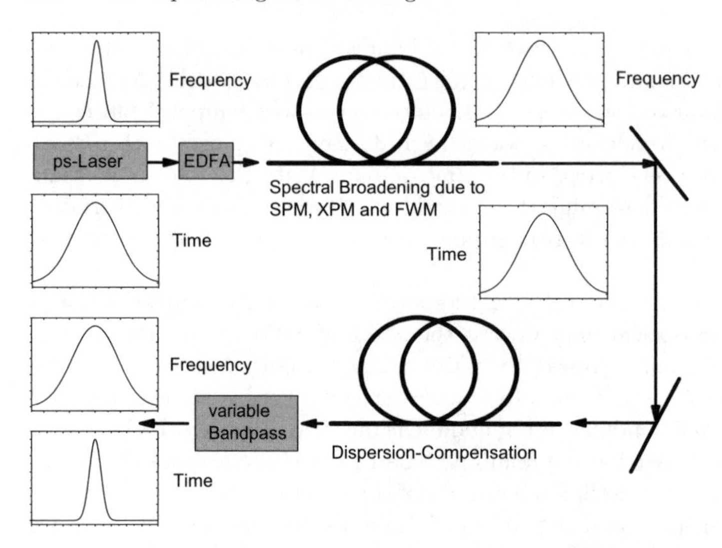

Fig. 12.1. Fundamental set-up for a spectrum-sliced source with tunable pulse length

intensive and it complicates the set-up and reliability of the system. Instead, nonlinear optical effects are able to provide all the required wavelength from one source only. This can, in principle, make the system much easier to handle.

According to the Fourier theorem, a very short pulse is a natural source of many different wavelengths. A pulse with a Gaussian shape and a temporal duration of 60 fs (FWHM) has, for instance, a spectral width of 4.4 THz. This would be sufficient for 44 WDM-channels with a channel spacing of 100 GHz. With nonlinear effects like SPM, the spectral width of the pulse can be further broadened by up to a few hundred nanometers. Accordingly, it is possible to use only one laser with nonlinear effects as a WDM source for the whole transmission range of an optical fiber.

An early experiment by Nuss et al. [117] used a femtosecond laser as a source and its pulses were injected into a 15-km long optical fiber. The pulses experienced a chirp due to the dispersion of the waveguide in the fiber. As a result, the spectral components of the pulse were propagated with different velocities, so the pulse was spread in time to around 15 ns. The different wavelengths of the pulse components arrived at the output of the fiber at different times. With a fast modulator it was possible to modulate spreaded spectral ranges of the pulse independently with a data signal. These spectral ranges were then used as WDM carriers, as shown in Fig. 12.2.

In [118], a pulse with a duration of 3.5 ps was used to spread its spectral width up to 200 nm in a 3-km long fiber. From this supercontinuum, the authors obtained 200 WDM channels. With a similar supercontinuum source, the study in [119] reported a data transmission rate of 3 Tbit/s over 40 km.

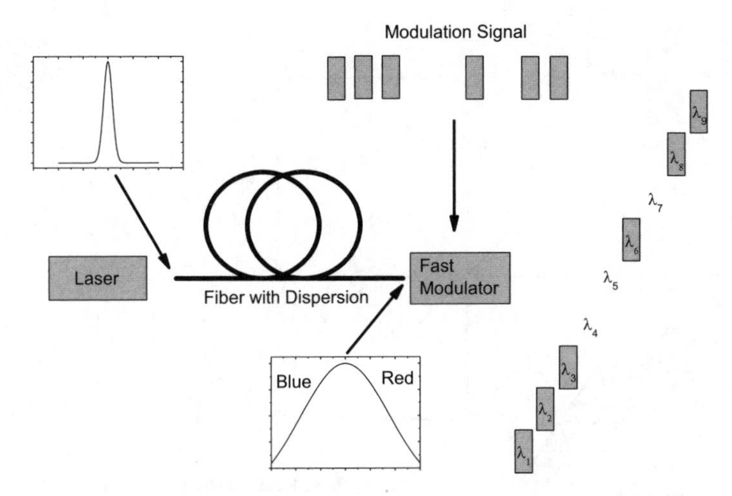

Fig. 12.2. Principle set-up of a WDM source that can exploit the spectral width of a ultrashort pulse and the dispersion of a fiber

In this experiment, each of 19 WDM channels were modulated with a data rate of 160 Gbit/s. Recently, the demonstration of more than 1000 WDM channels from one single laser source was shown [120].

For spectrum slicing coherent light sources like mode-locked erbium-doped fiber lasers can be used, they do not suffer from beat noise between different portions of the spectrum within a channel which can limit the transmission capacity [121], but incoherent sources like the amplified spontaneous emission of an EDFA can be exploited as well [122].

With pulses of a very high intensity – for instance from an amplified titanium sapphire fs-laser system – it is possible to produce a white light continuum in a transparent material like glass plate or a glass of water. This means the pulse, with a mean wavelength of 800 nm (wavelength of a titanium sapphire laser) and a spectral width according to its temporal duration, is spectrally broadened in the material over a range larger than the visible domain [123]. The infrared pulse produces a coherent white light continuum. The spectral width of the continuum depends on the thickness and the transparency range of the material.

Figure 12.3 shows the construction of such a white light source for possible applications in the telecommunications area. The laser delivers ultrashort pulses with very high intensities[1] (to the order of 10^{11} W/cm^2). Since nonlinear effects depend on the intensities but not the energy of the pulses, these

[1]Although these intensities are very high, their energy is not sufficient for the destruction of the material. The destruction depends mainly on the energy and not the intensity of the pulses. The crystal lattice of the material is very indolent, so it can act only after a few picoseconds. By this time the pulse is over already. The mechanism behind the destruction of a transparent material by a fs pulse is not very

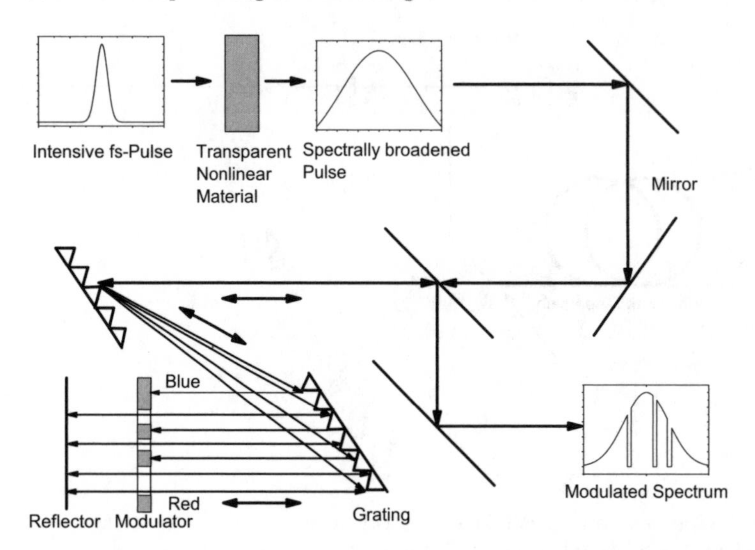

Fig. 12.3. Fundamental set-up for the spectral modulation of a white light source

high intensities are responsible for highly effective nonlinearities in the material. The pulse produces a white light continuum in a glass plate and due to the dispersion of the material, the pulse experiences a chirp. This chirp is compensated by a pair of gratings. The angle in which a grating reflects the spectral components of the white light continuum depends on their wavelengths. As shown in Fig. 12.3, the blue part of the pulse propagates a shorter way between the two gratings than the red one. This behavior is synonymous to a negative dispersion that can compensate the positive dispersion of the glass plate. The spectral components of the grating are spatially separated between the grating pair according to their wavelength, so a modulator is placed between the two gratings. The modulator absorbs or transmits defined wavelengths or WDM-channels according to the modulation signal.[2] On the other hand, this arrangement can be exploited to modulate each individual channel with a code sequence. In this case the difference between the channels is not determined by the time slot – like in OTDMA systems – and not the carrier wavelength, like in DWDM systems. Here, all channels use the whole time and bandwidth of the transmission system but they can be separated by the code sequence (code division multiple access CDMA)[124]. For CDMA the different codes must be orthogonal to each other. With such multiple access techniques it might be possible to increase the transmission capacity of the channel.

well understood [126, 127]. The destruction takes place probably via multiphoton absorption in the material for intensities above 10^{12} W/cm^2.

[2]According to the Fourier theorem, the pulses will experience an additional temporal broadening if parts of its spectrum are sliced.

The biggest part of the white light continuum cannot be propagated in standard single mode fibers because their cut-off wavelength is 1.2 μm. Therefore, other waveguides like photonic fibers or different laser sources – which can deliver mean wavelengths in the telecommunications range like fs fiber lasers – have to be used for this technique.

Due to the fact that the supercontinuum generation is a result of nonlinear effects like SPM, XPM and FWM, the efficiency of this process can be improved by the enhancement of the nonlinearity of the waveguide. Therefore, in tapered fibers supercontinuum generation with nanojoule femtosecond pulses is possible [10]. Photonic crystal or microstructured fibers offer additionally the possibility to tailor the dispersion characteristics. Since the supercontinuum generation involves the generation of new frequency components, the dispersion has an important influence on the efficiency and stability of the generated wavelengths [125].

12.2 Wavelength Conversion

In a simple WDM system, the initial wavelengths remain the same all over the network. The assigned wavelength will not be changed in the links along the route. This approach reduces the capacity of the network and can lead to a blocking of wavelengths if a failure occurs. Wavelength converters are devices that convert the data from an input wavelength into another output wavelength. Therefore, dynamically reconfigurable WDM networks are possible with wavelength converters, thus improving the network blocking performance [128].

Electrical signal processing has, beside many disadvantages, one big advantage for WDM systems. It is very easy to change the wavelength between the incoming and outgoing signal of a cross connection. In the case of a failure, some of the wavelengths could be reserved in a link but this has no influence on other links. Therefore, all wavelengths between two cross connects or switches are available to use.

In a network with pure optical signal processing, the signals are switched between the input and output port without changing their wavelength, making it impossible to adapt them to the relationships of the next link. As a result, this leads to a continuous wavelength carrier for the signals in the whole optical network from the beginning to the end. If a failure occurs, such as a DFB laser diode breaking down somewhere in the network, the net has to evade to another wavelength. For the system, this means it is no longer possible to use the former wavelength in the whole net although in fact only one link had a failure. This virtuality limits the capacity of the network. Therefore, for dynamically reconfigurable networks with a high performance optical wavelength converters are essential.

The simplest devices for changing the wavelength in a switch or cross connect are optoelectronic regenerators. They demodulate the optical carrier down to an electrical baseband signal with the help of a photodiode. This electrical signal is then used to modulate another DFB laser diode delivering a different frequency or wavelength. Optoelectronic regenerators depend on the modulation format of the signals and their switching speed is limited. Furthermore, they increase the hardware costs and the complexity of the network. Pure optical wavelength conversion, such as FWM in a highly nonlinear fiber, is instead completely independent of the modulation format and provides the possibility to handle data rates far beyond 100 Gbit/s.

In this section, different nonlinear principles for an all-optical conversion of wavelength are treated; first, FWM in optical fibers for wavelength conversion is considered in detail; after that, the principles of exploiting XPM for the changing of frequencies are described; in the last subsection, different nonlinear effects in semiconductor optical amplifiers (SOA) are mentioned.

12.2.1 Wavelength Conversion with FWM

In the case of linear optics, wavelength conversion is impossible. If an optical wave intrudes a material, the phase of the wave will be changed but not its wavelength in the linear case. Therefore, only nonlinear effects, like FWM in optical fibers, can change the frequency of the waves. For wavelength conversion degenerate – both pump waves have the same frequency – FWM is used. If the old wavelength is called the signal, for the new idler frequency is

$$f_{\text{idler}} = 2f_{\text{pump}} - f_{\text{signal}}. \tag{12.1}$$

In order to build such a wavelength converter, in principle, only a pump-laser delivering a sufficiently strong pump wave with the appropriate wavelength shift and an optical fiber are required. Fibers with a nonlinearity coefficient much higher than those in standard optical waveguides are very convenient for this purpose. According to Sect. 7.3, phase matching must be fulfilled in order to convert the wavelength efficiently, so these fibers have to be operated in their dispersion minimum. The natural dispersion minimum of silica glass is at around 1.3 μm, but the wavelength of the telecommunications C-band is around 1.5 μm, so the dispersion minimum must be shifted up to this wavelength. Such waveguides are called highly nonlinear dispersion-shifted fibers (HNL-DSF). Figure 12.4 shows the possible set-up of a FWM wavelength converter.

If the extensive waveguide is considered as a compact device, its efficiency can be calculated as the ratio between the power of the converted wave at the output and the power of the signal wave at the input:

$$G = \frac{P_{\text{idler}}(z = L)}{P_{\text{signal}}(z = 0)}. \tag{12.2}$$

For the case of degenerate FWM, if the attenuation in the fiber is neglected and under the consideration that the intensity of the pump is much

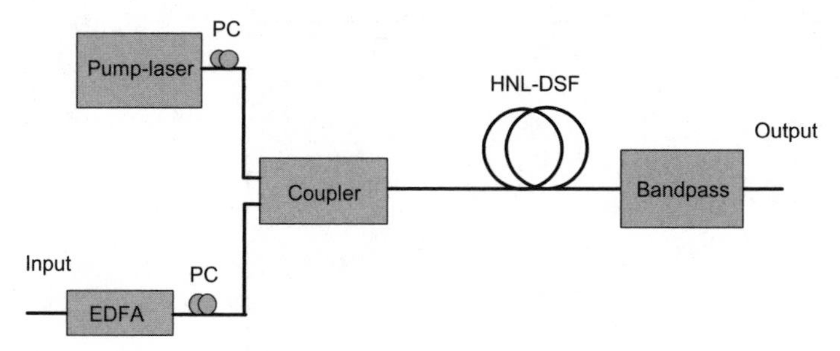

Fig. 12.4. Fundamental set-up of an FWM wavelength converter (HNL-DSF=highly nonlinear dispersion-shifted fiber, PC=polarization controller, EDFA=erbium-doped fiber amplifier)

higher than the signal and idler intensity, using the differential equations (7.13) to (7.16) with $A_1 = A_2 = A_{\text{Pump}}$; $A_3 = A_{\text{Signal}}$ and $A_4 = A_{\text{Idler}}$ yields

$$\frac{\partial A_{\text{pump}}}{\partial z} = \mathrm{j}\gamma P_{\text{pump}} \left|A_{\text{pump}}\right|^2 A_{\text{pump}}, \tag{12.3}$$

$$\frac{\partial A_{\text{signal}}}{\partial z} = \mathrm{j}2\gamma P_{\text{pump}} \left[\left|A_{\text{pump}}\right|^2 A_{\text{signal}} + A_{\text{pump}}^2 A_{\text{idler}}^* \mathrm{e}^{-\mathrm{j}\Delta kz}\right], \tag{12.4}$$

and

$$\frac{\partial A_{\text{idler}}}{\partial z} = \mathrm{j}2\gamma P_{\text{pump}} \left[\left|A_{\text{pump}}\right|^2 A_{\text{idler}} + A_{\text{pump}}^2 A_{\text{signal}}^* \mathrm{e}^{-\mathrm{j}\Delta kz}\right], \tag{12.5}$$

with $\Delta k = k_{\text{Signal}} + k_{\text{Idler}} - 2k_{\text{Pump}}$ as the phase mismatch. According to (12.3), only the SPM acts on the pump wave, whereas the phases of the signal and idler waves are altered by the XPM with the pump. Its amplitudes grow through the last term in (12.4) and (12.5), which is responsible for the FWM process. The increases of the amplitudes and, therefore, the intensity depend on the phase matching. If $D \gg \mathrm{d}D/\mathrm{d}\lambda$ the phase mismatch in the degenerate case is (7.36):

$$\Delta k = \frac{2\pi c}{f_{\text{pump}}^2} D(f_{\text{pump}}) \Delta f^2 \tag{12.6}$$

with $\Delta f = (f_{\text{Probe}} - f_{\text{Pump}})$. The efficiency of an FWM wavelength converter in the range of normal dispersion is (see Sect. 13.4)

$$G = \gamma^2 P_{\text{pump}}^2 L^2 \left(\frac{\sin(gL/2)}{gL/2}\right)^2, \tag{12.7}$$

where g is the gain, dependent on the phase mismatch, which can be calculated with

$$g = \sqrt{\frac{1}{4}\Delta k \left(\Delta k + 4\gamma P_{\text{Pump}}\right)}. \tag{12.8}$$

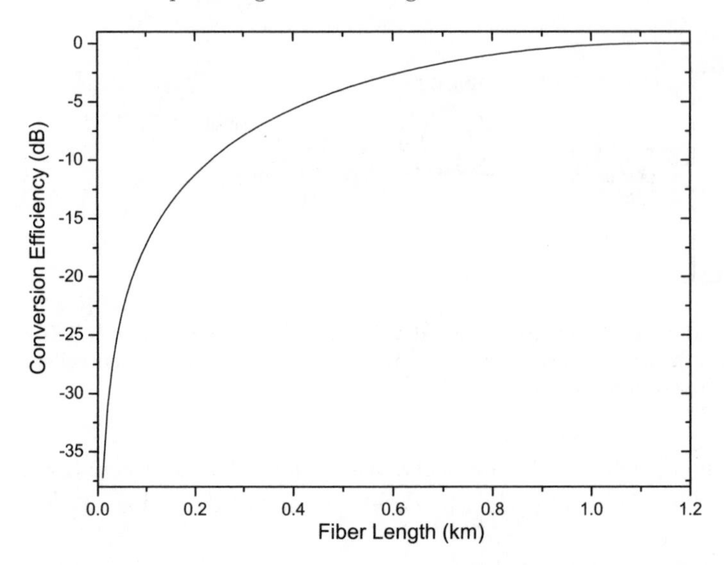

Fig. 12.5. Conversion efficiency of an optical parametric wavelength converter against the length of the waveguide. Higher order dispersion terms are neglected ($g = g_{\max} = 1.38 \times 10^{-3}$ m^{-1}, $P_{\text{Pump}} = 100$ mW, $\gamma = 13.8$ W^{-1}km^{-1})

The maximum gain for the parametric process is (see Fig. 13.12) $g_{\max} = \gamma P_{\text{pump}}$. With a nonlinearity coefficient for HNL fibers of $\gamma = 13.8$ W^{-1}km^{-1} and a pump power of 100 mW, the maximum gain becomes $g_{\max} = 1.38$ km^{-1}. Figure 12.5 shows the conversion efficiency for this example as a function of the fiber length.

Based on the equations (12.6), (12.7) and (12.8) the dispersion decreases very fast if the frequency of the signal wave is changed because the phase mismatch will be increased. For a large bandwidth of the FWM converter, the pump wavelength has to be at the zero dispersion of the fiber. In this case, higher order dispersion terms will play a significant role (see Sect. 13.4).

For maximum efficiency and bandwidth, the length of the converter should be shorter than the coherence length in the fiber, i.e.,

$$L \leq \frac{2\pi}{\Delta k} = \frac{f^2}{c\left|D\left(f_{\text{pump}}\right)\right|\Delta f^2}. \tag{12.9}$$

Aso et al. [145, 146] found an optimum converter length of about 200 m for an HNL-DS fiber with a nonlinearity coefficient of $\gamma = 13.8$ W^{-1}km^{-1}. With a pump power of 100 mW and a signal power of 1 mW, they achieved in this special fiber a conversion efficiency of -15 dB in a wavelength range of more than 20 nm. The pump wavelength coincides with the wavelength of the fiber's zero dispersion at 1565.5 nm.

With a 20-m long HNL-DSF ($\gamma = 18$ W^{-1}km^{-1}), Ho et al. [190] showed that optical parametric amplifiers and, therefore, wavelength converters with a bandwidth of more than 200 nm are possible (see Sect. 13.4). They used a pump wavelength of 1542.4 nm for an optimum gain spectrum. The dispersion minimum of the fiber was at 1540.2 nm. The pump source was a 13-dBm DFB laser followed by two EDFAs. It was pulse modulated to provide about 10 W peak power.

Watanabe [147] showed the simultaneous wavelength conversion between the C- and L-band of 32 × 10-Gb/s WDM signals with a polarization-maintaining HNL-DSF.

Due to the fact that electronic polarization is responsible for the FWM in optical fibers, it is very fast, to the order of 10^{-15} s. Therefore, wavelength converters based on this effect are not limited to the modulation format or bit rate of the signals. The conversion is transparent. Problems occur due to the polarization dependence of the FWM. In order to circumvent this, polarization-maintaining fibers, polarization controllers at the beginning of the converter, or long interaction length used to average out the dispersion dependence (Sect. 6.2.1) can be employed.

12.2.2 Wavelength Conversion with XPM

If two pulses with different carrier frequency propagate in a nonlinear material, they cannot be considered independently. Due to the third order susceptibility, they will interact with each other. As mentioned in Sect. 6.2, the effect of cross-phase modulation (XPM) leads to an alteration of the phase of one pulse by the intensity of the other. Since the time derivation of the phase is the frequency, this phase alteration will lead to a frequency change at the same time. According to (6.32), the spectrum of the pulse will be broadened:

$$\Delta B_{\mathrm{XPM}} = 2\gamma L_{\mathrm{eff}} \frac{\partial P_2}{\partial t}. \quad (12.10)$$

This broadening depends on the effective length in the fiber L_{eff}, its nonlinearity coefficient γ, and the temporal change of the power of the other pulse. The broadening is symmetric around the original spectrum of the pulse. Hence, the pulse will get sidebands if the other wave is amplitude-modulated and its temporal status represents a logical one. Otherwise – in the case of a logical zero – the sidebands are not present because the other wave is not present in this timeslot. If a filter separates only this new sidebands, a wavelength conversion occurs.

This basic idea was used by Rau et al. [148] for the simultaneous all-optical demultiplexing of a 40-Gbit/s optical time division multiplexed (OTDM) stream to four 10-Gbit/s wavelength division multiplexed (WDM) streams. Since every new WDM channel had another wavelength, this was a wavelength conversion as well. The basic set-up is shown in Fig. 12.6.

The input signal is an amplitude-modulated (return-to-zero) data stream with a bit rate of 40 Gbit/s. In order to demultiplex this OTDM stream into

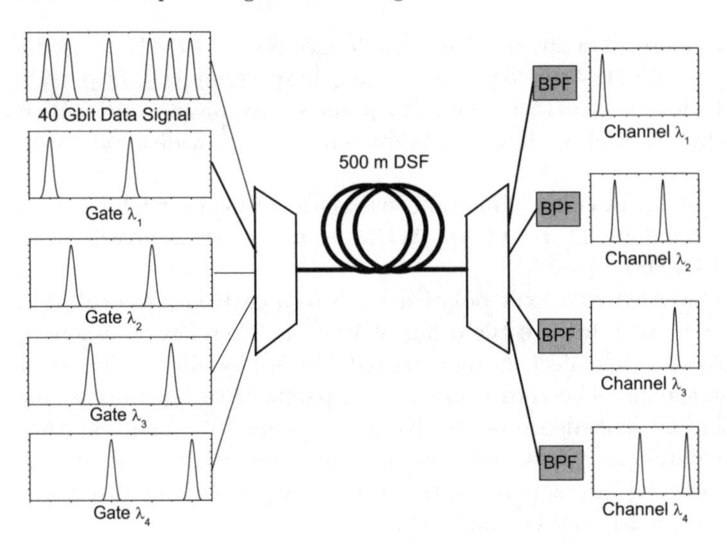

Fig. 12.6. Simultaneous demultiplexing and wavelength conversion of a 40-Gbit OTDM signal in a DSF due to the effect of XPM [148]. (DSF=dispersion-shifted fiber, BPF=bandpass filter)

4 WDM data streams with a bit rate of 10 Gbit/s each, four gate pulses – each of which has a different wavelength and time slot – copropagate with the OTDM signal in 500 m of dispersion-shifted fiber. Each gate stream overlaps with another 1/4 of the pulses in the original OTDM stream.

In the fiber, the pulses are spectrally-broadened through SPM. If the OTDM stream contains a logical one in the considered time slot, an additional broadening through the effect of XPM happens on both sides of the original spectrum of the gate pulse. This additional broadening is, therefore, dependent on the data in the OTDM stream. At the end of the fiber, one of the sidebands is filtered out with a narrowband bandpass filter. This sideband represents the new demultiplexed and wavelength-converted data stream.

The usable bandwidth of this device depends on the wavelength difference between the gate and signal waves and on the dispersion in the fiber. Contrary to FWM, XPM does not require the matching of the phases between the interacting waves, but – due to the effect of walk-off – the signal and gate pulses copropagate only for a certain distance in the waveguide. As shown in (6.31), the conversion efficiency is a function of the effective length of the waveguide and the peak power of the signal. For the set-up in [148], a peak power requirement for maximum efficiency of 3 W and a range of wavelength conversion of about 20 nm on either side of the zero dispersion wavelength was found. The authors pointed out that a shorter length of highly nonlinear fiber could be used to decrease the power and to scale the system to higher bit rates (potentially up to 160 Gbit/s) and a larger number of WDM channels.

12.2.3 Wavelength Conversion with SOA

The wavelength conversion with a semiconductor optical amplifier (SOA) has the advantage – contrary to FWM in optical fibers – that a conversion in a small compact device that can be monolythically integrated with more functionality is possible. Therefore, this approach has much more similarity with conventional electronics than nonlinear optics in fibers. On the other hand, the nonlinearity in SOAs is a resonant effect, so the switching speed is limited and the conversion is not transparent.

SOAs are, in principle, laserdiodes without a resonator in order to prevent lasing. The reflection of the facets is suppressed with an anti-reflection (AR) layer and by their angular alignment. Basically, two forms of SOA with different ARs exist. A resonant SOA has a medium AR and, therefore, a resonance at a distinct wavelength. The bandwidth is thus rather narrow (2–10 GHz). On the other hand, a travelling wave SOA (TW-SOA) has a low AR and hence a higher bandwidth of more than 40 nm. The whole device has a length of less than 1 mm. In order to provide polarization-independent operation, it has a near quadratic active region. Originally, SOAs were developed for linear amplification of optical signals, the maximum output power of an SOA is on the order of 10 dBm

Current electrons are created in the conduction band and holes in the valence band of the semiconductor via injection – for instance InGaAsP – resulting in a population inversion in the material. If an optical wave with a photon energy $\hbar\omega$, which is bigger than the band gap of the material but smaller than the difference between the quasi fermi energies of the conduction and valence band, propagates in the semiconductor, the wave will be amplified by stimulated emission. If the wave's photon energy is smaller than the band gap, then the material is transparent to the wave; if it is bigger than the energy difference between the two bands, then an absorption of the photons occurs.

The nonlinearity in an SOA is a resonant effect. Due to the fact that the band-structure of a semiconductor is rather complicated, transitions can occur between miscellaneous states and the nonlinearities in a semiconductor can have many different and partly complicated origins. They can, for instance, result from free carrier motions in the conduction band, from the inter- and intraband carrier relaxation, and from the carrier transition from the valence to the conduction band.

The rate equation for the charge density in an excited semiconductor is [149]:

$$\frac{\mathrm{d}N}{\mathrm{d}t} = \frac{I}{eV} - \frac{N}{\tau} - \frac{g\left(N,z\right)}{\hbar\omega} cn\varepsilon_0 \left\langle \left|E\left(t\right)\right|^2 \right\rangle, \tag{12.11}$$

The first term on the right side of (12.11) describes the increase in the charge density due to the electrical pumping of the device, where I is the current, e is the elementary charge, and V is the active volume. The second term is connected to the decrease in charge density due to a spontaneous recombination of the load carriers, where τ is the effective lifetime of the interband and the Auger recombination and it itself depends on N. The third term describes the stimulated emission, with c as the speed of light in vacuum, n as the refractive index of the medium, and ε_0 as the electrical field constant.

The last term in (12.11) depends on the intensity of the electric field of the wave, the brackets denote the averaging over the rapid fluctuations. The intensity is proportional to the amount of the electric field squared. Consequently, the carrier density and, therefore, the stimulated emission in the SOA depend nonlinearly on the electric field of the wave propagating in the device. This is the origin of the nonlinearity in SOAs. If, for instance, two waves with different frequencies, ω_1 and ω_2, propagate in the device, the charge density not only depends on the field of the two frequencies alone, as would be the result in linear case, it also depends on the field of the sum and difference frequencies and on their second harmonic. Therefore, the amplification behavior of the device is a nonlinear function of the electric field.

Due to the fact that nonlinearity in the SOA has a resonant origin, the propagation equation is a little more complicated. But the nonlinear polarization in the material can be described, like in an optical fiber, with an effective third order susceptibility. In contrast to the nonresonant nonlinearity in optical fibers, the resonant nonlinearity in a SOA is very strong, but it depends on the lifetime of the resonant effect, and is, therefore, not instantaneous.

Beside the already known effects of self-phase modulation, cross-phase modulation, and four-wave-mixing occurring from $\chi^{(3)}$ in a semiconductor amplifier, another effect called cross gain modulation can occur. All these effects will be described shortly in the following before their exploitation for wavelength conversion will be discussed.

The origin of the SPM in the SOA is, as in the case of optical fibers, the self-induced change of the refractive index by the intensity variation of the pulse and the associated modification of its phase and, therefore, its frequency. If the power of the pulses is rather low, they will be amplified by the device distortionless. But if the pulse power exceeds the saturation power of the amplifier, a distortion of the pulse will take place. Contrary to the instantaneous response of the nonresonant nonlinearity in fibers, the resonant nonlinearity in a SOA has time constants, for instance, the lifetime of the charge carriers (τ_{eff}). Therefore, the distortion of the pulse depends on the relationship between the pulse duration and τ_{eff}.

Four-wave-mixing takes place if waves with different frequencies propagating in the device are mixed over the third order susceptibility and produce a wave with a new frequency according to (7.1). Contrary to the case of optical fibers, the interaction length in an SOA is on the order of 1 mm, so phase matching between the waves plays no significant role.

Cross-phase modulation and cross-gain modulation are the result of the mutual interaction between different waves which propagate at the same time in the device. Which of the effects will occur depends on the energy and, therefore, the frequency of the waves. Wave number one (pump) lies in both cases within the gain spectrum of the SOA and its intensity may be much higher than that of wave number 2 (signal). If the energy of the signal is smaller than the band gap of the semiconductor, the material is transparent for the wave. Therefore, only the refractive index that the wave experiences can be significant. The refractive index is changed by the pump. This refractive index modification results in a phase and an associated frequency change of the signal. So the pump alters – via the refractive index – the phase of the signal, this is the effect of cross-phase modulation.

If the signal is within the gain spectrum of the device as well, then both waves have an influence on the carrier density in (12.11) and, therefore, the gain of the amplifier. If the intensity of the pump is high enough, then the carrier concentration in the active region of the device is depleted. As a result, the gain that the signal experiences is reduced by the pump. This mutual influence on the gain is called cross-gain modulation.

Figure 12.7 shows two possible set-ups for a wavelength conversion with semiconductor amplifiers [150]. Figure 12.7a shows a wavelength converter exploiting the effect of cross-gain modulation. An intensity-modulated data signal together with a CW wave of the desired frequency is coupled into the SOA. The ones of the data signal drive the amplifier into saturation, so the bit sequence determines the amplification of the CW wave via cross-gain modulation. If the data signal has a logical one, the gain of the CW wave with a different frequency is small; otherwise it is large. Therefore, the CW wave shows the inverted replica of the bit sequence. A filter behind the device separates the converted wave from the original one and from the additional spontaneous emission of the amplifier.

If both waves counterpropagate in the device, the filter at the end can be saved, but with this set-up, additional speed limitations will occur because the interaction length between the pulses is decreased.

The advantage of wave conversion with the XGM in SOAs is its simplicity, but due to the fact that the effect exploits a resonant nonlinearity, the reaction is not instantaneous. It especially suffers from the relatively slow recovery time of the carriers. Nevertheless, data rates of 40 Gbit/s [151] and – with some degradation – up to 100 Gbit/s [152] were possible with the XGM set-up. The wavelength conversion is not modulation-transparent. Therefore, only intensity-modulated signals can be converted, additionally, the signals will experience a chirp due to the process. The chirp in the device results in a red-shift of the leading edge of the inverted pulse and a blue-shift on its trailing wing. The chirp is quite large since the conversion relies on a maximized carrier density modulation. With XGM only up to 5 converters

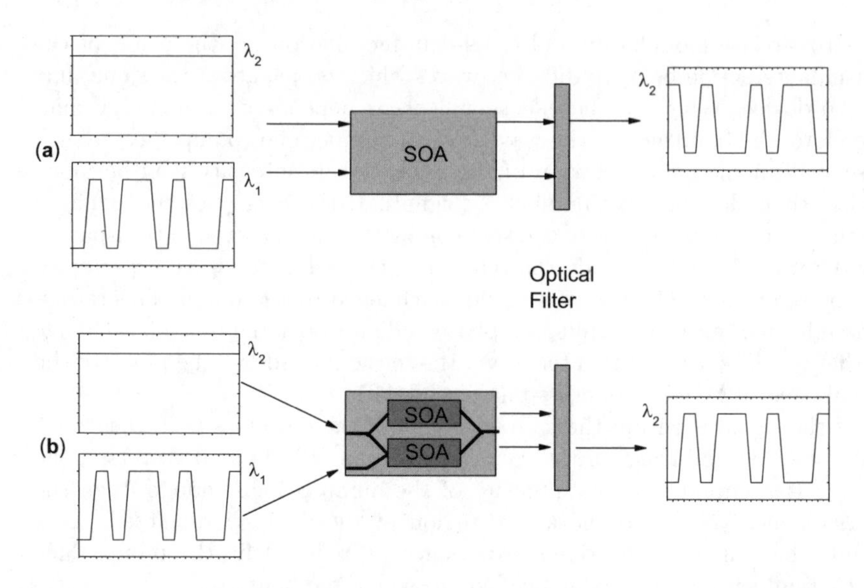

Fig. 12.7. Fundamental set-up of a wavelength conversion with semiconductor optical amplifiers (**a** Cross-gain modulation. **b** Cross-phase modulation) [150]

are cascadable. This is caused by the fact that XGM wavelength converters possess very limited regenerative capabilities.

Figure 12.7b shows a wavelength converter based on the effect of XPM on SOAs. Both SOAs are placed in the arms of an interferometer. The CW wave with the desired wavelength is injected into the interferometer and propagates through both SOAs. The data signal is injected into the lower SOA additionally. The data signal changes the refractive index in the SOA as a function of the bit sequence. The modulation of the refractive index alters the phase of the CW wave propagating in the lower branch of the interferometer via XPM. If the phase change is π, the CW waves in the two branches interfere destructively at the end of the interferometer, resulting in an intensity modulation of the wave. Here again, a filter behind the device separates the desired wavelength from the original and from ASE.

The duration of the trailing edges of the converted pulses depends on the effective carrier lifetime of the lower SOA. Therefore, for high bit rates, the trailing edges can overlap with the successive time slots which leads to an intersymbol interference between adjacent bits. This speed limitation can be avoided by a method called differential mode operation [153]. Here, a delayed and attenuated copy of the data signal is launched into an additional arm of the upper SOA (not shown in the figure). The copy changes the refractive index of the SOA in the upper branch and, therefore, the phase of the wave as well. Hence, the delay between the data copies in both arms defines a switching window for the interferometer.

In the XPM configuration, data rates of 100 Gbit/s with a much better performance than with the XGM configuration are achieved [154]. Variations of the differential interferometer scheme are possible as well. One approach uses only a single SOA followed by a passive interferometer and is referred as "delayed interference signal converter" (DISC). With these devices, wavelength conversion up to 168 Gbit/s [155] is possible in a hybrid and 100 Gbit/s in a fully integrated device [156].

The effect of FWM in SOAs can be exploited for frequency conversion as well. Contrary to the FWM in optical fibers, the phase matching of the waves plays no significant role, due to the short length of interaction.

12.3 Optical Switching

The cross connects in the optical backbone have to switch the signals from fibers with hundred and more wavelengths and data rates of 10 Gbit/s each. In the near future, this bit rate will be increased beyond 40 Gbit/s. The switches have to be equipped, therefore, with hundreds or even thousands of ports. Until now, electro-optical switching is used in this area where the optical signals are translated with a fast photodiode into the electrical domain, then the signals are electrically switched and translated back to the optical domain with laser diodes (optical-electronic-optical or oeo). But electro-optical switching will arrive at its limits.

First, the complexity of the switching fabrics increases dramatically with the number of channels and the bit rate. This is because every channel from an incoming fiber has to be translated even though they possibly only have to be connected to an outgoing fiber. Second, the switching speed of electrical signal processing is limited due to the limited carrier mobility in semiconductor devices. Even the fastest laboratory electronic switch operates at a speed of only 176 GHz [157].

Optical switching has, instead, a number of advantages over electrical switching [160, 187]. Due to the fact that the complexity of electrical switches will strongly increase with increasing bit rate and number of channels, the costs for electro-optical switches will dramatically increase as well. Switching in the optical domain, on the other hand, will save costs, power and required space. The new configuration of the switch in the case of a failure could be executed very rapidly. Optical switches can switch entire fibers, single wavelength, bands of wavelengths, or even packets. This granularity would be impossible with electro-optic switches.

Many different effects have been proposed for optical switching. Over the last few years, silicon-based micro-electromechanical systems (MEMS) have become very popular. MEMS are micro-mechanical structures fabricated on a silicon substrate with standard semiconductor processing techniques. Therefore, mass production with fairly low costs for each device should be possible.

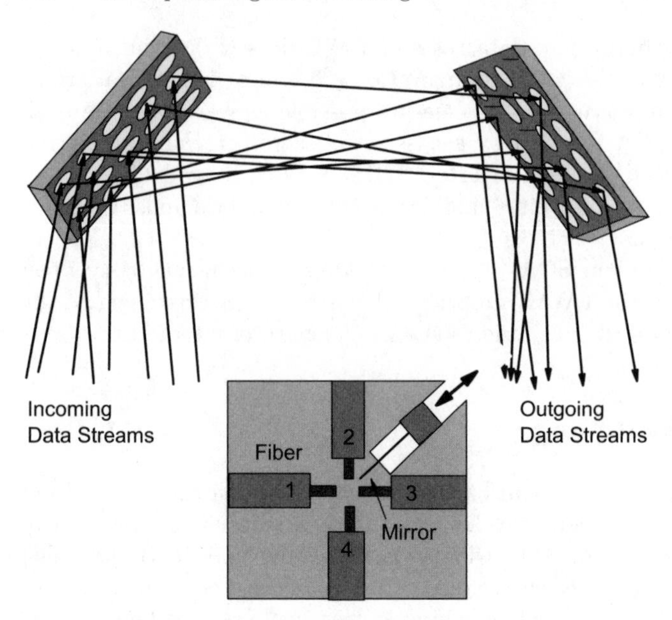

Fig. 12.8. Schematic illustration of a large scale switch with MEMS technology. The lenses for the mapping on the mirrors and the incoming and outgoing fiber bundles are not drawn. A small scale switch is shown on the bottom [160]

The structures on the silicon substrate range from a few hundred microns to millimeters. For the fabrication of the structures, two distinct methods are possible [160]. In bulk micromachining, the components of the device were crafted out of the bulk of the silicon wafer. In the surface micromachining technique, additional layers are added to the surface of the substrate which are then partially freed. An anchor attaches the free top layers to the surface underneath it. In most cases, the structures worked out from the substrate are flat plates used as mirrors. In order to increase the reflectivity of the plates, they were coated with a thin layer of gold. The application of a voltage to the substrate actuates the mirrors, steering them mechanically. If the mirrors are in the light path, they can reflect the beam, or the channels, from an incoming fiber to a defined outgoing fiber.

With this approach, different optical switches from small scale (2×2) up to large scale (a few thousand) are possible. Figure 12.8 shows a schematic illustration for a large and a small scale switch [160].

The data signals from the incoming fiber bundles on top of Fig. 12.8 are collimated with an array of lenses (not shown in the figure) onto the first MEMS in free space. The mirrors can be steered in an upward and downward direction as well from left to right, so every mirror on the left MEMS can switch the beam onto every mirror on the right MEMS. Then the light is again collimated with a lens array and injected into the outgoing fiber bundle. Every incoming and every outgoing fiber has its own corresponding mirror, but due to the fact that the angles of the mirrors can be adjusted independently in both directions, the data signal of every incoming fiber can be switched to every outgoing fiber. The size of the switch scales with the number of fibers, for N fibers an N $\times$ N switch needs 2N mirrors accordingly.

At the bottom of Fig. 12.8, a small scale 2 $\times$ 2 switch is shown [161]. This set-up does not need any lens so the fibers must be as close as possible. In the normal case, the data signals propagate in free space between the fibers 1 - 3 and 4 - 2. If the mirror is slid between the fibers, the propagation path will be switched to 1 - 2 and 4 - 3.

An advantage of the MEMS technology is that the carrier material for the switches is silicon. Therefore, it will be easy to integrate additional functionality into only one device. A disadvantage is the switching time, which is on the order of a millisecond. With this technology, transparent switching of high speed data signals from one fiber to another on many different wavelengths is possible in the case of a failure, for instance, when the network must change its configuration in order to bypass an interrupted link. But MEMS are much too slow to switch individual packets which would be essential for IP routers.

A faster technology based on nonlinear optics is the use of photorefractive switches.

12.3.1 Photorefractive Optical Switches

Due to the linearity of the wave equation (2.13) the switching of light by light is only possible in a material that reacts nonlinearly to the applied optical field. In such a material, it is possible to exploit different physical effects for the switching process. One of these is, for instance, the field dependence of the refractive index due to the so called photorefractive effect which is briefly treated here.

Let us assume two plane monochromatic pump waves are propagating in a crystal with a small angle 2θ between each other, as shown in Fig. 12.9. Both waves will interfere in the material, forming an intensity modulation of the electric field. In Fig. 12.9, the configuration of the wave vectors of the two waves is depicted as well.

The interference of the two waves with the wave vectors $\boldsymbol{k}_0$ and $\boldsymbol{k}_1$ creates an intensity modulation perpendicular to their propagation direction with the grating vector $\boldsymbol{q}$. According to the wave-vector diagram in Fig. 12.9, the vectors fulfill the relation

$$\boldsymbol{k}_1 = \boldsymbol{k}_0 + \boldsymbol{q} \tag{12.12}$$

or, under consideration of the angle,

$$|\boldsymbol{q}| = 2\,|\boldsymbol{k}_0|\sin\theta. \tag{12.13}$$

The absolute value of the wave vectors of the incident waves in the material is:

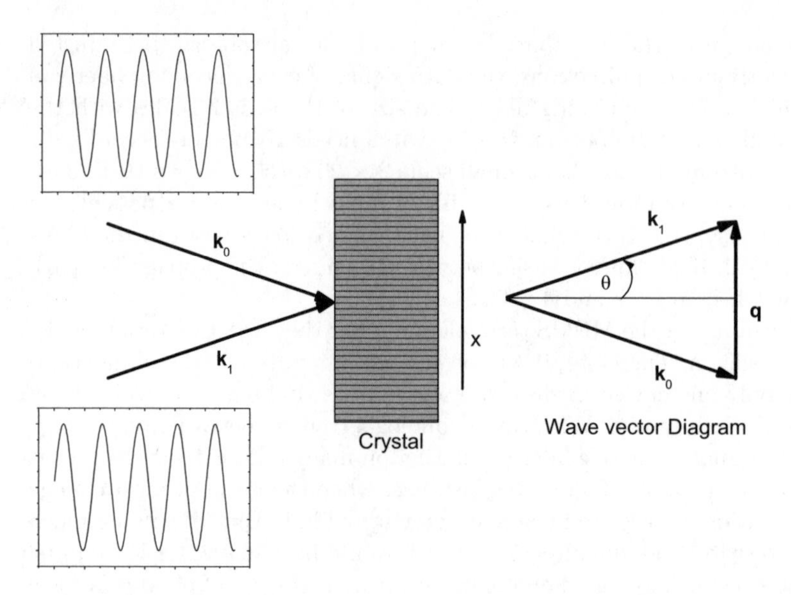

Fig. 12.9. Interference of two monochromatic plane waves with corresponding wave vector diagram

$$|\boldsymbol{k}_0| = |\boldsymbol{k}_1| = |\boldsymbol{k}| = \frac{2\pi n}{\lambda}, \tag{12.14}$$

with λ as the wavelength of the monochromatic waves writing the interference. The absolute value of the grating wave vector is $|\boldsymbol{q}| = 2\pi n/g$, with g as the wavelength of the created perpendicular interference pattern. With the equations (12.13) and (12.14), this wavelength is

$$g = \frac{\lambda}{2\sin\theta}, \tag{12.15}$$

where g is called the grating constant. If we determine the direction of the gratings wave vector as the x-axes, as shown in Fig. 12.9, the intensity interference in the crystal can be written as

$$I = I_0 \cos(qx) = I_0 \cos(2k \sin\theta x) . \tag{12.16}$$

Until now the interference pattern is only an intensity modulation, but if the intensity modulation can be translated into a refractive index modulation of the material, a refractive index transmission grating will be formed. At this grating a third beam, for instance, the data signal from an incoming optical fiber, can be diffracted. The diffraction of the third beam is shown in Fig. 12.10.

Due to the fact that the grating contrast[3] is in most cases rather slight the portions of the beam that are scattered at different maxima of the grating

[3]The grating contrast is the refractive index difference between maximum and minimum.

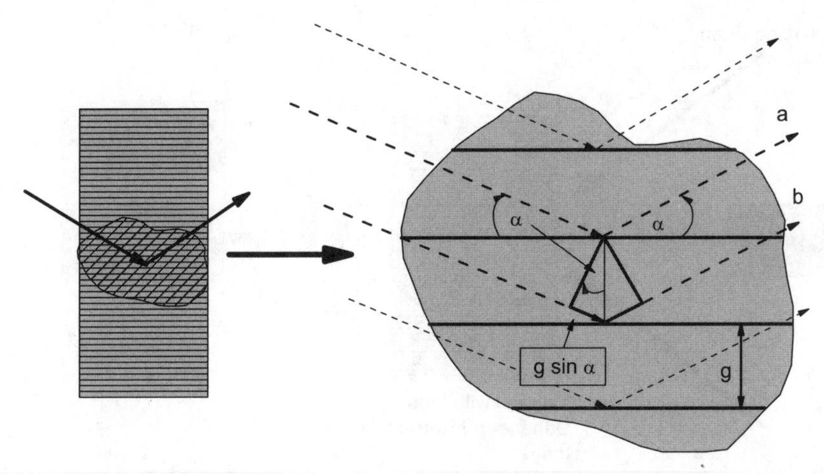

Fig. 12.10. Diffraction of an incident beam on a refractive index grating for the derivation of the Bragg condition

are relatively small. Therefore, for an intense diffracted beam, the portions from successive maxima have to interfere constructively with each other. This condition depends on the wavelength of the incident beam λ and is only fulfilled in designated directions. As a result, the difference in the propagation distance between part "a" and "b" of the diffracted beam has to be an integer m multiplied by the wavelength λ. With Fig. 12.10 this leads to:

$$m\lambda = 2g \sin \alpha \tag{12.17}$$

Equation (12.17) is known as the Bragg condition which is used to describe the diffraction of x-rays in the layers of an atom. In acousto-optic switches and modulators, it describes the scattering of light at sound waves. If the thickness of the device is large, only one diffracted beam is possible and therefore $m = 1$. The angle of the diffracted beam is then

$$\alpha = \arcsin\left(\frac{\lambda}{2g}\right) . \tag{12.18}$$

If the translation is fast – this means (1) there is no delay between the interference pattern and the creation of the grating and (2) there is a fast decay time of the grating when the writing beams are switched off – then the data signals can be switched with the grating, as shown in Fig. 12.11.

For a switching process like that in Fig. 12.11, a material with a light dependent refractive index is required. This condition is, for instance, fulfilled in photorefractive materials. The change of the refractive index in photorefractive materials is based on the shifting of charge carriers in the bulk of the crystal.

Suppose there are two interfering laser beams like those in Fig. 12.10. Due to photoionization, free charge carriers will be created in the intense, bright

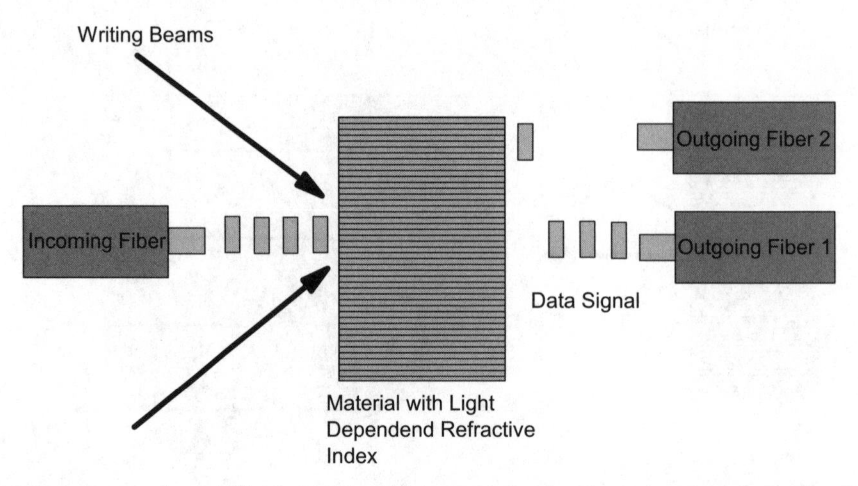

Fig. 12.11. A possible set-up for the switching of a data signal with a refractive index grating. The grating is written with two interfering laser pulses and is only present in the material for a short time. At this time, one channel is switched to the outgoing fiber 2, out of this time the data signal is transmitted to the outgoing fiber 1

parts of the interference pattern of the two beams. The electrons can migrate into the darker parts of the interference, and there it is possible for them to recombine. This spatial charge carrier difference results in a spatially varying charge distribution (ρ). If the intensity modulation in the crystal and the charge carrier distribution are in phase, the distribution follows the intensity modulation. According to (12.16), this leads to

$$\rho = \rho_0 \cos(qx) = \rho_0 \cos(2k \sin\theta \; x), \tag{12.19}$$

where ρ_0 is a constant. The charge carrier distribution is responsible for an internal electric field in the crystal. The Maxwell equation for this internal electric field (2.3) lead to

$$\mathrm{div}\boldsymbol{E} = \frac{\rho}{\varepsilon}. \tag{12.20}$$

If only the distribution in one spatial direction is considered we can deduce

$$\frac{\partial E}{\partial x} = \frac{\rho(x)}{\varepsilon}, \tag{12.21}$$

where ε is an effective dielectric constant that is relevant for the interaction. In anisotropic crystals, it depends on the relative orientation between the electric field and the crystal. The charge carrier distribution is given in (12.19). Therefore, the electric field in the material is

$$E = \frac{\rho_0}{\varepsilon q} \sin(qx) = \frac{\rho_0}{2\varepsilon k \sin\theta} \sin(2k \sin\theta \; x). \tag{12.22}$$

The local derivation in (12.21) leads to a spatial phase shift of 90° between the space load and the electric field. At the positions in the material, where the space load density has its maximum, the electric field is zero. Conversely, the electric field has its maximum at the positions where the space load density is zero.

The static field distribution can be translated into a refractive index modulation of the material via a second order nonlinear effect, the so called Pockels effect. Due to the symmetry condition, mentioned in Sect. 4.3, a second order nonlinear effect can only take place in crystals without an inversion center. Another requirement to the material is the ability to provide good photoionization at rather low intensities of the field. Therefore, special materials like the ferroelectric oxides barium titanate, lithium niobate, and lithium tantalate are required for good photorefractive performance.

Although a second order nonlinear effect is responsible for the translation of the spatial electric field modulation into a refractive index modulation, the effect is called a linear electro-optic effect. This linearity is related to the dependence of the refractive index on the static electric field. As shown in Sect. 4.7, the origin of an alteration of this refractive index lies in the nonlinear polarization of the crystal. For a second order nonlinear effect this has to be:

$$P_{\mathrm{NL}}(\omega) = \chi^{(2)} E(\omega) E_{\mathrm{dc}}(0) \tag{12.23}$$

where $E(\omega)$ is the electric field of the writing beams and $E_{\mathrm{dc}}(0)$ is the static electric field in the material due to the charge separation. According to (12.23), the change of the refractive index is in fact linearly dependent on the field.

Contrary to the linear electro-optic effect, the Kerr effect (Sect. 4.7) is called a nonlinear electro-optic effect since its refractive index alteration depends quadratically on an electric field:

$$P_{\mathrm{NL}}(\omega) = \chi^{(3)} E(\omega) E_{\mathrm{dc}}(0) E_{\mathrm{dc}}(0) . \tag{12.24}$$

Both effects could be steered by an external electric field applied to the crystal. This is, for example, the basis of a Pockels cell used to switch the Q-factor of a laser resonator.[4]

The depth of the index modulation due to the Pockels effect depends on the electro-optic coefficient and can be written as [162]

---

[4]In a laser resonator, the beams are highly polarized due to the brewster angle of the crystal. The Pockels-cell changes the polarization of the laser beam in the resonator. In the normal case, a polarizer switches the light out of the resonator. In the laser crystal, a high inversion will be built due to the pumping.

If the pockels cell is switched on, it changes the polarization of the beam to 90°. The light will stay in the resonator, building up a standing wave which can lead to a sudden decay of the inversion in the crystal, resulting in a short pulse with high intensity.

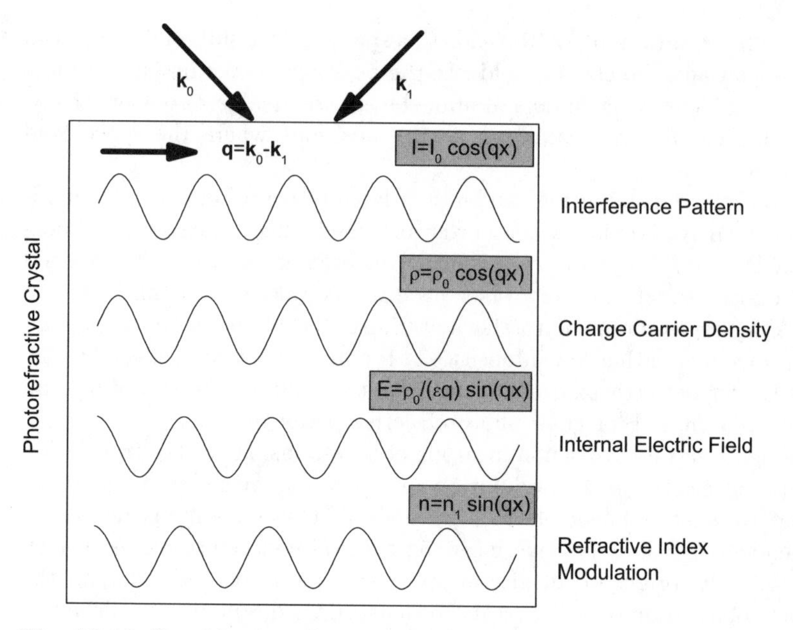

Fig. 12.12. Resulting interference pattern, charge carrier density, internal electric field, and refractive index in a photorefractive material if the crystal is excited by two interfering laser pulses. Note the phase shift between the charge carrier density and the internal electric field

$$n_1 = \frac{1}{2} n^3 r \frac{\rho_0}{\varepsilon q}, \tag{12.25}$$

where r is the effective electro-optic coefficient and n is the relevant refractive index. The index grating written by the Pockels effect is:

$$n = n_1 \sin(qx) = n_1 \sin(2k \sin\theta\ x) . \tag{12.26}$$

Figure 12.12 shows schematically the described distributions in a photorefractive material.

In order to provide packet switching as shown in Fig. 12.11, a very fast change of the refractive index is required. The grating has to be built up in a very short time and it must decay rather instantaneously. But the switching speed of the photorefractive effect is limited by the mobility of the charge carriers in the crystal. The charge carriers must be produced by the photoionization process and then they have to diffuse into the darker regions of the intensity modulation. Following an estimation by Yeh [162], it is possible to calculate the required intensity and, therefore, the time to form a volume index hologram in a photorefractive crystal.

If a pair of charge carriers should be formed due to photoionization, this requires the absorption of at least one photon. For an efficient beam coupling in a barium titanate crystal, a charge carrier density of $\approx 10^{16}$ cm^{-3} is essential. Therefore, in a volume of 1 cm^3 in the crystal, 10^{16} photons have to be absorbed. If a wave in the infrared with a power of 1 W is uniformly illuminating the considered volume, this corresponds to a photon flux of $\approx 10^{19}$/s. Therefore, with the assumption of a photoexcitation absorption of 10%, a period of at least 10 ms is required to create the charge separation. If the intensity is increased, the photon flux is increased as well and the time will be decreased.

If a higher switching speed is required, the intensity has to be increased but this is limited by the demolition threshold of the material. Another possibility is to switch to a nonresonant, nonlinear mechanism where the intensity modulation is translated instantaneously into a refractive index modulation. Such ultrafast optical switches are described in Sect. 12.3.4.

12.3.2 Optical Switching with the XPM Effect

As mentioned in Sect. 6.2, the effect of cross-phase modulation (XPM) will lead to an alteration in the phase of one optical pulse due to the intensity of another pulse propagating at the same time in the nonlinear material. If this phase change can be translated into an amplitude variance, it is possible to switch one pulse by the other. In XPM, one pulse changes the refractive index the other pulse will experience. The pulses can differ in their polarization, direction, or wavelength.

For a high switching speed, nonresonant nonlinearities are required but they are very weak, so a rather large interaction length – provided by optical waveguides only – will be necessary. One possibility of exploiting the XPM effect in fibers for optical switching is the Kerr shutter method. The basic set-up of a Kerr shutter is shown in Fig. 12.13.

A Kerr shutter consists of an optical fiber and a polarizer. The signal and gate pulses are both linearly polarized at the fiber's entry point, their polarizations have an angle of 45° with respect to each other. The polarizer at the fiber output is adjusted perpendicular to the signal pulse. Therefore, if no gate pulse is present, no light can pass through it, as shown in the upper part of Fig. 12.13. On the other hand, if a gate pulse is present in the fiber at the same time, it generates – due to its intensity – a higher refractive index in the direction of its polarization. This is the so called self-induced birefringence [132].

As mentioned in Sect. 4.10, a birefringent crystal consists of different refractive indices in two fundamental axes. Along the slower axis, the refractive index is higher and, therefore, the phase velocity of the wave is lower. Along the faster axis the behavior is converted. The electric field of the signal pulse consists of a part parallel to the fast axis and one parallel to the slow axis. Due to the different phase velocities of the components in the fiber, as a result of the birefringence induced by the gate pulse, the polarization state of the signal will be turned. For this turned fraction, it is now possible to pass the polarizer. The signal has been switched on. Since the fraction of the

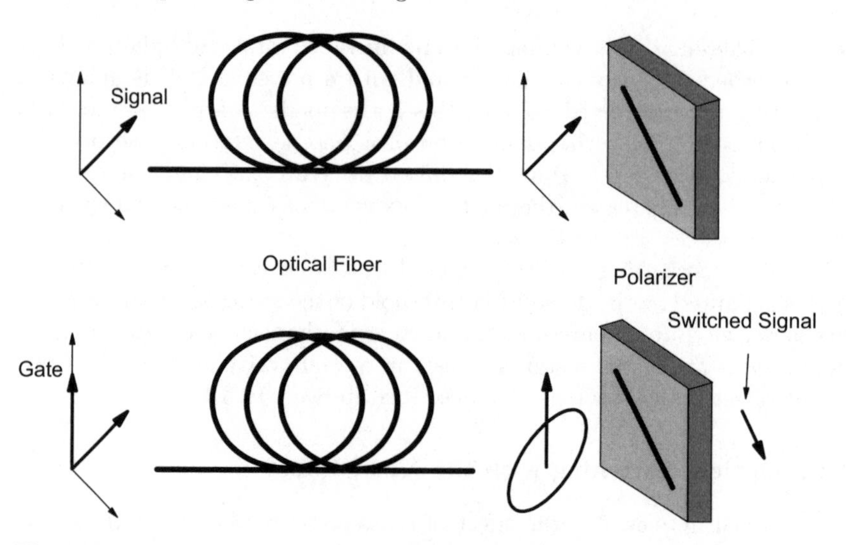

Fig. 12.13. Optical switching with XPM in an optical fiber [51]

signal pulse that can pass through the polarizer depends on the intensity of the gate pulse, the signal can be modulated. This set-up is also known as a Kerr modulator.

The signal and gate pulse are different in their wavelengths. Contrary to FWM, no phase matching condition has to be fulfilled, but if the wavelengths differ too much, the interaction length in a dispersive medium is limited. The switching speed of this method depends on the duration of the pulses and it is possible to switch and manipulate optical signals in the ps domain. For a high switching contrast (on-off ratio), rather long fibers are needed to increase the interaction length between the signal and gate pulses. Therefore, the duration of the pulses will be increased due to the dispersion in the fiber, and the signal and gate pulses will walk-off. Hence, switching contrast and speed are limited.

Using a short interaction length together with very high intensities can increase the switching speed. Figure 12.14 shows the experimental result of such an approach [163].

A thin barium fluoride (BaF_2) (111) crystal in air at room temperature was used as the nonlinear material. The output from an amplified titanium sapphire laser (wavelength ≈810 nm, pulse duration <100 fs, repetition rate = 1 kHz) was split into two beams. One was used as the signal, the other one as the gate.

The signal beam was s-polarized in front of the sample crystal. Both beams were focused non-collinearly (angle between the two beams 1.2°) onto the sample with a spot area ($1/e^2$ intensity) of 0.1 mm^2 and a confocal parameter larger than the sample thickness. The energy of both beams was ≈150 μJ. We made sure that the overall intensity was always below the

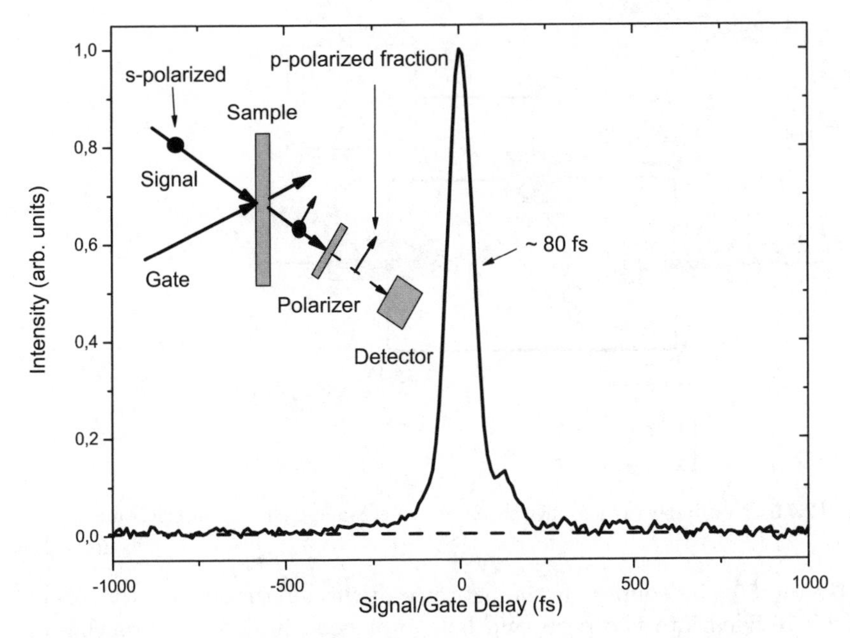

Fig. 12.14. P-polarized fraction of an s-polarized fs pulse. The inset shows the experimental set-up

threshold for white light continuum generation, which we determined to be $\approx 10^{12}$ W/cm^2.

Behind the sample, a polarizer was adjusted in a way that only p-polarized light could pass through. If only the signal or the gate pulse was present in the sample[5], no light could pass through the polarizer because the pulses were s-polarized. Only when both pulses arrive at the same time was birefringence induced in the crystal and hence a p-polarized fraction of the signal was seen by the detector. This behavior took place for a very short time. As shown in the figure, the response of the material was instantaneous. Therefore, with this set-up, it was possible to switch in the fs time domain.

In the two set-ups of optical switches based on the XPM between the data signal and a control pulse, the phase change due to the interaction leads to a change in the polarization of the data signal. This alteration in polarization is translated into a change of the amplitude via a polarizer. Interferometers have, instead, the possibility to translate the phase alteration directly into an amplitude change. Therefore nonlinear optical switches based on interferometers are very popular and versatile. The basic idea of an interferometric based switch is shown with a Mach-Zehnder interferometer in Fig. 12.15.

[5] A negative delay means that the signal pulse arrives a short time before the gate pulse at the sample, whereas for a positive delay the gate will arrive before the signal.

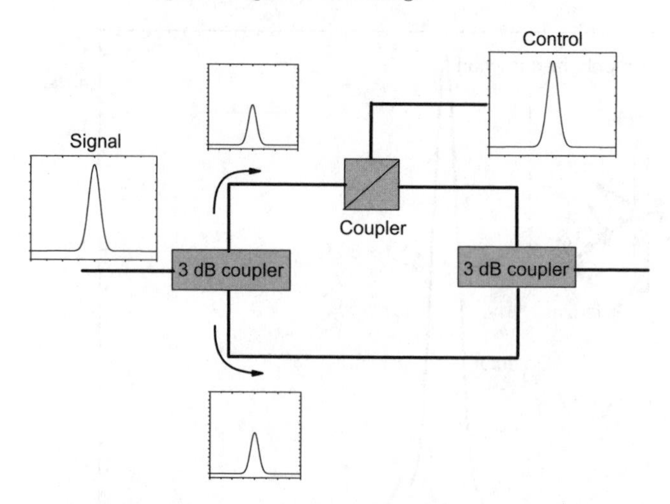

Fig. 12.15. Nonlinear optical switch based on a Mach-Zehnder interferometer

By the 50 : 50 coupler at the entrance of the interferometer, the signal pulse is divided into two parts and both propagate in different branches of the interferometer. Assuming there is no control pulse present, both branches will have the same properties and the two partial pulses will interfere constructively in the 3 dB coupler at the end of the interferometer. The SPM affects the optical pulses in both branches in the same way so, the phase distortions due to nonlinear effects are the same and will, therefore, have no influence on the interference.

An additional control pulse can be injected via a coupler into one branch of the device. This control pulse will be able to change the refractive index in the material, $n = n_0 + n_2 I$, via the third order susceptibility $\chi^{(3)}$. The second branch can consist of an ordinary optical fiber, or an SOA to decrease the volume of the device [164], or a highly nonlinear fiber. The refractive index alteration changes the phase of the signal pulse and a cross-phase modulation between control and signal pulse will take place. The output intensity of the interferometer is related to the input intensity via [165]

$$I_{\mathrm{out}} = I_{\mathrm{in}} \cos\left(\frac{\varphi_{\mathrm{L}} + \Delta\varphi_{\mathrm{NL}}}{2}\right), \tag{12.27}$$

where φ_{L} is the initial phase difference between both pulses and $\Delta\varphi_{\mathrm{NL}}$ is the additional phase difference due to the control pulse. If the whole phase difference is π, according to (12.27), the light in the output coupler will interfere destructively. The signal is then switched off. Assuming $\varphi_{\mathrm{L}} = 0$, the phase shift in the upper branch has to be π. According to (6.30), the phase shift due to XPM is

$$\Delta\varphi_{\mathrm{NL}} = \pi = 2\frac{2\pi}{\lambda} n_2 I L_{\mathrm{eff}}. \tag{12.28}$$

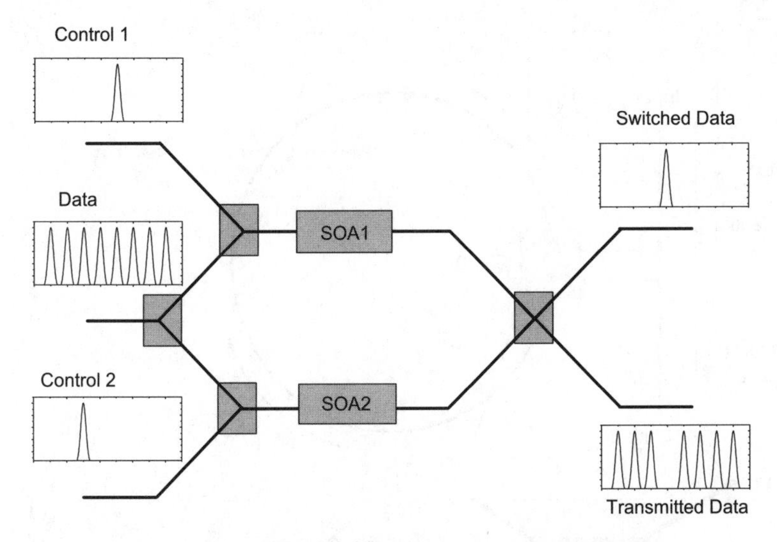

Fig. 12.16. Mach-Zehnder interferometer with two SOAs as optical elements for all-optical switching [149]

If an ordinary single mode fiber is used, the nonlinear refractive index n_2 is 2.6×10^{-20} m^2/W. A phase shift of π requires therefore an intensity of 1.5×10^{10} W/m^2 in a fiber of 1 km effective length and a wavelength of 1.55 μm. With an effective fiber area of 80 μm^2, this corresponds to an input power of 1.2 W for the control pulse.

If a high-bit-rate data signal is introduced at the signal input of the interferometer and a clock sequence with a lower bit rate is used at the control input, the set-up in Fig. 12.15 can be used as a demultiplexer.

If the optical fiber is replaced by a SOA, the required intensities are much lower because resonant, nonlinear effects are exploited. But, on the other hand, an SOA suffers from time constants related to the resonances in the semiconductor. The relaxation or recovery time, after a control pulse has changed the refractive index of a SOA, is rather long. In order to circumvent this drawback – which considerably limits the usable bit rate – it is possible to use two SOAs in a Mach-Zehnder arrangement with one in each arm of the interferometer. The basic set-up is shown in Fig. 12.16.

The control bit sequence is divided into two parts, each of which is introduced into one arm of the interferometer. If no control pulse is present in the device, both arms of the interferometer will behave in the same manner. Therefore, the parts of the signal pulse can interfere constructively. The first control pulse introduces, due to XPM, a phase shift of π for the parts of the pulses propagating in the lower branch. The second control pulse, introduced with a delay in the other arm of the interferometer, switches this arm as well in a manner that a phase shift of π will occur. Therefore, if both arms

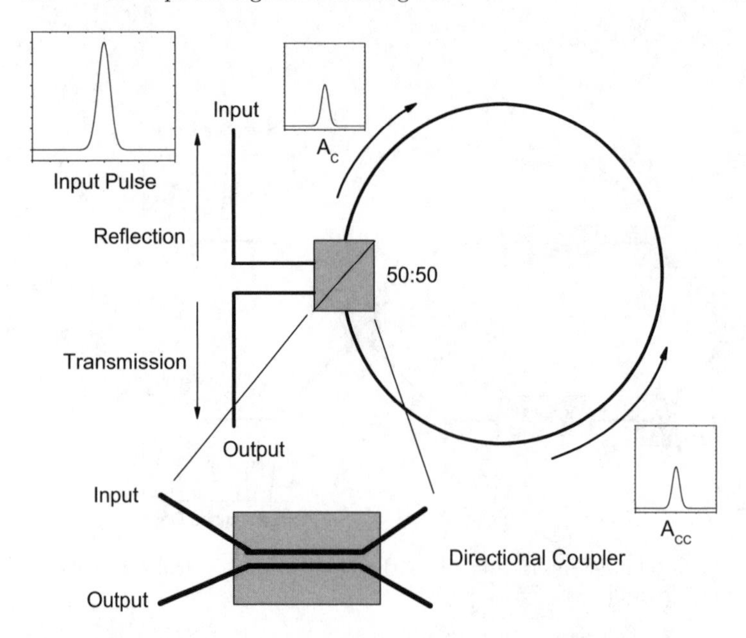

Fig. 12.17. Sagnac interferometer with an optical fiber loop

are switched, the pulses will interfere again constructively at the end of the interferometer.

The data stream is switched off only in the time interval between the two control pulses. If the last coupler introduces a constant phase shift of π for one part of the signals in its upper output, then the inverse of its lower output – or in other words the switched data pulse – can be seen.

Another often used set-up for optical switches is the Sagnac interferometer shown in Fig. 12.17. The pulse from the input is divided in the directional coupler into two beams counter-propagating in the fiber loop. The coupler divides the power of the input pulse equally, therefore the amplitude of the clockwise propagating part is $A_c = \sqrt{0.5}A_0$ where A_0 is the amplitude of the input pulse with power $P = A_0^2$.

The directional coupler consists of two fiber cores that are very tightly arranged. Therefore, the light field in one core will be injected into the other as well. The waves in the two cores have different wavenumbers and, therefore, a phase shift between them is introduced. If the wave changes the fiber core in the coupler the phase shift will be $\pi/2$. The counter-clockwise propagating part of the pulse changes the fiber in the directional coupler (see Fig. 12.17) and therefore its amplitude can be written as $A_{cc} = \mathrm{j}\sqrt{0.5}A_0$.

After one round trip, both pulses arrive at the same time at the coupler. According to their relative phase shift, they can interfere constructively or destructively in the device. The part that changes the fiber core in the coupler will again experience a phase shift of $\pi/2$. The amplitude A_r at the fiber input

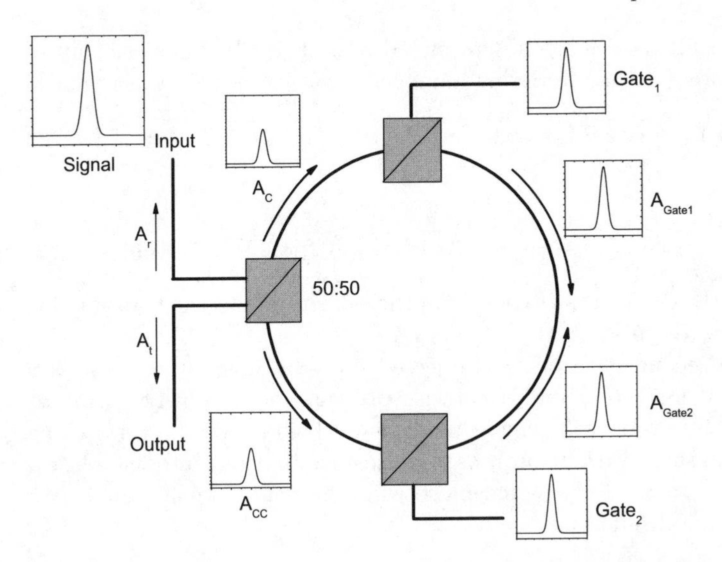

Fig. 12.18. Sagnac interferometer as an optical switch and XOR gate

is therefore

$$A_{\mathrm{r}} = \sqrt{0.5}A_{\mathrm{cc}} + j\sqrt{0.5}A_{\mathrm{c}} = \sqrt{0.5}\mathrm{j}\sqrt{0.5}A_0 + \mathrm{j}\sqrt{0.5}\sqrt{0.5}A_0 = \mathrm{j}A_0, \quad (12.29)$$

whereas for the transmitted amplitude at the fiber output, one yields

$$A_{\mathrm{t}} = \sqrt{0.5}A_{\mathrm{c}} + j\sqrt{0.5}A_{\mathrm{cc}} = \sqrt{0.5}\sqrt{0.5}A_0 + \mathrm{j}\sqrt{0.5}\mathrm{j}\sqrt{0.5}A_0 = 0. \quad (12.30)$$

The part of the signal that appears at the input connection will be reflected, whereas the part that appears at the output connection of the coupler will be transmitted. The Sagnac interferometer serves as an ideal mirror with a 100 % reflection.[6] Due to its properties, this set-up is also known as a nonlinear optical loop mirror (NOLM) [166, 167, 168].

Figure 12.18 shows a Sagnac interferometer with two additional couplers in the loop. First, only the coupler in the upper part is considered. With this device, it is possible to inject a gate pulse into the loop. This gate pulse propagates at the same time in the same direction as the clockwise propagating pulse A_{c}. Due to the XPM in the fiber, the pulse with amplitude A_{c} experiences a phase shift while the counter-clockwise propagating pulse remains unaffected.

[6]If the fiber loop is turned, then the part of the pulse that propagates in the same direction will arrive earlier at the coupler, whereas the part in the opposite direction will arrive a little later. Therefore the phase difference, or the interference fringes, can be used as a measure of the velocity of the rotation. Sagnac interferometers are used in navigation systems for airliners, ships, spacecraft, and in many other applications. The best devices are capable of detecting rotation rates as small as a fraction of a degree per hour.

The phase shift here is again determined by (12.28). If the phase shift of A_c is π, it follows for the interfering amplitudes at the input coupler that

$$A_r = \sqrt{0.5}A_{cc} + j\sqrt{0.5}A_c = \sqrt{0.5}j\sqrt{0.5}A_0 - j\sqrt{0.5}\sqrt{0.5}A_0 = 0 \qquad (12.31)$$

and

$$A_t = \sqrt{0.5}A_c + j\sqrt{0.5}A_{cc} = -\sqrt{0.5}\sqrt{0.5}A_0 + j\sqrt{0.5}j\sqrt{0.5}A_0 = -A_0. \qquad (12.32)$$

Therefore, the device is switched from the reflecting into the transmitting state with the gate pulse.

If only the second gate pulse is injected into the fiber, it introduces a phase shift of π for the counter-clockwise propagating amplitude, whereas the clockwise amplitude will remain unaffected. This results in a switching of the signal pulse as well. If both gate pulses are injected into the optical fiber, then both parts of the signal pulse will experience a phase shift of π the interfering amplitudes are

$$A_r = \sqrt{0.5}A_{cc} + j\sqrt{0.5}A_c = -\sqrt{0.5}j\sqrt{0.5}A_0 - j\sqrt{0.5}\sqrt{0.5}A_0 = -jA_0 \qquad (12.33)$$

and

$$A_t = \sqrt{0.5}A_c + j\sqrt{0.5}A_{cc} = -\sqrt{0.5}\sqrt{0.5}A_0 - j\sqrt{0.5}j\sqrt{0.5}A_0 = 0. \qquad (12.34)$$

In this case, the device will reflect the signal pulse again. The NOLM with two input gates acts, according to this behavior, like an XOR gate. With a similar set-up, all-optical pattern generation and matching at 10 Gbit/s were studied [165, 169].

As shown in Sect. A, the XPM effect is polarization-dependent. Therefore, special considerations of the polarization states of the pulses must be taken into account. One possibility is the use of additional polarization controllers for every gate, or the application of polarization-maintaining fibers for the loop.

The nonlinearity in an ordinary optical fiber is instantaneous, so very fast switching speeds are possible. But due to the fact that the nonlinear refractive index is rather small a long interaction or high powers of the gate pulse are required [170]. According to (12.28), fibers of several kilometers are essential for a gate pulse power in the mW range. If the signal and gate pulses are different in their wavelengths then a large interaction length can lead to a walk-off between both, limiting the switching speed. In order to circumvent this, different polarization states between the signal and gate pulses instead of different wavelengths are possible, but this approach can lead to problems with the birefringence of the fiber [171]. The walk-off in the loop element and the pulse width of the gate pulse determine the switching speed and the maximum bit rate for demultiplexing [172].

For a suppression of the walk-off effect, it is possible to exploit a kind of dispersion management in the loop of the NOLM. In [173], nine sections of dispersion flattened fibers with different dispersion were used to build a walk-off-free NOLM, but this device is rather complicated. Sotobayashi et al. [174] used, instead, a 100-m, highly nonlinear dispersion-shifted fiber for error free demultiplexing of 320 Gbit/s TDM-signals down to 10 Gbit/s. The HNL-DSF had a zero dispersion wavelength at 1554.5 nm, a dispersion slope of 0.028 (ps/nm^2)/km, and a nonlinear coefficient of $\gamma = 15\ \mathrm{W}^{-1}\mathrm{km}^{-1}$. The walk-off was suppressed in the HNL-DSF by adjusting the signal and gate wavelength symmetrically around the zero dispersion in the fiber (1544.5 and 1564.0 nm), therefore both experienced the same group velocity. The bit rate of the mentioned experiment was limited by the pulse width of the control pulse, so the authors ensured that an operation beyond Tbit/s was feasible.

Another possibility of decreasing the necessary interaction length and pump power requirement is the exploitation of photonic fibers. Due to the small core area in these devices, the effective area is decreased and, therefore, the intensity is increased. Another advantage of photonic fibers is that it is possible to use completely different wavelengths because photonic fibers can be fabricated to exhibit single mode behavior in a much wider range.

Sharping et al. [175] used a microstructured fiber with a core diameter of 1.7 μm surrounded by a hexagonal array of 1.4 μm diameter air holes. An asymmetric fabrication of the core maintained the polarization in the fiber. A length of 5.8 m of the fiber was arranged in a Sagnac interferometer set-up to demonstrate all-optical switching of 2.6 ps (FWHM) signal pulses at wavelengths near 1550 and 780 nm.

For shorter dimensions of the device, it is possible to replace the fiber with a SOA [176]. In order to allow the two parts of the data signal to reach the nonlinearity at different times, the SOA must be placed asymmetrically in the loop. Therefore, this arrangement is also known as terahertz optical asymmetric demultiplexer (TOAD) or semiconductor laser amplifier in a loop mirror (SLALOM) [177] [178].

The repetition rate of the TOAD suffers from the recovery time of the SOA, but with optical pumping techniques, it was possible to reach repetition rates of more than 100 GHz [179]. Contrary to NOLMs with an optical fiber, a TOAD uses an active device as the nonlinearity. This active device introduces – due to its spontaneous emission – noise into the data signal, degrading the signal to noise ratio of the system.

On the other hand, the active device requires only a very low control pulse power for its switching process. Switching energies of only 250 fJ are possible [180]. The great advantage of switching elements with active devices as the nonlinearity is its possibility to be integrated on a small chip with standard semiconductor techniques [181, 182].

Other studies reported the combination of arrayed waveguides with optic Sagnac interferometers [183]. The device was formed by wavelength selective light paths, provided by two arrayed waveguides, and parallel, independent Sagnac loops. Hence, this device had the potential of performing independent signal processing functions in a parallel manner over a number of WDM-channels.

12.3.3 FWM for Optical Switching

The principle of optical switching with four-wave-mixing is relatively easy. The basic set-up is shown in Fig. 12.19 where the signal is, together with a gate bit sequence injected into an optical fiber. Due to the equation

$$f_{\text{switch}} = 2f_{\text{gate}} - f_{\text{signal}}, \tag{12.35}$$

the signal is shifted in wavelength or switched only when a signal and a gate pulse are present together in the fiber. The term f_{Switch} is the frequency of the switched signal and f_{Gate} and f_{Signal} are the frequencies of the gate and signal pulse, respectively. The set-up could be understood as a logical AND gate. With a filter at the end of the fiber, the switched signal can be separated from the gate and the data sequence.

A disadvantage of this set-up is that the switched signal has a wavelength that is different from the original data sequence. The usable bandwidth of the device suffers from the phase matching condition of the FWM effect, but with a careful choice of the wavelength and modulation of the gate pulses, a bandwidth of up to 200 nm is possible [190]. Since the FWM is polarization-dependent, special attention has to be paid to the polarization states of the waves. One possibility is the use of polarization-maintaining fibers or polarization controllers. If the fiber is long enough, the polarization dependence will be averaged out (see Sect. A).

The nonlinearity in an optical waveguide is nonresonant and, therefore, very fast, but because the data and gate bit sequence different wavelengths,

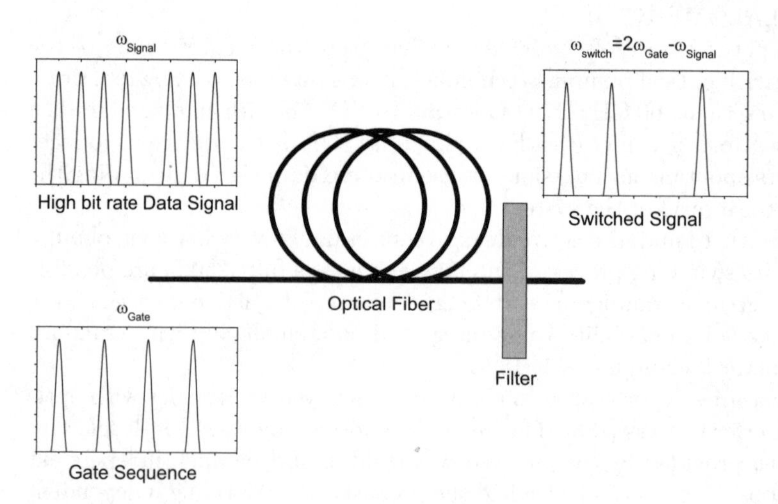

Fig. 12.19. Optical switching with the FWM effect in fibers

the walk-off in the dispersive wave guide leads to a limitation of the switching speed.

A 16-Gbit/s demultiplexer based on degenerate FWM was demonstrated by Andrek et al. [144]. They used 14 km of dispersion-shifted fiber as the nonlinear medium. The wavelength difference between the signal and the gate was 8 Å (0.8 nm).

Another possibility is to use a SOA instead of the optical fiber for the nonlinear medium. With such a set-up, Morioka showed polarization-independent, all-optical demultiplexing up to 200 Gbit/s [158, 159].

12.3.4 Ultrafast Optical Switching

Resonant nonlinearities have their origins in the excitation of carriers into another energy level. Since resonances in the material are exploited, only relatively low intensities are required. On the other hand, these processes are accompanied by time constants, such as the recovery time in a SOA which degrades the switching speed.

Nonresonant nonlinearities, like that in an optical fiber, result from the disturbance of the electron orbit due to the applied optical field. Due to the fact that only virtual levels are involved, the response and recovery time is very fast. The time of one round trip around the nucleus is on the order of $\approx 10^{-16}$ s [132].

Unfortunately, the strength of resonant nonlinear effects is very small. In order to increase the switching contrast, for instance, a large interaction length is required. This approach can lead to a walk-off between the signal and gate pulses due to the dispersion of the fiber, which limits the switching speed. The other basic possibility to increase the efficiency lies in the use of much higher intensities. These high intensities can be delivered by amplified femtosecond laser systems.

The basic set-up for ultrafast optical switching is shown in Fig. 12.20. Two femtosecond laser pulses from an amplified titanium sapphire laser system (mean wavelength 800 nm, pulse duration ≈ 120 fs, repetition rate 1 kHz) interfere under a small angle of 1.2° in a transparent material. The interference is shown in the figure as well. According to (12.15) the distance between the maxima of the pattern is 38 μm. Due to the fact that the pulses are very short, the interference is 24 μm wide for a constant time. The interference pattern can be translated into a refractive index grating via the nonlinear Kerr effect (4.91):

$$n_{\text{ges}} = n_0 + n_2 I. \qquad (12.36)$$

In the experiments, thin slides of cleaved barium fluoride at room temperature were used [184]. Due to the fact that the nonlinear refractive index (n_2) in barium fluoride is very weak ($\approx 2 \times 10^{-20} \text{m}^2/\text{W}$), very high intensities are required to achieve a sufficient grating contrast. The maxima and minima of the intensity are governed by the interference equation:

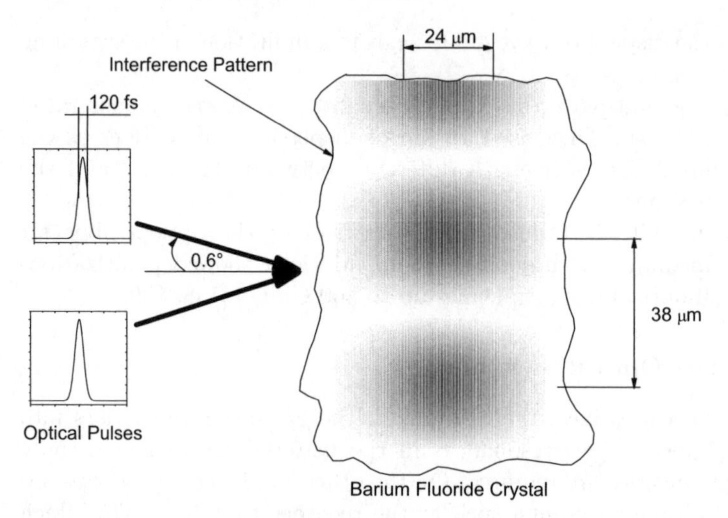

Fig. 12.20. Basic set-up for ultrafast optical switching with a travelling grating

$$I_{\text{Max/Min}} = I_1 + I_2 \pm 2\sqrt{I_1 I_2}. \tag{12.37}$$

Assuming an equal intensity of both pulses of $1.25 \times 10^{12}\text{W/cm}^2$, this leads to a grating contrast (difference between maxima and minima of the refractive index) of 10^{-3}.

The origin of the refractive index grating is the nonlinearity of the material, or – as mentioned earlier – the disturbance of the electron orbits due to the electric field of the interference. This nonlinearity is very fast and therefore, no lifetimes are detectable. The interference is instantaneously translated into the refractive index grating. If the interference disappears, the grating decays at the same time. Since the interference pattern is a thin slide with a thickness of around 24 μm, the refractive index grating has the same dimensions. The refractive index grating moves with the group velocity of the pulses through the crystal, as shown in Fig. 12.21.

The writing beams of the grating will be self-diffracted from this grating at the same time. Due to the fact that the refractive index grating is very thin, no phase matching is required and, therefore, not only the first diffracted order – as in Fig. 12.10 – can be seen behind the sample.

Figure 12.21 shows a typical self-diffraction pattern. The quoted percentages refer to the total input energy, i. e., the sum of the combined pump and probe energies in front of the sample. The diffraction is very efficient. Since only a completely nonresonant nonlinearity is involved, this self diffraction pattern is only present if the pulses have a sufficient intensity. The dynamic behavior of the self diffraction pattern, and therefore the grating, is shown in Fig. 12.22.

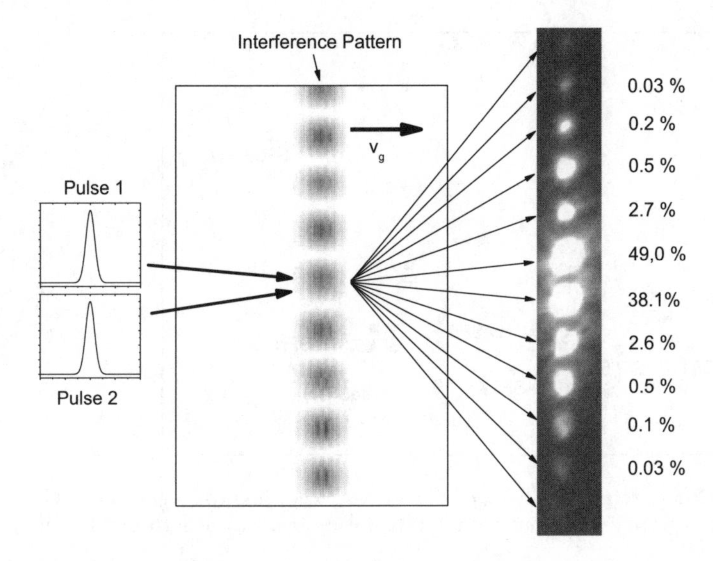

Fig. 12.21. Moving refractive index grating and self-diffraction pattern as a result of the interference of two fs pulses in a transparent material. The percentage values are related to the total input

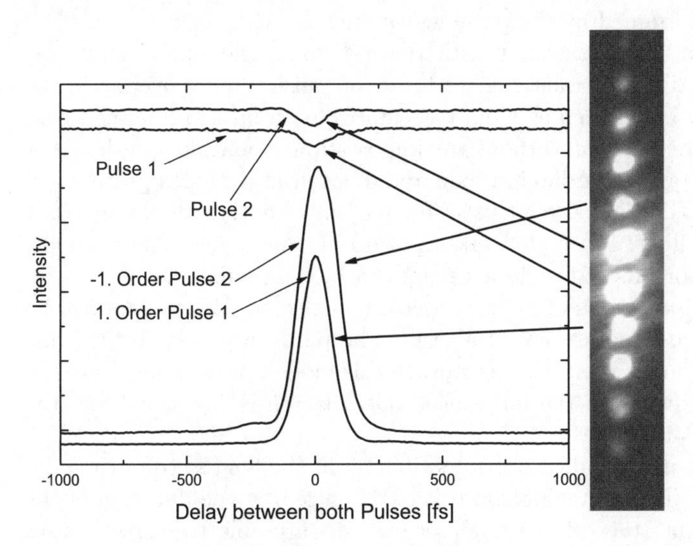

Fig. 12.22. Intensity of the first two orders of the self-diffraction pattern and the two input waves as a function of the relative delay between both pulses. A negative delay means that the second pulse reaches the sample before the first one and vice versa for a positive delay

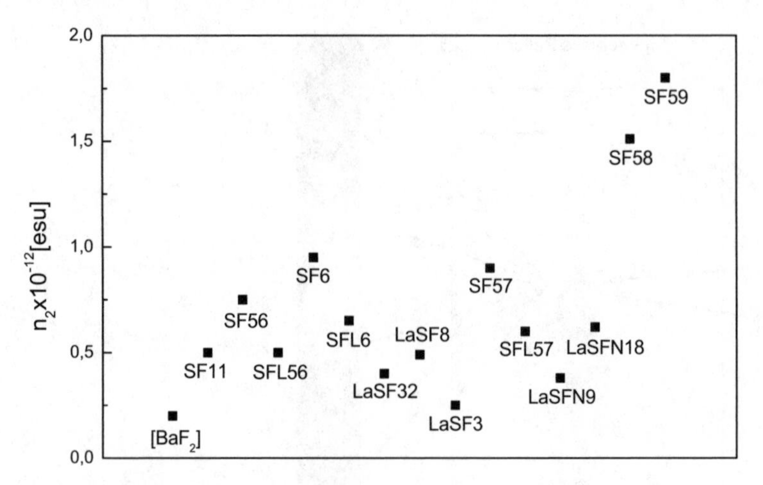

Fig. 12.23. Nonlinear refractive index of silica glass, according to Aber et al. [185]. The measurements were carried out with 130 fs pulses at a wavelength of 638 nm

The interference pattern is present only if both pulses overlap in the material. Therefore, the pulses will be diffracted only at this time and the diffraction is determined by the pulse width only.

Switching can be carried out with this set-up in the same manner as with photorefractive materials, Fig. 12.11. If the pulses do not overlap in the material, a third beam such as a data stream from an optical fiber can pass the thin transparent crystal without any important attenuation. On the other hand, if the two pulses overlap in the material, a refractive index grating will be written and the data stream is diffracted into another direction , i.e., another optical fiber. Since this takes place only for a few femtoseconds, discrete packets or bits of the data stream can be manipulated.

If the diffracted beams of higher order are neglected, the process can be seen as a partly degenerate FWM as well. The pump waves have the same frequency but differ in their wave vector. In this model, pulse 1 and 2 are the two pump waves whereas the introduced data stream is the signal and the switched beam is the idler.

With this set-up, ultrafast optical switching in the femtosecond time domain is possible. The switching speed is only limited by the duration of the pulses and current state of the art laser systems are able to deliver pulses with a duration of less than 5 fs. A drawback is the requirement of rather large table top laser systems for such a switch, but compact fs fiber laser systems are already commercially available.

In order to decrease the required intensity, other materials with a higher nonlinear refractive index but the same dynamic behavior are possible. Figure 12.23 shows the nonlinear refractive index of different commercial silica glass from SCHOTT.

According to the Figure, SF59 has a nonlinear refractive index about one order of magnitude higher than that of barium fluoride. Therefore, a drastic decrease in the required intensities should be possible in this materials. But it is not clear yet if the glass can withstand these intensities without an alteration of its optical properties. Harbold et al. [186] have synthesized a series of chalcogenide glasses from the As-S-Se system with nonlinearities greater than 400 times that of fused silica at 1.55 μm.

12.4 Retiming, Reshaping, and All-Optical Clock Recovery

For the minimization of intensity variations and accumulated timing jitter leading to a signal degradation, so called 3R (reamplify, reshape, retime) regenerators are essential. Regeneration of an optical signal can be done by a saturable absorber, for example. Such a device shows a low transmittance for the optical signal if the power level is below a certain threshold. Above the threshold, the transmittance increases rapidly and saturates asymptotically to transparency. Due to this behavior, the accumulated noise in the spaces is reduced and the contrast between marks and spaces increased. A suppression of the noise in the marks is possible by optical filtering with soliton pulses. The spectral width of the soliton depends on its temporal width and intensity. Hence, the filter shows higher losses for higher intensities and lower losses if the intensity is decreased. With a saturable absorber and such a filter a passive 2R regenerator can be build [319].

Another possibility for reamplification and reshaping is a Mach-Zehnder arrangement as shown in Fig. 12.15. Assume a CW signal with wavelength λ_2 is injected into the signal input port and the input data signal with wavelength λ_1 is launched into the upper branch which contains an SOA. In dependence on the power of the data signal, a phase shift via XPM is induced. This phase shift is translated into an amplitude variation of the CW signal λ_2. The maximum power of the output signal depends on the input power of λ_2. Furthermore the output signal shows an improvement in the signal extinction ratio due to the relatively fast nonlinearity of the SOA. Hence, the device performs a 2R regeneration. If the CW signal is substituted by a pulse clock signal a full 3R regeneration is possible [320].

In a 3-R regenerator clock recovery is a key element for the retiming and reshaping of the data signal. For clock recovery, the original clock is extracted from the incoming data signal. Clock recovery can, for instance, be reached with polarization switching in a TOAD arrangement or with a mode-locked ring laser configuration using a monolithically-integrated array of SOAs as an active mode-locking element [321].

A completely different approach, which will described here in more detail, is based on active Brillouin filters [322, 323, 324]. These filters exploit

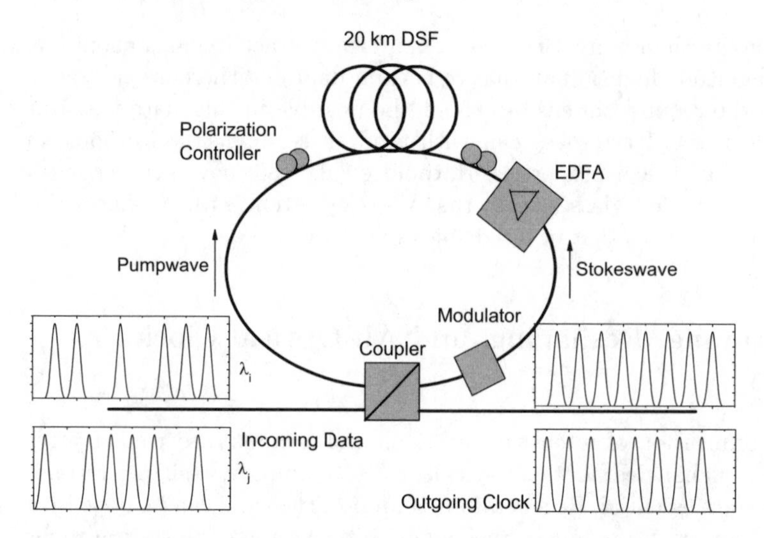

Fig. 12.24. Basic principle of optical clock recovery with SBS [324]

the very narrow bandwidth of the Brillouin scattering and the fact that different frequency components are scattered independently for the extraction of the clock component out of the signal spectrum. Figure 12.24 shows the experimental set-up for an optical clock recovery with stimulated Brillouin scattering (SBS).

With the directional coupler, the data signal is divided into two directions and injected into the loop. The counter-clockwise propagating data stream is used as the pump, whereas the clockwise propagating stream is the Stokes wave. For SBS pump and Stokes waves must have a frequency shift, therefore the data stream is amplitude-modulated with a Mach-Zehnder modulator. The modulation frequency complies with the Brillouin shift in the fiber and the modulation carrier is suppressed.

The two counter-propagating waves are injected into a dispersion-shifted fiber (DSF). Since the waves have a frequency shift that corresponds to the maximum of the Brillouin gain in the fiber, the Stokes wave will be amplified. The amplification is chosen in a way that only the strongest frequency component is higher than the required threshold for stimulated Brillouin scattering. Therefore only this frequency component will increase exponentially in the fiber. The spectrum is shown in Fig. 3.8 for a return-to-zero modulation. It consists of equally spaced frequency components enveloped by a $\sin(x)/x$ function. These frequency components represent the clock, shifted by multiples of the bit rate. The strongest frequency components are the first two sidebands with a frequency shift that corresponds to the bit rate. Due to the narrow bandwidth of the Brillouin scattering, one of these frequency components can be separated, to represent the clock of the data signal. The Brillouin active filter converts the data signal with an arbitrary distribution of ones and zeros into a clock signal. The clock output is obtained by effectively dumping some of the energy from the bit slots containing logical ones to the bit slots with logical zeros.

Since the pump and Stokes waves are generated from the original data stream, the set-up is independent of the bit rate and it is possible to regenerate the clock of several WDM channels at the same time. This is due to the fact that the bandwidth of the Brillouin gain is much narrower than the bandwidth of the WDM channels. Therefore, the scattering for each WDM channel takes place independently.

A restore-reshape (2R) regenerative optical switch based on SPM in a polarization-maintaining highly nonlinear optical fiber was shown by Petropoulos et al. [32]. The basic principle was suggested by Mamyshev [15], where the noisy data stream was amplified and then spectrally broadened by SPM in the highly nonlinear fiber. After that, a narrow band filter with a central wavelength offset relative to the peak wavelength of the incident signal regenerated the signal. If the input signals had low efficiency, the spectral broadening by SPM was not large enough and were rejected by the filter. On the other hand, high intensity pulses experienced a strong broadening and hence, transmitted a significant amount of light in the filter passband. Therefore, if the initial signal power was below a certain threshold, the bits were rejected by the device. Whereas, if it was above the threshold, the pulses were transmitted and limited in their maximum power. For appropriately designed systems the intensity of the output pulses became independent of the input pulse intensity if it was above a certain threshold.

The nonlinear device used by Petropoulos et al. [32] consisted of an HNL with a nonlinear coefficient of 35 $\text{W}^{-1}\text{km}^{-1}$ (SSMF $\gamma \approx 1.2\ \text{W}^{-1}\text{km}^{-1}$). Therefore, due to the $\approx$30 times higher nonlinearity, just 3.3 m of the fiber was sufficient.

12.5 Optical Filters with SBS

The very narrow bandwidth of the Brillouin gain can be exploited for narrow tunable optical filters [256]. If the bandwidth of each channel in a WDM system is smaller then the bandwidth of the Brillouin gain, it is possible to separate and amplify each channel by tuning a counter-propagating pump-laser. For high-bit-rate WDM channels, the bandwidth of the intrinsic narrow SBS gain is not sufficient. On the other hand, modulating the pump wave can increase the bandwidth drastically [257].

As mentioned in Sect. 11.5, a modulation of the pump wave results in a decrease and a broadening of the Brillouin gain. With a defined modulation pattern, one can adjust – to a certain limit – a desired filter function.

The normalized power density spectrum of a binary phase shift-keyed (BPSK) modulated wave is [259]

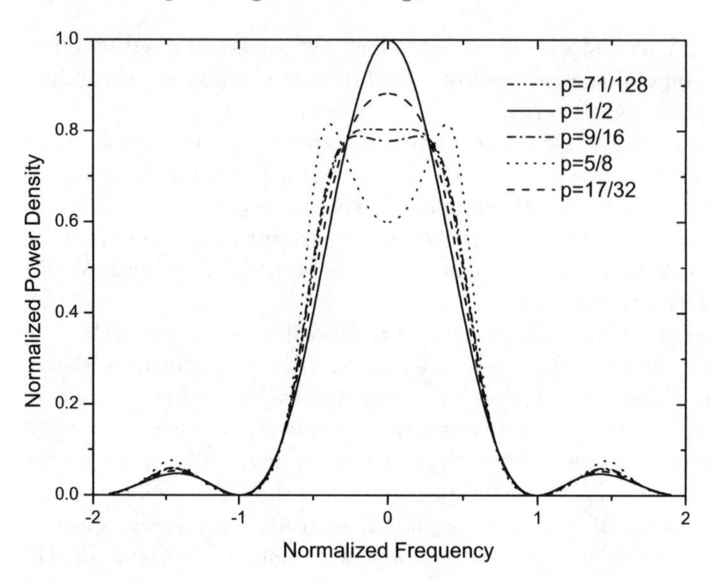

Fig. 12.25. Normalized power density spectrum of a BPSK modulated signal for different values of p [259]

$$P(f) = \frac{1}{B}\frac{\sin^2(\pi f/B)}{(\pi f/B)^2}\frac{4p(1-p)}{1+(1-2p)^2 - 2(1-2p)\cos(2\pi f/B)}, \qquad (12.38)$$

where B is the bit rate of the modulated signal. The power density spectrum depends on the statistical probability of the bit pattern. According to [259], p is the probability that two adjacent bits change its phase between 0 and π. A defined bit pattern of the form 0101010.... has a probability of $p = 1$, whereas a completely random bit pattern shows the highest uncertainty and therefore a probability of $p = 0.5$.

Figure 12.25 shows the normalized power density spectrum for different bit patterns. If the spectrum of the modulated pump is much broader than the intrinsic bandwidth of the SBS, the SBS gain follows directly the spectrum of the pump. Therefore, as shown in Fig. 12.25, a BPSK modulated pump with a parameter p of 9/16 or 71/128 can form a bandpass filter with flat transfer function.

The bandwidth of the filter is determined by the bit rate of the modulating signal. For increasing bit rates the bandwidth increases as well, but at the same time insufficiencies of the modulator can be seen. For high bit rates, the nonzero rise-fall times of the modulation data pattern cannot be neglected. Therefore sidebands at multiples of the clock frequency of the modulation signal are detectable in the power density spectrum. These sidebands grow rapidly for increasing bit rates [259].

As shown in Sect. 11.3, the normal operating mode of SBS is the amplification of a down-shifted wave, but if an upshifted wave with the Brillouin frequency and the pump wave are present in the material at the same time, this upshifted wave acts like a pump for the original pump wave and is, therefore, attenuated. The pump wave amplifies the Stokes wave with $f_P - f_A$ and, at the same time, it can attenuate the anti-Stokes wave with $f_P + f_A$, thus allowing the construction of an optical Notch filter with SBS [258, 259].

Tanemura et al. [259] realized with the described method a bandpass filter with a 3-dB bandwidth of 1.5 GHz and a 10-dB bandwidth of 2 GHz. They modulated externally the pump light of a laserdiode (wavelength 1.55 μm) with a lithium niobate phase modulator. The modulation signal had a bit rate of 2 Gbit/s with a parameter of $p = 9/16$ and was delivered by a pulse pattern generator. The modulated optical signal was amplified to 20 dBm by an EDFA and then injected into a 10 km DSF. The fiber was used as the nonlinear medium in which the data signal counter propagates.

Summary

- All-optical signal processing is the only way to overcome the physical limits of electronics in optical telecommunications.
- The alteration of light by light is impossible with linear optics.
- The most important fields for all-optical signal processing in all-optical networks are WDM sources, wavelength conversion in WDM cross connects, all-optical switches and demultiplexer, and elements for the retiming and reshaping of the data signal.
- Several nonlinear optical methods are suggested for these tasks. These methods can be divided in a partition which exploits resonant effects – for instance in photorefractive materials and SOAs – and a part that exploits nonresonant effects, especially in optical fibers.
- Resonant effects have several advantages. They only need very small intensities and they bear the potential for an integration into monolithical devices. Its disadvantage is that the resonances are connected with several time constants, such as the electron mobility in photorefractive media, or the decay time in SOAs, which limit the possible speed of resonant devices.
- Nonresonant effects are able to process optical signals in the femtosecond time domain. On the other hand they require high intensities or large interaction lengths because they exploit very weak nonlinearities.

Exercises

12.1 Determine the power of a pulse from a titanium sapphire laser with a duration of 80 fs and an energy of 1 mJ.

12.2. Calculate the power of the converted wave if a parametric wavelength converter with maximum gain and an input power of 10 mW is used. The converter consists of a standard singlemode fiber ($\gamma = 1.3\ \mathrm{W^{-1}km^{-1}}$) with a length of 500 m and the pump power is 1 W.

12.3. Derive the grating constant of the moving interference perpendicular to $\boldsymbol{q}$ in Fig. 12.9.

12.4. Compute the diffraction angle of a data signal in the C-band. The signal is diffracted at a photorefractive grating that was written with two beams from a Nd:YAG-Laser (1064 nm) interfering in a photorefractive crystal under an angle of 8°.

12.5. According to the interference equation $I_{\text{int}} = I_1 + I_2 + 2\sqrt{I_1 I_2}\cos\Delta$, two monochromatic waves with the same intensity ($I_1 = I_2$) and a phase shift $\Delta = 0$ have a past-interference intensity which is 4 times that of each wave. Assume a Mach-Zehnder interferometer like the one in Fig. 12.15 is used. What intensity will the pulse have behind it if the phase shift between the two branches is zero?

12.6. Determine the required pump power for switching off a signal in the C-band with XPM if an HNL fiber with a length of 500 m, a nonlinearity coefficient of 15 $\mathrm{W^{-1}km^{-1}}$, an attenuation constant of $\alpha = 0.3$ dB/km, and an effective area of 13.9 $\mu\mathrm{m}^2$ is used in a Mach-Zehnder arrangement.

12.7. Show that the NOLM in Fig. 12.18 behaves like an optical switch for the input signal as well if only the second gate pulse (gate2) is injected into the loop.

12.8. Determine the spatial dimensions of a moving grating for pulses from a titanium sapphire system (wavelength=800 nm, pulse duration=5 fs) if two beams of this system interfere in a nonresonant, nonlinear material with a refractive index of 1.5 under an angle of 1°.

12.9. Calculate the refractive index contrast for Exc. 12.8 if the intensity of one writing beam in the grating area is $5 \times 10^{12}\ \mathrm{W/cm^2}$ and the nonlinear refractive index of the material $n_2 = 2 \times 10^{-19}\ \mathrm{m^2/W}$.

12.10. Compute the normalized power density spectrum for $p = 111/200$ according to (12.38).

13. Nonlinear Lasers and Amplifiers

Current optical transmission systems incorporate a mix of two multiplex techniques. Many different low-bit-rate channels are interlaced in the time domain (OTDM) to form a high-bit-rate channel. These high-bit-rate channels with data rates of 10 or 40 Gbit/s are then again multiplexed in the wavelength domain (WDM). For an increase of the transmission capacity in an OTDM-WDM system, in principle, three ways are possible:

1. A higher bit rate in each individual WDM channel
   This approach requires the introduction of fast and expensive electronics. Furthermore, the switching speed of electronic systems is limited due to physical reasons. As mentioned in the former chapter, ultrafast optical switching could circumvent these limitations but such switches are still in the experimental stage. Additional problems are the dispersion and nonlinear effects like SPM and XPM, and intrachannel effects like IFWM and IXPM. Due to the strong intensity gradient, all of these effects are very efficient for short pulses.
2. A decrease of the channel spacing between WDM channels
   The spacing between two channels is determined by the selectivity of the filters in the receiver because the signals must be separated after transmission. Furthermore, the relationship between frequency and temporal duration is governed by the uncertainty relation. High-bit-rate time domain channels require a large bandwidth. Hence, if the channel spacing is small, the bit rate of each channel is limited. At the same time, the phase matching between the different channels is increased if the channel spacing is decreased. Therefore, the efficiency of FWM is increased as well.
3. Enlargement of the available bandwidth in optical transmission systems
   The bandwidth of optical fibers was already treated in Sect. 3.2.1. If no OH^- ion absorption occurs in the fiber, the available bandwidth is determined by the cut-off wavelength and the attenuation in the fiber material, mainly due to Rayleigh scattering. Therefore, the transmission bandwidth is ≈ 400 nm ($1.2 - 1.6$ μm) in a standard single mode fiber. On the other hand, an optical transmission system consists not only of fibers, it incorporates other devices like optical amplifiers. The EDFA has a bandwidth of only 20 nm in the C-band. Filters can increase it to 35 nm

and with high additional effort, a bandwidth of 80 nm is possible, as can be seen from Fig. 2.5. So EDFAs can be used in the C- and L- band of an optical fiber in a wavelength range between 1530 and 1610 nm. The amplification bandwidth of an EDFA is determined by the energy levels of the erbium ions and can not be adapted to other requests.

Hence, other physical mechanisms for an optical amplification are required for the enlargement of the transparency range of optical transmission systems. Candidates for this purpose are nonlinear effects like SBS, SRS, or FWM, all of which transfer power from a pump wave to a signal wave with a different frequency. They offer several advantages in comparison to an EDFA.

Contrary to its erbium ions, nonlinear amplification mechanisms are independent of the frequency and possible in every standard single mode fiber. The incorporation of highly nonlinear fibers like HNL or holey fibers allow the construction of small compact devices. Amplification at any wavelength can be achieved if a suitable pump source is available. With SRS amplifiers, for instance, the amplification at 1.3 μm [285], at 1.4 μm [286], in the C-band [287], and in the C- and L-band together [288] was shown [293]. If multiple pump wavelengths are combined, the SRS effect can be used to build broadband amplifiers [289, 299]. Since the last few years, pump laser diodes that deliver the required wavelength with a high power are available, so nonlinear amplifiers are a serious alternative to current EDFAs.

With only one pump laser, Ho et al. [190] built an optical parametric amplifier with a bandwidth of 200 nm. On the other hand, amplifiers that exploit the effect of SBS have a very narrow bandwidth of only a few MHz, but this small bandwidth can be advantageous for small bandwidth applications like radio-over-fiber.

This chapter discusses the principles of nonlinear amplifiers and lasers. The first two sections deal with the exploitation of the Raman effect for the amplification of an optical signal and for the construction of a laser. Section 13.3 shows the basic principles of Brillouin amplifiers and lasers and gives an overview on the radio-over-fiber technique. The last section shows how the parametric effect of FWM can be used for optical amplification.

13.1 Raman Amplifier

Before the invention of the EDFAs, much attention was given to the development of Raman amplifiers as a possibility for all-optical amplification. In those early years, Raman scattering in optical fibers was, for instance, used for the proof of soliton transmission in optical fibers by Mollenauer et al.[294]. A major drawback of the Raman effect is the very high pump power required for stimulated scattering. Therefore, after the development of EDFAs in 1987 [3], which required only laser diodes with a few mW power for

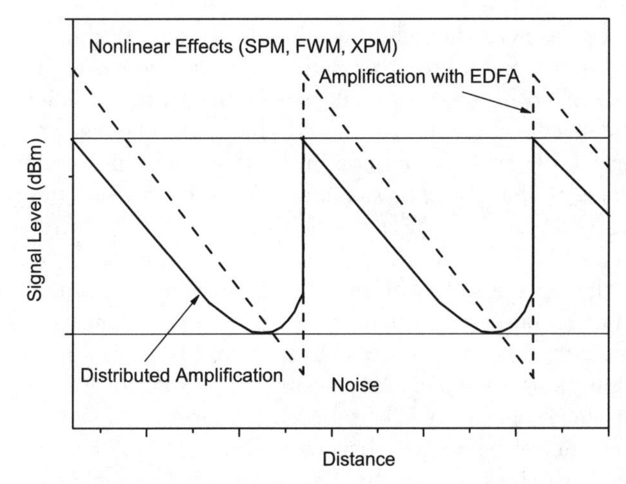

Fig. 13.1. Signal level against distance for a conventional lumped and a distributed Raman amplification pumped from both sides [298]

pumping [4], the Raman amplification lost attention. Today there is a strong renewed interest in Raman amplification, for several reasons:

- Laser diodes which deliver the required power at several pump wavelengths were developed in the last few years and are commercially available today.
- Using Raman amplification, the whole transmission bandwidth of an optical fiber can be exploited. For instance, studies proposed the realization of an U-band (1625 – 1675 nm) amplifier with the Raman effect [292].
- In an optical fiber, multiple Raman processes can be carried out simultaneously. This means that a broadband amplification is possible if many pump lasers are combined. With such a concept, a Raman amplifier with a gain bandwidth of more than 100 nm was demonstrated [299].
- A particular advantage of the Raman amplification is the distributed amplification inherent in the process.

Amplifier devices like EDFA and SOA are used for a so called lumped amplification between the spans. This means the signal loses power during its propagation in the fiber and this power loss is compensated at a distinct point with an amplifier device. In contrast to that, it is possible to compensate the losses over the entire transmission distance with a Raman amplifier.

Figure 13.1 shows the signal level for a distributed amplification in comparison to a conventional lumped amplification. As can be seen the conventional amplification can have mainly two problems. First, the signal level is very high directly behind the amplifier. Hence, the optical intensity in the fiber can be so high that nonlinear effects like SRS, SPM, XPM, and FWM are very effective. The nonlinear effects lead to a distortion of the signals and an increase of the noise in the system. Second, the signal is so low at the end of the fiber span that it comes near the noise floor in the system. The small SNR leads to detection failures in the receiver and, therefore, increases the BER. A possible solution of both problems could be reducing the amplifier spacing or increasing the spacing between the WDM channels. The last approach reduces the strength of intrachannel nonlinear effects as well as the system transmission capacity. The first approach increases the costs and, due to the amplified spontaneous emission (ASE, see Sect. 2.5), the noise in the system.

On the other hand, the incorporation of distributed amplifiers in transmission spans gives a better noise performance than a lumped amplification between the spans [290]. The distributed Raman amplification allows a nearly flat amplification along the span. As a result, the signal level does not exceed the range in which nonlinear effects become important nor does it get low enough to become overshadowed by noise. Raman amplification can be carried out from both sides of the fiber. Therefore, it is possible to use rather large amplifier distances. For instance, transmission experiments in the Tbit/s range over several thousand kilometers would be impossible without the incorporation of distributed Raman amplification. Distributed Raman amplification lowers the required signal power and, therefore, the impact of fiber nonlinearities. In 40-Gbit/s systems, studies showed that the amplifier spacing can be doubled if Raman amplifiers instead of EDFAs are used at almost the same transmission performance [291].

Another possible application field is the amplification of analog signals in hybrid fiber-coax cable TV (HFC-CATV) systems. The powers that must be transmitted via these fibers are on the order of 10 dBm and, therefore, relatively high. If the radio and TV channels are multiplexed in the wavelength domain, several nonlinear effects degrade the system performance. With a distributed Raman amplification, the power and the cross talk between the channels can be reduced drastically. In an experiment, Marhic et al. [295] showed a reduction of the nonlinear cross talk in a 25-km fiber link with measured improvements ranging from 8 to 15 dB.

If Raman scattering is used to amplify a weak signal, in principle, only a second pump wave with a frequency shift of ≈ 13 THz (for the maximum of the Raman shift) has to be launched together with the signal into the fiber input. For the amplification, the entire transparency range of the optical fiber can be used. Therefore, contrary to the EDFA, any wavelength can be amplified if the signal and the pump lie in the transparency range of the fiber and if a corresponding pump source is available. For instance, for an amplification between 1520 and 1620 nm, a pump source with a wavelength between 1410 and 1510 nm is required. Due to the relatively high bandwidth of the Raman gain, several wavelength channels can be amplified simultaneously with one pump source.

A power transfer between the pump wave and the signal wave happens if the pump wave propagates in the direction of the signal or opposite to it.

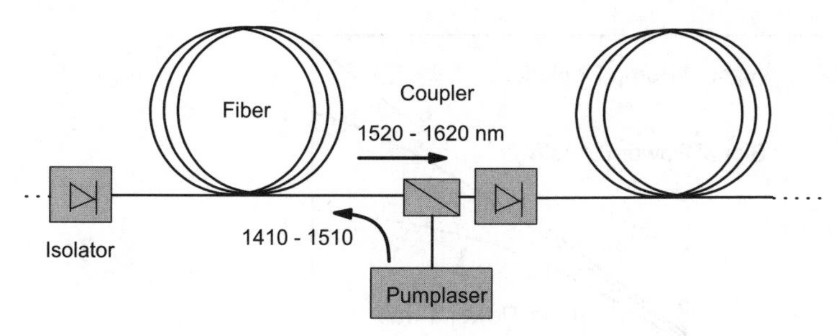

Fig. 13.2. Schematic set-up of a backward-pumped Raman amplifier

Hence, a Raman amplifier can be pumped in forward, backward, or both directions. The basic set up of a Raman amplifier pumped backwards is shown in Fig. 13.2. The basic configuration is the same as that for an EDFA. The gain medium is an optical fiber, one or two lasers serve as the pump source, and the wavelength-sensitive coupler directs the pump light in one and the signal in the other direction. The optical isolator, e.g., a Faraday rotator, prevents destruction of the transmitter. Nevertheless, two important differences to an EDFA exist. First, the gain medium can be an ordinary fiber and second, the pump sources have to deliver unpolarized light because the SRS is polarization-dependent.

The maximum length or gain beyond which Raman amplifiers no longer improve system performance is determined by the so called double Rayleigh backscattering (DRBS) multiple-path interference (MPI) noise [298]. Suppose a signal, propagating through an optical fiber, is amplified by a co-propagating pump wave. Due to Rayleigh scattering, a part of the signal is scattered back. The SRS amplifies the forward travelling signal as well as the backscattered part of the light. The latter thus becomes strong enough to generate a new backscattered wave again. This time, the double-backscattered wave propagates in the same direction as the pump wave itself and will be amplified by it. The superposition between the signal and the time-delayed, double-backscattered Rayleigh light leads to a phase-dependent noise. Contrary to the original wave, the double-backscattered wave moves twice through the gain medium. Therefore, the signal to noise ratio decreases if the gain or the effective length of the Raman amplifier is increased.

The gain of an amplifier is the ratio between the signal power at its output and input. The gain of the Raman process alone can be defined as the ratio between the intensity (or power) of the amplified signal, $I_{\mathrm{SR}}(L)$, and the signal intensity (or power) without amplification at the fiber output, $I_{\mathrm{S}}(L)$. The gain can be calculated with (10.13), (10.14), and (10.17). Without Raman amplification, the signal experiences attenuation in the fiber and its intensity at the fiber output where $z = L$ is

$$I_{\mathrm{s}}(L) = I_{\mathrm{s}}(0)\mathrm{e}^{-\alpha_{\mathrm{s}} L}. \tag{13.1}$$

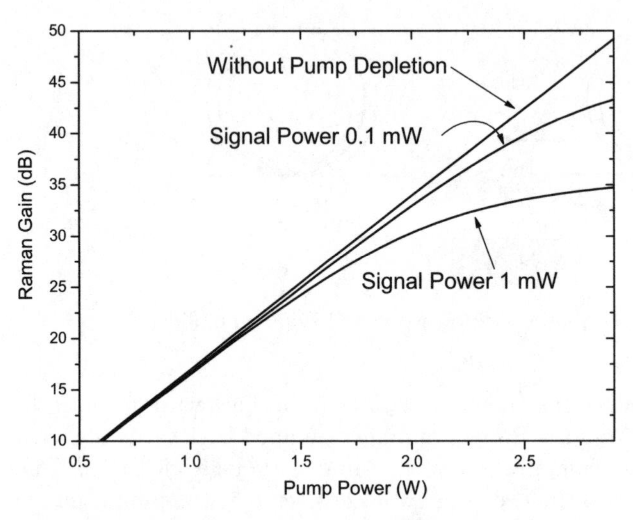

Fig. 13.3. Raman gain for two different signal powers against the pump power in a fiber with a length of 5 km (Attenuation $\alpha = 0.2$ dB/km, Raman coefficient $g_R = 7 \times 10^{-14}$ m/W, $A_{\text{eff}} = 80$ µm^2, $K_S = 1$)

If, at the same time, a strong wave with a frequency shift that corresponds to the Raman shift is launched into the fiber input, an amplification due to the Raman process occurs. If the pump depletion is neglected, then (10.17) is valid for the Raman-amplified signal wave at the fiber output. The ratio between both cases – (10.17) and (13.1) – is the gain that will be achieved by the Raman process:

$$G_R = \frac{I_{SR}(L)}{I_s(L)} = \frac{I_s(0)\exp(g_R/K_s I_p(0) L_{\text{eff}} - \alpha_s L)}{I_s(0)\mathrm{e}^{-\alpha_s L}} = \mathrm{e}^{g_R/K_S I_p(0) L_{\text{eff}}}. \quad (13.2)$$

Alternatively, with $P_0 = I_p(0) A_{\text{eff}}$ as the power of the pump at the fiber input, (13.2) can be written as

$$G_R = \mathrm{e}^{g_R/K_S P_0 L_{\text{eff}}/A_{\text{eff}}}, \quad (13.3)$$

where, the gain for the Raman amplifier is

$$G_{RA} = \frac{I_{SR}(L)}{I_s(0)} = \exp(g_R/K_S P_0 L_{\text{eff}}/A_{\text{eff}} - \alpha_s L). \quad (13.4)$$

Figure 13.3 shows the Raman gain against the pump power at the fiber input for a fiber with a length of 5 km. If the pump depletion is not considered, according to (13.3), the gain grows exponentially. As can be seen from the figure, this is only approximately valid for small input powers of signal and pump. If the powers are increased from a distinct value, a saturation of the gain can be seen. In this case (13.3) is no longer valid. The saturation

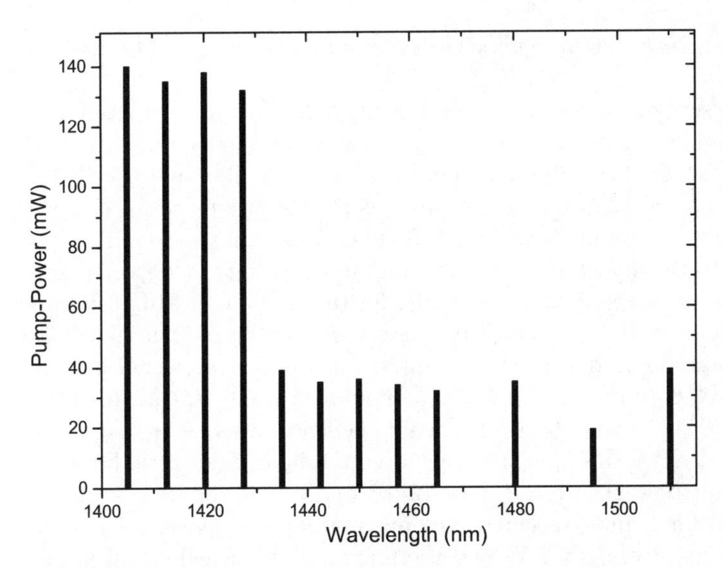

Fig. 13.4. Distribution of the power of 12 pump laser diodes for a 100-nm Raman amplifier [299]

can be considered if the differential equations (10.13) and (10.14) are solved numerically. Figure 13.3 shows the result for two different signal powers of 1 and 0.1 mW. Evidently the saturation depends on the signal power, where the amplifier is saturated earlier if the signal power is higher. Therefore, (13.3) is only valid for low pump and signal powers.

A Raman amplifier with one pump source is restricted to the rather small range of the Raman gain maximum, shown in Fig. 10.5. Pumping the fiber simultaneously with multiple laser diodes at different wavelengths enable the construction of broadband amplifiers. These laser diodes are tuned in wavelength and output power and then multiplexed together into the fiber. The maximum bandwidth of such an amplifier is limited by the SRS frequency shift to 110 nm. Rottwitt et al. built a Raman amplifier with a bandwidth of 92 nm [289]. The output spectrum of a pumping unit for a Raman amplifier in a 25-km DSF with a bandwidth of 100 nm is shown in Fig. 13.4.

For such a broadband pump source, the SRS acts not only on the signal but between the pump wavelengths as well. In order to achieve the flattest gain curve possible each pump laser must be fine-tuned. Due to the SRS, the pump waves with a smaller wavelength transfer a part of their power to the pump waves with the higher wavelength. Hence, the pump lasers with the smaller wavelength require a higher power.

In order to compensate the polarization dependence of SRS, the pump unit consists of two lasers with crossed polarization for each pump wavelength. For the limitation of the linewidth and the time stability, a fiber Bragg grating is used. The whole required pump power for the DSF fiber was 790 mW whereas, the power for each individual laser (20 - 140 mW) is relatively small.

With the described pump unit, Emori et al. [299] built a broadband amplifier with a bandwidth of 100 nm and a gain variation of only ± 0.5 dB. The average gain was 2 dB for a 25-km standard single mode fiber and 6.5 dB for a DSF with the same length. The gain here is defined as the ratio of input to output signals. Therefore, it includes the fiber losses that were 5 dB.

In contrast to the distributed amplification in the former example, Yusoff et al. [246] built a compact Raman amplifier for the L^+ band (1610-1640 nm) in a highly nonlinear, holey fiber with a length of only 75 m. The fiber was polarization-maintaining due to an asymmetry in the core area and had an effective area of 2.85 μm^2 at 1550 nm. The nonlinear coefficient in this fiber was $\gamma = 31\ W^{-1}km^{-1}$ and the Raman gain coefficient was estimated to be 7.6×10^{-14} m/W. With this fiber, they achieved a Raman gain (ratio between input and output power) of more than 42 dB and a noise figure of ≈ 6 dB under a forward single pump scheme. The maximum pump power they could launch into the fiber was ≈ 6.7 W and the maximum launched signal power was ≈ -10 dBm.

13.2 Raman Laser

A Raman laser is, in principle, a Raman amplifier in a resonator cavity. Due to the cavity, the threshold of stimulated scattering is decreased. For fiber lengths of 10 m, SSMF pump powers of 1W are sufficient. Very low threshold powers for Raman lasing were shown by Min et al. [305]. They used an ultrahigh-Q, silica microsphere resonator coupled to a tapered fiber. This device required less than 900 µW of pump power near 980 nm to generate five cascading Stokes laser lines. The threshold for the first-order Stokes wave was only 56.4 µW.

Due to the spectral width of the Raman gain in an optical fiber, a fiber Raman laser can be tuned to different wavelength. Figure 13.5 shows a fiber Raman laser tunable by a grating in Littman configuration.

The pump wave is coupled into the fiber via a coupler. The Raman scattering in the fiber generates new waves with different frequencies. Depending on the particular wavelength, the grating diffracts the distinct beams into different directions. Only the beam, or the wavelength, that hits the tuning mirror perpendicularly forms a Fabry-Perot resonator together with the first dichroic mirror. Hence, only for this beam with a narrow bandwidth the set up in Fig. 13.5 serves as a laser. All other beams are reflected out of the cavity without amplification. The mirror can be turned to tune for another wavelength, as shown in the Figure.

For compact laser devices, the mirrors can be replaced by a periodic alteration of the refractive index in the core of the fiber (fiber Bragg grating). Figure 13.6 (a) shows a fiber Raman laser with a resonator built by a pair

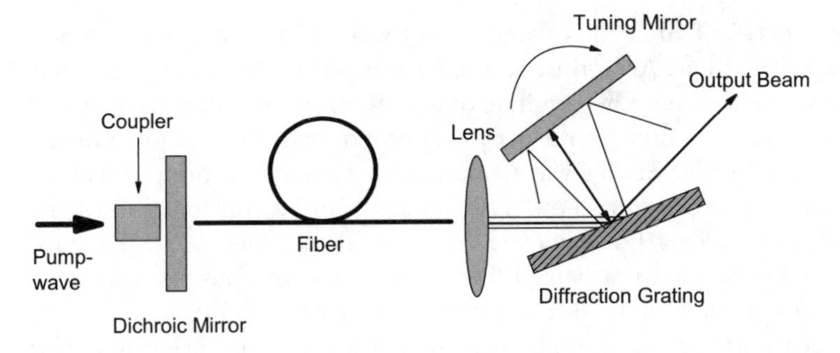

Fig. 13.5. Schematic of a tunable fiber Raman laser in Littman configuration

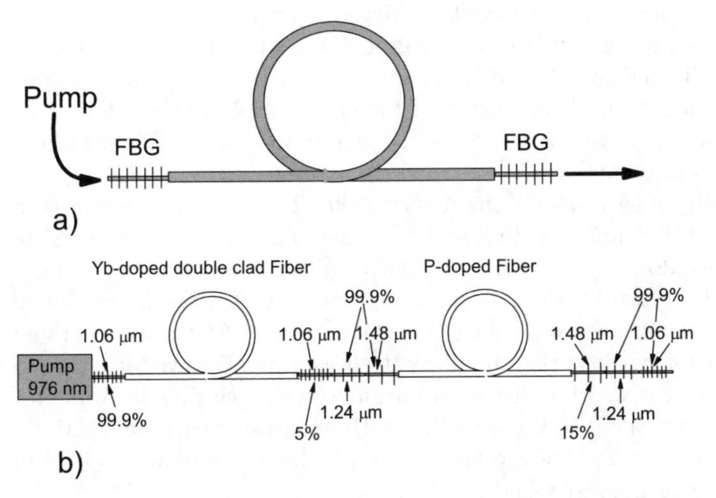

Fig. 13.6. Fiber Raman laser with a resonator composed of fiber Bragg gratings (FBG = fiber Bragg grating, Yb = Ytterbium, P = Phosphosilicate). **a** Basic set up. **b** Diode-pumped fiber laser

of Bragg gratings at the input and output of the fiber. The Bragg grating can be written directly into the core of the fiber. A comparison to Fig. 13.5 shows that the geometry is extremely reduced.

For Raman lasers in telecommunications, pump sources with an output power in the Watt range and a wavelength of 1.5 μm are required. Raman lasers can be used as pump sources for Raman amplifiers or as remote pump sources for EDFAs. In the latter case, they must deliver a wavelength of 1.48 μm. Until recently, the output power of semiconductor lasers was too small for pumping and conventional solid state lasers did not deliver the correct wavelengths. Solid state lasers with a high output power that can be used as pump sources include Nd:Yag lasers but, their output wavelength is 1064 nm. Only with a multistage, or cascaded Raman process can they deliver pump wavelengths required for the telecommunications range [300, 301].

Via the process of SRS the output pump wave of the laser generates a Stokes wave in the fiber. According to the Raman shift, this Stokes wave has a higher wavelength than the original pump. In a second step, the Stokes wave itself is used as a pump to generate a second order Stokes wave in the fiber. This second order Stokes wave can again be used as a pump wave for the next Raman process and so on. This process can be continued until the required output wavelength is achieved. Since the device has to be a laser for each individual wavelength, a pair of fiber Bragg gratings must be used as a resonator for each wavelength (see the right side of Fig. 13.6b).

On the other hand, the Raman shift in optical fibers is rather small. If the laser is pumped at a wavelength of 1.06 μm, a Raman shift of 105 nm requires a fourth order Stokes wave to achieve an output wavelength of 1.48 μm. Therefore, a large number of Bragg gratings has to be involved, making the laser complicated and less reliable. In contrast, phosphosilicate-doped fibers have a much higher Raman shift of 1330 cm^{-1} [302]. In these fibers the second order Stokes wave has a wavelength of 1.48 μm already if the laser is pumped at 1.06 μm.

Figure 13.6b shows a fiber Raman laser [303]. The first pump wave with a wavelength of 976 nm is delivered by a laser diode array. It pumps an ytterbium-doped double clad fiber laser which emits a wave with a wavelength of 1.06 μm. This wave is now used as the pump wave for the fiber Raman laser. A first order Stokes wave with a wavelength of 1.24 μm is generated via Raman scattering. It is then used, by itself as a pump wave for a second order Stokes process which delivers an output wavelength of 1.48 μm. This wavelength is coupled out of the laser by the Bragg grating with a reflectivity of 15%. The output power of the laser is 1W, which corresponds to 34% of the pump power of the laser diodes.

Demidov et al. [304] presented a cascaded Raman laser with a chain of Raman cavities. In principle, it consists of a set up that is similar to the one in Fig. 13.6b followed by two additional stages based on Ge-doped fibers. They used a different pump wavelengths (P-doped fiber output 1278 nm) and the reflectivity of the FBGs was controlled by a piezoelectric bender. The laser provided three wavelengths (1425, 1454, and 1463 nm) simultaneously for a flat Raman gain profile in the C-band and had an efficient power control and large dynamic range for all generated wavelengths. The control was achieved by an alteration of the reflectivity of the Bragg gratings used as output couplers.

13.3 Brillouin Amplifiers and Lasers

Stimulated Brillouin scattering in a fiber means that power is transferred from a strong pump to a weak Stokes wave. If no Stokes wave is present, it can be generated from noise. But, if a weak wave with the corresponding frequency shift is launched from the other fiber side, this wave will be amplified by the

SBS. Hence, in principle, a Brillouin amplifier is an optical fiber in which the pump and Stokes waves propagate in opposite directions. The only required condition is that the pump and signal must have a frequency shift relative to each other that corresponds to the Brillouin shift in the used fiber.

Since the Brillouin frequency shift is only a few GHz, the required pump wavelength is in the telecom range. Furthermore, because stimulated scattering occurs readily at relatively small pump powers, cheap pump sources are available without any problems.

The SBS has only a small bandwidth, however, and the utilizable frequency range of a Brillouin amplifier is very small (< 100 Mhz). For the amplification of high-bit-rate digital channels, this bandwidth is not sufficient. Furthermore, the spontaneous emission noise of a Brillouin amplifier is, due to the large number of temperature excited phonons, relatively large. Therefore, Brillouin amplifiers cannot be applied in the field of digital optical communications.

On the other hand, if the signals are not digitized but transmitted in their analog form, their bandwidth is much smaller. For analog signals, Brillouin amplifiers have some important advantages, as will be shown in this Section. But first, the basic amplification process in a Brillouin amplifier is treated in detail.

If we assume that the pump wave propagates from the amplifier output to its input, then the gain of a Brillouin amplifier is the ratio between the signal power at the amplifier output to the signal power at its input. Due to the constant effective area in a single mode fiber this ratio holds for the intensities as well

$$G_{\mathrm{BA}} = \frac{P_{\mathrm{s}}(L)}{P_{\mathrm{s}}(0)} = \frac{I_{\mathrm{s}}(L)}{I_{\mathrm{s}}(0)}. \tag{13.5}$$

If the depletion of the pump due to the Brillouin process is neglected, the signal intensity at the fiber output can be calculated with (11.13). If output and input are changed it can be written

$$I_{\mathrm{s}}(L) = I_{\mathrm{s}}(0)\mathrm{e}^{g_{\mathrm{B}}P_0 L_{\mathrm{eff}}/(K_{\mathrm{B}}A_{\mathrm{eff}})-\alpha L} \tag{13.6}$$

with g_{B} as the peak value of the Brillouin gain coefficient and K_{B} as a factor that includes the polarization dependence of pump and Stokes wave. Therefore, the gain of a Brillouin amplifier is

$$G_{\mathrm{BA}} = \frac{I_{\mathrm{s}}(0)e^{g_{\mathrm{B}}P_0 L_{\mathrm{eff}}/(K_{\mathrm{B}}A_{\mathrm{eff}})-\alpha L}}{I_{\mathrm{s}}(0)} = \mathrm{e}^{g_{\mathrm{B}}P_0 L_{\mathrm{eff}}/(K_{\mathrm{B}}A_{\mathrm{eff}})-\alpha L}. \tag{13.7}$$

For the derivation of (11.13), we assumed that the depletion of the pump can be neglected. If the pump intensities exceed the threshold, this simplification is no longer valid. Therefore, both equations of the differential equation system, (11.7) and (11.8), have to be considered. An analytical solution is

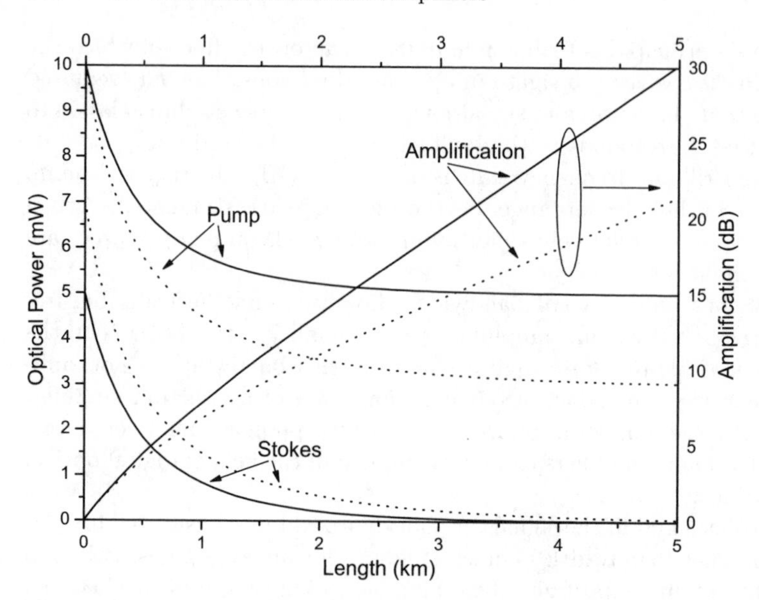

Fig. 13.7. Optical power of the pump and Stokes waves and amplifier gain in a Brillouin amplifier against the fiber length ($A_{\text{eff}} = 80\ \mu\text{m}^2$, $\alpha = 0$, $g_{\text{B}} = 2 \times 10^{-11}$ m/W, $P_{\text{p}}(0) = 10$ mW, $P_{\text{s}}(0) = 7$ mW and 5 mW)

complicated and leads to an integral for which no closed form solution was found [306]. For the strong simplification that the attenuation in the fiber is neglected, the intensity of pump and Stokes wave [307] can be written as

$$I_{\text{p}}(z) = A \frac{I_{\text{p}}(0)\text{e}^{g_{\text{B}}Az}}{I_{\text{p}}(0)\text{e}^{g_{\text{B}}Az} - I_{\text{s}}(0)} \tag{13.8}$$

and

$$I_{\text{s}}(z) = A \frac{I_{\text{s}}(0)}{I_{\text{p}}(0)\text{e}^{g_{\text{B}}Az} - I_{\text{s}}(0)}, \tag{13.9}$$

with

$$A = I_{\text{p}}(0) - I_{\text{s}}(0), \tag{13.10}$$

Figure 13.7 shows the power of the pump and Stokes waves in a Brillouin amplifier with a pump power of 10 mW for two different output powers of the Stokes wave ($I_{\text{s}}(0)$), while neglecting the effects of attenuation ($A_{\text{eff}} = 80\ \mu\text{m}^2$, $\alpha = 0$). The power exchange between the pump and the Stokes wave is 50 and 70%, respectively. The amplification factor or amplifier gain ($G_{\text{BAdB}} = 10 \log (P_{\text{s}}(0)/P_{\text{S}}(L))$) against the fiber length is shown in the figure as well.

A Brillouin laser can be built in the same way as a Raman laser. This means a Brillouin amplifier has to be placed into the cavity of a resonator.

The threshold of a Brillouin laser is orders of magnitude smaller than that of a Raman laser. A few 100 μW can be sufficient if the fiber is long enough. For higher pump powers, cascaded Brillouin scattering can lead to higher order Stokes waves. At the same time, due to FWM, anti-Stokes waves can be generated in the laser.

A Brillouin laser with a highly nonlinear holey optical fiber in a Fabry-Perot resonator was shown by Lee et al. [315]. The fiber had an effective area of 2.85 μm^2 and a nonlinear coefficient of 31 $W^{-1}km^{-1}$. In a fiber length of 73.5 m, they achieved a laser power conversion efficiency of 70% and a threshold of 125 mW. The Brillouin shifted output power was 110 mW at a wavelength of 1552.18 nm. According to the authors, this rather high threshold was caused by structural variations along the imperfect fiber they used. Hence, it may be possible to obtain much lower thresholds and (or) reduced device lengths in holey fibers.

13.3.1 Radio-over-Fiber

Due to the small bandwidth of a Brillouin amplifier, such devices are only suitable for small bandwidth applications. One such application is the amplification of analog or narrow band digital signals. Especially important is the radio-over-fiber technique [308]. In wireless communications, mobile phones, or wireless LANs, carriers with wavelengths in the centimeter or even millimeter range are used. For these short wavelengths, coax cables have very high attenuation so, the signals can not be transmitted over long distances. The information has to be delivered to the base station in the baseband which provides the carrier, modulates it with the baseband signal, and radiates the modulated radio frequency (RF) via its antennas. Hence, every base station can be very complex requiring a link to the power grid, an air conditioning unit, and possibilities for a remote control. Wireless communication networks are cellular, this means the network consists of a large number of cells and the cell number increases with the number of subscribers. Each cell is equipped with its own base station.

In contrast to coax cables, optical fibers have extremely low losses. This opens the possibility to generate the RF signal in a control station far away from the base station. Each control station can serve several base stations and transmits the RF signal via optical fibers. Hence, the complexity of a large number of base stations is concentrated in just one control station. This bears the possibility of an enormous cost reduction and makes the control of the whole system much easier.

For radio-over-fiber, the RF signal (or RF modulated baseband signal) is generated in the control station, which is again modulated to an optical carrier and transmitted over large distances in an optical fiber. The base station transfers the optical signal back to the RF domain, amplifies, and radiates it. If the cell around the base station is smaller than 100 m, the amplification can be dropped if electro-absorption modulators[1] (EAM) are used. In this case, the base station is completely passive and consists only of a demodulator and an antenna [309].

In principle, three techniques can be used for the modulation of the optical carrier with the RF signal [311]: the direct, the external, and the heterodyne modulation. The frequency of the carrier is much higher than the bandwidth of the base band signal and can lead to problems for the optical modulation. In the case of a direct modulation, the intensity of the emitted light from a laser diode is directly controlled by its current. This method is very easy and cheap but in addition to the intensity, the frequency of the emitted light is changed as well. The additional generated frequency modulation (chirp) leads to several problems.

In the case of an external modulation, the laser stays tuned, the current is constant, and an external modulator (EAM, Mach-Zehnder) modulates the light. At high frequencies, problems with the nonlinear characteristic curve of the modulator can occur. This leads to higher harmonics and mixing frequencies.

A method which can be used for very high frequencies is the heterodyne technique. In principle, this method is based on the fact that only the intensity of light can be measured and not even a fast photodiode can follow the optical frequencies. For a heterodyne superposition, two waves with different frequencies $E_1(t, \omega_1)$ and $E_2(t, \omega_2)$ are required. The frequency difference between both corresponds to the carrier frequency of the RF signal, $\omega_{\mathrm{RF}} = \omega_2 - \omega_1$. The two waves are represented by

$$E_1(t, \omega_1) = m(t) E_1(z) \cos(\omega_1 t) \tag{13.11}$$

and

$$E_2(t, \omega_2) = E_2(z) \cos\left[(\omega_1 + \omega_{\mathrm{RF}})\, t + \varphi\right], \tag{13.12}$$

where $m(t)$ is the base band signal which is amplitude modulated and φ denotes an arbitrary phase shift. A photodiode is – like all optical detectors – unable to detect the electric field. Only the intensity of the whole field leads to its output current, which depends on the square of the superposition between both waves, i.e., $I \sim (E_1(t, \omega_1) + E_2(t, \omega_2))^2$. The intensity contains components with twice the frequencies of the fundamental waves as well as sum and difference frequencies. The diode is fast enough only for the difference frequency. Therefore, the output current of the photo diode I_D is

$$I_D = C\; m(t)\; \cos\left(\omega_{\mathrm{RF}} t + \varphi\right), \tag{13.13}$$

[1] An electro-absorption modulator exploits the Franz-Keldysch effect. If a semiconductor is placed in an electric field it can change its bandgap. If the bandgap is reduced, the semiconductor absorbs wavelengths at which it was formerly transparent. Therefore, the effect is called electro-absorption. It is possible to modulate an optical signal by an external electric field.

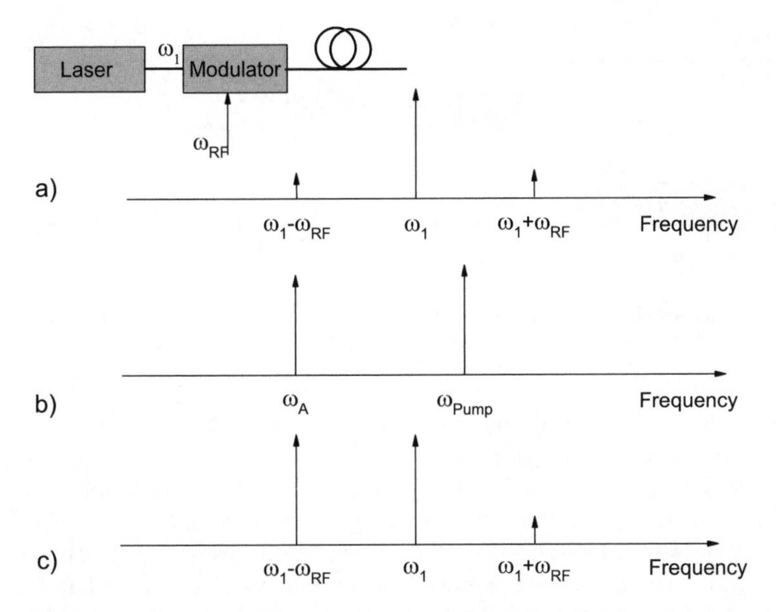

Fig. 13.8. Spectrum of an amplitude modulated optical carrier (**a**), a pump wave and corresponding Brillouin scattered wave (**b**) and a Brillouin amplified signal (**c**)

with C as a constant. The output current of the diode follows the frequency of the RF signal and the modulated base band information stays unaffected. A phase shift (φ) in the optical domain stays the same in the electrical domain. The output signal of the photodiode can now be amplified and radiated via an antenna.

An external amplitude modulation with a narrow band signal is shown in Fig. 13.8a where most of the power is contained in the carrier. In Fig. 13.8b, the spectrum of a pump wave and a Stokes wave, generated and frequency-shifted via Brillouin scattering, is shown. If the amplitude-modulated signal propagates in the opposite direction towards a pump wave, and if the frequency difference between the pump and its lower sideband corresponds to the Brillouin shift in the fiber, then a frequency-selective, narrow-band amplification of the lower sideband occurs, as shown in Fig. 13.8c. The heterodyne superposition of the spectrum in Fig. 13.8c in a photodiode leads to the RF signal. A schematic sketch of such an Brillouin amplifier is shown in Fig. 13.9.

If the set up is compared to other possible optical amplification mechanisms, the method above is much more efficient because the total power of the scattered pump wave is transferred to the sideband. Furthermore, since the strong carrier is not amplified, a saturation of the photodiode is suppressed. Cheap diodes can serve as pump laser sources. An additional advantage lies in the suppression of fiber dispersion [312]. In the case of a conventional demodulation of an amplitude-modulated signal, both sidebands and the carrier are mixed in the receiver. Due to the fiber dispersion, the upper sideband

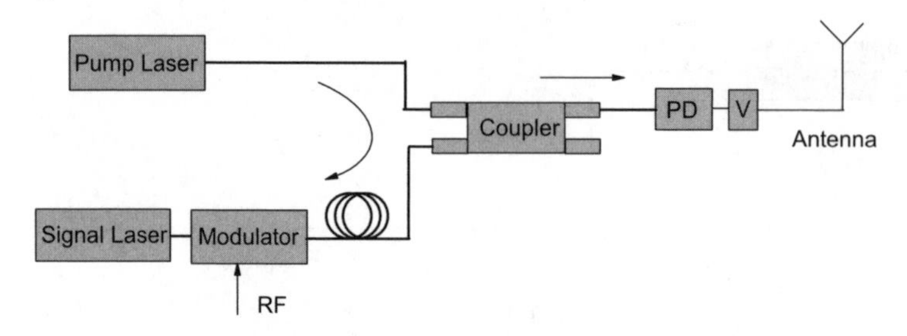

Fig. 13.9. Principle sketch of the generation of an RF signal with a Brillouin amplifier [312] (PD=photodiode, V=electrical amplifier)

propagates with velocity other than the lower. Therefore, both accumulate different phases which will lead to a fading of the mixed signal in dependence on the fiber length. For a solution of the problem, an one-sideband modulation [313][2] and the incorporation of an optical filter for the suppression of one of the sidebands were proposed [314]. The above set-up works, in principle, in a similar way – the upper side band is not amplified and has a negligible influence on the demodulated signal when compared to the lower sideband.

13.4 Parametric Amplifiers

In the cases of Raman and Brillouin scattering, the medium is directly involved in the nonlinear process because an energy transfer between the optical waves and the material takes place. Whereas for SPM, XPM and FWM, the refractive index of the material becomes intensity-dependent due to the nonlinear interaction. Since the refractive index is a parameter of the material, these effects are considered parametric.

Traditionally, optical parametric amplification is a $\chi^{(2)}$ process treated in Sect. 4.5. An incident pump wave is divided in a signal and an idler wave. If a wave with the signal frequency is already present in the material, it will be amplified. A similar behavior for a $\chi^{(3)}$ process takes place in the case of FWM – a part of the power of the pump waves is transferred to the new generated idler wave, while the signal wave is amplified in the medium at the same time. Hence, it is possible to exploit the FWM as a nonlinear optical amplification mechanism and FWM amplifiers are commonly called optical parametric amplifiers (OPA) as well. An OPA can, like a Raman amplifier, be tailored to operate at any wavelength. On the other hand, the parametric process does not depend not on energy transitions between energy states like in Raman and erbium-doped amplifiers. Therefore, it enables a wideband and a flat gain profile [310].

[2]The spectrum of the modulated signal contains only the carrier and one of the side bands.

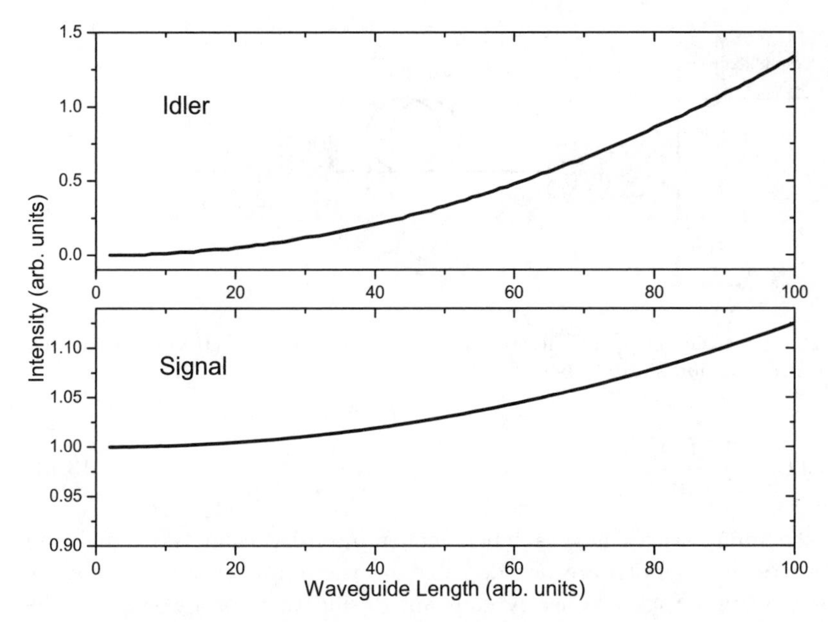

Fig. 13.10. Signal and idler waves generated by FWM in a waveguide in the case of complete phase matching

Figure 13.10 shows the behavior of the signal and idler waves along a waveguide if the phases between the waves are phase-matched. As can be seen from the figure, both waves increase quadratically.

The principle set up of a parametric amplifier is shown in Fig. 13.11. Since degenerate FWM is exploited hence, a pump laser delivers the pump waves which are then launched into the fiber together with the signal. During the propagation in the fiber, the pump waves transfer a part of their power to the signal due to the nonlinear interaction. At the same time, a new idler wave with the frequency $\omega_i = 2\omega_p - \omega_s$ is generated. Since only the amplified signal wave is of interest the pump and idler waves are blocked at the fiber output. The only difference to the parametric wavelength converter shown in Fig. 12.4 is the bandpass which blocks the signal and the pump frequency. Therefore, every OPA can be used as a wavelength converter and vice versa if the bandpass is changed.

The efficiency of the amplifier depends on the nonlinearity coefficient (γ) and the phase mismatch between the involved waves (Δk). Hence, for the construction of short compact devices, fibers with a high γ and a small Δk are required. These conditions are fulfilled by HNL-DSF (highly nonlinear dispersion-shifted fiber) or by highly nonlinear holey fibers. The gain of an amplifier is the ratio between the signal intensity (or power) at the fiber output to the signal intensity at the fiber input:

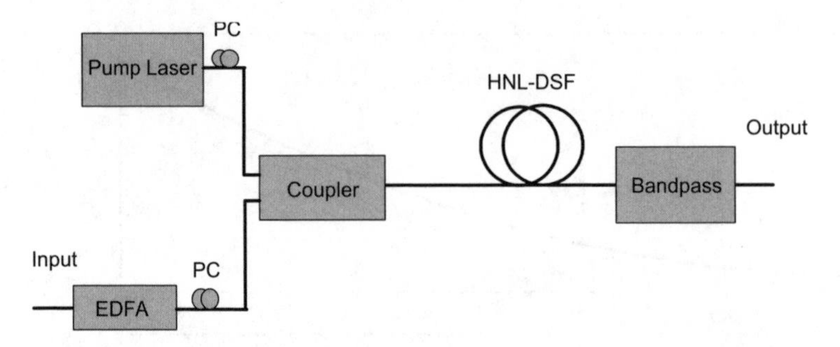

Fig. 13.11. Basic set-up of an optical parametric amplifier (HNL-DSF=highly nonlinear dispersion-shifted fiber)

$$G_s(L) = \frac{|E_s(L)|^2}{|E_s(0)|^2} \tag{13.14}$$

If the pump waves have a much higher intensity than the signal and idler waves and if both are not depleted by the amplification process, we can assume that they are nearly constant during their propagation in the fiber. Therefore, in the coupled equation system (7.13) to (7.16), the equations which describe the pump propagation ($\frac{\partial A_1}{\partial z} = \frac{\partial A_2}{\partial z} = 0$) and the FWM terms responsible for the pump can be neglected. The analytical solution of the residual coupled equations together with (13.14) gives the gain of the amplifier when the attenuation is neglected [296][295]:

$$G_s(L) = 1 + \left[\frac{\gamma P_0}{g} \sinh(gL)\right]^2, \tag{13.15}$$

with L as the length of the waveguide, γ as the nonlinearity coefficient, P_0 as the power of the degenerate pump, and g as a parametric gain coefficient [189]:

$$g^2 = -\Delta k \left(\frac{\Delta k}{4} + \gamma P_0\right). \tag{13.16}$$

Here, Δk denotes again the phase mismatch between the distinct waves due to the dispersion in the fiber. With k_s, k_i, and k_p as the wavevectors of the signal, idler, and pump, respectively, we have

$$\Delta k = k_s + k_i - 2k_p. \tag{13.17}$$

Figure 13.12 shows the parametric gain coefficient against the phase mismatch Δk for different pump powers. The maximum of the gain coefficient does not coincide not with the minimum of the phase mismatch, $\Delta k = 0$. This is caused by the influence of SPM and XPM, which alters the phases. As denoted in the figure, the maximum value of the parametric gain for a phase mismatch of $\Delta k = -2\gamma P_0$ is

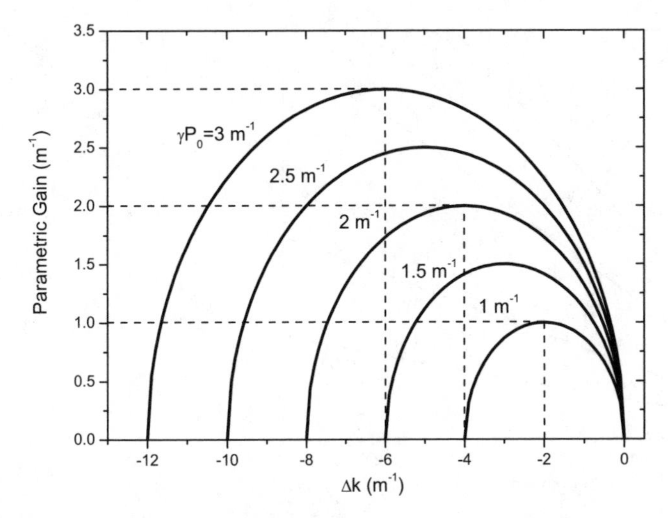

Fig. 13.12. Parametric gain coefficient g due to FWM against the phase mismatch for different pump powers

$$g_{\max} = \gamma P_0. \tag{13.18}$$

For a comparison between the gain coefficient of the parametric process and the gain coefficients for scattering processes (SRS and SBS), the nonlinearity coefficient γ has to be expressed with (5.29). Using (13.18) we have

$$g_{\max} = g_{\mathrm{p}} P_0 / A_{\mathrm{eff}}, \tag{13.19}$$

where $g_{\mathrm{p}} = 2\pi n_2/\lambda$. For an average wavelength of 1.55 μm and a nonlinear refractive index of $n_2 = 2.6 \times 10^{-20}$ m^2/W, the gain coefficient is $g_{\mathrm{p}} \approx 1.05 \times 10^{-13}$ m/W. Therefore, it is $\approx$ 3.4 times larger than the gain coefficient of stimulated Raman scattering ($g_{\mathrm{R}} \approx 3.1 \times 10^{-14}$ m/W), but $\approx$ 190 times smaller than the gain factor of SBS ($g_{\mathrm{B}} \approx 2 \times 10^{-11}$ m/W).

Figure 13.13 shows the maximum gain of a parametric amplifier against the fiber length for ($g = \gamma P_0$) and different input powers, assuming the fiber attenuation is negligible.

As can be seen from the figure, a pump power of 1W and 1.8 km of dispersion-shifted fiber are required for an amplification of 25.4 dB. For small scale compact devices, highly nonlinear fibers like HNL or holey microstructured fibers have to be incorporated. Tang et al. [297] achieved a peak net gain of 25.4 dB with a peak pump power of 12.7 W in a microstructured fiber with a length of only 12.5 m and a nonlinearity coefficient of $\gamma = 24$ W^{-1}km^{-1}. The gain was higher than 20 dB over a wavelength band of 30 nm in the telecom range around 1540 nm.

If (13.18) is introduced into (13.15) and the pump power is high ($\gamma P_0 L \gg 1$), the maximum amplifier gain is

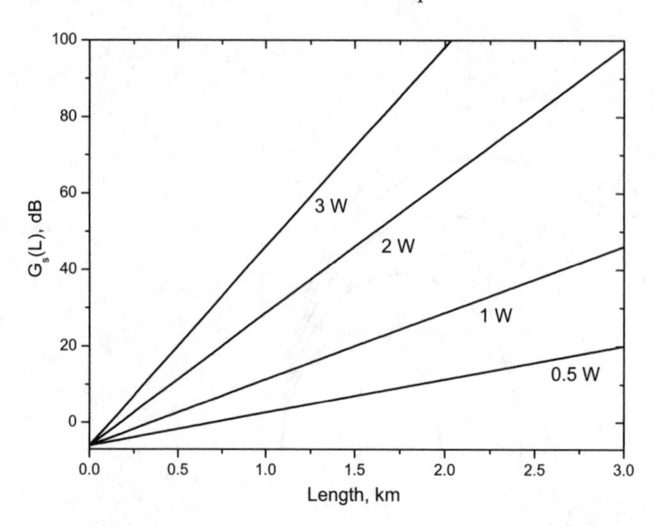

Fig. 13.13. Amplification gain of a parametric amplifier against the fiber length for a maximum gain coefficient ($\gamma = 2\ \mathrm{km}^{-1}\mathrm{W}^{-1}$, $g = \gamma P_0$)

$$\begin{aligned} G_{\mathrm{smax}}(L) &\approx \sinh^2(\gamma P_0 L) \\ &= \frac{\left(\mathrm{e}^{\gamma P_0 L} - \mathrm{e}^{-\gamma P_0 L}\right)^2}{4} \\ &\approx \frac{1}{4}\mathrm{e}^{2\gamma P_0 L}. \end{aligned} \tag{13.20}$$

If (13.20) is rewritten in decibel units, a very simple equation for the gain maximum of the amplifier results [275]:

$$\begin{aligned} G_{\mathrm{s\,max}}(dB) &= 10\log_{10}\left(\frac{1}{4}\mathrm{e}^{2\gamma P_0 L}\right) \\ &= P_0 L S_P - 6, \end{aligned} \tag{13.21}$$

with $S_P = 10\log_{10}\left(\mathrm{e}^2\right)\gamma \approx 8.7\gamma$. For 200 m of a dispersion-shifted fiber with a nonlinearity coefficient of $\gamma = 2\ \mathrm{W}^{-1}\mathrm{km}^{-1}$ and a pump power of 7 W a maximum gain of $G_{\mathrm{s\,max}} \approx 18.4$ dB can be calculated.

If the signal wavelength coincides with the pump wavelength, then the phase mismatch is zero and the parametric gain coefficient is, according to (13.16), zero as well. Nevertheless, there is still amplifier gain. If $\sinh(x)$ in (13.15) is developed in a Taylor series the amplifier gain can be expanded as follows [275]:

$$G_{\mathrm{s}}(L) = 1 + (\gamma P_0 L)^2 \left[1 + \frac{gL^2}{6} + \frac{gL^4}{120} + \ldots\right]^2. \tag{13.22}$$

Therefore, in the case where $\lambda_s = \lambda_p$, the amplifier gain is

$$G_{\mathrm{s}}(L)_{\lambda_s=\lambda_p} = 1 + (\gamma P_0 L)^2. \tag{13.23}$$

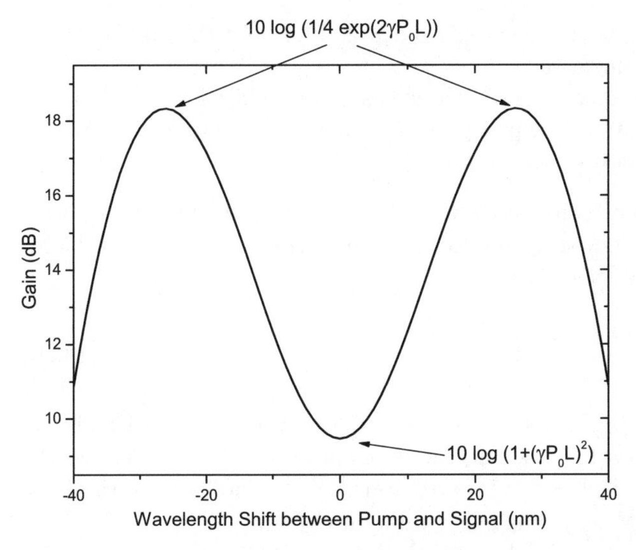

Fig. 13.14. Calculated gain spectrum of a fiber OPA for $\lambda_p - \lambda_0 = 0.8$ nm ($\gamma = 2\times10^{-3}$ m^{-1}W^{-1}, $P_0 = 7$ W, $L = 200$ m, $k_2 = k_3\,(\omega_p - \omega_0)$, $k_3 = 1.2\times10^{-40}$ s^3/m, $k_4 = 2.5 \times 10^{-55}$ s^4/m)

Under the assumptions for (13.21), we have $G_{\rm s}(L)_{\lambda_s=\lambda_p} = (\gamma P_0 L)^2$. Using the values from the example yields $G_{{\rm s}\lambda_s=\lambda_p} \approx 9.46$ dB. Therefore, the gain spectrum varies between 9.46 dB and 18.4 dB. Figure 13.4 shows the amplifier gain versus the shift between the pump and signal wavelengths.

The distribution of the amplifier gain depends on the phase matching between the involved waves. According to Fig. 13.12 outside the inequality,

$$-4\gamma P_0 \le \Delta k \le 0, \tag{13.24}$$

the gain coefficient is zero. The bandwidth of the amplifier as a function of the phase mismatch is therefore $4\gamma P_0$. Thus the bandwidth of the amplifier can be increased with a higher nonlinearity coefficient (γ) and a higher input power (P_0). On the other hand, the frequency bandwidth of the amplifier depends on the fiber dispersion. Near the zero dispersion of the fiber, higher order dispersion terms play an important role. If k is expanded in a power series around the frequency of the pump wave, the phase mismatch is [189]:

$$\Delta k = 2\sum_{m=1}^{\infty} \frac{k_{2m}}{(2m)!}\left(\omega_{\rm s} - \omega_{\rm p}\right)^{2m}, \tag{13.25}$$

where k_{2m} denotes the $2m$th derivation of k at $\omega_{\rm p}$. According to (13.25), the phase mismatch and, therefore, the amplification depends on even powers of the frequency difference between the pump and signal waves. Hence, the amplification bandwidth of a parametric amplifier is symmetrical around the pump frequency (ω_p), as can be seen from Fig. 13.4, and the odd orders of the dispersion can be neglected.

According to Marhic et al. [189], three cases can be distinguished. These cases differ in the position of the pump frequency (ω_p) relative to the zero-crossing of the dispersion in the fiber (ω_0):

1. The pump frequency coincides with ω_0. In this case, the phase mismatch in (13.25) is determined by the fourth order dispersion (k_4). The bandwidth of the amplifier is then

$$\Delta\omega_{2m} = \left[-\frac{2\,(2m)!\gamma P_0}{k_{2m}}\right]^{1/2m}, \tag{13.26}$$

   where $m = 2$, the pump wavelength is 1.55 μm, the fourth order dispersion is $k_4 = -4.93 \times 10^{-55}$ s^4/m, the input power is $P_0 = 15$ W, and the nonlinearity coefficient is $\gamma = 2 \times 10^{-3}$ m^{-1}W^{-1}. The bandwidth of the amplifier ($\Delta\lambda \approx \frac{\lambda^2}{2\pi c}\Delta\omega$) thus has a value of ≈ 53 nm.
2. The pump frequency is so close to ω_0 that in (13.25), the second order dispersion (k_2) as well as the fourth order dispersion (k_4) has to be considered. In this case, the sign and the value of k_2 can be chosen in a manner that together with k_4, a bandwidth optimum occurs. Hence, the bandwidth can be higher than in the first case. Figure 13.15 shows the bandwidth of an optical parametric amplifier for three different frequency shifts between λ_0 and λ_p. If the pump wavelength is 0.55 nm higher than the wavelength for dispersion zero-crossing in the fiber, the amplifier has a 3 dB bandwidth of more than 40 nm. Note that since the parameters for pump power, nonlinear coefficient, and fiber length are the same as in Fig. 13.4, the same values for maxima and local minimum can be seen.
3. The pump frequency is far away from ω_0. Only the second order dispersion (k_2) has to be considered in the equation for the phase mismatch. The bandwidth of the amplifier is determined by (13.26) with $m = 1$.

In a highly nonlinear fiber ($\gamma = 18$ km^{-1}W^{-1}) with a length of 20 m, Ho et al. [190] showed optical parametric amplification with a bandwidth of more than 200 nm for a pulsed 10-W pump. The zero-crossing of the dispersion in this fiber was at 1540.2 nm. The pump wavelength was chosen at 1542.4 nm so that the gain spectrum showed its optimum. The parametric amplifier had a gain of more than 10 dB in the range between 1443 and 1651 nm. The shape of the spectrum was similar to that in Fig. 13.4 but, of course, much broader. According to the values for pump power, length and nonlinearity coefficient, the maxima of the distribution were ≈ 25.3 dB and the local minimum was ≈ 11.5 dB. Therefore, the gain difference was 13.8 dB. For such broadband amplifiers, the bandwidth of the amplifier exceeded the Raman shift of ≈ 110 nm. According to the authors, the influence of the Raman scattering was a weak asymmetry of the gain spectrum; for shorter wavelengths, the gain decreased while it increased for longer wavelengths.

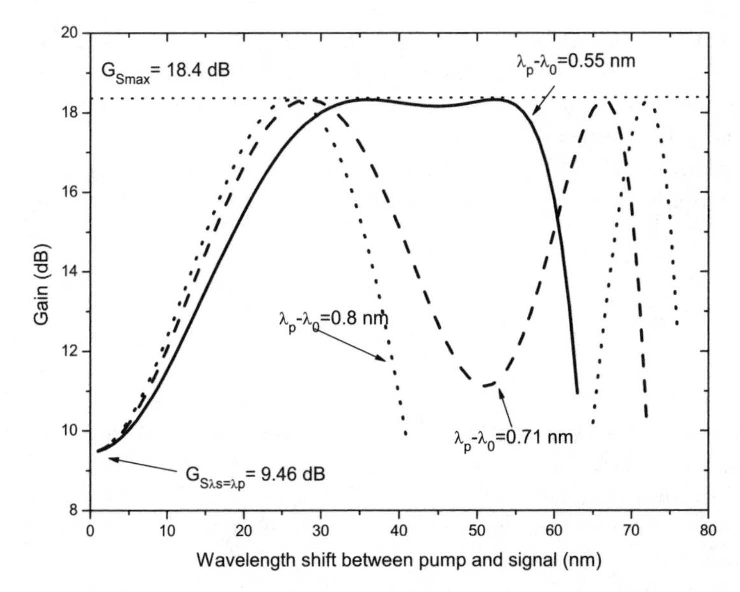

Fig. 13.15. Simulated gain spectrum of an optical parametric amplifier. Note, that the Figure shows only one side of the symmetric spectrum ($\gamma = 2 \times 10^{-3}$ m^{-1}W^{-1}, $P_0 = 7$ W, $L = 200$ m, $k_2 = k_3\,(\omega_p - \omega_0)$, $k_3 = 1.2 \times 10^{-40}$ s^3/m, $k_4 = 2.5 \times 10^{-55}$ s^4/m) [189]

Summary

- Nonlinear amplification mechanisms are independent of the frequency and possible in every standard single mode fiber. For small compact devices, fibers with a high nonlinearity can be incorporated.
- Raman amplification can be carried out in the entire transmission bandwidth of an optical fiber. If many pump lasers are combined, a broadband amplification is possible. A particular advantage of the Raman amplification is the distributed amplification that is inherent in the process.
- Distributed Raman amplification with co-propagating pump provides very low equivalent noise figures [191]. If combined with counter-propagating pumping, it reduces multipath interference noise induced by double Rayleigh signal scattering [192].
- A Raman laser is a Raman amplifier in a resonator cavity. For the resonator fiber, Bragg gratings directly written in the fiber core are possible. Raman laser can be built as compact, tunable small scale fiber devices.
- Brillouin amplifiers require a very low pump power. On the other hand, due to their small bandwidth, they are only suitable for narrow band applications like radio-over-fiber.
- Radio-over-fiber is a technique to transmit radio signals over large distances in optical fibers.

- Optical parametric amplifiers exploit the effect of degenerate FWM for amplification. Therefore, they do not depend on energy transitions between energy states like in Raman and erbium-doped amplifiers and enable a wideband and a flat gain profile
- If the pump wavelength is slightly shifted to the fiber zero-dispersion with parametric amplifiers, extremely wideband amplification with only one pump source is possible.

Exercises

13.1. Determine the gain (ratio between input and output power) of a Raman amplifier pumped with a power of 5 W in decibel units (dB) and the gain of the Raman process alone in a microstructured fiber with a length of 100 m, a Raman gain coefficient of $g_{\mathrm{R\,max}} = 7.6 \times 10^{-14}$m/W, and a loss of 40 dB/km. The effective area in the fiber at a wavelength of 1550 nm is $A_{\mathrm{eff}} = 2.85\ \mu\mathrm{m}^2$.

13.2. Suppose a distributed Raman amplification in an SSMF ($A_{\mathrm{eff}} = 80\ \mu\mathrm{m}^2$; $g_{R\,\mathrm{max}} = 3.5 \times 10^{-14}$m/W; $L = 30$ km; $\alpha = 0.2$ dB/km). Estimate the number of WDM channels from the C-band (100 GHz channel spacing) that can be amplified with a single 1 W pump if the maximum alteration of the amplification gain should not exceed ± 1.5 dB.

13.3. Calculate the gain of a Brillouin amplifier in decibel units in an SSMF with a length of 30 km if the pump depletion is neglected. The fiber has an effective area of 80 µm^2, an attenuation constant of 0.2 dB/km and a Brillouin gain of 2×10^{-11} m/W. The pump power is 7 mW.

13.4. An universal mobile telecommunications standard (UMTS) signal (carrier frequency $f_c = 2$ GHz) has to be transmitted in the fiber C-band with a radio-over-fiber system. Calculate the required wavelength difference if a heterodyne superposition is used to produce the carrier.

13.5. Estimate the fiber length that is required for a parametric amplifier with a pump power of 2 W and a maximum gain of 50 dB if a DSF ($\gamma = 2\ \mathrm{W}^{-1}\mathrm{km}^{-1}$) or a holey fiber with $\gamma = 35\ \mathrm{W}^{-1}\mathrm{km}^{-1}$ is used as the nonlinear medium.

13.6. Determine the maximum gain and the gain for coincidence between the pump and signal wavelengths in a parametric amplifier in an experiment carried out by Tang et al. [297] ($\gamma = 24\ \mathrm{W}^{-1}\mathrm{km}^{-1}$, $L = 12.5$ m, $P_0 = 12.7$ W).

13.7. Determine the bandwidth of an OPA if the pump frequency coincides with the zero-crossing of the fiber dispersion. The pump power is 5 W and the fiber parameters are $\lambda_0 = 1.55\ \mu$m, $k_4 = -4.93 \times 10^{-55}\ \mathrm{s}^4/\mathrm{m}$, and $\gamma = 2 \times 10^{-3}\ \mathrm{m}^{-1}\mathrm{W}^{-1}$).

14. Nonlinear Optical Phase Conjugation

A very fascinating application of nonlinear optical effects is the nonlinear phase conjugation (NPC) which will be discussed in this Chapter. Under particular circumstances, it is possible to compensate optical distortions with NPC and optical pulses in a fiber can be regenerated. Therefore, in the range of optical telecommunications the NPC, can be exploited for the compensation of group velocity dispersion [193] as well as nonlinear effects like SPM [194] or Raman scattering [195].

Other interesting fields of application include phase-conjugating mirrors in laser systems for a correction of beam quality [196, 197] and the optical control of laser beams [198].

A phase-conjugated replica of the incident wave can be generated with FWM and the backscattered waves of stimulated scattering processes like SRS and SBS can also be exploited for NPC.

In this chapter the basic difference between a conventional and a phase-conjugating mirror is treated first. Section 14.2 then shows the background of distortion compensation with a phase-conjugated replica. In Sect. 14.3 the connections between NPC and nonlinear effects like FWM and stimulated scattering are treated followed by an introduction to a holographic model for the description of NPC in Sect. 14.4. The last section deals with the application of NPC for distortion compensation in optical transmission systems using a technique called "mid-span spectral inversion".

14.1 Phase-Conjugating Mirrors

The NPC was mentioned for the first time in the early seventies [199, 200]. Figure 14.1 shows the basic experimental set-up for the first verification of phase conjugation where the beam of a ruby laser was send through a distorting medium, before being focused into a nonlinear material (CF_6). The intensity of the laser was high enough to initiate stimulated Brillouin scattering in the nonlinear medium. The backscattered beam then passed the distorting medium again and was coupled out with a beam splitter.

As was shown by a comparison between the incident and the backscattered beam, although the backscattered beam moved through the distorting

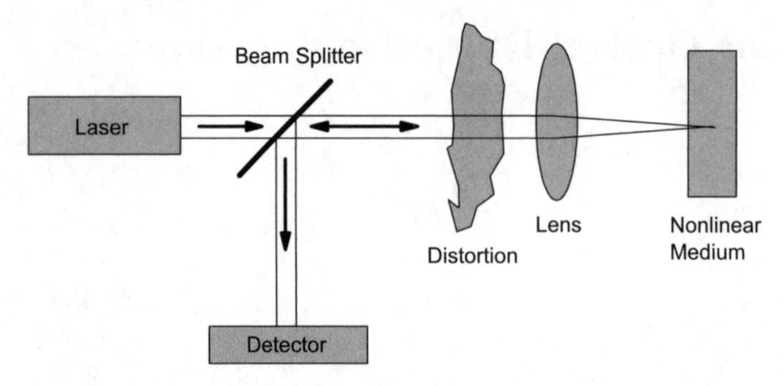

Fig. 14.1. Basic experimental set-up for the first proof of nonlinear optical phase-conjugation [199]

medium twice, it had the same divergence as the incident beam. Evidently, the distortion influence was compensated.

Two optical waves are phase-conjugated to each other if both have the same spatial wave front but move in opposite directions. Hence, the incident wave front will be reflected in a special manner by the phase-conjugating medium. Therefore, the nonlinear medium can be seen as a kind of a mirror. As will be shown later, due to phase conjugation by the original wave, a second wave is generated in the medium and the position-dependent part of this second wave is a complex conjugate of that of the original wave.

The basic difference between a conventional and a phase-conjugating mirror is shown in Fig. 14.2. In the case of a conventional mirror due to reflection, the incident wave front is turned over in Fig. 14.2a, while a phase-conjugating

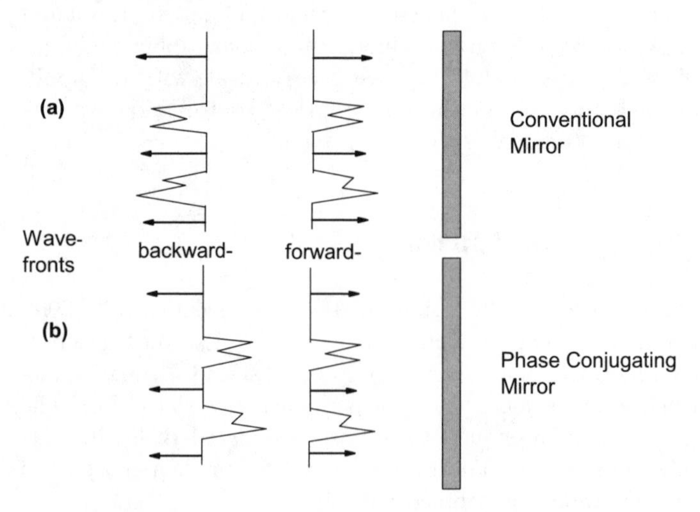

Fig. 14.2. Reflection of an arbitrary wavefront on a conventional (**a**) and a phase-conjugating mirror (**b**)

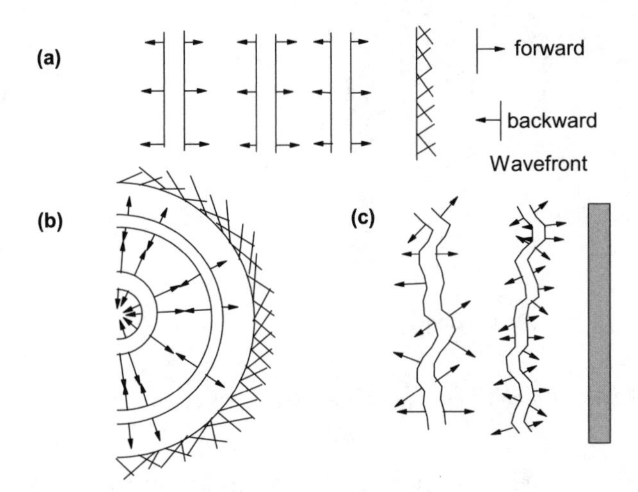

Fig. 14.3. Generation of a phase-conjugated wave for plane waves at a plane mirror (**a**), for spherical wavefronts at a spherical mirror (**b**), and for arbitrary wavefronts at a phase-conjugating mirror (**c**)

mirror generates the exact copy of this wave front. The propagation path of the copy is opposite to the original wave front. The easiest way to fulfill this condition is a perfectly plane wave reflected at a perfectly plane mirror, as shown in Fig. 14.3a.

Another possibility is the reflection of an ideal spherical wave front at an ideal spherical mirror with the same radius (Fig. 14.3b). Both of these cases are very special. Real phase-conjugating mirrors can generate a reverse copy of each arbitrary wave front (Fig. 14.3c). A conventional mirror can be used to generate a phase-conjugated replica only if the mirror has the same shape as the incident wave front. Therefore, phase-conjugating reflection can be seen as a kind of adaptation of the mirror shape to the wave front. A nonlinear material is required as a medium for this kind of mirror.

Another interesting difference between NPC and a conventional mirror is shown in Fig. 14.4. As already mentioned, the phase-conjugated wave is the exact replica of the incident wave front propagating in exactly the opposite direction. The behavior of a phase-conjugating mirror can be understood with the following simplified consideration. Suppose the incident wave is a superposition of a huge number of wave fronts with different orientations and the mirror is divided into the same number of parts, each of which is parallel to the corresponding wave front. Each subwave front will be reflected at its corresponding part of the mirror in the same way as in Fig. 14.2a. The superposition of the reflected wave fronts gives rise to the original wave moving in the opposite direction. From a beam optics point of view, this means that each individual beam is reflected into the direction where it came from. Therefore, as shown in the figure, originally divergent beams will converge at the point

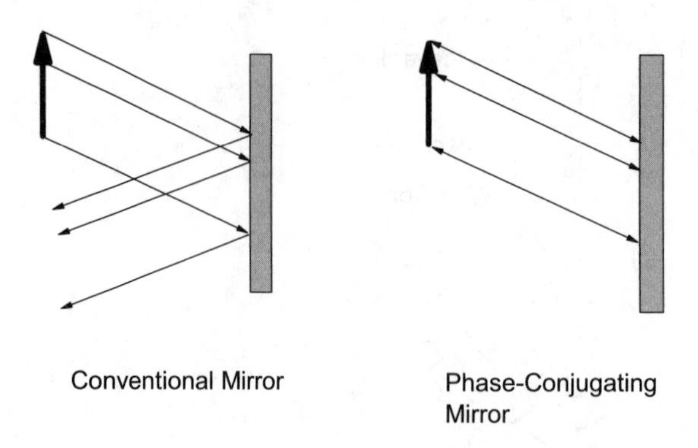

Fig. 14.4. Comparison between a conventional and a phase-conjugating mirror

where they were emitted after reflection on a phase-conjugating mirror. Contrary to that, a conventional mirror's divergent beams stay divergent after a reflection. This means that an observer watches his or her own mirror image in a conventional mirror. Some of the beams originally emitted by the ears, for example, will hit the eye after reflection. Whereas, after reflection on a phase-conjugating mirror, the beams emitted by the ears will always hit the ears and never the eye of the observer. Only the beams emitted by the eye will be directed back to it. As a consequence, the phase-conjugating mirror appears to be an evenly illuminated plane to the observer.

14.2 Distortion Compensation due to Phase Conjugation

The important potential of NPC for optical telecommunications lies in the possibility to compensate distortions in the fiber caused by linear effects, such as the chromatic dispersion, or by nonlinear effects like SPM and Raman scattering. The behavior of a wave propagating through a distorting medium, for instance a wet windshield, is shown in Fig. 14.5.

If the wave passes the windshield for the first time, its wave front will be distorted by the water droplets. If the wave is reflected at a conventional mirror, the wave front is turned over and the wave propagates in the opposite direction (Fig. 14.5a). After reflection, the wave passes the windshield a second time and its wave front will again be distorted by the droplets. On the other hand, if the wave is reflected at a phase-conjugating mirror, the distortion will be compensated successively during the second propagation through the medium, as shown in Fig. 14.5b. If the wave leaves the medium, its original wave front is completely regenerated.

If the behavior of the forward propagating wave (E^i) is filmed by a camera and recorded by a video recorder, the same behavior as in Fig. 14.5b would

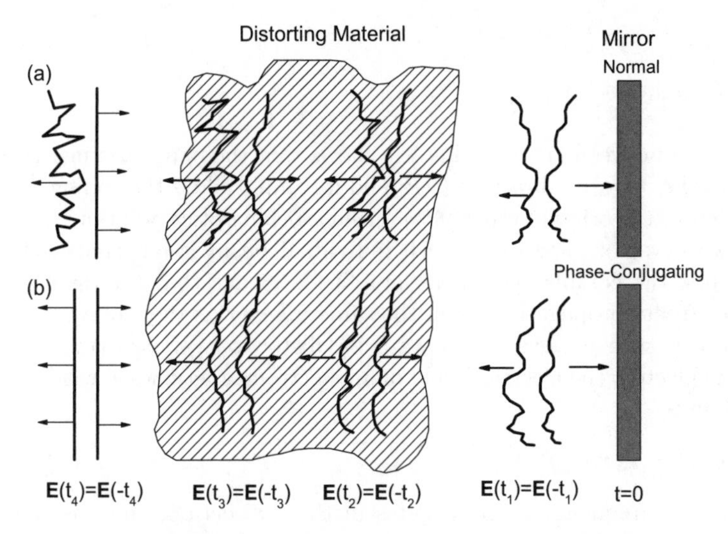

Fig. 14.5. Wave fronts distorted by a medium. **a** Reflection at a conventional mirror. **b** Distortion compensation due to a phase-conjugating mirror

be observed if the recorded tape is played in the opposite direction (from the end to the beginning). The back-propagating wave front (E^r) at the time t_4 has the same shape as the forward propagating wave front 8 time units earlier ($-t_4$). After reflection one moves further back into the past of the wave front. Therefore, the phase conjugation allows a view back in time, but only as long as the properties of the distorting medium will not change. Hence, the phase-conjugated wave is the time-reversed copy of the original wave.[1]

A monochromatic wave which propagates to the phase-conjugating mirror can be described as

$$E^i = \frac{1}{2}\left(A^i \mathrm{e}^{\mathrm{j}(kz-\omega t)} + \mathrm{c.c.}\right) = \frac{1}{2}\left(A^i \mathrm{e}^{\mathrm{j}(kz-\omega t)} + A^{i*}\mathrm{e}^{-\mathrm{j}(kz-\omega t)}\right). \quad (14.1)$$

According to (2.18), (14.1) can be written as

$$E^i = \left|A^i\right| \cos\left(kz - \omega t + \varphi\right). \quad (14.2)$$

If the wave is reflected at the phase-conjugating mirror, it moves into the opposite direction and it has a negative phase angle φ. Hence, for the phase-conjugated wave we have

$$E^r = r\left|A^i\right| \cos\left(-kz - \omega t - \varphi\right), \quad (14.3)$$

where r is an arbitrary reflection factor. If (14.3) is expressed in exponential form it can be written

[1] The same behavior occurs for acoustic waves as well [201].

$$E^r = \frac{r}{2}\left(A^i \mathrm{e}^{-\mathrm{j}\varphi}\mathrm{e}^{\mathrm{j}(-kz-\omega t)} + A^i \mathrm{e}^{+\mathrm{j}\varphi}\mathrm{e}^{-\mathrm{j}(-kz-\omega t)}\right) \\ = \frac{r}{2}\left(A^{i*}\mathrm{e}^{\mathrm{j}(-kz-\omega t)} + \text{c.c.}\right). \tag{14.4}$$

A comparison between (14.4) and (14.1) shows that the phase-conjugated wave moves in the opposite direction ($-k$) and its amplitude is the conjugate complex of that of the original wave (A^{i*}). From a beam optics point of view, the inverse wave vector shows that each individual beam is reflected back into itself, while the complex conjugate amplitude means that the distinct reflected wave fronts propagate back in the same manner as they arrived.

But, how can such a wave compensate for distortions? If the complex conjugate is neglected, the electrical field of an electromagnetic wave with an arbitrary phase is

$$E^i(z, x, y, \omega) = A(z, x, y)\mathrm{e}^{\mathrm{j}(kz-\omega t)}. \tag{14.5}$$

The wave has the frequency ω, propagates in the z-direction, and has the wave number k. The term $A(z, x, y)$ is the complex amplitude of the wave that is dependent on the coordinates x, y and z and contains the complex phase

$$A(z, x, y) = A_0\mathrm{e}^{\mathrm{j}(\varphi(z,x,y)+\varphi_0)}, \tag{14.6}$$

where $\varphi(z, x, y)$ is a phase function which describes the difference of the considered wave front to an ideal plane wave, A_0 is a real amplitude, and φ_0 is a phase angle which is assumed to be zero in the following. For an ideally plane wave, the phase function is $\varphi(z, x, y) = 0$. If this wave is reflected at an ideal plane mirror (Fig. 14.3a), the reflected wave is

$$E^r(z, x, y, \omega) = rA_0\mathrm{e}^{\mathrm{j}(-kz-\omega t)}, \tag{14.7}$$

with r as the reflection factor of the mirror. According to the definition, (14.7) is in fact the phase-conjugated wave to (14.5) because both have the same spatial wave front and they move into opposite directions. But this only holds as long as the wave front is really a plane wave.

If the plane wave front penetrates an attenuation-free distorting medium (see Fig. 14.5), its wave front and its phase will be changed, i.e., $\varphi(z, x, y) \neq 0$. Equation (14.5) then becomes [129]:

$$E^v(z, x, y, \omega) = A(z, x, y)\mathrm{e}^{\mathrm{j}(kz-\omega t)} = A_0\mathrm{e}^{\mathrm{j}(kz+\varphi(z,x,y)-\omega t)}. \tag{14.8}$$

After reflection at the same perfectly plane mirror, the reflected wave at the input of the distorting medium is

$$E^r(z, x, y, \omega) = rA(z, x, y)\mathrm{e}^{\mathrm{j}(-kz-\omega t)} = rA_0\mathrm{e}^{\mathrm{j}(-kz+\varphi(z,x,y)-\omega t)}. \tag{14.9}$$

If the reflected wave propagates a second time through the distorting medium, its wave front will be changed further. At the output of the medium, the wave will then be

$$E^{rv}(z,x,y,\omega) = rA(z,x,y)\mathrm{e}^{\mathrm{j}(-kz-\omega t)} = rA_0\mathrm{e}^{\mathrm{j}(-kz+2\varphi(z,x,y)-\omega t)}. \quad (14.10)$$

Since the wave propagates twice through the same distorting medium, the influence of the distortion on the wave front is twice as strong.

On the other hand, if the mirror is a phase-conjugating one, the distorted wave is phase-conjugated to (14.8) after reflection. The wave at the input of the distorting medium is then

$$E^{r}(z,x,y,\omega) = r^{p}A^{*}(z,x,y)\mathrm{e}^{\mathrm{j}(-kz-\omega t)} = r^{p}A_0\mathrm{e}^{\mathrm{j}(-kz-\varphi(z,x,y)-\omega t)}, \quad (14.11)$$

with r^p as the reflection factor of the phase-conjugating mirror and A^* as an amplitude that is the complex conjugate of the original amplitude:

$$A = A_0\mathrm{e}^{\mathrm{j}\varphi(x,y,z)} \quad (14.12)$$

and

$$A^{*} = A_0\mathrm{e}^{-\mathrm{j}\varphi(x,y,z)}. \quad (14.13)$$

If this wave propagates through the same distorting medium again, it becomes

$$E^{rv}(z,x,y,\omega) = r^{p}A_0\mathrm{e}^{\mathrm{j}(-kz-\omega t)} \quad (14.14)$$

at the output of the medium.

A comparison between (14.14) and (14.5) shows that the original wave appears again if it propagates twice through the same distorting medium. Therefore, phase-conjugating mirrors can in fact compensate for phase distortions of the wave.

14.3 Theoretical Description of NPC

Suppose the NPC is generated by four-wave-mixing in a nonlinear medium in an arrangement similar to that shown in Fig. 14.7. The waves E_1 and E_2 are pump waves and E_3 is a signal wave propagating in the forward direction. Due to FWM, a phase-conjugated idler wave E_4, propagating in the backward direction is generated. The distinct waves E_1 to E_4 are plane and monochromatic and have the amplitudes A_1 to A_4. Hence, the signal wave can be approximated as follows:

$$E_3 \sim A_3\mathrm{e}^{\mathrm{j}(kz-\omega_3 t)}. \quad (14.15)$$

The FWM can be described by the coupled equations (7.13) to (7.16). Only the FWM terms are responsible for a change of the amplitudes, while SPM and XPM lead to an alteration of the phases. At the same time, the medium is rather short and the intensities of the pump and signal waves are relatively high. Therefore, we can assume that the amplitudes of the pump as well as that of the signal are approximately constant during propagation, i. e., $\frac{\partial A_1}{\partial z} = \frac{\partial A_2}{\partial z} = \frac{\partial A_3}{\partial z} = 0$. Therefore, the amplitude of the idler wave is approximately (see Sec. 7.3)

$$\frac{\partial A_4}{\partial z} = 2j\gamma P_1 A_1 A_2 A_3^* \mathrm{e}^{\mathrm{j}\Delta k_{1,2,-3,-4} z}, \tag{14.16}$$

$$\Rightarrow E_4 \sim A_1 A_2 A_3^* \mathrm{e}^{\mathrm{j}\Delta k_{1,2,-3,-4} z}.$$

According to (7.5), the frequency of E_4 is $\omega_4 = \omega_1 + \omega_2 - \omega_3$. If all three waves have the same frequency ω (degenerate FWM), the frequency of the idler is then $\omega_4 = \omega + \omega - \omega = \omega$. The wave vector of E_4 is a result of the phase matching condition $\Delta k_{1,2,-3,-4} = k_1 + k_2 - k_3 - k_4$. Since a degenerate FWM is incorporated, all four waves have the same absolute value. The direction of the wave vectors is shown in Fig. 14.7 where the process is in fact phase-matched $\Delta k_{1,2,-3,-4} = 0$. As a result, the wave vector of the signal wave is

$$\boldsymbol{k}_4 = \boldsymbol{k}_1 + \boldsymbol{k}_2 - \boldsymbol{k}_3. \tag{14.17}$$

According to Fig. 14.7, the two pump waves E_1 and E_2 propagate in opposite directions, so the wave vectors are equal and opposite, i.e., $\boldsymbol{k}_1 = -\boldsymbol{k}_2$. Therefore, (14.17) together with Fig. 14.7 yields $\boldsymbol{k}_4 = -\boldsymbol{k}_3$ for the wave vector of the idler. If the counter-propagating pump waves have plane wave fronts and the same amplitude ($A_2 = A_1^*$) and the product between both is a constant ($|A|^2 = r$). The idler wave is represented by

$$E_4 \sim r A_3^* \mathrm{e}^{\mathrm{j}(-kz - \omega t)}. \tag{14.18}$$

A comparison between (14.18) and (14.15) shows that the idler wave (E_4) is the phase-conjugated wave to the signal wave (E_3).

The mathematical derivation of the phase conjugation with stimulated scattering effects is much more complicated and can be found elsewhere (see, for instance, [129]). Only the most important points will be considered here. Let's consider an experimental arrangement similar to that shown in Fig. 14.6 where a strong pump wave, E_0, propagates through a distorting medium. The resulting wave, E_i, penetrates into the nonlinear medium and consists of two parts: the plane undistorted wave, E_1, and the distorted contingent, E_2:

$$E_i = E_1 + E_2 = A_1 \mathrm{e}^{\mathrm{j}\varphi_1} + A_2 \mathrm{e}^{\mathrm{j}\varphi_2} = \mathrm{e}^{\mathrm{j}\varphi_1}\left[A_1 + A_2 \mathrm{e}^{\mathrm{j}(\varphi_2 - \varphi_1)}\right]. \tag{14.19}$$

The pump wave is a plane monochromatic wave propagating in the negative z-direction (see Fig. 14.6). Therefore, its phase is

$$\varphi_1 = -(\omega t + kz). \tag{14.20}$$

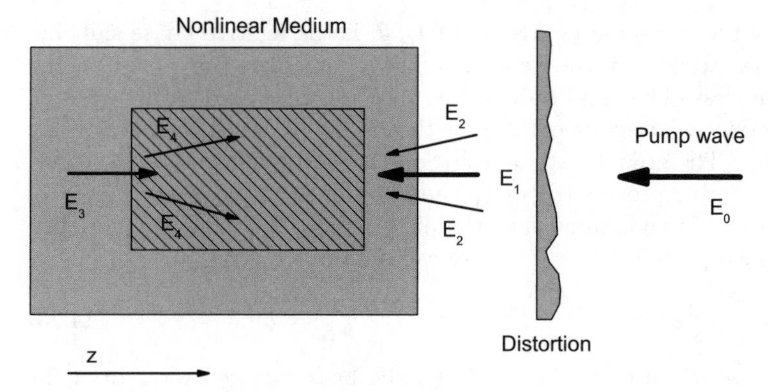

Fig. 14.6. Schematic sketch for the description of phase conjugation due to stimulated scattering processes [129]

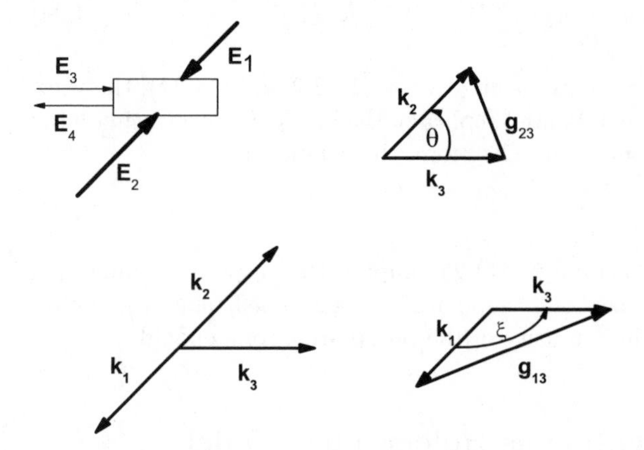

Fig. 14.7. Interfering waves and wave vector diagrams for the derivation of the grating constant

The phase alteration due to the influence of the distortion is $\varphi_2 - \varphi_1 = \theta(z)$. Hence, the penetrating pump can be written as

$$E_i(\omega, z) = \mathrm{e}^{-\mathrm{j}(\omega t + kz)} \left[A_1(z) + A_2(z)\mathrm{e}^{\mathrm{j}\theta(z)} \right] . \tag{14.21}$$

The whole backscattered field propagates into the positive z-direction. According to Fig. 14.6, it can be expressed as a superposition of two contingents as well:

$$E_r(\omega', z) = \mathrm{e}^{-\mathrm{j}\left(\omega' t - k' z\right)} \left[A_3(z) + A_4(z)\mathrm{e}^{\mathrm{j}\theta'(z)} \right] , \tag{14.22}$$

where $A_3(z)$ is a real amplitude function of the read out wave E_3 which was generated by stimulated scattering, $A_4(z)$ is the part of this wave which was diffracted by the hologram (see Sect. 14.4), $\theta^{'}$ is the relative phase shift between the waves, and $\omega^{'}$ is the frequency shift according to the corresponding scattering process. The amplitude of the wave E_3, generated at the position $z = 0$, experiences an exponential growth due to the stimulated scattering (SBS or SRS). The growth rate depends on the gain coefficient of the corresponding scattering process (g_R or g_B) and on the intensity, or the power, of the pump wave. If we assume that the distorted part is much smaller than the undistorted portion ($A_2 \ll A_1$), we can obtain

$$A_3(z) = A_3(0)\mathrm{e}^{\frac{1}{2}gA_1^2 z}. \tag{14.23}$$

Together with the total wave (14.21), the backscattered wave from the nonlinear wave equation is [129]

$$E_r = \frac{A_3(0)}{A_1}\mathrm{e}^{\frac{1}{2}gA_1^2 z}\left[A_1 + A_2 e^{-\mathrm{j}\theta(z)}\right]\mathrm{e}^{-\mathrm{j}\left(\omega^{'}t - k^{'}z\right)}. \tag{14.24}$$

As can be seen from a comparison between (14.21) and (14.24), the wave (14.24) is only the phase-conjugated replica of the incident wave if stimulated scattering occurs in the medium. Therefore, the condition

$$gA_1^2 z \gg 1 \tag{14.25}$$

has to be fulfilled. The inequality (14.25) implies that a phase conjugation by stimulated scattering can only happen if the gain coefficient (g), the interaction length (z) or the intensity of the pump are strong enough.

14.4 Phase Conjugation as Holographic Model

Independent of its actual origin, optical phase conjugation can be seen as a kind of real-time holography. The NPC due to FWM as well as phase conjugation due to stimulated scattering processes has, therefore, a kind of a higher common level. The connecting element is a grating generated by the interference of waves in the volume of the nonlinear material.

In the case of conventional holography, a film is simultaneously illuminated by an object and a reference wave. The object wave comes from the source and will be reflected at the object of interest. The reference wave, on the other hand, reaches the film directly. Both will interfere with each other in the film material. A complex interference pattern is generated and it depends on the intensities and the relative phase between the waves. Due to chemical reactions, this pattern can be stored in the film. In order to read the film, it has to be illuminated by the reference wave which is now propagating in the opposite direction. The image of the object then appears at its original position. Contrary to that, conventional image storage techniques, such as those in a camera, require only the object wave. Therefore, they can only store the intensity distribution of the reflected light. The spatial information – transmitted by the phase of the waves – is lost.

In the case of phase conjugation, interference patterns that depend on the intensity and phase of the involved waves are generated as well. But now the interference pattern is not permanently stored in the material. Due to the nonlinearity of the material, the interference pattern is directly translated into a hologram. Since the nonlinearity is very fast, the hologram vanishes if the waves are no longer present in the material. Therefore, the effect is called real time holography – two waves write the hologram while another wave acts as reference and generates the phase-conjugated wave simultaneously.

First, the phase conjugation due to FWM will be treated. The upper left corner of Fig. 14.7 shows the fundamental direction of the distinct wave vectors involved in the phase conjugation process. The wave $\boldsymbol{E}_3$ can interfere with the first pump wave $\boldsymbol{E}_1$ as well as with the second pump wave $\boldsymbol{E}_2$. Hence, two different interference patterns can be generated and both can be responsible for a signal wave reflection. The wave $\boldsymbol{E}_4$ reads out the hologram and is, therefore, the phase-conjugated wave to $\boldsymbol{E}_3$.

Since the distinct wave vectors have different angles, different interference patterns are generated. If the waves $\boldsymbol{E}_3$ and $\boldsymbol{E}_2$ interfere with each other, an interference pattern with the grating vector $\boldsymbol{g}_{23}$ will be generated. The angle between $\boldsymbol{k}_3$ and $\boldsymbol{k}_2$ is θ (see Fig. 14.7, top right) and the two vectors are related via the vector summation

$$\boldsymbol{k}_2 = \boldsymbol{k}_3 + \boldsymbol{g}_{23}. \tag{14.26}$$

Since all waves have the same frequency (ω), the wave vectors $\boldsymbol{k}_3$ and $\boldsymbol{k}_2$ have the same length and absolute value. The grating vector of the interference grating is perpendicular to the half angle between the vectors. The sine of the half angle $\theta/2$ is

$$\sin\frac{\theta}{2} = \frac{|\boldsymbol{g}_{23}/2|}{|\boldsymbol{k}_2|} \qquad \Longrightarrow \qquad |\boldsymbol{g}_{23}| = 2\,|\boldsymbol{k}_2|\sin\frac{\theta}{2}. \tag{14.27}$$

The absolute values of the wave vectors can be written as $|g_{23}| = 2\pi/d_{23}$ and $|\boldsymbol{k}_2| = 2\pi/\lambda$, with d_{23} as the distance between the maxima of the interference pattern and λ as the wavelength of the interfering waves in the material. Therefore, the distance of the interference maxima is

$$d_{23} = \frac{\lambda}{2\sin\theta/2}. \tag{14.28}$$

On the other hand, if the waves $\boldsymbol{E}_3$ and $\boldsymbol{E}_1$ interfere with each other, the generated interference grating has the grating vector $\boldsymbol{g}_{13}$. The angle between the vectors $\boldsymbol{k}_3$ and $\boldsymbol{k}_1$is ξ and the generated interference grating is perpendicular to the half angle $\xi/2$. As can be seen from Fig. 14.7 (bottom left) the sum of ξ and θ is 180° so $\xi = \pi - \theta$. With the derivation above, the grating constant of the interference pattern between the waves $\boldsymbol{E}_3$ and $\boldsymbol{E}_1$ is then

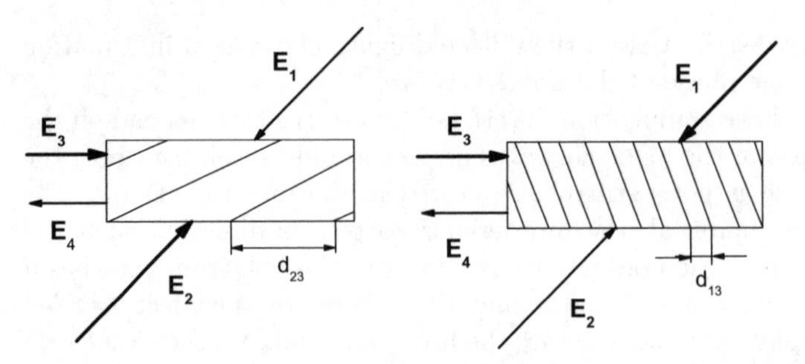

Fig. 14.8. Interference grating between the distinct wave vectors

$$d_{13} = \frac{\lambda}{2\cos\theta/2}. \tag{14.29}$$

It follows that both holograms have different directions and grating vectors, as shown in Fig. 14.8.

The phase conjugation due to nonlinear scattering processes can be described with a real-time hologram generated in the volume of the material as well [129]. Figure 14.6 shows a schematic sketch of this process. The pump wave $\boldsymbol{E}_0$ is a plane monochromatic wave. It penetrates a distorting medium where only a small fraction ($\boldsymbol{E}_2$) will be really distorted and most of the wave ($\boldsymbol{E}_1$) experiences no change. The modificated wave $\boldsymbol{E}_2$ can be different in its wave vector (as shown in Fig. 14.6), its phase, or in both. The superposition between $\boldsymbol{E}_1$ and $\boldsymbol{E}_2$ leads to the generation of an interference pattern in all cases, respectively to a hologram in the volume of the nonlinear material. Only the wave $\boldsymbol{E}_1$ has sufficient intensity to stimulate Raman or Brillouin scattering. Therefore, only $\boldsymbol{E}_1$ can generate a backscattered wave $\boldsymbol{E}_3$. This backscattered wave reads out the hologram, i.e., it is diffracted at the grating and generates a wave $\boldsymbol{E}_4$. Since $\boldsymbol{E}_4$ has the same frequency as the wave $\boldsymbol{E}_3$, originally generated by stimulated scattering, it will be amplified by SBS or SRS during its propagation in the medium. The two backscattered waves $\boldsymbol{E}_3$ and $\boldsymbol{E}_4$ are phase-conjugated to the incident waves $\boldsymbol{E}_1$ and $\boldsymbol{E}_2$. In the case of Brillouin scattering, $\boldsymbol{E}_3$ and $\boldsymbol{E}_4$ have nearly the same frequency as $\boldsymbol{E}_2$ and $\boldsymbol{E}_3$. For SRS, the frequency is correspondingly smaller.

14.5 Mid-span Spectral Inversion

Due to the possibility to compensate for distortions in optical transmission systems, NPC is of particular importance in the field of telecommunications. The usual method to compensate for group velocity dispersion is the dispersion management, discussed in Sect. 5.4. In high-bit-rate transmission systems, this method cannot completely compensate for chromatic dispersions. In particular the transmission of pulses with a temporal duration of less than

1 ps suffers from higher order dispersion terms, which are very complicated to handle by dispersion management. Nonlinear distortions, such as those due to Raman scattering, cannot be compensated by dispersion management either. A method that bears the potential to compensate for all of these distortions is nonlinear phase conjugation; The temporal NPC in its standard form can compensate for second and fourth order chromatic dispersions as well as nonlinear distortions due to SPM and Raman scattering [333]. A scheme that compensates for all effects (odd and even order dispersion, SPM, SRS, and self-steepening) is combining temporal NPC with a suitable chosen dispersion map, as proposed by Pina et al. [334, 336]. Another very interesting method with the same potential was investigated by Tsang et al. [335] and was based on a temporal NPC together with a spectral phase conjugation of the signals. In the case of a temporal NPC, the time envelope of the pulse is conjugated whereas for spectral phase conjugation, the combination of conjugation and time reversal of the envelope is required. The latter corresponds to the conjugation of the optical pulse in the frequency domain [335]. We will only treat the temporal NPC here.

Kunimatsu et al. [337] have shown the transmission of pulses with a temporal duration of 600 fs and a repetition rate of 10 GHz over a distance of 144 km. The second and fourth order dispersions were compensated by NPC in a lithium niobate ($LiNbO_3$) wavelength converter while the third order dispersion was compensated by a combination of a negative slope DSF with a DSF. The temporal duration of the input pulses was 680 fs. As shown by these laboratory results, due to NPC, the transmission of high-bit-rate fs pulses over large distances is possible. But NPC is used in other interesting fields as well, such as for an improvement of the beam quality in laser systems [338]. The phase-conjugating mirror can be introduced in an amplifier that follows the laser [196] or directly in the laser's oscillator [197]. Another application is carrier suppression in micrometer and millimeter-wave radio-over-fiber systems [339, 340] or distortion compensation in multimode fibers [342] which, due to their large core area, are used for the transmission of high-power laser fields over short distances.

The basic idea of distortion compensation with phase conjugation in a transmission system (mid-span spectral inversion) is shown in Fig. 14.9. A pulse propagating in an optical fiber will be spread in time due to the group velocity dispersion and the pulse shape is distorted. Directly in the middle of the transmission system, the phase-conjugated replica of the pulse is generated by nonlinear effects. If the second part of the fiber shows exactly the same behavior as the first, the pulse will be compressed and, therefore, the distortion will be compensated. As a result, the pulse shape at the output of the transmission system should be the same as the pulse shape at the fiber input if attenuation losses are neglected.

Figure 14.10 shows an ideal (left) and a frequency modulated (right) "chirped" pulse in the time (top) and frequency domain (bottom). Such a fre-

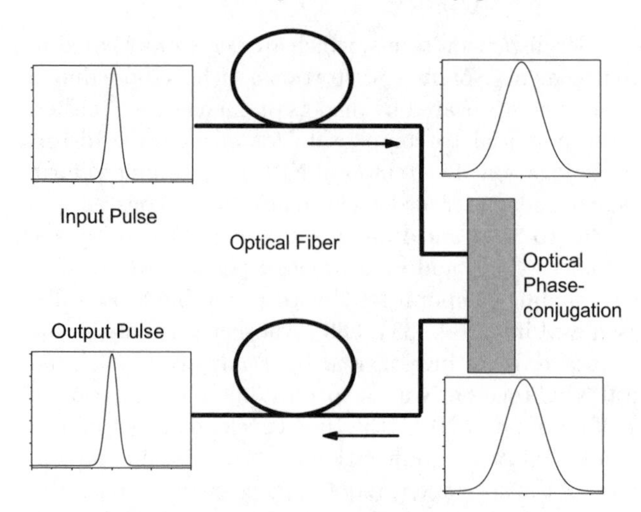

Fig. 14.9. Principle sketch of the compensation of pulse distortions in optical fibers (only dispersion here) with nonlinear optical phase conjugation

quency modulation follows, for instance, from a self-phase modulation during the transmission in the nonlinear waveguide but, it can be the consequence of a laser diode modulation as well.

As a result of the group velocity dispersion, the ideal as well as the chirped pulse shape spread in time during propagation in the fiber. The effect is much stronger for the chirped pulse because its spectral width is larger. Therefore, we will study the behavior of the chirped pulse closely in the following section.

If the pulse has an average wavelength of $\lambda_{\mathrm{m}} = 1.542\ \mu\mathrm{m}$, then the propagation is in the range of positive group velocity dispersion in standard single mode fibers (the dispersion parameter D is positive). In this case, the group velocity decreases with increasing wavelength. Consequently, the red-shifted frequency component $(\omega_{\mathrm{m}} - \Delta\omega)$ at the rear side of the pulse propagates slower than the component at the rising edge which is blue-shifted $(\omega_{\mathrm{m}} + \Delta\omega)$. In the middle of the propagation, the blue-shifted edge arrives earlier than it would be in the case without the chirp and the red-shifted wing arrives a little later. Therefore, the pulse shape is further spread in time. A phase-conjugating mirror is placed exactly in the middle of the system. This can be a highly nonlinear fiber or a semiconductor optical amplifier (SOA) working in its nonlinear regime, for example. These nonlinear elements can phase-conjugate the signal by degenerate FWM. If the two pump waves have a wavelength of $\lambda_{\mathrm{p}} = 1.544\ \mu\mathrm{m}$, then the conservation of energy and momentum lead to $\omega_{\mathrm{mPC}} = 2\omega_{\mathrm{p}} - \omega_{\mathrm{m}}$ for the average wavelength of the converted signal and $k_{\mathrm{mPC}} = 2k_{\mathrm{p}} - k_{\mathrm{m}}$ for the wave number. According to these values, the average wavelength of the converted signal is $\lambda_{\mathrm{mPC}} \approx 1.546\ \mu\mathrm{m}$.

The blue chirped range of the pulse, which arrived first, has a frequency of $(\omega_{\mathrm{m}} + \Delta\omega)$. After the phase conjugation, this part is in the conjugated

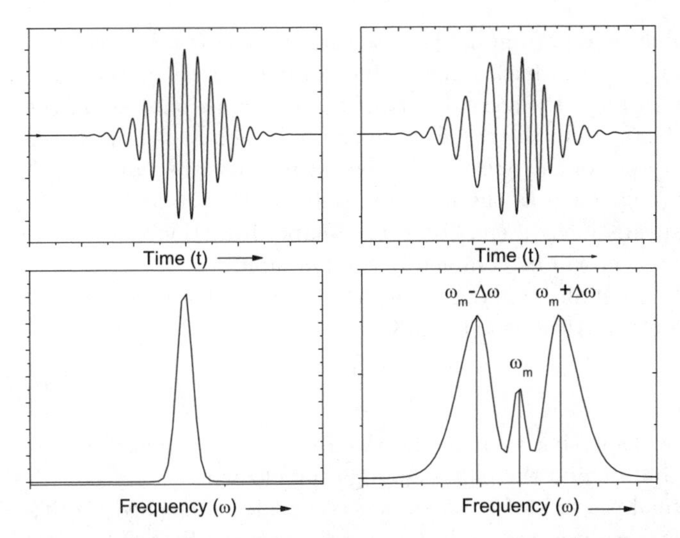

Fig. 14.10. Comparison between a normal and a chirped pulse in the time and frequency domain. Note that the chirp of a laser diode is slightly asymmetric

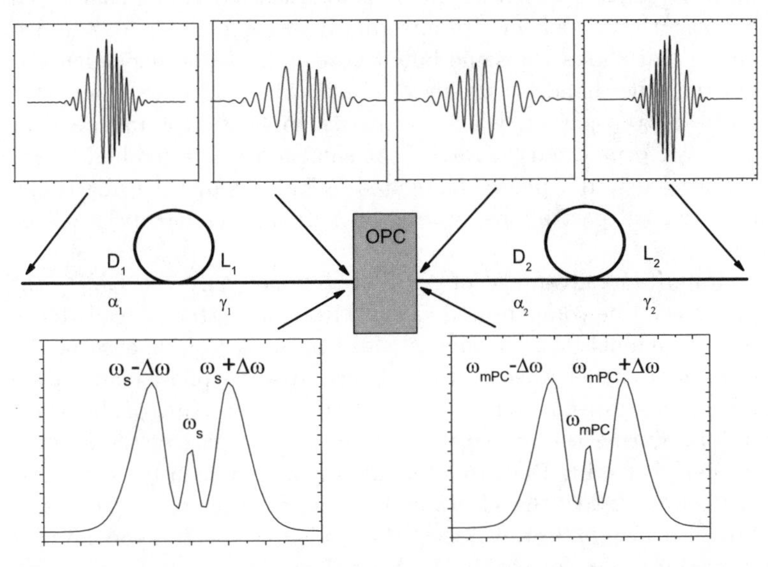

Fig. 14.11. Dispersion compensation of a chirped pulse due to optical phase conjugation (*Top* Pulse in time domain. *Bottom* Pulse in the frequency domain)

pulse, as defined by $2\omega_{\mathrm{p}} - (\omega_{\mathrm{m}} + \Delta\omega) = (\omega_{\mathrm{mPC}} - \Delta\omega)$. For the red-shifted part that arrives later a similar consideration leads to $2\omega_{\mathrm{p}} - (\omega_{\mathrm{m}} - \Delta\omega) = (\omega_{\mathrm{mPC}} + \Delta\omega)$. Therefore, the pulse is spectrally inverted where every spectral component from the lower frequency range is shifted to the higher range and vice versa. Figure 14.11 shows a schematic sketch of this behavior.

The formerly blue edge in front of the pulse is now red, at the same time the wing at the pulse in the back is now blue (formerly red). If the pulse propagates further through the second part of the transmission system, the blue part is again faster than the red. But since the blue part is now at the back and the red in front of the pulse, if the second part has the same length and group velocity dispersion as the first, the pulse will be compressed. At the end of the transmission system, the pulse shape distortion due to the group velocity dispersion will be completely compensated.

The compensation condition is approximately achieved if the total GVD in both fibers is identical, this leads to [343, 344, 345]

$$D_1 L_1 = D_2 L_2. \tag{14.30}$$

with D_1, D_2 as the dispersion parameters and L_1, L_2 as the length of the fiber spans before and behind the phase-conjugating element.

If only one channel is considered, a compression could be reached via dispersion management, as discussed in Sect. 5.4. But for very short pulses and DWDM systems with many channels, the slope of the GVD is very important. The chromatic dispersion is slightly different for each individual channel. If fibers with a negative dispersion are introduced to compensate for the dispersion, a fiber that shows the same but negative dispersion dependence is required. On the other hand, according to (14.30) the dispersion compensation with NPC is transparent, which means its application is independent of the modulation format and bit rate of the signals. Each individual WDM channel is converted into a phase-conjugated WDM channel. During transmission in the second part of the system, all pulses in all channels will be compressed.

Another important advantage of NPC is that not only the GVD, but other distortions can be compensated as well. Raman scattering leads to an amplification of channels with a higher wavelength at the expense of the channels with lower carrier wavelength (see Sect. 10.4). A phase conjugation in the middle of the transmission distance inverts the spectrum of the whole band. Therefore, during the propagation in the second part of the system, the same Raman scattering leads to a balance of the power in the channels.

Another nonlinear effect that degrades system performance in WDM systems is SPM. As shown in Sect. 6.1, the SPM leads to a spectral spreading of the pulses that can be detrimental in WDM systems. The resulting distortion of the pulses can be compensated by NPC as well if the path averaged SPM induced phase shift along both fibers is almost the same [346, 347, 348]

$$\gamma_1 \bar{P}_1 L_1 = \gamma_2 \bar{P}_2 L_2 \tag{14.31}$$

where $\bar{P}_{\mathrm{j}} = 1/L_{\mathrm{j}} \int P_{\mathrm{j}}(z) dz$ $(j = 1, 2)$ is the path averaged power. But the problem is that in conventional transmission systems, the SPM is asymmetrical. Due to its intensity dependence, the distortion is very strong if the intensities are strong, i.e., shortly behind the amplifiers. If the pulse propagates further, the intensity decreases and, therefore, the influence of SPM decreases as well. If NPC is chosen for a compensation of the distortions caused by SPM, an exact converse amplification distribution is required for the second part of the transmission. Hence, the SPM influence can be completely compensated by phase conjugation only with additional constructive changes [202].

In early MSSI experiments as nonlinear media for optical phase conjugation, dispersion-shifted fibers were used. In practical applications the output phase-conjugated signal has to be insensitive to the polarization of the input signal. Polarization independent FWM can be offered by two orthogonally polarized pump waves [142, 349] (Sect. 7.7). Nevertheless, with this approach the conversion efficiency is not completely polarization independent. This behavior is caused by polarization sensitive cross-talk FWM waves, asymmetrical depletion of the two pumps and the input signal and polarization sensitive phase-conjugated FWM waves [350].

The disadvantage of NPC in dispersion-shifted fibers is the relatively large required interaction length. In order to build small compact devices optical fibers with a high nonlinearity (HNL or microstructured fibers) can be used. Other possibilities are offered by compact elements like SOAs. On the other hand, SOAs lead to an increase of ASE noise in the system.

The efficiency of NPC by FWM depends on the phase matching between the involved waves. A possibility to circumvent this is the incorporation of quasi phase matching offered by periodically-poled structures in crystals like lithium niobate (PPLN, see Sect. 4.10). Such a PPLN can be very compact (< 5 cm), has a very low noise, and has a conversion bandwidth of more than 140 nm [341].

Summary

- Phase conjugation can compensate for distortions caused by linear and nonlinear effects in optical transmission systems.
- Phase-conjugating reflection can be seen as a kind of adaptation of the mirror shape to the wave front. The phase-conjugated wave is the exact replica of the incident wave front propagating in exactly the opposite direction. Each individual beam is reflected back to its origin.
- Two optical waves are phase-conjugated to each other if both have the same spatial wave front but move in opposite directions.
- Phase conjugation requires a nonlinear medium. In this medium, FWM, SBS, or SRS can generate a phase-conjugated replica of the incident wave.
- NPC is a kind of real time holography. Two waves write the hologram, while another wave acts as reference and generates the phase-conjugated wave at the same time.

- For temporal NPC, the time envelope of the pulse is conjugated, i.e., the pulse is spectrally inverted.
- In the case of a spectral phase conjugation, the combination of conjugation and time reversal of the envelope is required. This corresponds to the conjugation of the optical pulse in the frequency domain.
- With temporal NPC, the effects of even order dispersion, self-phase modulation and Raman scattering can be compensated.
- A spectral NPC can compensate for even and odd order dispersion, self-phase modulation, and self-steepening of the pulses.
- With the introduction of NPC in the middle of an optical transmission system, the propagation of high-bit-rate fs pulses over large distances is possible. Other application fields are the carrier suppression in micrometer and millimeter wave radio-over-fiber systems, an improvement of the beam quality in laser systems, or the distortion compensation in multimode fibers.
- For the nonlinear medium required for NPC, optical fibers with a high nonlinearity (HNL or microstructured fibers), SOAs, or PPLNs can be used.

Exercises

14.1. Suppose we have an NPC arrangement for the C-band like the one shown in Fig. 14.12. Show that the phase matching condition is fulfilled for this arrangement and determine the distances between the grating maxima of the generated interferences.

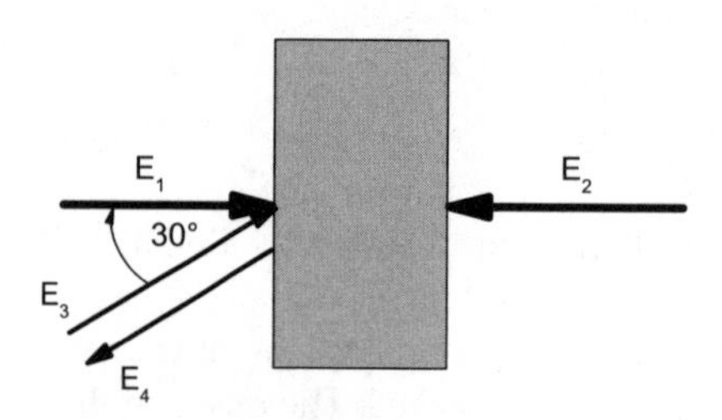

Fig. 14.12. NPC arrangement

14.2. Estimate the power of the phase-conjugated wave if FWM in a highly nonlinear microstructured fiber with a length of 50 m, a nonlinearity coefficient of $\gamma = 35\ \mathrm{W}^{-1}\mathrm{km}^{-1}$, and an attenuation constant of 0.8 dB/km is used for the NPC. The signal has a carrier wavelength of 1.55 μm and a power of 1 mW. The pump has a carrier wavelength of 1.551 μm and a power of 1 W. The process is completely phase-matched.

14.3. Assume the first ten channels of the C-band with a channel spacing of 100 GHz (see Table 14.1) will be phase-conjugated with a pump wavelength of 1536.61 nm. Determine the carrier wavelength of the distinct channels after phase conjugation.

Table 14.1. Wavelengths of the first 10 WDM channels in the C-band for a channel spacing of 100 GHz

Channel #	Wavelength (nm)
1	1528.77
2	1529.55
3	1530.33
4	1531.12
5	1531.9
6	1532.68
7	1533.47
8	1534.25
9	1535.04
10	1535.82

A. Appendices

Derivation of the Polarization Dependence of FWM and XPM

The nonlinear wave equation serves as the starting point for the investigation of the polarization dependence of the interacting waves. Due to the inversion symmetry of silica glass and the relatively weak intensities of the applied fields, only the third order susceptibility is involved. Hence, the nonlinear wave equation can be written as

$$\Delta\boldsymbol{E} - \frac{n_0^2}{c^2}\frac{\partial^2 \boldsymbol{E}}{\partial t^2} = \frac{\chi^{(3)}}{c^2}\frac{\partial^2}{\partial t^2}\boldsymbol{E}\boldsymbol{E}\boldsymbol{E}. \tag{A.1}$$

A wave with arbitrary polarization can be expressed as a superposition of two orthogonal parts with different phases. Therefore, the electrical field vector can be written as

$$\boldsymbol{E} = \frac{1}{2}\left[\left(E_x(\boldsymbol{r})\mathrm{e}^{\mathrm{j}(k_x z-\omega t)} + \text{c.c.}\right)\boldsymbol{e}_x + \left(E_y(\boldsymbol{r})\mathrm{e}^{\mathrm{j}(k_y z-\omega t)} + \text{c.c.}\right)\boldsymbol{e}_y\right] \tag{A.2}$$

With k_x and k_y as the wave vectors in $z-$direction for the perpendicular polarized waves. For the simplification of the derivation, the same assumptions as for the derivation of the NSE in Sect. 5.1 are made. The only difference is that different polarizations of the waves are allowed now and the dispersion length exceeds the fiber length, i.e., the influence of the dispersion on the pulses can be neglected.

In this section, the derivation of the polarization dependence of FWM and XPM will be shown together. For non-degenerate FWM, four distinct waves interact with each other in the fiber. The total electric field is then

$$\boldsymbol{E} = \frac{1}{2}\sum_{m=1}^{4}\left[\left(E_{xm}(\boldsymbol{r})\mathrm{e}^{\mathrm{j}(k_{xm} z-\omega_m t)} + \text{c.c.}\right)\boldsymbol{e}_x \right. \tag{A.3}$$
$$\left. + \left(E_{ym}(\boldsymbol{r})\mathrm{e}^{\mathrm{j}(k_{ym} z-\omega_m t)} + \text{c.c.}\right)\right]\boldsymbol{e}_y,$$

where c.c. is the complex conjugate, k_m denotes the wavenumber of the mth wave and ω_m its frequency. As in Sect. 5.1, the complex amplitude of the mth wave can again be separated into a transverse and a longitudinal component, A_{im} and F_{im}.

$$E_{im}(\boldsymbol{r}) = F_{im}(r,\varphi)A_{im}(z) \qquad i = x, y \tag{A.4}$$

The intensity of the mth vector component is proportional to its absolute value squared, as follows:

$$I_{im} = \frac{1}{2}\varepsilon_0 n_0 c \left|\hat{E}_{im}\right|^2 . \tag{A.5}$$

In a single mode fiber, the radiated area is the same for all waves. Hence, the power of the distinct waves is

$$P_{im} = \int\int I_{im} r \mathrm{d}r \mathrm{d}\varphi . \tag{A.6}$$

If (A.4) and (A.5) are introduced into (A.6), and if it is assumed that the field distribution is independent on φ we have

$$\begin{aligned} P_{im} &= \frac{1}{2}\varepsilon_0 n_0 c \left|A_{im}\right|^2 2\pi \int \left|F_{im}(r)\right|^2 r\mathrm{d}r \\ &= N_{im}^2 \left|A_{im}\right|^2 , \end{aligned} \tag{A.7}$$

with N_{im}^2 as the transverse part of the component im, which corresponds to the intensity distribution of its mode:

$$N_{im}^2 = \frac{1}{2}\varepsilon_0 n_0 c 2\pi \int \left|F_{im}(r)\right|^2 r\mathrm{d}r. \tag{A.8}$$

Now it is possible to introduce a longitudinal amplitude

$$B_{im}(z) = N_{im}A_{im}(z) \tag{A.9}$$

so that the power of the component im is

$$P_{im} = \left|B_{im}\right|^2 . \tag{A.10}$$

With the equations (A.4) to (A.10), (A.3) can be written as

$$\begin{aligned} \boldsymbol{E} = \frac{1}{2}\sum_{m=1}^{4} \Big[&\left(\Gamma_x \mathrm{e}^{\mathrm{j}(k_{xm}z-\omega_m t)} + \mathrm{c.c.}\right)\boldsymbol{e}_x \\ &+ \left(\Gamma_y \mathrm{e}^{\mathrm{j}(k_{ym}z-\omega_m t)} + \mathrm{c.c.}\right)\boldsymbol{e}_y \Big] , \end{aligned} \tag{A.11}$$

where

$$\Gamma_{x,y} = \frac{B_{x,ym}(z)}{N_{x,ym}} F_{x,ym}(r). \tag{A.12}$$

The propagating wave leads to a polarization of the material. This polarization is a wave as well:

$$\boldsymbol{P} = \frac{1}{2}\sum_{m=1}^{4}\left[\left(P_{xm}\mathrm{e}^{\mathrm{j}(k_{xm}z-\omega_m t)} + \text{c.c.}\right)\boldsymbol{e}_x + \left(P_{ym}\mathrm{e}^{\mathrm{j}(k_{ym}z-\omega_m t)} + \text{c.c.}\right)\boldsymbol{e}_y\right]. \tag{A.13}$$

The wave equation (A.1) can now be separated into 8 parts for the different involved terms, i.e., four frequencies with two vector components each. As an example in the following, only the x-vector component of the wave with frequency ω_1 will be considered. For all other waves and polarizations, a completely analogous derivation is valid. For the longitudinal part of the left side of the wave equation, E_{1x} can be represented by

$$\frac{\partial^2 E_{1x}}{\partial z^2} = \frac{1}{2}\left[\frac{\partial^2 U_{1x}}{\partial z^2} + 2jk_{1x}\frac{\partial U_{1x}}{\partial z} - U_{1x}k_1^2\right] F_{1x}(r)\mathrm{e}^{\mathrm{j}(k_{1x}z-\omega_1 t)}. \tag{A.14}$$

The temporal derivation is then

$$\frac{n_0^2}{c^2}\frac{\partial^2 E_{1x}}{\partial t^2} = -\frac{1}{2}\frac{n_0^2}{c^2}\omega_1^2 U_{1x}F_{1x}(r)\mathrm{e}^{\mathrm{j}(k_{1x}z-\omega_1 t)}, \tag{A.15}$$

where $U_{1x} = B_{1x}(z)/N_{1x}$. The whole left side of the wave equation becomes:

$$\triangle E_{1x} - \frac{n_0^2}{c^2}\frac{\partial^2 E_{1x}}{\partial t^2} = \frac{1}{2}\left[\triangle_{\mathrm{T}} E_{1x} + \frac{\partial^2 U_{1x}}{\partial z^2} + 2jk_{1x}\frac{\partial U_{1x}}{\partial z} - U_1 k_{1x}^2 + \frac{n_0^2}{c^2}\omega_1^2 U_{1x}\right] \times F_{1x}(r)\mathrm{e}^{\mathrm{j}(k_{1x}z-\omega_1 t)}. \tag{A.16}$$

With the slowly varying amplitude approximation, the second spatial derivation can be neglected. The influence of the nonlinearity to the perpendicular field components should be negligibly small as well. Contrary to the derivation of the NSE in Sect. 5.1, we can assume now that the dispersion has no influence on the pulses. Hence, quasi-monochromatic waves are considered and the wavenumbers k_{1x} and $\kappa_1 = \frac{n_0^2}{c^2}\omega_1^2$ should be approximately the same. With these simplifications, the left side of the wave equation becomes

$$\triangle E_{1x} - \frac{n_0^2}{c^2}\frac{\partial^2 E_{1x}}{\partial t^2} \approx jk_{1x}\frac{\partial U_{1x}}{\partial z}F_{1x}(r)\mathrm{e}^{\mathrm{j}(k_{1x}z-\omega_1 t)}. \tag{A.17}$$

The right side of the wave equation is

$$\frac{\chi^{(3)}}{c^2}\frac{\partial^2}{\partial t^2}\boldsymbol{EEE} = \frac{1}{8}\frac{\chi^{(3)}}{c^2}\frac{\partial^2}{\partial t^2}\left[\left(E_x^3 + E_y^2 E_x\right)\boldsymbol{e}_x + \left(E_y^3 + E_x^2 E_y\right)\boldsymbol{e}_y\right]. \tag{A.18}$$

If (A.3) is introduced into (A.18), a large number of terms follow. The solution for the nonlinear polarization in x-direction is

$$\frac{\chi^{(3)}}{c^2}\frac{\partial^2}{\partial t^2}\boldsymbol{EEE} = \frac{\chi^{(3)}}{c^2}\frac{\partial^2}{\partial t^2}\frac{1}{8}\Bigg(\sum_{k,m,n} E_{xk}E_{xm}E_{xn}\mathrm{e}^{\mathrm{j}(\Delta kz-\Delta\omega t)} + \sum_{k,m,n} E_{yk}E_{ym}E_{xn}\mathrm{e}^{\mathrm{j}(\Delta kz-\Delta\omega t)}\Bigg). \tag{A.19}$$

With Δk as the wavenumber and $\Delta\omega$ as the frequency superposition between the distinct involved waves. The indices k, m, and n can have the values 1-4 and 1^*-4^*. The terms for which all three indices are equal ($k = m = n$) denote the third harmonic of the original waves. According to the phase matching condition, these terms can be neglected. For the SPM, terms of the form $3\left|E_i\right|^2 E_i$ with i and $j = k, m, n$ are responsible, whereas the XPM is connected to $6\left|E_i\right|^2 E_j$. All other terms are related to FWM. According to Sect. 7.2, it is possible to separate the relevant terms with (7.5). The frequency f_1 is for instance $f_3 + f_4 - f_2$. If the frequency differences between the components are equal, f_1 can be a result of $2f_2 - f_4$ and $f_2 + f_3 - f_1$ as well. For convenience, it should be assumed that only the condition $f_1 = f_3 + f_4 - f_2$ remains in the following.

Furthermore, $N_{\mathrm{x,y}i}$ and $F_{\mathrm{x,y}i}(r)$ should be identical for all considered waves. Therefore, the right side of the wave equation for the component ω_1 in the x-direction is with (A.12)

$$\begin{aligned}\frac{\chi^{(3)}}{c^2}\frac{\partial^2}{\partial t^2}\boldsymbol{EEE} =& \Psi\Bigg[\Bigg(\frac{1}{2}\left|\Gamma_{1x}\right|^2 + \sum_{i\neq 1}\left|\Gamma_{ix}\right|^2 + \frac{1}{3}\sum_{i}\left|\Gamma_{iy}\right|^2\Bigg)\Gamma_{1x}\mathrm{e}^{\mathrm{j}(k_1 z-\omega_1 t)} \\ &+\Gamma_{2x}^*\Gamma_{3x}\Gamma_{4x}\mathrm{e}^{\mathrm{j}(k_{3x}+k_{4x}-k_{2x})z-\mathrm{j}(\omega_3+\omega_4-\omega_2)t} \\ &+\frac{1}{3}\Gamma_{2x}^*\Gamma_{3y}\Gamma_{4y}\mathrm{e}^{\mathrm{j}(k_{3y}+k_{4y}-k_{2x})z-\mathrm{j}(\omega_3+\omega_4-\omega_2)t} \\ &+\frac{1}{3}\Gamma_{2y}^*\Gamma_{3x}\Gamma_{4y}\mathrm{e}^{\mathrm{j}(k_{3x}+k_{4y}-k_{2y})z-\mathrm{j}(\omega_3+\omega_4-\omega_2)t} \\ &+\frac{1}{3}\Gamma_{2y}^*\Gamma_{3y}\Gamma_{4x}\mathrm{e}^{\mathrm{j}(k_{3y}+k_{4x}-k_{2y})z-\mathrm{j}(\omega_3+\omega_4-\omega_2)t}\Bigg],\end{aligned} \tag{A.20}$$

where $\Psi = -\frac{6}{8}\frac{\chi^{(3)}\omega_1^2}{c^2}$ and $i = $ 1-4.

In order to eliminate the transverse field distribution both sides are multiplied by $2\pi\int\left|F_{im}(r)\right|^2 r\mathrm{d}r$. Using A_{eff} (4.93), n_2 (4.90), and (A.8) both sides of the wave equation can be expanded. If the distinct amplitudes are normalized to the input power of wave 1

$$B_m = \sqrt{P_1}A_m, \tag{A.21}$$

it can be written

$$\frac{\partial A_{1\mathrm{x}}}{\partial z} = 2j\gamma P_1 \Bigg[\left(\frac{1}{2} |A_{1x}|^2 + \sum_{i \neq 1} |A_{ix}|^2 + \frac{1}{3} \sum_{i} |A_{iy}|^2 \right) A_{1x} \tag{A.22}$$

$$+ A_{2x}^* A_{3x} A_{4x} \mathrm{e}^{\mathrm{j}\Delta k_{x1} z} + \frac{1}{3} \Big(A_{2x}^* A_{3y} A_{4y} \mathrm{e}^{\mathrm{j}\Delta k_{xy1} z}$$

$$+ A_{2y}^* A_{3x} A_{4y} \mathrm{e}^{\mathrm{j}\Delta k_{xy2} z} + A_{2y}^* A_{3y} A_{4x} \mathrm{e}^{\mathrm{j}\Delta k_{xy3} z} \Big) \Bigg]$$

$i = 1\text{-}4.$

Equation (A.22) contains the phase terms

$$\Delta k_{x1} = k_{3x} + k_{4x} - k_{2x} - k_{1x}, \tag{A.23}$$

$$\Delta k_{xy1} = k_{3y} + k_{4y} - k_{2x} - k_{1x}, \tag{A.24}$$

$$\Delta k_{xy2} = k_{3x} + k_{4y} - k_{2y} - k_{1x}, \tag{A.25}$$

$$\Delta k_{xy3} = k_{3y} + k_{4x} - k_{2y} - k_{1x}. \tag{A.26}$$

If we assume that the frequency difference is small, (A.24) can be written as

$$|\Delta k_{xy1}| = 2 |k_y - k_x| . \tag{A.27}$$

Due to the phase mismatch, the coherence length of this term is then

$$L_{\mathrm{coh}} = \frac{\pi}{|\Delta k_{xy1}|} = \frac{\pi}{2 |k_y - k_x|} = \frac{\lambda}{4\Delta n}. \tag{A.28}$$

Using $\Delta n/n \approx 10^{-7} - 10^{-5}$, the coherence length for this mixing product can be calculated to be between 3 m and 3 cm. Hence, this term can be neglected in the case of a small frequency difference.

Equation (A.23) is determined by the wavelength dependence of the refractive index $n(\lambda)$, i.e., the material dispersion k_{M} so that

$$\Delta k_{x1} = \frac{1}{c} \left(n_{3x}\omega_3 + n_{4x}\omega_4 - n_{2x}\omega_2 - n_{1x}\omega_1 \right) = k_{\mathrm{M1}}. \tag{A.29}$$

Under the assumption of a small frequency difference for the other two phase terms, the waveguide dispersion is stronger. Hence, the refractive indices can be seen as approximately equal. With $k_{x,yi} = n_{x,y}\omega_i/c$, it can be written that

$$\Delta k_{xy2} \approx \frac{1}{c} \left[n_x (\omega_3 - \omega_1) - n_y (\omega_2 - \omega_4) \right] \tag{A.30}$$

and

$$\Delta k_{xy3} \approx \frac{1}{c} \left[n_x (\omega_4 - \omega_1) - n_y (\omega_2 - \omega_3) \right] . \tag{A.31}$$

If the frequency difference is $\omega_1 - \omega_3 = \omega_2 - \omega_1 = \omega_4 - \omega_2 = \Delta\omega$, it follows from Fig. 7.4 that

$$\Delta k_{xy2} = -\Delta n \Delta\omega / c = -\Delta k_{\mathrm{W}} \tag{A.32}$$

and

$$\Delta k_{xy3} = 2\Delta n \Delta\omega / c = 2\Delta k_{\mathrm{W}}. \tag{A.33}$$

The other 7 equations for the description of the polarization dependence of XPM and FWM can be derived completely analogously [142], as follows.

$$\frac{\partial A_{1x}}{\partial z} = 2\mathrm{j}\gamma P_1 \begin{bmatrix} \left(\frac{1}{2}|A_{1x}|^2 + \sum_{i\neq 1}|A_{ix}|^2 + \frac{1}{3}\sum_i |A_{iy}|^2\right) A_{1x} \\ +A_{2x}^* A_{3x} A_{4x} \mathrm{e}^{\mathrm{j}\Delta k_{\mathrm{M1}} z} \\ +\frac{1}{3}\left(A_{2y}^* A_{3x} A_{4y} \mathrm{e}^{-\mathrm{j}\Delta k_{\mathrm{W}} z} + A_{2y}^* A_{3y} A_{4x} \mathrm{e}^{\mathrm{j}2\Delta k_{\mathrm{W}} z}\right) \end{bmatrix}, \tag{A.34}$$

$$\frac{\partial A_{2x}}{\partial z} = 2\mathrm{j}\gamma P_1 \begin{bmatrix} \left(\frac{1}{2}|A_{2x}|^2 + \sum_{i\neq 2}|A_{ix}|^2 + \frac{1}{3}\sum_i |A_{iy}|^2\right) A_{2x} \\ +A_{1x}^* A_{3x} A_{4x} \mathrm{e}^{\mathrm{j}\Delta k_{\mathrm{M2}} z} \\ +\frac{1}{3}\left(A_{1y}^* A_{3x} A_{4y} \mathrm{e}^{-\mathrm{j}2\Delta k_{\mathrm{W}} z} + A_{1y}^* A_{3y} A_{4x} \mathrm{e}^{\mathrm{j}\Delta k_{\mathrm{W}} z}\right) \end{bmatrix}, \tag{A.35}$$

$$\frac{\partial A_{3x}}{\partial z} = 2\mathrm{j}\gamma P_1 \begin{bmatrix} \left(\frac{1}{2}|A_{3x}|^2 + \sum_{i\neq 3}|A_{ix}|^2 + \frac{1}{3}\sum_i |A_{iy}|^2\right) A_{3x} + \\ A_{4x}^* A_{1x} A_{2x} \mathrm{e}^{\mathrm{j}\Delta k_{\mathrm{M3}} z} \\ +\frac{1}{3}\left(A_{4y}^* A_{1x} A_{2y} \mathrm{e}^{\mathrm{j}\Delta k_{\mathrm{W}} z} + A_{4y}^* A_{1y} A_{2x} \mathrm{e}^{\mathrm{j}2\Delta k_{\mathrm{W}} z}\right) \end{bmatrix}, \tag{A.36}$$

$$\frac{\partial A_{4x}}{\partial z} = 2\mathrm{j}\gamma P_1 \begin{bmatrix} \left(\frac{1}{2}|A_{4x}|^2 + \sum_{i\neq 4}|A_{ix}|^2 + \frac{1}{3}\sum_i |A_{iy}|^2\right) A_{4x} \\ +A_{3x}^* A_{1x} A_{2x} \mathrm{e}^{\mathrm{j}\Delta k_{\mathrm{M4}} z} \\ +\frac{1}{3}\left(A_{3y}^* A_{1x} A_{2y} \mathrm{e}^{-\mathrm{j}2\Delta k_{\mathrm{W}} z} + A_{3y}^* A_{1y} A_{2x} \mathrm{e}^{-\mathrm{j}\Delta k_{\mathrm{W}} z}\right) \end{bmatrix}, \tag{A.37}$$

$$\frac{\partial A_{1y}}{\partial z} = 2\mathrm{j}\gamma P_1 \begin{bmatrix} \left(\frac{1}{2}|A_{1y}|^2 + \sum_{i\neq 1}|A_{iy}|^2 + \frac{1}{3}\sum_i |A_{ix}|^2\right) A_{1y} \\ +A_{2y}^* A_{3y} A_{4y} \mathrm{e}^{\mathrm{j}\Delta k_{\mathrm{M5}} z} \\ +\frac{1}{3}\left(A_{2x}^* A_{3y} A_{4x} \mathrm{e}^{\mathrm{j}\Delta k_{\mathrm{W}} z} + A_{2x}^* A_{3x} A_{4y} \mathrm{e}^{-\mathrm{j}2\Delta k_{\mathrm{W}} z}\right) \end{bmatrix}, \tag{A.38}$$

$$\frac{\partial A_{2y}}{\partial z} = 2\mathrm{j}\gamma P_1 \begin{bmatrix} \left(\frac{1}{2}|A_{2y}|^2 + \sum_{i\neq 2}|A_{iy}|^2 + \frac{1}{3}\sum_i |A_{ix}|^2\right) A_{2y} \\ +A_{1y}^* A_{3y} A_{4y} \mathrm{e}^{\mathrm{j}\Delta k_{\mathrm{M6}} z} \\ +\frac{1}{3}\left(A_{1x}^* A_{3y} A_{4x} \mathrm{e}^{2\mathrm{j}\Delta k_{\mathrm{W}} z} + A_{1x}^* A_{3x} A_{4y} \mathrm{e}^{-\mathrm{j}\Delta k_{\mathrm{W}} z}\right) \end{bmatrix}, \tag{A.39}$$

$$\frac{\partial A_{3y}}{\partial z} = 2\mathrm{j}\gamma P_1 \begin{bmatrix} \left(\frac{1}{2}|A_{3y}|^2 + \sum_{i\neq 3}|A_{iy}|^2 + \frac{1}{3}\sum_i |A_{ix}|^2\right) A_{3y} \\ +A_{4y}^* A_{1y} A_{2y} \mathrm{e}^{\mathrm{j}\Delta k_{\mathrm{M7}} z} \\ +\frac{1}{3}\left(A_{4x}^* A_{1y} A_{2x} \mathrm{e}^{-\mathrm{j}\Delta k_{\mathrm{W}} z} + A_{4x}^* A_{1x} A_{2y} \mathrm{e}^{-\mathrm{j}2\Delta k_{\mathrm{W}} z}\right) \end{bmatrix}, \tag{A.40}$$

and

$$\frac{\partial A_{4y}}{\partial z} = 2\mathrm{j}\gamma P_1 \left[\begin{array}{c} \left(\frac{1}{2}|A_{4y}|^2 + \sum_{i\neq 4}|A_{iy}|^2 + \frac{1}{3}\sum_i |A_{ix}|^2\right) A_{4y} \\ +A_{3y}^* A_{1y} A_{2y} \mathrm{e}^{\mathrm{j}\Delta k_{\mathrm{M8}} z} \\ +\frac{1}{3}\left(A_{3x}^* A_{1y} A_{2x} \mathrm{e}^{\mathrm{j}2\Delta k_{\mathrm{W}} z} + A_{3x}^* A_{1x} A_{2y} \mathrm{e}^{\mathrm{j}\Delta k_{\mathrm{W}} z}\right) \end{array} \right]. \quad (A.41)$$

where $i = 1\text{-}4$.

The differential equation system, (A.34) to (A.41), describes the SPM, XPM, and FWM in optical fibers where the polarization state of the waves are taken into consideration. If all waves have an identical polarization, then

$$A_{1x} = A_{2x} = A_{3x} = A_{4x} = 1$$

and

$$A_{1y} = A_{2y} = A_{3y} = A_{4y} = 0.$$

The differential equation system derived from the NSE in Sect. 7.2 (7.13) to (7.16) is then

$$\frac{\partial A_{1x}}{\partial z} = 2\mathrm{j}\gamma P_1 \left[\left(\frac{1}{2}|A_{1x}|^2 + \sum_{i\neq 1}|A_{ix}|^2\right) A_{1x} + A_{2x}^* A_{3x} A_{4x} \mathrm{e}^{\mathrm{j}\Delta k_{\mathrm{M1}} z}\right], \quad (A.42)$$

$$\frac{\partial A_{2x}}{\partial z} = 2\mathrm{j}\gamma P_1 \left[\left(\frac{1}{2}|A_{2x}|^2 + \sum_{i\neq 2}|A_{ix}|^2\right) A_{2x} + A_{1x}^* A_{3x} A_{4x} \mathrm{e}^{\mathrm{j}\Delta k_{\mathrm{M2}} z}\right],$$

$$\frac{\partial A_{3x}}{\partial z} = 2\mathrm{j}\gamma P_1 \left[\left(\frac{1}{2}|A_{3x}|^2 + \sum_{i\neq 3}|A_{ix}|^2\right) A_{3x} + A_{4x}^* A_{1x} A_{2x} \mathrm{e}^{\mathrm{j}\Delta k_{\mathrm{M3}} z}\right],$$

$$\frac{\partial A_{4x}}{\partial z} = 2\mathrm{j}\gamma P_1 \left[\left(\frac{1}{2}|A_{4x}|^2 + \sum_{i\neq 4}|A_{ix}|^2\right) A_{4x} + A_{3x}^* A_{1x} A_{2x} \mathrm{e}^{\mathrm{j}\Delta k_{\mathrm{M4}} z}\right].$$

If only two waves with an identical polarization are present and if the phase-dependent terms are neglected, the equations (A.34) to (A.41) describe the effects of SPM and XPM only, which are derived in Sect. 6.2:

$$A_{1x} = A_{2x} = 1, \quad (A.43)$$

$$A_{1y} = A_{2y} = A_{3y} = A_{4y} = A_{3x} = A_{4x} = 0,$$

$$\frac{\partial A_{1x}}{\partial z} = 2\mathrm{j}\gamma P_1 \left(\frac{1}{2}|A_{1x}|^2 + |A_{2x}|^2\right) A_{1x}, \quad (A.44)$$

and

$$\frac{\partial A_{2x}}{\partial z} = 2j\gamma P_1 \left(\frac{1}{2} |A_{2x}|^2 + |A_{1x}|^2 \right) A_{2x}.$$

If in the case of XPM the waves have different polarizations, the Equations (A.34) to (A.41) then become

$$A_{1x} = A_{2x} = A_{1y} = A_{2y} = 1, \tag{A.45}$$

$$A_{3y} = A_{4y} = A_{3x} = A_{4x} = 0,$$

$$\frac{\partial A_{1x}}{\partial z} = 2j\gamma P_1 \left(\frac{1}{2} |A_{1x}|^2 + |A_{2x}|^2 + \frac{1}{3} \sum_i |A_{iy}|^2 \right) A_{1x}, \tag{A.46}$$

$$\frac{\partial A_{2x}}{\partial z} = 2j\gamma P_1 \left(\frac{1}{2} |A_{2x}|^2 + |A_{1x}|^2 + \frac{1}{3} \sum_i |A_{iy}|^2 \right) A_{2x},$$

$$\frac{\partial A_{1y}}{\partial z} = 2j\gamma P_1 \left(\frac{1}{2} |A_{1y}|^2 + |A_{2y}|^2 + \frac{1}{3} \sum_i |A_{ix}|^2 \right) A_{1y},$$

and

$$\frac{\partial A_{2y}}{\partial z} = 2j\gamma P_1 \left(\frac{1}{2} |A_{2y}|^2 + |A_{1y}|^2 + \frac{1}{3} \sum_i |A_{ix}|^2 \right) A_{2y},$$

where $i = 1, 2$.

A.1 div, grad, curl, rot

In the electromagnetic field theory described by the Maxwell equations, abbreviations for mathematical operations like the Laplace and Nabla operator are often used. The essential concepts are reviewed here

- Divergence **div**
  The divergence of a field of vectors $\boldsymbol{E}$ is a measure of the sources of this field. The result is a field of scalars:

  $$\mathrm{div}\boldsymbol{E} = \frac{\rho}{\varepsilon_0}.$$

- Gradient **grad**
  With the gradient operation it is possible to generate a field of vectors from a scalar field. If, for instance, sources are present in a considered volume, they will produce a potential function φ. Hence, the corresponding electric vector field is

  $$\boldsymbol{E} = -\mathrm{grad}\varphi.$$

  The result is a source field and the field lines have a beginning and an end.

- Rotation **curl** (**rot**)
  If the field lines of a field of vectors have neither a beginning nor an end, the result is a field of vortices. With the rotation, it is possible to determine the origin of these vortices. The result is again a field of vectors:

$$\mathrm{curl}\boldsymbol{E} = -\frac{\partial \boldsymbol{B}}{\partial t}.$$

- Nabla Operator
  The Nabla operator is a vector and a regulation for a mathematical operation. If a Cartesian coordinate system with x, y, and z is used, it can be written as

$$\nabla = \frac{\partial}{\partial x}\boldsymbol{e}_x + \frac{\partial}{\partial y}\boldsymbol{e}_y + \frac{\partial}{\partial z}\boldsymbol{e}_z.$$

With the help of the Nabla operator the mentioned operations of the field theory can be expressed as

$$\mathrm{div}\boldsymbol{E} = \nabla \cdot \boldsymbol{E} = \frac{\partial E_x}{\partial x} + \frac{\partial E_y}{\partial y} + \frac{\partial E_z}{\partial z},$$

$$\mathrm{grad}\varphi = \nabla\varphi = \frac{\partial \varphi}{\partial x}\boldsymbol{e}_x + \frac{\partial \varphi}{\partial y}\boldsymbol{e}_y + \frac{\partial \varphi}{\partial z}\boldsymbol{e}_z,$$

and

$$\mathrm{curl}\boldsymbol{E} = \nabla \times \boldsymbol{E} = \begin{vmatrix} \boldsymbol{e}_x & \boldsymbol{e}_y & \boldsymbol{e}_z \\ \frac{\partial}{\partial x} & \frac{\partial}{\partial y} & \frac{\partial}{\partial z} \\ E_x & E_y & E_z \end{vmatrix}.$$

In cylinder coordinates, useful for wires and fibers, the Nabla operator has the form

$$\nabla = \frac{\partial}{\partial r}\boldsymbol{e}_{\mathrm{r}} + \frac{1}{r}\frac{\partial}{\partial \varphi}\boldsymbol{e}_{\varphi} + \frac{\partial}{\partial z}\boldsymbol{e}_z$$

In spherical coordinates, the Nabla operator can be expressed as

$$\nabla = \frac{\partial}{\partial r}\boldsymbol{e}_{\mathrm{r}} + \frac{1}{r}\frac{\partial}{\partial \vartheta}\boldsymbol{e}_{\vartheta} + \frac{1}{r\sin\vartheta}\frac{\partial}{\partial \varphi}\boldsymbol{e}_{\varphi}.$$

- Laplace Operator
  Twice the cross product between vectors can be developed in

$$\boldsymbol{a} \times \boldsymbol{b} \times \boldsymbol{c} = \boldsymbol{b}\,(\boldsymbol{a} \cdot \boldsymbol{c}) - \boldsymbol{c}\,(\boldsymbol{a} \cdot \boldsymbol{b})\,.$$

With $\boldsymbol{a} = \nabla$, $\boldsymbol{b} = \nabla$, and $\boldsymbol{c} = \boldsymbol{E}$ it follows that

$$\nabla \times \nabla \times \boldsymbol{E} = \nabla\,(\nabla \cdot \boldsymbol{E}) - (\nabla \cdot \nabla)\,\boldsymbol{E}.$$

Using the above operations, it can be written as

$$\text{curl curl}\boldsymbol{E} = \text{grad div}\boldsymbol{E} - (\nabla \cdot \nabla)\,\boldsymbol{E}$$

if the considered field is free of sources it follows div$\boldsymbol{E}$ =0 and, hence, curl curl$\boldsymbol{E}$ = $-(\nabla \cdot \nabla)\,\boldsymbol{E}$. The scalar product between the Nabla operators is the Laplace operator. In a right angled coordinate system with x, y and z it means that

$$\nabla \cdot \nabla = \begin{pmatrix} \frac{\partial}{\partial x} \\ \frac{\partial}{\partial y} \\ \frac{\partial}{\partial z} \end{pmatrix} \cdot \begin{pmatrix} \frac{\partial}{\partial x} \\ \frac{\partial}{\partial y} \\ \frac{\partial}{\partial z} \end{pmatrix} = \frac{\partial^2}{\partial x^2} + \frac{\partial^2}{\partial y^2} + \frac{\partial^2}{\partial z^2} = \Delta .$$

Hence, contrary to the Nabla operator the Laplace operator is not a vector. If it is applied on a vector, one yields

$$\Delta \boldsymbol{E} = \left(\frac{\partial^2}{\partial x^2} + \frac{\partial^2}{\partial y^2} + \frac{\partial^2}{\partial z^2} \right) \begin{pmatrix} E_x \\ E_y \\ E_z \end{pmatrix} .$$

The Laplace operator in the cylindrical coordinates r, φ and z can be written as

$$\Delta = \frac{1}{r}\frac{\partial}{\partial r} + \frac{\partial^2}{\partial r^2} + \frac{1}{r^2}\frac{\partial^2}{\partial \varphi^2} + \frac{\partial^2}{\partial z^2} .$$

A.2 The Gaussian Pulse

For analytical descriptions especially, pulses with a Gaussian shape are of special interest because the shape remains the same if the pulse is Fourier transformed. Hence, the mathematical derivation can be simplified in most cases. A Gaussian pulse can be expressed as

$$A = A_0 \exp\left(-\frac{2 \ln 2 t^2}{\tau_{\text{FWHM}}^2} \right) ,$$

where τ_{FWHM} is the full width of the pulse at half maximum amplitude ($1/2A_0$). Another possibility is offered by the description over its half width at 1/e intensity

$$A = A_0 \exp\left(-\frac{t^2}{2\tau_0^2} \right) .$$

Both are related via

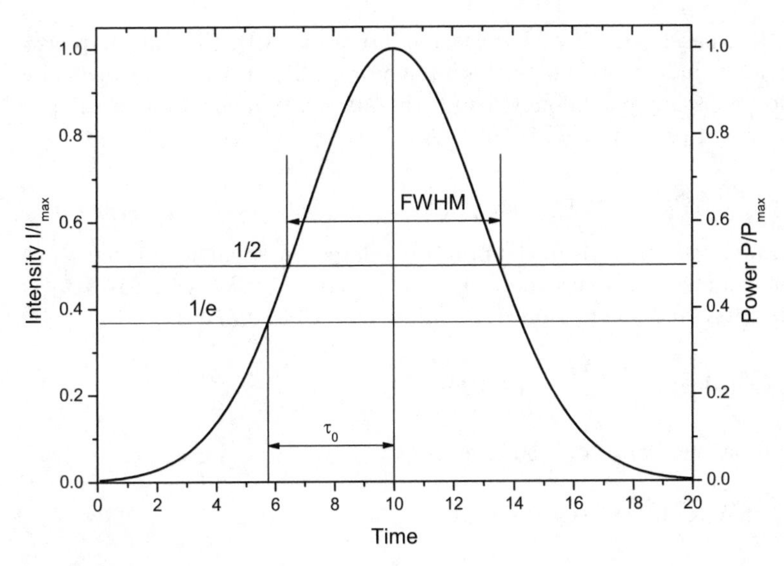

Fig. A.1. Normalized Gaussian pulse

$$\frac{2\ln 2t^2}{\tau_{\mathrm{FWHM}}^2} = \frac{t^2}{2\tau_0^2}$$
$$\tau_{\mathrm{FWHM}} = 2\sqrt{\ln 2}\tau_0 \approx 1.665\tau_0.$$

Figure A.1 shows a Gaussian pulse as well as the typical values for the description of the width. If the Gaussian pulse is bandwidth-limited, i.e., it has no additional chirp, the relation between the temporal and spectral widths is determined by

$$\Delta\omega = \frac{1}{\tau_0} \approx \frac{1.665}{\tau_{\mathrm{FWHM}}}.$$

A.3 Logarithmical Units

Logarithmical units are often used for convenience in telecommunications. The gain of an amplifier is, for instance,

$$G_{\mathrm{dB}} = 10\log_{10}\frac{P_{\mathrm{out}}}{P_{\mathrm{in}}}.$$

Hence, a gain of 30 dB means that the output power of the device is 1000 times stronger than the input power:

$$\frac{P_{\mathrm{out}}}{P_{\mathrm{in}}} = 10^{\frac{G_{\mathrm{dB}}}{10}}.$$

Whereas, if a passive device has an attenuation of 3 dB, the output power of the device is half as strong as the input power. The relation between the attenuation constant per kilometer and the attenuation constant in dB per kilometer was already derived in Sect. 2.4. It is:

$$\alpha_{\mathrm{km}^{-1}} = 0.23026\ \alpha_{\mathrm{dB/km}}.$$

The power values in decibel units are related to a particular reference value. For instance, the expression dBm is related to 1 mW, whereas dBµ is related to 1 µW and so on. Hence, an input power of 10 mW is

$$P_{\mathrm{dBm}} = 10 \log_{10} \frac{10\ \mathrm{mW}}{1\ \mathrm{mW}} = 10\ \mathrm{dBm}$$

whereas an output power of −30 dBm is

$$P = 1\ \mathrm{mW} \times 10^{\frac{-30}{10}} = 1\ \mu\mathrm{W}{=}0\ \mathrm{dB}\mu$$

References

1. M. Soljačić, C. Luo, J. D. Joannopoulos, S. Fan: Opt. Lett. **28**, 637 (2003)
2. J. F. Nye: *Physical Properties of Crystals: Their representation by Tensors and Matrices* (Oxford University Press, Oxford 1985)
3. R. J. Mears, L. Reekie, I. M. Jauncey, D. N. Payne: Elect. Lett. **23**, 1026 (1987)
4. E. Desurvire, J. R. Simpson, P. C. Becker: Opt. Lett. **12**, 888 (1987)
5. Y. Sun, A. K. Srivastava, J. Zhou, J.W. Sulhoff: Bell Labs Techn. J., Jan.-March, 187 (1999)
6. F. und L. Pedrotti: *Optik eine Einführing (*Prentice Hall, München 1996)
7. I. H. Mlitson: J. Opt. Soc. Am. **54**, 628 (1964)
8. I. N. Bronstein, K. A. Semendjajew: *Taschenbuch der Mathematik* (Teubner, Leipzig 1979)
9. J.R. Thompson, R. Roy: Phys. Rev. A, **43**, 4987 (1991)
10. J. Teipel et al: Appl. Phys. B **77**, 245 (2003)
11. J. Hiroishi, R. Sigizaki, O. Aso, M. Tadakuma, T. Shibuta: Furukawa Review **23**, 21 (2003)
12. K. W. Chung, S. Kim, S. Yin: Opt. Lett. **28**, 2031 (2003)
13. E. Yablonovitch: Phys. Rev. Lett. **58**, 2059 (1987)
14. S. John: Phys. Rev. Lett. **58**, 2486 (1987)
15. P. V. Mamyshev: ECOC'98, Madrid, Spain (1998)
16. A. R. Parker et al: Nature **409**, 36 (2001)
17. U. Kilian: Physik Journal **1**, 58 (2002)
18. P. Vukusic, J. R. Sambles: Nature, **424**, 852 (2003)
19. S. D. Cheng et al: Apl. Phys. Lett. **67**, 3399 (1995)
20. M. P. Kesler, J. G. Maloney, B. L. Shirley, G. S. Smith: Microwave and Opt. Technol. Lett. **11**, 169 (1996)
21. E. Yablonovitch, T. J. Gmitter, K. M. Leung: Phys. Rev. Lett. **67**, 2295 (1991)
22. E. Ozbay et al: Phys. Rev. B **50**, 1945 (1994)
23. T. A. Birks, J. C. Knight, P.St. J. Russell: Opt. Lett. **22**, 961 (1997)
24. B. Temelkuran, S. D. Hart, G. Benoit, J. Joannopoulos, Y. Fink: Nature **420**, 650 (2002)
25. D. W. Hewak, Y. D. West, N. G. R. Broderick, T. M. Monro, D. J. Richardson: in *Optical Fiber Communications*, 2001 OSA Technical Digest Series, paper TuC4 Washington D. C. (2001)
26. P. Kaiser, H. W. Astle: Bell Syst. Tech. J. **53**, 1021 (1974)
27. J. C. Knight: Nature **424**, 847 (2003)
28. V. V. R. K. Kumar et al: Opt. Express **10**, 1520 (2002)
29. D. Kominsky, G. Pickrell, R. Stolen: Opt. Lett. **28**, 1409 (2003)
30. N. G. R. Broderick, T. M. Monroe, P. J. Bennett, D. J. Richardson: Opt. Lett. **24**, 1395 (1999)
31. J. Leagsgaard, A. Bjarklev: Opt. Lett. **28**, 783 (2003)

32. P. Petropoulos, T. M. Monro, W. Belardi, K. Furusawa, J. H. Lee, D. J. Richardson: Opt. Lett. **26**, 1233 (2001)
33. J. Broeng, D. Mogilevstev, S. E. Barkou, A. Bjarklev: Optical Fiber Technology **5**, 305 (1999)
34. J. Broeng, S. E. Barkou, T. Sondergaard A. Bjarklev: Opt. Lett. **25**, 96 (2000)
35. J. K. Ranka, R. S. Windeler, A. J. Stentz: Opt. Lett. **25**, 25 (2000)
36. W. J. Wadsworth, J. C. Knight, A. Ortigosa-Blanch, J. Arriaga, E. Silvestre, P. St. J. Russell: Elect. Lett. 36, 53 (2000)
37. T. A. Birks, D. Mogilevtsev, J. C. Knight, P. St. J. Russell: IEEE Photon. Technol. Lett. **11**, 674 (1999)
38. K. Tajima, J. Zhou, K. Nakajima, K. Sato: OFC 2003 (2003)
39. M. A. van Eijkelenborg et al: Opt. Express **9**, 319 (2001)
40. P. St. J. Russell: Science 299, 385 (2003)
41. Y. Guo, C. K. Kao, E. H. Li, K. S. Chiang: *Nonlinear Photonics, Nonlinearities in Optics, Optoelectronics and Fiber Communications (*Springer Berlin Heidelberg New York 2002)
42. G. Mourou, D. Umstadter: Spektrum der Wissenschaft, 70, Juli (2002)
43. P. A. Norreys e. al.: Phys. Rev. Lett. **76**, 1832 (1996)
44. M. Hegelich et al: Phys. Rev. Lett. **89**, 085002 (2002)
45. C. E. Shannon: Bell. Syst. Tech. J. **27**, 379-423, 623-656 (1948)
46. P. P. Mitra, J. B. Stark: Nature **411**, 1027 (2001)
47. E. E. Narimanov, P. Mitra: J. Lightw. Technol. **20**, 530 (2002)
48. J. M. Kahn, K. P. Ho: Optical Fiber Communications Conference, Anaheim, CA, March 17-22 (2002)
49. M. Eiselt: J. Lightw. Technol. **17**, 2261 (1999)
50. J. M. Kahn, K. P. Ho: OptoElectronics and Communications Conference, Yokohama, Japan, July 8-12 (2002)
51. G.P. Agrawal: *Nonlinear Fiber Optics* (Academic, New York 1995)
52. G. P. Agrawal: *Fiber Optic Communication Systems* (Wiley, New York 1997)
53. J. A. J. Fells et al: Proc. 26th Optical Fiber Communications Conference, OFC 2001, Anaheim Cal. Paper PD11 (2001)
54. W. Glaser: *Photonik für Ingenieure (*Verlag Technik, Berlin 1997)
55. L. F. Mollenauer, P. V. Mamyshev, M. J. Neubelt: Opt.Lett. **21**, 1724 (1996)
56. K. Nakajima, M. Ohashi, M. Tateda: J. Lightw. Technol. **15**, 1095 (1997)
57. J. Gripp, L. F. Mollenauer: Opt. Lett. **23**, 1603 (1998)
58. E. Poutrina, G. P. Agrawal: J. Lightw. Technol. **21**, 990 (2003)
59. Y.R. Shen: *The Principals of nonlinear Optics* (Wiley, New York 1984)
60. J. W. Nicholson et al: Opt. Lett. **28**, 643 (2003)
61. D. J. Richardson, T. M. Monro, N.G.R. Broderick: Proc. ECOC'2000, P 10.2.2, Munich, Germany (2000)
62. J. J. Refi: Bell Labs Technical Journal, Jan-March, 246 (1999)
63. N. Bloembergen, P.S. Pershan: Phys. Rev., **128**, 606 (1962)
64. J.Reif, P.Tepper, E.Matthias, E,Westin, A.Rosèn: Appl. Phys. B, **46**, 131, (1988).
65. J. Hohlfeld, U. Conrad, E. Matthias: Appl. Phys. B, **63**, 541 (1996)
66. J.M. Hicks, L. E. Urbach, E. W. Plummer, H.L. Dai: Phys.Rev.Lett., **61**, 2588 (1988)
67. K.Sokolowski-Tinten, J.Bialkowski, D.von der Linde: Phys. Rev. B, **51**, 14186, (1995).
68. H.W.K.Tom, G.D.Aumiller, C.H.Brito-Cruz: Phys. Rev. Lett., **60**,1438, (1988).
69. Y.M. Chang, L.Xu, H.W.K. Tom: Phys.Rev.Lett.,**78**, 4649 (1997)
70. J. Reif: Opt. Engin., **28**, 1122 (1989)

71. J. E. Sipe, D. J. Moss, H. M. van Driel: Phys. Rev. B, **35**, 1129 (1987)
72. D. A. Kleinman: Phys. Rev. **126**, 1977 (1962)
73. R. W. Boyd: *Nonlinear Optics* (Academic, Boston 1992)
74. A. Laubereau: In *Ultrashort Laser Pulses-Generation and Applications*, Topics in Applied Physics, Vol.60, 2nd Edition, ed by W. Kaiser, (Springer-Verlag, Berlin 1993)
75. M. A. Arbore et al: Optics Letters, **22**, 13 (1997)
76. J. W. Shelton, Y.R. Shen: Phys.Rev.Lett., **25**, 23 (1970)
77. P. P. Bey, J. F. Giuliani. H. Rabin: Phys. Letters **26A**, 128 (1968)
78. Y. Tamaki, K. Midorikawa: Appl. Phys. B, **67**, 59 (1998)
79. Th. Schneider, R. P. Schmid, J. Reif: Appl. Phys. B, **72**, 563 (2001)
80. R. Billington: Effective Area of Optical Fibres-Definition and Measurement Techniques, National Physical Laboratory (1999)
81. R. C. Wittman, M. Young: Technical Digest of the Symposium on Optical Fiber Measurements, Boulder, Colorado, 141 (1998)
82. Y. Namihira: Electron. Lett., **30**, 262 (1994)
83. ITU COM 15-273-E Appendix on Nonlinearities for G.650 (1996)
84. R. H. Stolen, J. E. Bjorkholm: IEEE J. Quant. El., **QE-18**, 1062 (1982)
85. F. Forghieri, R.W. Tkach, A.R. Chraplyvy: Fiber Nonlinearities and Their Impact on Transmission Systems, In: Optical Fiber Telecommunications Vol.IIIA, ed by I.P. Kaminov, T.L.Koch: (Academic, San Diego 1997)
86. A. H.Gnauck, A.R. Chraplyvy, R.W Tkach.,R.M. Derosier: Electron. Lett. **30**, 1241 (1994)
87. R. -J. Essiambre, B. Mikkelsen, G. Raybon: Electron. Lett. **35**, 1576 (1999)
88. A. Mecozzi, C. B. Clausen, M. Shtaif, S. G. Park, A. H. Gnauck: IEEE Photon. Technol. Lett. **13**, 445 (2001)
89. A. Mecozzi, C. B. Clausen, M. Shtaif, S. G. Park, A. H. Gnauck: IEEE Photon. Technol. Lett. **12**, 392 (2000)
90. P. V. Mamyshev, N. A. Mamysheva: Opt. Lett. **24**, 1454 (1999)
91. R. I. Killey, H. J. Thiele, V. Mikhailov, P. Bayvel: IEEE Photon. Technol. Lett. **12**, 1624 (2000)
92. J. Martensson, A. Berntson, M. Westlund, A. Danielsson, P. Johannisson, D. Anderson, M. Lisak: Opt. Lett. **26**, 55 (2001)
93. X. Liu, X. Wei, A. H. Gnauck, C. Xu, L. K. Wickham: Opt. Lett. **27**, 1177 (2002)
94. F. Merlaud, S. K. Turitsyn: Proc. ECOC'2000, P 7.2.4, Munich, Germany (2000)
95. M. J. Ablowitz, T. Hirooka: Opt. Lett. **25**, 1750 (2000)
96. P. Johannisson, D. Anderson, A. Berntson, J. Mårtensson: Opt. Lett. **26**, 1227 (2001)
97. S. Kumar: IEEE Photon. Technol. Lett. **13**, 800 (2001)
98. K. S. Cheng, J. Conradi: IEEE Photon. Technol. Lett. **14**, 98 (2002)
99. Y. Miyamoto, A. Hiramno, K. Yonenaga, A. Sano, Hi Toba, K. Murata, O. Mitomi: Electron. Lett. 35, 2041 (1999)
100. M. Suzuki, N. Edagawa: J. Lightw. Technol. **21**, 916 (2003)
101. S. N. Knudsen, M. O. Pedersen, L. Grüner-Nielsen: Electron. Lett. **36**, 2067 (2000)
102. N. Hanik, C. Caspar, F. Schmidt, R. Freund, L. Molle, C. Peucheret: ECOC'00, München, Vol.3, pp.195-197 (2000)
103. C. Caspar, R. Freund, N. Hanik, L. Molle, C. Peucheret: In *New Trends in Optical Network Design and Modelling* (Kluwer Academic Publishers, Boston 2001)
104. T.Morioka, K.Okamoto, M. Ishii, M. Saruwatari: El. Lett., **32**, 836 (1996)

105. R. Alfano: ed., *The Supercontinuum Laser Source*, (Springer, Berlin 1989)
106. A. Proulx, J.M. Ménard, N. Hô, J. M. Laniel, R. Vallée, C. Paré: Optics Express **11**, 3338 (2003)
107. B. Mikkelsen, G. Raybon, R. J. Essiambre, J. E. Johnson, K. Dreyer, L. F. Nelson: Electron. Lett. **35**, 1866 (1999)
108. N. Shibata, R. B. Braun, R. G. Waarts: IEEE J. Quant. Electron. **QE-23**, 1205 (1987)
109. R. W. Tkach, A. R. Chraplyvy, F. Forghieri, A. H. Gnauck, R. M. Derosier: J. Lightw. Technol. **13**, 841 (1995)
110. K. Inoue: Opt. Lett. **17**, 801 (1992)
111. K. Inoue: J. Lightw. Technol. **11**, 2116 (1993)
112. K. Inoue: IEEE Photon. Technol. Lett. **4**, 1301 (1992)
113. K. Inoue: J. Lightw. Technol. **11**, 455 (1993)
114. D. Marcuse: J. Lightw. Technol. **8**, 1816 (1990)
115. W. C. Babcock: Bell Syst. Tech. J. **31**, 63 (1953)
116. D. G. Schadt: Electron. Lett. **27**, 1805 (1991)
117. M. C. Nuss, W. H. Knox, U. Koren: El. Lett. **32**, 1311 (1996)
118. T. Morioka, K.Uchiyama, S. Kawanishi, S.Suzuki, M. Saruwatari: El. Lett., **31**, 1064 (1995)
119. S. Kawanishi, H. Takara, K. Uchiyama, I. Shake, K. Mori: El. Lett. **35**, 826 (1999)
120. H. Takara et al: Electron. Lett. **36**, 2089 (2000)
121. B. Mikulla, L. Leng, S. Sears, B. C. Collings, M. Arend, K. Bergman: IEEE Phot. Technol. Lett. **11**, 418 (1999)
122. J. Hee Han, J. W. Ko, J. S. Lee, S. Y. Shin: IEEE Phot. Technol. Lett. **10**, 1501 (1998)
123. A.Brodeur, S.L.Chin: J. Opt. Soc. Am. B, **16**, 637 (1999)
124. H. P. Sardesai, C. C. Chang, A. M. Weiner: J. Lightw. Technol. **16**, 1953 (1998)
125. A. B. Fedotov et al: J. Opt. Soc. Am. B **19**, 2156 (2002)
126. B. C. Stuart, M. D. Feit, S. Herman, A. M. Rubenchik, B. W. Shore, M. D. Perry: Phys. Rev. Lett. **74**, 2248, (1995)
127. D. Du, X. Liu, G. Mourou: Appl. Phys. B **63**, 617, (1996)
128. J. M. Yates, M. P. Rumsewicz, J. P. R. Lacey: IEEE Comm. Surveys, Second Quarter '99, 2 (1999)
129. G. S. He, S. H. Liu: *Physics of Nonlinear Optics* (World Scientific Publishing Co., Singapore 1999)
130. R. S. Vodhanel: Opt. Fiber Conf. Housten TX 1989 (cited after [248])
131. P. V. Mamyshev, L. F. Mollenauer: Opt. Lett. **24**, 448 (1999)
132. R.W.Hellwarth: Third-Order optical susceptibilities of liquids and solids In: Prog. Quant. Electr., 5 pp.1-68, Pergamon Press (1977)
133. K. O. Hill, D. C. Johnson, B. S. Kawasaki, R. I. MacDonald: J. Appl. Phys. **49**, 5098 (1978)
134. F. Forghieri, R.W. Tkach, A.R. Chraplyvy, D. Marcuse: IEEE Photon. Tech. Lett. **6**, 754 (1994)
135. F. Forghieri, R. W. Tkach, A. R. Chraplyvy: J. Lightw. Technol. **13**, 889 (1995)
136. J. Zweck, C.R. Menyuk: Opt. Lett. **27**, ,(2002)
137. J. Zweck, C.R. Menyuk: IEEE Photon. Technol. Lett. **15**, 323 (2003)
138. C. Xiang, J. F. Young: IEEE Lasers & Electro-Optics Society Annual Meeting Proc., Paper WZ2, 609, San Francisco (1999)
139. C. R. Menyuk: J. Opt. Soc. Am. B **5**, 392 (1988)
140. I. P. Kaminow: IEEE J. Quant. Electr. **QE-17**, 15 (1981)

141. M. Wolf: Diploma Thesis *Untersuchung und Bewertung von Verfahren zur Signalqualitätsanalyse in DWDM-Systemen* Deutsche Telekom AG Fachhochschule Leipzig (2002)
142. T.Schneider, D. Schilder: J. Opt. Commun.,**19**, 115 (1998)
143. P. K. A. Wai, C. R. Menyuk, H. H. Chen: Opt. Lett. **16**, 1231 (1991)
144. P. A. Andrekson, N. A. Olsson, J. R. Simpson, T. Tanbun-Ek, R. A. Logan, M. Haner: El. Lett, **27**, 922 (1991)
145. O. Asu, M. Tadakuma, S. Namiki: Furukawa Review, 19, 63 (2000)
146. O. Aso, S.-I. Arai, T. Yagi, M. Tadakuma, Y. Suzuki, S. Namiki: Proc. ECOC'99, 2, 226, Nice, France (1999)
147. S. Watanabe: COST 266, 1, (2000)
148. L. Rau, W. Wang, B. E. Olsson, Y. Chiu, H. F. Chou, D. J. Blumenthal, J. E. Bowers: IEEE Phot. Technol. Lett. 14, 1725 (2002)
149. H. G. Weber: Nichtlineare Optik und optische Signalverarbeitung In E. Voges, K. Petermann (Hrsg.) *Optische Kommunikationstechnik* (Springer, Berlin, Heidelberg, New York 2002)
150. D. Nesset, T. Kelly, D. Marcenac: IEEE Comm. Mag., 56, Dec.(1998)
151. S. L. Danielsen et al: Electron. Lett. **32**, 1688 (1996)
152. A. D. Ellis et al: Electron. Lett. **34**, 1955 (1998)
153. J. Mork, M. L. Nielsen, T. W. Berg: Optics and Photonics News, **14**, 42 (2003)
154. J. Leuthold et al: Electron. Lett. **36**, 1129 (2000)
155. S. Nakamura, Y. Ueno, K. Tajima: IEEE Photon. Technol. Lett. **13**, 1091 (2001)
156. J. Leuthold et al: Optical and Quantum Electron. **33**, 939 (2001)
157. H. Nakajima, K. Kurishima, S. Yamahta, T. Kobayashi, Y. Matsuoka: Electron. Lett. **29**, 1887 (1993)
158. S. Kawanishi, T. Morioka, O. Kamatani, H. Takara, J. M. Jacob, M. Saruwatari: Electron. Lett. **30**, 981 (1994)
159. T. Morioka, H. Takara, S. Kawanishi, K. Uchiyama, M. Saruwatari: Electron. Lett. **32**, 840 (1996)
160. A. Neukermans, R. Ramaswami: IEEE Comm. Mag., 62, Jan. (2001)
161. C. Marxer et al: IEEE J. Micro. Mech. Sys., **6**, 277 (1997)
162. P. Yeh: Appl. Opt. **26**, 602 (1987)
163. Th. Schneider, J. Reif: Phys. Rev. A 65, 023801-1 (2002)
164. M. Vaa, B. Mikkelsen, K. S. Jepsen, K. E. Stubkjaer: ECOC'97, Edinbourgh Scotland, U. K., pp. WE1B-4-1438 (1997)
165. V. W. S. Chan, K. L. Hall, E. Modiano, K. A. Rauschenbach: J. Lightw. Technol. **16**, 2146 (1998)
166. K. Uchiyama, H. Takara, S. Kawanishi, T. Morioka, and M. Saruwatari: Electron. Lett. **28**, 1864 (1992)
167. J. D. Moores, K. Bergmann, H. A. Haus, E. P. Ippen: Opt. Lett. **16**, 138 (1991)
168. N. A. Whitaker Jr. et al: Opt. Lett. **16**, 1838 (1991)
169. K. L. Hall, K. A. Rauschenbach: Electron. Lett. **32**, 1214 (1996)
170. D. Wang, E. A. Golovchenko, A. N. Pilipetskii, C. R. Menyuk, A. F. Arend: J. Lightw. Technol. **15**, 642 (1997)
171. M. F. Arend. M. L. Dennis, I. N. Dulling III, E. A. Golovchenko, A. N. Pilipetskii, C. R. Menyuk: Opt. Lett. **22**, 886 (1997)
172. K. Uchiyama, H. Takara, T.Morioka, S. Kawanishi, M. Saruwatari: Electron. Lett. **29**, 1313 (1993)
173. T. Yamamoto, E. Yoshida, M. Nakazawa: Electron. Lett. **34**, 1013 (1998)
174. H. Sotobayashi, C. Sawagushi, Y. Koyamada, W. Chujo: Opt. Lett. **27**, 1555 (2002)

175. J. E. Sharping, M. Fiorentino, P. Kumar, R. S. Windeler: IEEE Photon. Technol. Lett. **14**, 77 (2002)
176. D. Phillips, A. D. Ellis, H. J. Thiele, R. J. Manning, A. E. Kelly: Electron. Lett. **34**, 2340 (1998)
177. J. P. Sokoloff, P. R. Prucnal, I. Glesk, M. Kane: IEEE Photon. Technol. Lett. **5**, 787 (1993)
178. I. Glesk, J. P. Sokoloff, P. R. Prucnal: Electron. Lett. **30**, 339 (1994)
179. R. J. Manning, D. A. O. Davies, D. Cotter, J. K. Lucek: Electron. Lett. **30**, 787 (1994)
180. K. L. Deng, R. J. Runser, P. Toliver, C. Coldwell, D. Zhou, I. Glesk, P. R. Prucnal: Electron. Lett. **34**, 2418 (1999)
181. S. Diez, R. Ludwig, H. G. Weber: IEEE Photon. Technol. Lett. **11**, 60 (1999)
182. S. Nakamura et al: IEEE Photon. Technol. Lett. **12**, 425 (2000)
183. J. Capmany, P. Muñoz, S. Sales, D. Pastor, B. Ortega, A. Martinez: Opt. Lett., **28**, 197 (2003)
184. Th. Schneider, D. Wolfframm, R. Mitzner, J.Reif: Appl. Phys. B **68**, 749 (1999)
185. J. E. Aber, M. C. Newstein, B. A. Garetz: J. Opt. Soc. Am. B **17**, 120 (2000)
186. J. M. Harbold, F. Ö. Ilday, F. W. Wise, J. S. Sanghera, V. Q. Nguyen, L. B. Shaw, I. D. Aggarwal: Opt. Lett. **27**, 119 (2002)
187. J. Gruber, P. Roorda, F. LaLonde: Proc. NFOEC, 678, (2000)
188. A. H. Gnauck, S. G. Park, J. M. Wiesenfeld, L. D. Garrett: Proc. LEOS Annu. Meeting, San Francisco, CA, PD1.2 (1999)
189. M. E. Marhic, N. Kagi, T.-K. Chiang, L. G. Kazovsky: Opt. Lett. **21**, 573 (1996)
190. M. C. Ho, K. Uesaka, M. Marhic, Y. Akasaka, L. G. Kazovsky: J. Opt. Lightw. Technol. **19**, 977 (2001)
191. P. M. Krummrich, R. E. Neuhauser, H. Bock, W. Fischler, C. Glingener: Paper TuA1.4, ECOC 2001, Amsterdam, Holland (2001)
192. M. Nisson, K. Rottwitt, H. D. Kidorf, M. X. Ma: Electron. Lett. **35**, 997 (1999)
193. A. Yariv, D. Fekete, D. M. Pepper: Opt. Lett. **4**, 52 (1979)
194. R. . Fisher, B. R. Suydam, D. Yevick: Opt. Lett. **8**, 611 (1983)
195. S. Chi, S. F. Wen: Opt. Lett. **19**, 1705 (1994)
196. D. Rockwell: IEEE J. Quantum Opt. **24**, 1124 (1988)
197. M. Ostermeyer, R. Menzel: Opt. Commun. **171**, 85 (1999)
198. A. M. Scott, G. Cook, A. P. G. Davies: Appl. Opt. **40**, 2461 (2001)
199. B. Ya.Zel'dovich, V.I. Popovichev, V.V. Ragul'skii, F.S. Faisullov: JETP Lett., **15**, 109 (1972)
200. O. Yu. Nosach, V.I. Popovichev, V.V. Ragul'skii, F.S. Faisullov: JETP Lett., **16**, 435 (1972)
201. M. Fink: Scientific American, 67, November (1999)
202. S. Watanabe, S. Kaneko, T. Chikama: Opt.Fib.Techn., **2**, 169, (1996)
203. M. Remoissenet: *Waves Called Solitons* (Springer, Berlin Heidelberg New York 1996)
204. J. S. Russel: Rep. 14th Meet. Brit. Assoc. Adv. Sci., York, 311 (1844)
205. F. X. Kärtner: Physik in unserer Zeit, **26**, 152 (1995)
206. A. Hasegawa, F. Tappert: Appl. Phys. Lett. **23**, 142 (1973)
207. F. Mitschke, M. Böhm: Physikalische Blätter, 56, 25 (2000)
208. C. S. Gardener, J. M. Greene, M. D. Kruskal, R. M. Miura: Phys. Rev. Lett. **19**, 1095 (1967)
209. L. F. Mollenauer, R. H. Stolen, J. P. Gordon: Phys. Rev. Lett. 45, 1095, (1980)

210. L. F. Mollenauer, J. P. Gordon, P. V. Mamyshev: *Solitons in High Bit-Rate, Long-Distance Transmission* In: Optical Fiber Telecommunications Vol.IIIA, ed by I.P. Kaminov, T.L.Koch (Academic, San Diego 1997)
211. V.E. Zakharov, A.B. Shabat: Sov. Phys. JETP **34**, 62 (1972)
212. W. J. Tomlinson, R. J. Hawkins, A. M. Weiner: J.P. Heritage, R.N. Thurston, J. Opt. Soc. Am. B, **6**, 329 (1989)
213. W. Zhao, E. Bourkoff: Opt. Lett. 14, 703 (1989)
214. W. Zhao, E. Bourkoff: Opt. Lett. 14, 808 (1989)
215. M. Suzuki, I. Morita, N. Edagawa, S. Yamamoto, H. Taga, S. Akiba: Electron. Lett. **31**, 2027 (1995)
216. N. J. Smith, F. M. Knox, N. J. Doran, K. J. Blow, I. Bennion: Electron. Lett. **32**, 54 (1996)
217. G. M. Carter, J. M. Jacob, C. R. Menyuk, E. A. Golovchenko, A. N. Pilipetskii: Opt. Lett. **22**, 513 (1997)
218. M. Suzuki, N. Edagawa, I. Morit, S. Yamamoto, S. Akiba: J. Opt. Soc. Am. B **14**, 2953 (1997)
219. P. V. Mamyshev, L. F. Mollenauer: Opt. Lett. **24**, 448 (1999)
220. M. Shirasaki: IEEE Photon. Technol. Lett. **9**, 1598 (1997)
221. C. K. Madesn, G. Lenz: IEEE Photon. Technol. Lett. **10**, 994 (1998)
222. X. Wei, X. Liu, C. Xie, L. F. Mollenauer: Opt. Lett. **28**, 983 (2003)
223. E. Poutrina, G. P. Agrawal: Opt. Commun. **206**, 193 (2002)
224. X. Liu, X. Wei, L. F. Mollenauer, C. J. McKinstrie, C. Xie: Opt. Lett. **28**, 1412 (2003)
225. L. F. Mollenauer et al: Opt. Lett. **25**, 704 (2000)
226. L. F. Mollenauer, A. Grant, X. Liu, X. Wei, C. Xie, I. Kang: Opt. Lett. **28**, 2043 (2003)
227. J.P. Gordon, H. A. Haus: Opt. Lett. **11**, 665 (1986)
228. E. M. Dianov, A.V. Luchnikov, A.N. Pilipetskii, A. M. Prokhorov: Soviet Lightw. Comm. **1**, 235 (1991)
229. E. M. Dianov, A.V. Luchnikov, A.N. Pilipetskii, A. M. Prokhorov: Appl. Phys. B **54**, 175 (1992)
230. L. F. Mollenauer, J. P. Gordon, S. G. Evangelides: Opt. Lett., **20**, 539 (1992)
231. M. Nakazawa, E. Yamada, H. Kubotra, K. Suzuki: Electron. Lett. **27**, 1270 (1991)
232. J. P. Gordon: Opt. Lett. **8**, 596 (1983)
233. T. H. Maiman: Nature **187**, 493 (1960)
234. E. J. Woodbury, W. K. Ng: Proc. I.R.E. **50**, 2367 (1962)
235. G. Eckhardt et al: Phys. Rev. Lett. **9**, 455 (1962)
236. N. Bloembergen: Am. J. Phys. **35**, 989 (1967)
237. M. G. Raymer, I. A. Walmsley: Progress in Optics, **28**, edited by E. Wolf, North Holland, Amsterdam (1990)
238. E. Garmire, F. Pandarese, C. H. Townes: Phys. Rev. Lett. **11**, 160 (1963)
239. R. G. Smith: Appl. Opt. **11**, 2489 (1972)
240. F. Benabid, J. C. Knight, P. St. Russell: Science **298**, 399 (2002)
241. R. H. Stolen: Proc. IEEE 68, 1232 (1980)
242. F.L. Gallener et al: Appl. Phys. Lett., **32**, 34 (1978)
243. R. H. Stolen, E. P. Ippen: Appl. Phys. Lett. 22, 276 (1973)
244. R. W. Hellwarth, J. Cherlow, T. T. Yang: Phys. Rev. B **11**, 964 (1975)
245. D. Mahgerefteh et al: Optics Lett., **21**, 2026 (1996)
246. Z. Yusoff, J. H. Lee, W. Belardi, T. M. Monro, P. C. Teh, D. J. Richardson: Opt. Lett. **27**, 424 (2002)
247. N. R. Newbury: Opt. Lett. **27**, 1232 (2002)
248. A. R. Chraplyvy: J. Lightw. Technol., **8**, 1548 (1990)

249. A. R. Chraplyvy: Electron. Lett. **20**, 58 (1984)
250. A. R. Chraplyvy, R. W. Tkach: IEEE Phot. Technol. Lett. **5**, 666 (1993)
251. M. O. van Deventer, A. J. Boot: J. Lightwave Technol. **12**, 585 (1994)
252. E. L. Buckland, R.W. Boyd: Opt. Lett. **22**, 676 (1997)
253. T. Hoeppner: Diploma Thesis *"Nutzung der SBS für Radio over Fiber Verfahren"* Deutsche Telekom AG Fachhochschule Leipzig (2003)
254. R. Billington: "Meausurement Methods for Stimulated Raman and Brillouin Scattering in Optical Fibres", National Physical Laboratory Report COEM 31 (1999)
255. T. Horiguchi, K. Shimizu, T. Kurashima, M. Tateda, Y. Koyamada: J. Lightwave Technol. **13**, 1296 (1995)
256. A. R. Chraplyvy, R. W. Tkach: Elektron. Lett. **22**, 1084 (1986)
257. N. A. Olsson, J. P. van der Ziel: J. Lightw. Technol. **5**, 147 (1987)
258. A. Loayssa, D. Benito, M. J. Garde: Opt. Lett. **25**, 197 (2000)
259. T. Tanemura, Y. Takushima, K. Kikuchi: Opt. Lett. **27**, 1552 (2002)
260. X. Bao, D. J. Webb, D. A. Jackson: Opt. Lett. **19**, 141 (1994)
261. X. Bao, J. Dhilwayo, N. Heron, D. J. Webb, D. A. Jackson: J. Lightwave Technol. **13**, 1340 (1995)
262. X. Bao, A. Brown, M. DeMerchant, J. Smith: Optics Lett. **24**, 510 (1999)
263. K. Shiraki et.al.: J. Lightw. Technol., **14**, 50 (1996)
264. J. Yu, I. B. Kwon, K. Oh: J. Lightw. Technol., **21**, 1779 (2003)
265. C. K. Jen, A. Saffai-Jazi, G. W. Farnell: IEEE Trans. Ultrason. Ferroelect. Freq. Contr. UFFC-**33**, 634 (1986)
266. L. Zou, X. Bao, L. Chen: Opt. Lett. **28**, 2022 (2003)
267. A. Yeniay, J. M. Delavaux, J. Toulouse: J. Lightwave Technol. **20**, 1425 (2002)
268. Y. Zel'dovich, N. F. Pilipetskii, N. Shkunov: *Principal of Phase Conjugation* (Springer, Berlin Heidelberg New York 1985)
269. R. W. Boyd, K. Rzazewski, P. Narum: Phys. Rev. A **42**, 5514 (1990)
270. X.P. Mao, R. W. Tkach, A. R. Chraplyvy, R. M. Jopson, R. M. Derosier: IEEE Photon. Tech. Lett. **4**, 66 (1992)
271. Y. Aoki, K. Tajima, I. Mito: J. Lightwave Technol. **6**, 710 (1988)
272. A. L. Gaeta, R. W. Boyd: Phys. Rev. A **44**, 3205 (1991)
273. V. I. Kovalev, R. G. Harrison: Phys. Rev. Lett. **85**, 1879 (2000)
274. A. A. Fotiadi, R. Kiyan, O. Deparis, P. Megret, M. Blondel: Opt. Lett. **27**, 83 (2002)
275. J. Hansryd, F. Dross, M. Westlund, P.A. Andreksen, S. N. Knudsen: J. Lightw. Technol. **19**, 1691 (2001)
276. W. B. Gardner: ITU Document COM 15-273-E (1996)
277. T. C. E. Jones: National Physical Laboratory Report COEM 10 (1998)
278. X. P. Mao et. al: Proc. OFC 41 (1991)
279. D. Cotter: Electron. Lett., **18**, 495 (1982)
280. R. D. Esman et al: Proc. OFC 227 (1996)
281. P. Bayvel, P.M. Radmore: Electron. Lett., **26**, 434 (1990)
282. M. Ohashi: ITU Document COM 15-187-E (1995)
283. L. Thevanez: COST 241 - Final Report, SG2.4 97, (1998)
284. E. Lichtman, R. G. Waarts, A. A. Friesem: J. Lightw. Technol. **7**, 171 (1989)
285. E. M. Dianov et al: Opt. Fiber Technol., **1**, 236 (1995)
286. A. K. Srivastava et al: Optical Amplifiers and their Applications, vol PD2, Vail, CO (1998)
287. T. N. Nielsen et al: Eur. Conf. Optical Communications (ECOC'99), PD2-2, Nice, France (1999)
288. T. N. Nielsen et al: Optical Fiber Communications Conf. Baltimore, MD, PD23 (2000)

289. K. Rottwitt, H. D. Kidorf: Optical Fiber Communications Conf., San Jose, CA, PD6 (1998)
290. P. B. Hansen et al: IEEE Photon. Technol. Lett., **9**, 262 (1997)
291. I. Morita, K. Tanaka, N. Edagawa: Proc. OFC'01, Paper TuF5-1-TuF5-3 (2001)
292. P. C. Reeves-Hall, D. A. Chestnut, C. J. S. De Matos, J. R. Taylor: Electron. Lett. **37**, 883 (2001)
293. K. Rottwitt. J. Bromage, A. J. Stentz, L. Leng, M. E. Lines, H. Smith: J. Lightw. Technol. **21**, 1652 (2003)
294. L. F. Mollenauer, J. P. Gordon, M. N. Islam: J. Quant. Electron. **QE-22**, 157 (1986)
295. M. E. Marhic, Y. Akasaka, K. Y. Wong, F. S. Yang, L. G. Kazovsky: Electron. Lett. **36**, 1637 (2000)
296. R. H. Stolen, J. E. Bjorkholm: IEEE J. Quant. Electron. **QE-18**, 1062 (1982)
297. R. Tang, J. Lasri, P. Devgan, J. E. Sharping, P. Kumar: Electron. Lett. **39**, DOI:10.1049/el:20030141 (2003)
298. S. Namiki, Y. Emori: Optics and Photonics News, 52, July (2002)
299. Y. Emori, S. Namiki: Furukawa Rev. **19**, 59 (2000)
300. E. M. Dianov et al: Quant. Electron. **24**, 749 (1994)
301. S. G. Grubb et al: SA4 in *Optical Amplifiers and Their Applications*, Vol.18 OSA Technical Digest Series (OSA Washington D.C. 1995)
302. V. V. Grigoryants et al: Opt. Quant. Electron. **9**, 351 (1977)
303. V. I. Karpov et al: Opt. Lett. **24**, 887 (1999)
304. A. A. Demidov, A. N. Starodunov, X. Li, A. Martinez-Rios, H. Po: Opt. Lett. **28**, 1540 (2003)
305. B. Min, T. J. Kippenberg, K. J. Vahala: Opt. Lett. **28**, 1507 (2003)
306. L. Chen, X. Bao: Opt. Comm. **152**, 65 (1998)
307. C. L. Tang: J. Appl,. Phys. **37**, 2945 (1966)
308. Project P816-PF, Implementation frameworks for integrated wireless optical access networks, Deliverable 4 EURESCOM (2000)
309. D. Wake et al: Electron. Lett. **33** (1997)
310. J. Hansryd, P. A. Andrekson, M. Westlund, J. Li, P. O. Hedekvist: IEEE J. Select. Top. in Q. Electron. **8**, 506 (2002)
311. Th. Schneider: Deutsche Telekom Unterrichtsblätter **55**, 356 (2002)
312. X. S. Yao: IEEE Photon. Technol. Lett. **10**, 138 (1998)
313. G. H. Smith, D. Novak, Z. Ahmed: Electron. Lett. **33**, 74, (1997)
314. K. Yonenanga, N. Takachio: IEEE Photon. Technol. Lett. **5**, 949 (1993)
315. J. H. Lee, Z. Yusoff, W. Belardi, M. Ibsen, T. M. Monro, D. J. Richardson: Opt. Lett. **27**, 927 (2002)
316. B. C. Rodgers, T. H. Russell, W. B. Roh: Opt. Lett. **24**, 1124 (1999)
317. T. H. Russell, W. B. Roh, J. R. Marciante: Opt. Express **8**, 246 (2001)
318. D. Cavendish: IEEE Comm. Mag., 164, June (2000)
319. N. J. Doran, D. Wood: Opt. Lett. **13**, 5658 (1988)
320. O. Leclerc, B. Lavigne, E. Balmefrezol, P. Brindel, L. Pierre, D. Rouvillain, F. Seguineau: J. Lightwave Technol., **21**, 2779, (2003)
321. V. Mikhailov, P. Bayvel: Electron. Lett. 37 (2001)
322. D. Butler, L. Douglas, J. Wey, M.Chbat, G. Burdge, J. Goldhar: Opt. Lett., **20**, 560 (1995)
323. H. Kawakami, T. Mitamoto, T. Kataoka, K. Hagimoto: IEICE Trans. Commun. **E78-B**, 694 (1995)
324. C. Johnson, K. Demarest, C. Allen, R. Hui, K. V. Peddanarappagari, B. Zhu: IEEE Photon. Technol Lett., **11**, 895 (1999)

325. M. Niklès, L. Thévenaz, Ph. Robert: Proc. 10th. OFS Conf. (Glasgow, Scotland), pp. 138-141 (1994)
326. M. Niklès, L. Thévenaz, Ph. Robert: Optics Letters, **21**, no.10, 758, (1996).
327. Y. Li, F. Zhang, T.Yoshino: J. Lightw. Technol. **21**, 1663 (2003)
328. M. Niklès, L. Thévenaz, Ph. Robert: J. Lightw. Technol., **15**, 1842 (1997)
329. A. Fellay, L. Thevanez, M. Facchini, M. Nikles, P. Robert: In *Optical Fiber Sensors* 16, 324, OSA Technical Digest Series, Washington, D.C. (1997)
330. T. Horiguchi, K. Shimizu, T. Kurashima, Y. Koyamada: Proc. SPIE **2507**, 126 (1995)
331. K. Shiraki, M. Ohashi, M. Tateda: Electron. Lett., **31**, 668 (1995)
332. N. Yoshizawa, T. Imai: J. Lightwave Technol., **11**, 1518, (1993)
333. S. Watanabe, M. Shirasaki: J. Lightwave Technol., **14**, 243, (1996)
334. J. Pina, B. Abueva, G. Goedde: Opt. Commun. **176**, 397 (2000)
335. M. Tsang, D. Psaltis: Opt. Lett. **28**, 1558, (2003)
336. T. Yamamoto, E. Yoshida, K. R. Tamura, M. Nakazawa: ECOC'99, Nice, France, vol 2, 38, (1999)
337. D. Kunimatsu, C. Q. Xu, M. D. Pelusi, X. Wang, K. Kikuchi, A. Suzuki: ECOC 2000, Munich, Germany, 8.3.1 (2000)
338. A. Heuer, C. Hänisch, R. Menzel: Opt. Lett. **28**, 34, (2003)
339. K. Kitayama, H. Sotobayashi: OFC'99 Technical Digest, paper WD4-1, 64 (1999)
340. F. Ramos, J. Marti, V. Polo: IEEE Photon. Technol. Lett. **11**, 1171 (1999)
341. C. Q. Xu, H. Okayama, T. Kamijoh: ECOC'98, Madrid, Spain, 173 , (1998)
342. S. C. Matthews, D. A. Rockwell: Opt. Lett. **19**, 1729 (1994)
343. A. Yariv, D. Fekete, D. M. Pepper: Opt. Lett. **4**, 52 (1979)
344. S. Watanabe, T. Naito, T. Chikama: IEEE Photon. Technol. Lett. **5**, 92 (1993)
345. S. Watanabe, S. Kaneko, T. Chikama: Opt. Fiber Technol. **2**, 169 (1996)
346. K. Kikuchi, C. Lorattanasane: Proc. OAA'93, Yokohama, 22, paper SuC1 (1993)
347. S. Watanabe, T. Chikama, G. Ishikawa, T. Terahara, H. Kuwahara: IEEE Photon. Technol. Lett. **5**, 1241 (1993)
348. S. Watanabe, G. Ishikawa, T. Naito, T. Chikama: J. Lightw. Technol. **12**, 2139 (1994)
349. R. Jopson, R. Tench: Electron. Lett. **29**, 2216 (1993)
350. X. Zhang, B. F. Jorgensen: Opt. Lett. **21**, 791 (1996)

Index